FRACTALS IN PHYSICS

© Elsevier Science Publishers B.V., 1990

All rights reserved. No part of this publication may be reproduced, stored in a retrieval system, or transmitted in any form or by any means, electronic, mechanical, photocopying, recording or otherwise, without the written permission of the Publisher, Elsevier Science Publishers B.V., P.O. Box 211, 1000 AE Amsterdam, The Netherlands.

Special regulations for readers in the USA: This publication has been registered with the Copyright Clearance Center Inc. (CCC), Salem, Massachusetts. Information can be obtained from the CCC about conditions under which photocopies of parts of this publication may be made in the USA. All other copyright questions, including photocopying outside of the USA, should be referred to the Publisher.

No responsibility is assumed by the Publisher for any injured and/or damage to persons or property as a matter of products liability, negligence or otherwise, or from any use or operation of any methods, products, instructions or ideas contained in the material herein.

REPRINTED FROM PHYSICA D Vol. 38, Nos. 1–3 (1989)

ISBN: 0 444 88646 X

North-Holland
Elsevier Science Publishers B.V.
P.O. Box 103
1000 AC Amsterdam
The Netherlands

Sole distributors for the USA and Canada:

Elsevier Science Publishing Company, Inc.
655 Avenue of the Americas
New York, NY 10010
USA

Library of Congress Cataloging-in-Publication Data

Fractals in physics: essays in honour of Benoit B. Mandelbrot:
   proceedings of the international conference honouring Benoit B. Mandelbrot on his 65th birthday, Vence, France, 1–4 October, 1989/editors, Amnon Aharony, Jens Feder.
     p. cm.
   "Reprinted from Physica D, volume 38" - - Verso t.p.
   Includes bibliographical references.
   ISBN 0-444-88646-X (U.S.)
   1. Fractals - - Congresses. 2. Mathematical physics - - Congresses. 3. Mandelbrot, Benoit B. - - Congresses. I. Mandelbrot, Benoit B. II. Aharony, Amnon. III. Feder, Jens.
   QC20.7.G44F73 1990
   530.1′5615 - - dc20   89-72204
                                                                               CIP

Printed in The Netherlands

Fractals in Physics

Essays in honour of Benoit B. Mandelbrot

Proceedings of the International Conference honouring Benoit B. Mandelbrot on his 65th birthday
Vence, France, 1–4 October, 1989

Editors:

Amnon Aharony
School of Physics and Astronomy
Tel Aviv University
Tel Aviv 69978, Israel

Jens Feder
Department of Physics
University of Oslo
N-0316 Oslo, Norway

1990

NORTH-HOLLAND

PREFACE TO THE BOOK EDITION

This book contains the integral text of the Proceedings of the conference "Fractals in Physics" held in Vence, France, 1–4 October, 1989. The conference was held on occasion of the 65th birthday of Benoit B. Mandelbrot, the man who gave us the word "fractal" and opened up a very exciting field of research.

Written by leading experts, the work is a very complete account of the history and current state of the science and the wide variety of applications of the fractal concept.

This collection of essays aims at the individual "affectionado". Although the book is of broad interest, it is not so much intended for the uninitiated or superficial reader as for those who want to penetrate further into the secrets of the use of this fascinating fractal geometry in Physics.

At the end of this book edition a few corrections and additions to the reprinted text of the journal edition (Physica D, Vol. 38, Nos. 1–3 (1989)) are given.

The Publisher

Benoit B. Mandelbrot

PREFACE

This volume contains the Proceedings of a conference held in Le Mas d'Artigny (Vence), France, on October 1–4, 1989, in honour of Benoit B. Mandelbrot's 65th birthday.

Benoit Mandelbrot is certainly responsible for the creation and development of 'fractals' into an active sub-discipline in mathematics as well as in physical and social sciences. In his three seminal books:

1975 Les Objets Fractals: Forme, Hazard et Dimension

1977 Fractals: Form, Chance and Dimension

1982 The Fractal Geometry of Nature

and in his many papers (see list at the end of this volume), Mandelbrot examined and classified in a mathematical way numerous spatial patterns in nature, which are described by *fractal geometry*. Mandelbrot coined the word 'fractal', revived and extended the relevant old mathematics, first initiated by Hausdorff and Besicovitch, and supplied most of the tools and fractal measures necessary for the analysis of fractal phenomena.

In this volume, as in the conference which it summarizes, we restricted ourselves to *fractals in physics*. Although self-similarity and scaling have come up in various independent contexts in statistical physics, it was Benoit Mandelbrot who taught us to think systematically about these concepts in a *geometrical way*. It was this geometrical approach that made many of the complex self-similar phenomena more transparent, and led to the exponential growth of activity since 1974 [#1].

Mandelbrot's specific important contributions to the field are listed in his list of publications, at the end of this volume. They are also highlighted via many references and discussions throughout this volume.

When we set out to organize the Mandelbrot birthday conference, we thought of a small intimate meeting with a few papers representing each of the fields in physics in which fractals have become an essential tool. We thus agreed with IBM on their kind sponsorship, as well as on the size of the conference. Arrangements were made for the hotel Le Mas d'Artigny (Vence) in the south of France and for the Proceedings (with North-Holland), and both set an upper bound of 60 participants. Following that, we started to work on the list of invited participants. It was then that we discovered that the community of physicists who work with fractals, who made significant contributions to our understanding of fractals in physics and who feel indebted to Benoit Mandelbrot and would like to honour him on his birthday, is much much larger. We were therefore forced to make difficult choices, and we deeply apologize to our many friends (hoping we have not lost their friendship) whom we were unable to accommodate.

All the invited participants were asked to contribute a manuscript for the Proceedings, with a tight deadline (aimed to have the book ready to present to Benoit Mandelbrot at the meeting). Although some of the authors required a few bitnets, faxes and phone calls with reminders, we were impressed by their prompt responses and by the fact that we were able to pass all the manuscripts to the publisher on the agreed date. We are grateful to the authors for their cooperation.

[#1] The cumulative number of papers identified by the word *fractal* ... in the INSPEC database is fit well by $\exp\{(t-1974)/1.74\}$, where t is the time in years, indicating that the activity increases by a factor 1.8 every year.

Owing to the tight schedule, we have not been able to transfer to some authors our refereeing comments on their papers, or to negotiate on appropriate changes. Perhaps because of its interdisciplinary nature, 'Fractals in Physics' has been a subject that has led to several heated controversies and disputes, and continues to be a field in which some people tend to put forward unjustified conjectures or to create unnecessary formalism to describe simple phenomena. It also happens that authors present as new results which can be found (sometimes well hidden) in Mandelbrot's books. We leave the reader the identification of papers in this volume which commit these 'crimes', and hereby declare that those do not express our own views.

As said above, we attempted to represent here the most active topics involving fractals in physics. Thus, this volume contains papers on fractal growth phenomena (including diffusion-limited aggregation, viscous fingering, cracks, etc.), on physical properties (e.g. diffusion, flow, vibrations, magnetism, etc.) of fractal structures (such as percolation clusters, polymers, porous media, etc.), on turbulence and on galaxies. It also contains papers attempting to understand the fractal geometry from basic principles, like self-organization. Some papers also discuss history, philosophy and the teaching of fractal concepts. Because of the large variety of topics, we chose to order the papers alphabetically (by contributor). The analytic subject index, based on keywords provided by the authors, should help identify the papers by topic.

Personally, we have both benefited greatly from Benoit Mandelbrot's broad insight, intuitive creativity and enthusiastic encouragements. During the years, we have had many fruitful discussions and collaborations with him, and we certainly owe him the fact that we have become active in this exciting field. We are therefore very pleased that we had the possibility to organize this conference, and we are sure that we represent all the participants in the conference and in this book in wishing Benoit a happy birthday and many more fruitful and stimulating active years.

We are very grateful for all the help Inger Lauvstad gave us in preparing this meeting, and to Dr. V. Sadagopan in helping us to arrange it. The financial support by IBM France and the IBM Research Division is gratefully acknowledged.

Amnon Aharony and Jens Feder
Oslo, June 18, 1989

CONTENTS

Preface to the book edition	v
Preface	vii
Contents	ix

Measuring multifractals
   A. Aharony — 1

The physics of fractals
   P. Bak and K. Chen — 5

Fractal colloidal aggregates: consolidation and elasticity
   R.C. Ball — 13

Ordered shapes in nonequilibrium growth
   E. Ben-Jacob and P. Garik — 16

Falling fractal flakes
   M.V. Berry — 29

The fractal structure of evolution
   G. Binnig — 32

Multiscaling and multifractality
   A. Coniglio and M. Zannetti — 37

Fractons observed
   E. Courtens, R. Vacher and E. Stoll — 41

Cantor set spectra and self-similar critical modes in a 1D-quasicrystal
   J.P. Desideri, O. Legrand, L. Macon and D. Sornette — 56

Of men and ideas (after Mandelbrot)
   C. Domb — 64

Fractals in two dimensions and conformal invariance
   B. Duplantier — 71

The transmission of stress in an aggregate
   S.F. Edwards and R.B.S. Oakeshott — 88

Dynamic structure factor of fractals
   O. Entin-Wohlman, U. Sivan, R. Blumenfeld and Y. Meir — 93

Fractal pattern formation in human retinal vessels
   F. Family, B.R. Masters and D.E. Platt — 98

Geometrical crossover and self-similarity of DLA and viscous fingering clusters
   J. Feder, E.L. Hinrichsen, K.J. Måløy and T. Jøssang — 104

Fractal and nonfractal shapes in two-dimensional vesicles
   M.E. Fisher — 112

The building blocks of random walks
    Y. Gefen and I. Goldhirsch   119

Fractal motion of mammalian cells
    I. Giaever and C.R. Keese   128

A light scattering study of turbulence
    W.I. Goldburg, P. Tong and H.K. Pak   134

Asymmetric random walk on a random Thue–Morse lattice
    S. Goldstein, K. Kelly, J.L. Lebowitz and D. Szasz   141

Aggregates, broccoli and cauliflower
    F. Grey and J.K. Kjems   154

Multifractal measures and stability islands in the anisotropic Kepler problem
    M.C. Gutzwiller   160

Fractals and percolation in porous media and flows?
    E. Guyon, C.D. Mitescu, J.-P. Hulin and S. Roux   172

Remarks on percolation and transport in networks with a wide range of bond strengths
    B.I. Halperin   179

Probability densities of random walks in random systems
    S. Havlin and A. Bunde   184

Fractal deterministic cracks
    H.J. Herrmann   192

Fractals and self-organized criticality in dissipative dynamics
    T. Hwa and M. Kardar   198

Boundary layer instability in a coupled-map model
    M.H. Jensen   203

Geometrical optics in fractals
    R. Jullien and R. Botet   208

Fractals and multifractals in avalanche models
    L.P. Kadanoff   213

Energetic and entropic elasticity of the Sierpiński gasket
    Y. Kantor   215

Growth of self-affine surfaces
    J. Kertész and D.E. Wolf   221

Johnson–Nyquist noise derived from quantum mechanical transmission
    R. Landauer   226

Applications of fractal concepts in petroleum engineering
    R. Lenormand   230

Fracture as a growth process
    E. Louis and F. Guinea   235

Fractal dimension of the fractured surface of materials
    C.W. Lung and S.Z. Zhang   242

On the self-affinity of various curves
    M. Matsushita and S. Ouchi   246

The growth of self-affine fractal surfaces
  P. Meakin — 252

Fractals and the ac conductivity of disordered materials
  G.A. Niklasson — 260

Fracton dynamics
  R. Orbach — 266

The fractal galaxy distribution
  P.J.E. Peebles — 273

Theoretical concepts for fractal growth
  L. Pietronero — 279

Superdiffusive transport due to random velocity fields
  S. Redner — 287

Luminescence decay in chain-like polymers using fractal concepts
  A.K. Roy and A. Blumen — 291

Experimental observation of local modes in fractal drums
  B. Sapoval — 296

Growth velocity of electrochemical deposition and its concentration dependence
  Y. Sawada and H. Hyosu — 299

Levy flights: variations on a theme
  M.F. Shlesinger — 304

Scattering from fractal structures
  S.K. Sinha — 310

Geometrical scaling of microsphere-deposited monolayers with holes
  A.T. Skjeltorp — 315

New results on the fractal and multifractal structure of the large Schmidt number passive scalars in fully turbulent flows
  K.R. Sreenivasan and R.R. Prasad — 322

Learning concepts of fractals and probability by "doing science"
  H.E. Stanley — 330

Hunting for the fractal dimension of the Kauffman model
  D. Stauffer — 341

Frustration and correlations in fractals
  R.B. Stinchcombe — 345

Phase transitions for polymers on fractal lattices
  J. Vannimenus — 351

Deterministic models of fractal and multifractal growth
  T. Vicsek — 356

Random fractals: self-affinity in noise, music, mountains, and clouds
  R.F. Voss — 362

Lagrangian chaos and small scale structure of passive scalars
  A. Vulpiani — 372

Hull-generating walks
   R.M. Ziff   377

Vita and publications of Benoit B. Mandelbrot   385
List of contributors   396
Analytic subject index   397
Errata   399

MEASURING MULTIFRACTALS

Amnon AHARONY

*Raymond and Beverly Sackler Faculty of Exact Sciences, School of Physics and Astronomy,
Tel Aviv University, Ramat Aviv, Tel Aviv 69978, Israel*

Dedicated to Benoit B. Mandelbrot on his 65th Birthday

Differences between typical and average moments of a multifractal distribution are discussed. Negative moments are often nominated by exponentially small terms, which lead to a breakdown of multifractal scaling. Particular attention is given to the current distribution on percolating resistor networks and to the growth probabilities in diffusion-limited aggregation.

Dedication: my encounters with fractals

It gives me personal pleasure to dedicate this article, as well as these proceedings, to my good teacher and friend Benoit B. Mandelbrot on the occasion of his 65th birthday.

My first encounter with fractals happened in 1975. I gave a seminar at IBM, Yorktown Heights, on the use of the ϵ-expansion, and I failed to satisfy Mandelbrot, who wanted me to tell him about the explicit geometry of a system in 3.99 dimensions. Benoit kept bugging me on this whenever we met, but we finally did something about it only in 1980, when we both spent sabbaticals at Harvard. Yuval Gefen, who visited me at Harvard, had to run back and forth between the physics and mathematics buildings, in order to match the physical properties we wanted to imitate with corresponding fractal geometries. Benoit was very surprised after a few weeks, when I felt it was already time to submit the results to Physical Review Letters [1]. That joint paper started a long list of papers in which physical problems were solved on exact ordered fractal structures. This certainly allowed systematic studies of the interplay between geometry and physics.

Our meeting at Harvard also led to a continuous collaboration [2-8], which eventually also returned to Benoit's question on describing Euclidean structures in non-integer dimensions [3], modeled percolation clusters [2] and introduced transfer matrix methods on fractals [7,8]. Benoit's geometrical and analytical insights were invaluable in these collaborations, and they continue to inspire us at Tel Aviv University in our ongoing research of physics on fractals.

Looking forward to many more years of fruitful discussions and collaborations, I use this occasion to say "**Happy Birthday, Benoit**"!

1. Introduction

Multifractals, first introduced by Benoit Mandelbrot in 1974 [9], represent infinite sets of exponents which describe the (power law) scaling of all the moments of a distribution of some quantities which are defined on a fractal structure. In many cases, specific members of these families of exponents coincide with the fractal dimensionalities of geometrical substructures of the underlying fractal. Although knowledge of the multifractal spectrum is completely equivalent to knowledge of the corresponding probability distribution, the literature contains many attempts to attach a much deeper significance to the former.

The aim of the present paper is to point at several

difficulties related to the measurement of the multifractal spectrum, when it exists, and to raise some doubts about its existence, in some specific cases. Particularly, I wish to emphasize the differences between "average" and "typical" random fractals, and the problems with negative moments.

2. Problems with averages

Following ref. [1], ordered fractal models have been very helpful for understanding the interplay between the geometrical structure of a cluster and its physical properties. In particular, such models allowed exact evaluations of the multifractal spectra, and led to useful approximants for these spectra, e.g. for percolation clusters.

A useful example, to illustrate the point, concerns the moments of the current distribution on a fractal cluster,

$$M_q(L) = \sum_b |i_b|^{2q}, \qquad (1)$$

where i_b is the current in bond b when a unit current is inserted at point x and removed at point x', with $|x'-x|=L$, and every bond on the fractal has unit resistance [10–16]. On an ordered hierarchical fractal, like the Mandelbrot–Given curve [17], all the currents iterate multiplicatively [11–13], and therefore all the moments M_q behave as powers of L,

$$M_q(L) \sim L^{\tilde{\zeta}(k)}. \qquad (2)$$

The set of exponents $\tilde{\zeta}(k)$ thus represents a multifractal spectrum.

In all the interesting physical applications, we are dealing with *random fractal structures*. For example, at the percolation threshold, percolating clusters exhibit *statistical self-similarity*. The actual results then depend on the procedure of *averaging* over the *distribution* of such clusters [18].

In what follows, we shall distinguish between the configurational average over *all* the possible structures, e.g. $[M_q(L)]_{av}$, and between measurements of *typical* clusters. As we shall see, the complete average is sometimes dominated by a very rare (and therefore un-representative) configuration, arising in the tail of the distribution. *Typical* results, on the other hand, should arise from the vicinity of the maximum in the distribution. This is often achieved by using averages like $[\log M_q(L)]_{av}$, or $[M_q(L)^{1/q}]_{av}$ [18].

In the example of the current distribution, problems arise for *negative* moments [14]. For any given cluster, at $q \to -\infty$, $M_q(L)$ is dominated by $|i_{\min}|^{2q}$, where $i_{\min}$ is the smallest current in the cluster. If we average over *all* clusters, then $(|i_{\min}|^{2q})_{av}$ is dominated by the cluster which has the smallest current. Ref. [14] showed that this cluster could have ladder configurations, such that the current in the topmost rung decays exponentially with the ladder's height, l. Since the probability for such a ladder is of order p^l, and since $|i_{\min}|^2 \sim e^{-Al}$, we found that $[M_q(L)]_{av} \sim p^l e^{-qAl}$, diverging *exponentially* with l for sufficiently negative q. Since $l \sim L^y$, $[M_q(L)]_{av}$ does not behave as a power of L, and multifractality breaks down.

In contrast, $[M_q(L)^{1/q}]_{av}$ is equivalent to $(|i_{\min}|^{2q})_{av}$, and this might still be dominated by the clusters which have the most probable minimal currents. These might well behave as powers of L [18]. I am not aware of systematic studies of these averages, and I hope this paper will stimulate some.

A similar problem of averaging arose recently in a discussion of random walks and self-avoiding walks on fractals [19–21]. The probability to find a random walker after t time steps at distance r from the origin, $P(r, t)$, has a different decay on *typical* clusters (when one calculates $[\log P(r, t)]_{av}$) and on the *average* cluster (calculating $[P(r, t)]_{av}$). It is the former, equivalent to the quenched average of the entropy, that is needed in order to evaluate a Flory approximant for self-avoiding walks on fractals [20,21].

3. Diffusion-limited aggregates

In diffusion-limited aggregation (DLA), the growth probabilities $\{p_i\}$ on the surface of the aggregate are related to the harmonic measure, associated with the Laplacian field of the distribution of random walkers around the aggregate [22]. This growth probability distribution is then characterized by its moments,

$$M_q = \sum_i p_i^q, \quad (3)$$

and one considers the dependence of M_q on the aggregate's linear size, L.

Again, results will depend on the averaging used. Recently, Lee and Stanley [23] used exact enumeration for $L \leq 5$, to predict a divergence of $[M_q]_{av}$ for sufficiently negative a. They associated this divergence with an exponential decay of the minimal growth probability (over all possible aggregates of size L) with L.

Looking at a narrow long fjord, of length l, it is easy to convince oneself that the growth probability at the bottom of the fjord decays exponentially with l [24]. If such fjords occur on *typical* aggregates, then one expects an exponential divergence of $M_q(L)$ for all $q < 0$, even on such aggregates. This scenario was recently analyzed in detail [24]. We have also analyzed the average $[M_q(L)]_{av}$, and found circumstances in which it might diverge exponentially only for $q < q_0 < 0$.

Similar to the current distribution, $[M_q(L)]_{av}$ will be dominated by the smallest growth probability, p_{min} (over all aggregates), only if the probability for p_{min} to appear does not decay with L faster than $1/p_{min}$. Recently, Harris [25] raised doubts if this is indeed the case.

At present, I consider the question of the breakdown of multifractality in DLA open. More detailed quantitative studies of the distribution function of the p_i's, and calculations of averages like $[\log M_q(L)]_{av}$ or $[M_q(L)^{1/q}]_{av}$, might help resolve this problem.

4. Conclusion

This paper is meant as a warning for average calculations on multifractals. Care should be given to the types of averages used. Care should also be given to negative moments of distributions which may involve exponentially small quantities. Although most of these difficulties may disappear by averaging over the logarithms of the corresponding moments, exponentially small growth probabilities may still represent difficulties in DLA and similar structures.

Acknowledgements

Many of the results discussed here arose from collaborations with R. Blumenfeld, Y. Meir and A.B. Harris. The support of the Israel Academy of Sciences and of the US–Israel Binational Science Foundation is also gratefully acknowledged.

References

[1] Y. Gefen, B.B. Mandelbrot and A. Aharony, Phys. Rev. Lett. 45 (1980) 855.
[2] Y. Gefen, A. Aharony, B.B. Mandelbrot and S. Kirkpatrick, Phys. Rev. Lett. 47(1981) 1771.
[3] Y. Gefen, Y. Meir, A. Aharony and B.B. Mandelbrot, Phys. Rev. Lett. 50 (1983) 145.
[4] Y. Gefen, A. Aharony and B.B. Mandelbrot, J. Phys. A 16 (1983) 1267.
[5] Y. Gefen, A. Aharony, Y. Shapir and B.B. Mandelbrot, J. Phys. A 17 (1984) 435.
[6] Y. Gefen, A. Aharony and B.B. Mandelbrot, J. Phys. A 17 (1984) 1277.
[7] Y. Gefen, B.B. Mandelbrot, A. Aharony and A. Kapitulnik, J. Stat. Phys. 36 (1984) 827.
[8] B.B. Mandelbrot, Y. Gefen, A. Aharony and J. Peyriere, J. Phys. A 18 (1985) 335.
[9] B.B. Mandelbrot, J. Fluid Mech. 62 (1974) 331; E. Cabib, C.G. Kuper and I. Reiss, eds., STATPHYS 13 (Hilger, Bristol, 1978).
[10] R. Rammal, C. Tannous and A.-M.S. Tremblay, Phys. Rev. A 31 (1985) 2662.
[11] L. de Arcangelis, S. Redner and A. Coniglio, Phys. Rev. B 31 (1985) 4725.
[12] R. Blumenfeld and A. Aharony, J. Phys. A 18 (1985) L433.
[13] R. Blumenfeld, Y. Meir, A.B. Harris and A. Aharony, J. Phys. A 19 (1985) L791.
[14] R. Blumenfeld, Y. Meir, A. Aharony and A.B. Harris, Phys. Rev. B 35 (1987) 3524.
[15] B. Fourcade, P. Breton and A.-M.S. Tremblay, Phys. Rev. B 36 (1987) 8925.
[16] A. Aharony, R. Blumenfeld, P. Breton, B. Fourcade, A.B. Harris, Y. Meir and A.-M.S. Tremblay, unpublished.
[17] B.B. Mandelbrot and J. Given, Phys. Rev. Lett. 52 (1984) 1853.

[18] Y. Meir and A. Aharony, Phys. Rev. B 37 (1988) 596.
[19] A.B. Harris and A. Aharony, Europhys. Lett. 4 (1987) 1355.
[20] A. Aharony and A.B. Harris, J. Stat. Phys. 59 (1980) 1091.
[21] A. Aharony and A.B. Harris, in: STATPHYS 17, Physica A, to appear.
[22] T.A. Witten and L.M. Sander, Phys. Rev. Lett. 47 (1981) 1400.
[23] J. Lee and H.E. Stanley, Phys. Rev. Lett. 61 (1988) 2945.
[24] R. Blumenfeld and A. Aharony, Phys. Rev. Lett. 62 (1989) 2977.
[25] A.B. Harris, Phys. Rev. B 39 (1989) 7292.

THE PHYSICS OF FRACTALS

Per BAK and Kan CHEN

Brookhaven National Laboratory, Upton, NY 11973, USA

Fractals in nature originate from self-organized critical dynamical processes.

1. Introduction

The importance of Mandelsbrot's discovery that fractals occur widespread in nature can hardly be exaggerated. Many things which we used to think of as messy and structureless are in fact characterized by well-defined power-law spatial correlation functions. By now, we are so used to seeing fractals that we are tempted to feel that we understand them. But do we simply have to accept their existence as "God-given" without further explanation or is it possible to construct a dynamical theory of the physics of fractals?

There is another ubiquitous phenomenon which has defied explanation for decades. The signal (water, electrical current, light, prices, ...) from a variety of sources has a power spectrum decaying with an exponent near unity at low frequencies. Typically, the accumulated signal follows Hurst's empirical law: the standard deviation measured over a duration τ increases as τ^H where the exponent H exceeds the value $1/2$ for random processes. This type of behavior is known as "$1/f$" noise, or flicker noise.

Strangely enough, just as those working on fractal phenomena in nature never seem to be interested in the temporal aspects of the phenomenon, but concentrate on further geometrical characterization of fractals (which does not bring us one inch closer to understanding the phenomenon), those working on "$1/f$" noise never bother with the spatial structure of the source of the signal. We believe that those two phenomena are often two sides of the same coin: they are the spatial and temporal manifestations of a self-organized critical state. Actually, for those (like us) who are brought up as condensed matter physicists it is hard to believe that long-range spatial and temporal correlations can exist independently. A local signal cannot be "robust" and remain coherent over long times in the presence of any amount of noise, unless stabilized by the interactions with its environment. And a large, coherent spatial structure cannot disappear (or be created) instantly. For an illustration, think of the temporal distribution of sunshine, which must be correlated with the spatial distribution of clouds, through the dynamics of meteorology.

In fact, there is one area of physics where the relation between spatial and temporal power-law behavior is well-established. At the critical point for continuous phase transitions, the correlation function for the order parameter decays spatially as $r^{2-d-\eta}$ and temporally as $t^{-d/z}$. But in order to arrive at the critical point, one has to fine-tune an *external* control parameter such as the temperature or pressure, in contrast to the phenomena above which occur universally without any fine-tuning. The explanation is that open, extended, dissipative dynamical systems may go automatically to the critical state as long as they are driven slowly: the critical state is self-organized. We see fractals as snapshots of systems operating at the self-organized critical state.

Our models are discrete in both space and time: they are cellular automata. More realistic models might involve infinities of coupled differential equations, but we believe that the discretization does not affect the asymptotic long-time and -space behavior

Essays in honour of Benoit B. Mandelbrot
Fractals in Physics – A. Aharony and J. Feder (editors)

that we are interested in. Moreover, cellular automata seem particularly suited to systems with many metastable states, as in many of the physical situations that we consider.

We (and our collaborators) have studied four dynamical models spanning a great variety of phenomena. The models are *not* realistic representations of any real systems. We have chosen to sacrifice realism for simplicity in order to obtain a general flavor of the mechanisms at work. The first model is an "earthquake" model or "sandpile" model [1]. We use the "sandpile" picture not because we are particularly interested in sandpiles, or because the models are particularly realistic for sandpiles (in fact they are probably not), but because it provides an intuitive feeling for the phenomena that we are discussing. The model evolves to a critical state with a scale-free fractal distribution of avalanches or earthquakes. The famous Gutenberg–Richter power law for size-distribution of earthquakes indicates that the crust of the earth is at a stationary critical state [2]. Simple generalizations of the model may explain the fractal distribution of earthquake epicenters. The second model is Conway's "Game of Life" which is a toy model for biological evolution. We find that the model evolves to a self-organized critical state with concentration $p_c \approx 0.03$ of live individuals. At the stationary state individuals are born or die on a fractal with dimension $D \approx 1.6$. The third model is a model of ballistic particles, created and annihilated when they hit each other. The model might be thought of as a toy model of interacting matter in the universe. It seems to belong to the same universality class as "Life", and was constructed in an effort to understand the mechanisms of "Life". Active bright matter exists on a fractal of dimension 1.6. In contrast to "Life", the model can be generalized to any dimension and might possibly be attacked by renormalization group methods. The fourth model is a "forest-fire" model which can be thought of as a model of spreading of disease or chemical activity. The model explicitly demonstrates how energy which is injected uniformly can be dissipated on a fractal, a scenario envisioned by Mandelbrot for turbulence. The fractal dimension is 2.5 in agreement with analysis of turbulence experiments. We suggest that turbulence indeed be viewed as a sustained "forest-fire".

2. The earthquake model

The standard model of self-organized criticality is a simple "integrate and fire" model. A discrete variable $Z = 0, 1, 2, 3...$ is defined on a d-dimensional lattice. The variable at a particular site and its $2d$ nearest neighbors are updated according to the simple rule

$$Z \to Z - 4; \quad Z_{nn} \to Z_{nn} - 1 \quad \text{if } Z > Z_{cr}, \tag{1}$$

and it is driven by letting

$$Z \to Z + 1 \tag{2}$$

at some random position. Z can be thought of as the local force at some position in a fault region, and (2) represents a gradually increasing force (energy injection) from tectonic plate motion. Rule (1) simulates stress release caused by a local slip. Actually, our model is pretty close to the generally accepted stick–slip picture of earthquakes [3]; this is the justification for making bold claims on the subject! The variable Z can also be thought of as the local slope of a sandpile. Rule (1) represents the tumbling of a particle when the slope is too steep, and (2) represents a tilting of the pile.

As the local forces increase by repeated application of (2), eventually at some site the force exceeds the critical value and rule (1) is applied again and again until the activity stops; then rule (2) is applied, and so on. The system is driven infinitely slowly compared with the relaxation process. Where will the system go as the process is continued? Or where will the system end up if initiated at a state where all forces exceed the critical value? Naively, one might suspect that the sandpile gets pumped up to a state where all the slopes or forces assume the critical value. This cannot be the case since the sandpile at this point is very sensitive to perturbations: any activity will spread by a chain reaction causing the system to collapse. Fig. 1 shows the final configuration of a sandpile initiated at a uniformly steep state. The various colors indicate various subcritical values of local forces. Note the emergence of structures of all lengthscales.

Hence the system must go to a different type of stationary state. As the pile grows, the earthquakes become bigger and bigger. We argue that the pile will continue to grow until avalanches are just able to propagate infinitely throughout the system: the stationary state is at a dynamical critical state. Indeed, we find that the distribution of earthquakes is given by the scale-free Gutenberg–Richter power law, with exponent 1 in two dimensions and 1.33... in three dimensions [2,4]. The Gutenberg–Richter law can be taken as evidence that the crust of the earth is at a stationary critical state!

The most striking feature of the self-organized critical state is its resilience. The system responds to any perturbation, randomness, etc., by returning to the critical state after a transient period. If "snowsceens" are added in an effort to prevent falling sand, the pile simply builds up to a steeper critical state, thus completely counteracting any effort to take the system away from criticality. Could this be the explanation for the resilience of biological, geological and economical systems?

3. The "Game of Life"

In 1970 the mathematician Conway discovered a solitaire game, the fascinating "Game of Life" [5], a cellular automaton showing complex static and dynamic configurations. The game simulates the rise, fall, and alterations of a society of living organisms. Despite its simplicity, the dynamics of the game are poorly understood. The main interest in "Life" has been on the generation of complexity in local configurations, which has been suggested to mimic a general scenario of the emergence of complexity nature. In contrast to the "sandpile" model (but in accordance with most realistic situations) "Life" has no local conservation laws. We show that local configurations in the "Game of Life" self-organize into a critical state with the critical density of live individuals $p_c \approx 0.03$ [6].

The "Game of Life" is defined on a square lattice. There are two states on each lattice site, representing the presence or absence of a live individual. The rules for the evolution of "Life" are very simple:

(i) The fate of a live individual depends on its nearest neighbors; it will die at the next time step if there are less than 2 (over-exposure) or more than 3 live neighbors (over-crowding); it will remain alive otherwise.

(ii) At a dead site, a new individual will be born at the next time step only if there are exactly 3 live neighbors.

In order to elucidate the collective behavior of the society we study the following process: Starting with a random distribution of live sites, the system evolves according to the rules (i) and (ii) until it comes to "rest" in a simple periodic state with a distribution of local still life and simple cyclic life. There is no propagating activity. The system is then perturbed at a randomly chosen local site, for instance by adding a live individual, or by adding a "Glider", and is allowed to evolve according to the rules until it comes to rest again. As the process is repeated, the system evolves into a statistically stationary state.

Indeed the distribution of "clusters" of size s averaged over thirty thousand perturbations is a power law, $D(s) \propto s^{-\tau}$, $\tau \approx 1.4$, and so is the distribution of the durations of the perturbations $D(T) \propto T^{-b}$, with $b \approx 1.6$. The fact that the activity does not decay, or explode exponentially (becomes chaotic) indicates that life and death are highly correlated in time and space to allow the activity to continue indefinitely: the system has evolved into a critical state. Fig. 2 shows a snapshot of the activity in the middle of a perturbation. Notice the clustering of the activity. The number distribution of active sites at a distance r from a given active site increases with r as $N(r) \propto r^{D-1}$, where the fractal dimension $D \approx 1.6$. Thus, life is sustained on a fractal!

If life (and its environment) is indeed at a critical state, the concept of a stable equilibrium of nature is meaningless. Nature is ever changing along consecu-

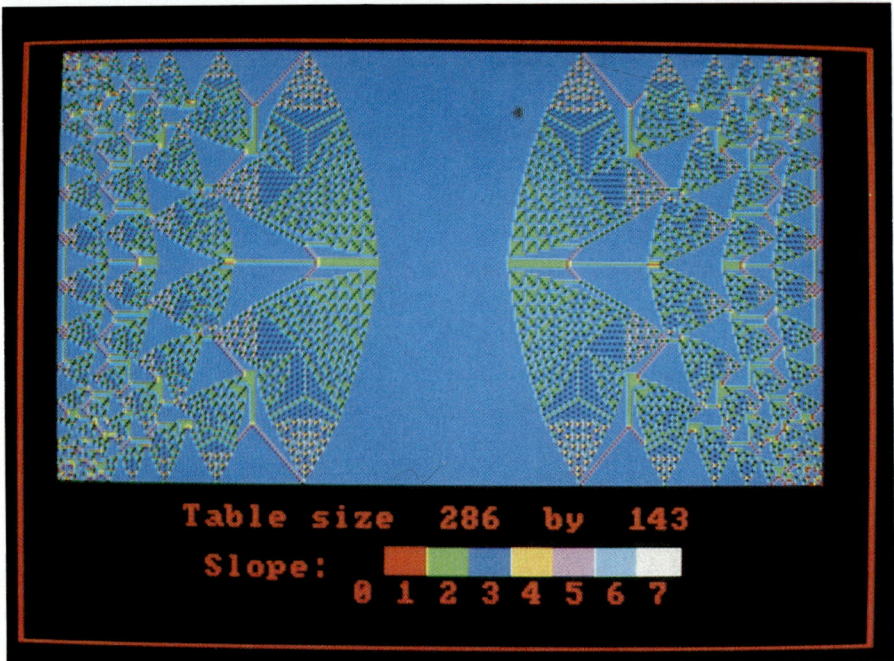

Fig. 1. Collapsed sandpile. The critical force (blue) is $Z=3$.

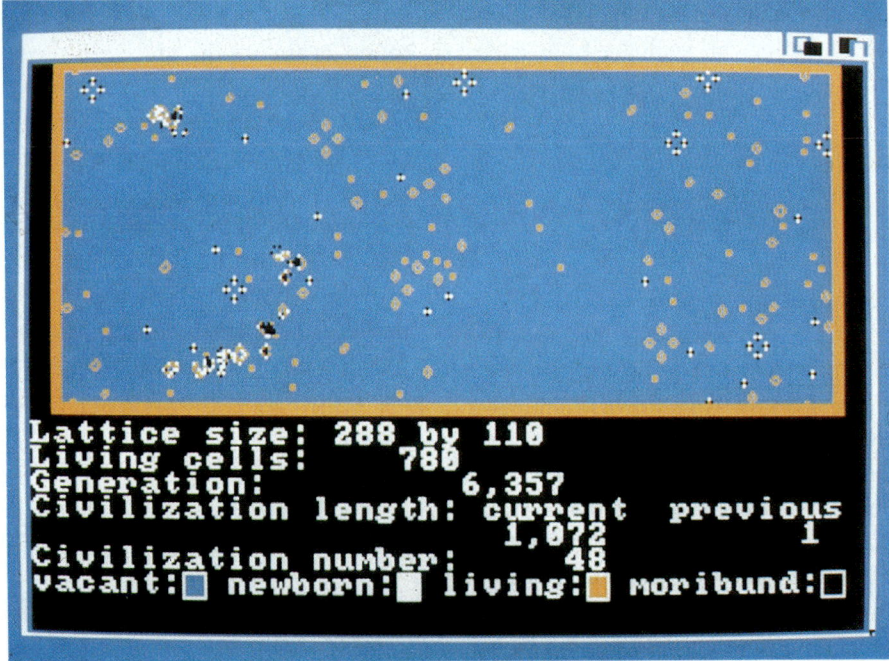

Fig. 2. Snaphsot of critical configuration of "Game of Life".

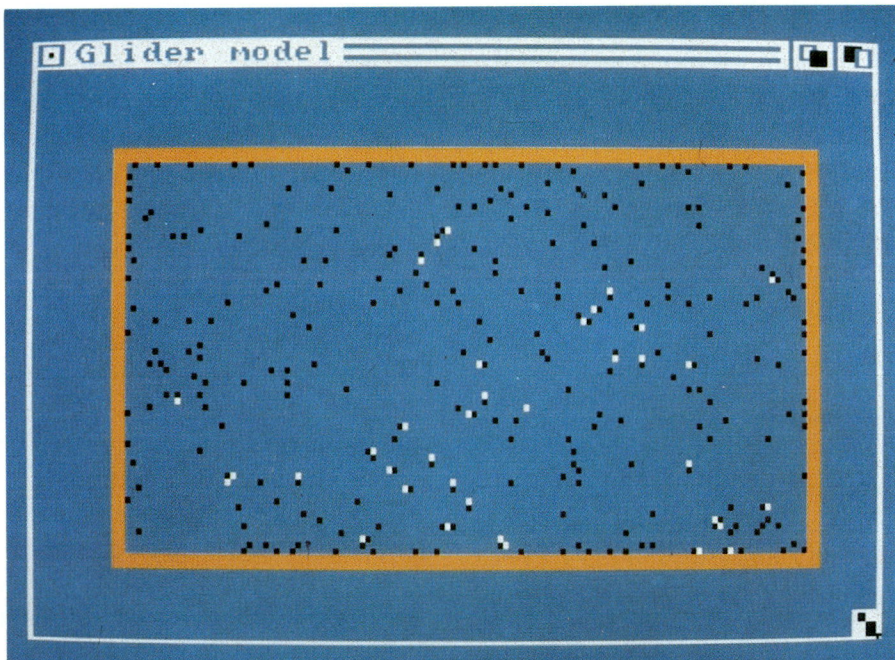

Fig. 3. The toy "universe model". The white sites are the active sites.

Fig. 4. The forest-fire model. The red sites are fires and the green sites are trees.

tive configurations of the critical state. The apparent logical connectivity does *not* indicate that nature is in balance. In analogy with the sand model, any effort to stop the clock and trying to "conserve nature" is a meaningless and certainly loosing battle. The more we try to place "snowscreens" around us to preserve ourselves, or the more we try to restructure nature, the bigger will be the apocalypse when the configuration of nature changes into one where we are no longer present (après nous le déluge)!

4. A toy model of the universe

There is now ample support for Mandelbrot's suggestion that the distribution of bright matter in the universe is fractal, with a dimension $D = 1.2$–1.5 [7]. The fact that the dimension is less than two suggests a way of circumventing Olber's paradox of divergent light intensity for a homogeneous brightness distribution. We have constructed a simple model of an extended system with active propagating particles interacting with other active or passive particles [8]. Starting from a homogeneous "mass" distribution, the model evolves towards a statistically stationary state with a fractal distribution of "bright" active sites.

The model is defined on a regular d-dimensional lattice, with "active" (white), "passive" (black) and empty sites. The rules for updating the system from time t to $t+1$ are:

(i) Active sites at t "burn out" and become passive at time $t+1$.

(ii) Passive sites are annihilated when they have one, and only one, active neighbor.

(iii) Active sites are created when they have one, and only one, active neighbor. This neighbor must have a passive site at the opposite neighbor position.

Rule (iii) allows a black site next to a white site to propagate ballistically along straight lines in free space, in analogy with the "Gliders" of the "Game of Life". Active sites are formed next to active sites, unless there is overcrowding causing an instant burnout. We have performed simulations in two and three dimensions. The correlation between active sites is studied in the following process: starting with a "hot" initial universe with a random distribution of active, passive and empty sites, the system evolves according to the rules (i)–(iii) until it comes to rest with no remaining active sites. The system is then perturbed by adding a single propagating particle or a single passive site. As the process is repeated, the universe evolves into a statistically stationary state. Fig. 3 shows a snapshot of the system at the stationary state. The background of passive particles has become organized into a complex structure which precisely allows the signal to propagate indefinitely; the ballistic motion of individual particles has been transformed into a collective diffusion-type propagation. One might think of the initial state as a "forest fire" which continues until the point where the "burned out" material just cuts the communication.

The bright sites are clearly grouped together in clusters indicating a fractal distribution. Indeed, by measuring the total number of active sites within a radius r from a chosen active site we find the fractal dimension $D \approx 1.6 \pm 0.2$ in two dimensions, $D \approx 1.7 \pm 0.2$ in three dimensions, which happens to be close to the one observed for the universe. The self-organization process solves Olber's paradox of divergent light intensity for this model by lowering the dimension to just below $D = 2$. The critical exponents including the fractal dimension appear to be the same as for the "Game of Life" indicating that the models are in the same universality class. This supports our expectation from the analogy with equilibrium critical phenomena that the scaling properties only depend on general features, such as the existence of dynamically generated propagating particles, and the dimension and symmetry, rather than particular details of the system.

5. A forest-fire model and some ideas on turbulence

The phenomenon of turbulence is essentially not understood. On the geometric aspects, Mandelbrot has suggested that in the turbulent state, energy is dissipated on a fractal set, with dimension slightly higher than two [9]. Some phenomenological models have been proposed [10], in which the fractal set is pre-assumed. It is, however, essential to understand the dynamical mechanism which generates the fractal itself: How can a uniform energy injection result in a fractal dissipation?

We have studied a simple "forest-fire" model. Specifically, we focus on the spatial distribution of dissipation (fire) and its dependence on the driving force (tree growth). The mechanism for fractal dissipation is demonstrated explicitly. Our lattice model is defined in any dimension, with the following simple rules:

(i) Trees grow with a small probability p from empty sites at each time step.

(ii) Trees on fire will burn at the next time step.

(iii) The fire on a site will spread to trees at its nearest-neighbor sites at the next time step.

There is only one parameter in the model, namely the growth rate of trees. We are interested in the limit where trees grow infinitely slowly. We find that the fire is characterized by a correlation length $\xi(p) \approx p^{-\nu}$ [11]. The critical point is at $p=0$. Because of the finite size of our lattices, the probability p has to be non-zero (such that the correlation length is smaller than the system) in order to prevent accidental extinction of the fire. Starting from a homogeneous distribution of trees and fires, the forest fire evolves to a stationary state. Fig. 4 shows a snapshot of the forest on fire, taken in the stationary state after an initial transient period. Note the coherent domains of trees separated by a fractal distribution of fires, indicating that the system is operating near a critical point. By measuring the number distribution $D(r)$ of fire at a distance r from a chosen site on fire, we obtain the fractal dimension: $D=1.0\pm0.2$ in 2d and $D=2.5\pm0.2$ in 3d. This value of D agrees with experimental observations for turbulence [10]. Of course, this could be accidental. If one prefers the language from traditional equilibrium critical phenomena, the fire–fire correlation function $G(x) = \langle f(x')f(x'+x) \rangle$ decays as $G(x) \propto x^{2-d-\eta}$, $\eta = 2 - D \approx 1.0$ in 2d, $\eta \approx -0.5$ in 3d.

In real turbulence, the Reynolds number is a combination of the size of the system and the driving force: $R = LV/\nu$. Similarly, we can define, by combining L and p, a "Reynolds number" R for the forest: $R = Lp^\nu$, which uniquely determines the behavior of the forest fire up to an overall scale. The transition to the steady "turbulent" state of the forest fire occurs at $R = R_c$, where the critical "Reynolds number" is $R_c \approx 1.8$ in 2d and $R_c \approx 0.6$ in 3d. Note that the transition described here is a finite-size crossover effect; so it is for real turbulence: the driving force goes to zero for infinite L. The dependence of the energy dissipation on the "Reynolds number" in the forest fire is also a power law: $E_d = p \approx R^\beta$ with $\beta = 1/\nu$, where the first equation expresses that the system is stationary. The scaling has been confirmed by means of a Monte Carlo renormalization group type calculation.

The model may be rather directly applied to spreading of diseases, propagation of chemical activity, such as real fire. We believe that the model is simple enough to allow for explicit theoretical analysis, for instance renormalization group theories based on expansions around the upper critical dimension. The limitation of the present model when applied to turbulence is obvious. It is a lattice model, thus the "Kolmogorov" length for energy dissipation is a fixed length scale, namely the lattice spacing. The dynamics is simple; for instance, the coherent domains are structureless. However, our study on a specific dynamical model shows explicitly that certain principles are viable: (a) Driven non-equilibrium systems may operate near critical points. This is *not* low-dimensional chaos and the fractal nature of turbulence cannot be described in terms of a strange attractor. (b) Homogeneously injected energy is dissipated on a fractal, as suggested by Mandelbrot.

Acknowledgement

We are grateful to Dr. M. Creuz for providing the color computer graphics. This work was supported by the Division of Materials Science, US Department of Energy, under contract DE-AC02-76CH00016.

References

[1] P. Bak, C. Tang and K. Wiesenfeld, Phys. Rev. Lett. 59 (1987) 381; Phys. Rev. A 38 (1988) 364;
C. Tang and P. Bak, Phys. Rev. Lett. 60 (1988) 2347; J. Stat. Phys. 51 (1988) 797;
P. Bak and C. Tang, Phys. Today 42 (1989) S27.

[2] P. Bak and C. Tang, J. Geophys. Res., submitted for publication;
K. Ito and M. Matsuzaki, J. Geophys. Res., submitted for publication;
A. Sornette and D. Sornette, Europhys. Lett., submitted for publication;
J.M. Carlson and J.S. Langer, Phys. Rev. Lett. 62 (1989) 2632.

[3] T. Mikumo and T. Miyatake, Geophys. J. Roy. Astron. Soc. 58 (1978) 417; 59 (1979) 497.

[4] S.P. Obukhov in: Random Fluctuations and Pattern Growth: Experiments and Models, H.E. Stanley and N. Ostrowsky, eds. (Kluwer, Dordrecht, 1989) p. 336;
Y-C. Zhang, Phys. Rev. Lett., in press;
L.P. Kadanoff, S.R. Nagel, L. Wu and S. Zhou, Phys. Rev., in press;
T. Hwa and M. Kardar, Phys. Rev. Lett. 62 (1989) 1813.

[5] E.R. Berlekamp, J.H. Conway and R.K. Guy, Winning Ways (Academic Press, New York, 1985).

[6] P. Bak, K. Chen and M. Creutz, Nature, submitted for publication.

[7] L. Pietronero, Physica A 144 (1987) 257;
P.H. Coleman, L. Pietronero and L.R.H. Sanders, Astron. Astrophys. 200 (1988) 32.

[8] K. Chen and P. Bak, unpublished.

[9] B.B. Mandelbrot, J. Fluid Mech. 62 (1974) 331; The Fractal Geometry of Nature (Freeman, San Francisco, 1982).

[10] U. Frisch, P. Sulem and M. Nelkin, J. Fluid Mech. 87 (1978) 719;
R. Benzi, G. Paladin, G. Parisi and A. Vulpiani, J. Phys. A 17 (1984) 3521.

[11] P. Bak, K. Chen and C. Tang, Europhys. Lett., submitted for publication.

FRACTAL COLLOIDAL AGGREGATES: CONSOLIDATION AND ELASTICITY

R.C. BALL

Cavendish Laboratory, Madingley Road, Cambridge CB3 0HE, UK

The assumption that colloidal aggregates are rigid must break down beyond a critical size; thermal flexibility then leads to internal contacts which under flocculating conditions become new permanent bonds in the structure. It is argued that these self-consistently raise the spectral dimension and the fractal dimension should also be expected to increase. This may explain the unexpectedly high values of both observed by Courtens and Vacher in Brillouin scattering from alcogels.

The established picture of the aggregation of small solid particles from a dilute colloid is based on computer models introduced independently by Meakin [1] and by Jullien and Kolb [2]. These presume rigid irreversible joining of clusters either at first contact (diffusion-limited) or with low sticking probability leading to more uniform sampling over possible contacts (reaction-limited). These give relatively open and almost loopless aggregates of fractal dimensions ≈ 1.8 [1,2] and 2.1 [3,4], respectively, the latter being higher due to the increased chance of interdigitation as well as high polydispersity. As first pointed out by Kantor and Webman [5], the elasticity of the loopless structure is particularly simple to calculate because the relations between angular displacement and local bending moment reduce in two dimensions to Kirchoff's equations; including twist and torque they can still be bounded both above and below by them in three dimensions as shown by Brown [6].

Simple colloidal flocculation studies confirm the fractal dimensions of the rigid model [7–9]. In most cases the vibrational motion has not been observed, but direct studies of the elasticity of compacted flocs have been shown to be consistent with the simplest theoretical picture [6].

Courtens and Vacher's study [10] of alcogels tests the theoretical picture down to aggregation on a rather smaller, almost molecular, scale and suggests that it breaks down: not only is the fractal dimension significantly higher, $d_f \approx 2.4$, but also the elasticity scales much less softly with length scale than for a rigid loopless structure. (The fact that the loopless elastic behaviour breaks down provides some reassurance that their system has not simply been somehow driven into particle–cluster aggregation.)

The simplest (but not necessarily conventional) approach to fractal elasticity is to consider how the spring constant K relating force δF to displacement δR scales with separation R for two (typical) points on the fractal,

$$\delta F = K \delta R \quad \text{with } K(R) \sim R^{-z}. \tag{1}$$

This then suggests that the frequency of vibration for motion on length scale R should scale as

$$\omega(R) \approx \left(\frac{K(R)}{m(R)}\right)^{-1/2} \sim R^{-(z+d_f)/2}. \tag{2}$$

We also expect the number of such modes [11] (essentially the integrated density of states) to scale as

$$N(\omega) \sim R^{-d_f} \tag{3}$$

so that if the scaling of N with ω is interpreted in terms of a spectral dimension d_s [11] we have

$$N(\omega) \sim \omega^{d_s} \quad \text{with } d_s = 1 + (d_f - z)/(d_f + z).$$

For loopless aggregates we have precisely [5] $z = 2 + b$, where b is the backbone dimension (≥ 1) which gives $z \geq 3$ so that $d_s < 1$ for $d_f < 3$. Realistic values from computer-simulated clusters give $d_s \approx 0.8$ [12], whereas the alcogel study [10] gave best fit to the Brillouin scattering data for $d_s \approx 1.3$. This value of d_s has been compared to estimates ($d_s \approx 4/3$) based on "scalar" or entropic elasticity [11]. This, however, assumes that entropy dominates the structure, that is it is thermally fully floppy. The key point of this paper is that floppy clusters would continually suffer internal collisions, leading under flocculating conditions to extra bonds, higher rigidity and consolidation of the floc. In general this will raise both d_s and d_f: the former effect we quantify below.

How floppy can a floc remain without further consolidation being provoked thermally? This question was first addressed by Kantor and Webman [5] and the simplest criterion is that the thermal strain

$$\epsilon(r) = \left(\frac{dR}{R}\right)_{\text{rms}} \approx \frac{(k_B T/K)^{1/2}}{R} \sim R^{z/2-1} \quad (4)$$

be of order unity, so that $z=2$. In practice, we would expect that at small enough scales (up to some length a_T) the loopless structure would still be sufficiently rigid so that

$$z_{\text{eff}} = 2 + b, \quad R < a_T, \quad (5)$$
$$= 2, \quad a_T < R < \xi,$$

where ξ is the correlation length scale above which the floc is constrained to be uniform by its non-zero density.

The above analysis assumes a truly harmonic elasticity, and also ignores the difficulty that for a harmonic oscillator the maximum amplitude explored increases logarithmically with time beyond the thermal amplitude. Thus the true criterion for no further consolidation becomes

$$\epsilon(R) < \epsilon_c(\omega(R)t), \quad (6)$$

where the right-hand side is both small and very weakly decreasing at large times.

The net result is that in the consolidated regime we will have $z=2$ and so $d_s > 1$ without the requirement that the structure be completely floppy. The actual value predicted for d_s using $z=2$ and the observed fractal dimension of 2.4 is $d_s = 1.3(2)$, in excellent agreement with experiment. Whilst there is little scope for the scalar elasticity value to be greatly different, there is no intrinsic reason for them to coincide.

There is no (model-independent) fixed relation between the increase in rigidity and the densification, i.e. increase in d_f. However, we can address the question as to whether the observed values of $d_f \approx 2.4$ are attainable by consolidation after the gel has been formed, given the limited degree of densification observed in the alcogel systems.

Assuming that the structure is unchanged on scales below a_T and that the mass in a correlation blob (i.e. up to scale ξ) is preserved, then

$$(\xi'/a_T)^{d_f'} = (\xi/a_T)^{d_f} \quad (7)$$

so that the degree of linear shrinkage is given by

$$\xi'/\xi = (\xi/a_T)^{(d_f - d_f')/d_f'}. \quad (8)$$

Thus for $d_f' - d_f \approx 0.3$ and assuming a linear shrinkage of no more than 10%, we could not have consolidated structure over a wider range of length scales than 3, which is far too small to be consistent with the experimental data. We are thus forced to conclude that much of the consolidation occurs before the aggregates have spanned into a gel, in this way it does not contribute to observed macroscopic shrinkage.

In conclusion, the hypothesis of consolidation and $z=2$ brings the spectral dimension into excellent agreement with experiment, but only provided the experimental value for the fractal dimension is used. To achieve this high value of d_f by densification after gellation would require an amount of shrinkage inconsistent with experiment. It is inferred that the consolidation must largely occur as the clusters aggregate and before the gel is complete.

The author wishes to acknowledge discussions with D.A. Weitz and P.B. Warren.

References

[1] P. Meakin, Phys. Rev. Lett. 52 (1983) 1119–1122.

[2] M. Kolb, R. Botet and R. Jullien, Phys. Rev. Lett. 51 (1983) 1123-1126.
[3] R. Jullien, M. Kolb and R. Botet, J. Phys. A 17 (1984) L75.
[4] W.D. Brown and R.C. Ball, J. Phys. A 18 (1985) L517-L519.
[5] Y. Kantor and I. Webman, Phys. Rev. Lett. 52 (1984) 1891.
[6] W.D. Brown, Ph.D. Thesis, University of Cambridge (1986).
[7] D.A. Weitz and M. Oliveria, Phys. Rev. Lett. 52 (1984) 1433-1436.
[8] D.W. Schaeffer, J.E. Martin, P. Wiltzius and D.S. Cannell, Phys. Rev. Lett. 52 (1984) 1433-1436.
[9] M.Y. Lin, H.M. Lindsay, D.A. Weitz, R.C. Ball, R. Klein and P. Meakin, Nature (1989), in press.
[10] E. Courtens and R. Vacher, Proc. Roy. Soc. A 423 (1989) 55-69.
[11] S. Alexander and R. Orbach, J. Phys. Lett. 43 (1982) L625.
[12] I. Webman and G. Grest, Phys. Rev. B 31 (1985) 1689.

ORDERED SHAPES IN NONEQUILIBRIUM GROWTH

Eshel BEN-JACOB and Peter GARIK

Department of Physics, University of Michigan, Ann Arbor, MI 48109, USA
and School of Physics and Astronomy, Tel Aviv University, 69978 Tel Aviv, Israel

Patterns observed during nonequilibrium growth display complex ordering on many length scales. We focus on ordered patterns which reflect the interplay of microscopic and macroscopic dynamics. The fundamental morphologies which result, and which are the building blocks of more complex patterns, include dendritic and tip-splitting growth. The latter gives rise to the two-dimensional dense-branching morphology (DBM). We review the current understanding of how dendritic growth and the DBM arise from the microscopic dynamics of surface tension and surface kinetics. We emphasize the open questions, with particular attention to the question of developing theory for morphology selection and transitions between dendritic and dense-branching growth. In this context, we review our hypotheses of the selection of the fastest growing morphology, and the existence of first- and second-order-like morphology transitions. Theoretical issues are illustrated using the Hele-Shaw and electrodeposition experiments.

1. Introduction

We are surrounded by a nature out of equilibrium, a nature which presents the scientist with a bewildering and mesmerizing universe of patterns. A principal challenge to physicists is to understand the geometry of these nonequilibrium patterns, whether they arise as physical objects – mountains, snowflakes, dust motes – or as mathematical abstractions, e.g. time correlations, or density fluctuations. The quantification of the geometrical properties of these nonequilibrium systems, and our understanding of the dynamics which gives rise to these geometries, has made large gains over the past decade. In large part this has followed from a recognition that nature supercedes Euclid's Elements, and that much of her patterning is best understood with the "fractal" geometry described by Mandelbrot [1].

Concurrent with the scientific community's recognition that nature is not restricted to patterns of Euclidean dimension, these past few years have seen the development of a new understanding of specific patterns that recur during nonequilibrium growth[#1]. These patterns, the dendrite and tip-splitting "fingering" growth, appear on system-dependent length scales varying by many orders of magnitude. In effect, they are the short-length-scale building block morphologies from which are composed the more complex patterns visible on larger length scales. While the larger patterns which develop during solidification, aggregation, or condensation still require new insights to explain the global morphology assemblage, significant progress has been made in understanding the determining physics of dendritic and tip-splitting growth, and their interrelationship as fundamental morphologies.

The cornerstone of the recent developments is the recognition of the interplay of microscopic interfacial dynamics with external macroscopic forces in the determination of growth patterns. Most of the research has focused on systems where the macroscopic dynamics are determined by a diffusion field. We now understand that for these systems, the patterns that form, result from competition between the

---

[#1] By 1611 the great astronomer Johannes Kepler [2] was already captivated by the beautiful shapes of snowflakes, which is perhaps the most striking example of pattern formation in inorganic systems. See also ref. [3]. For a review of the previous phase of the research on pattern formation during solidification see ref. [4].

diffusion field on the one hand, and the microscopic dynamics of the interface on the other. The patterns may be grouped into a small number of typical "essential shapes" or morphologies, observed in different systems and over many different length scales (from meters to micrometers). These are the faceted [5], dendritic [4], dense-branching [6] and fractal [7,8] [2] morphologies (see fig. 1). It is the purpose of this short review to provide a perspective on the advances in the field of morphology selection with an emphasis on general principles and the remaining open questions.

2. The selection problem for dendritic growth

The principal mystery in how microscopic dynamics, operative on the scale of angstroms, is amplified to the extent that in a system out of equilibrium it controls the macroscopic shape on a scale of centimeters. From a theoretical perspective herein lies the rub. The natural inclination is to attempt theories of growth emphasizing macroscopic dynamics and relegate the microscopic dynamics to subsequent refinements of theory. Indeed this was how the theory of dendritic growth initially evolved. In 1947 Ivantsov showed [10] that propagating solutions, with a parabolic shape, exist for a solid forming from an undercooled melt by assuming only diffusion control of the heat field but *neglecting surface tension and surface kinetics*. Both the parabolic shape and the predicted constant velocity fit well a semi-quantitative description of a dendrite. However, a conundrum comes with the Ivantsov solution: it specifies only the product of the dendrite tip's radius of curvature and velocity, but cannot predict either one alone. The 1976 experiments of Glicksman et al. [11] demonstrated that under controlled conditions, *for given undercooling* the same dendrite (i.e. same tip velocity and radius of curvature) is reproducibly observed. This implies a "selection problem": for given undercooling the

Ivantsov solution admits a continuous family of parabolic solutions, and yet for specified conditions only one is observed. Moreover, it was shown that these Ivantsov solutions were also "linearly unstable", meaning that they would be unable to maintain their shape during growth.

Sensibly the first attempts to resolve the stability problems were based on a hope that incorporation of surface tension would involve only a minor shape modification of Ivantsov's parabolic fronts, while stabilizing all parabolas below a characteristic length scale. However, the selection problem remained inherent in this. In 1973 Oldfield [12] proposed that the selected dendrite was the one moving with the minimum speed (or maximum radius of tip) for which the surface tension can stabilize the underlying needle-crystal. Oldfield's idea was revived and elaborated in 1977 by Langer and Müller-Krumbhaar [13], who performed extensive calculations in order to find this marginally stable operating point.

The real breakthroughs in understanding dendritic growth waited until this decade, and arrived with results of broader significance for morphology determination. The resolution required sufficient computing power and the subsequent application of more advanced mathematical methods, to properly incorporate the microscopic dynamics. The surprise was that despite their small size, surface tension and surface kinetics are singular perturbations in the dynamical equations for interface evolution. Singular perturbations, no matter how small, totally change the character of the solution. As such, the microscopic dynamics cannot be treated as small corrections to solutions initially determined from the macroscopic dynamics. What has emerged is that when surface tension and surface kinetics are isotropic, dendritic growth does not occur. Instead, tip-splitting fingers develop leading to the dense-branching morphology. Anisotropy is required in the interfacial dynamics to produce dendritic growth [3]. This picture is now con-

---

[2] For a review of the recent developments in the study of diffusion-limited aggregation (DLA) growth see ref. [9].

[3] This was first demonstrated in a local model of solidification – the boundary layer model [14].

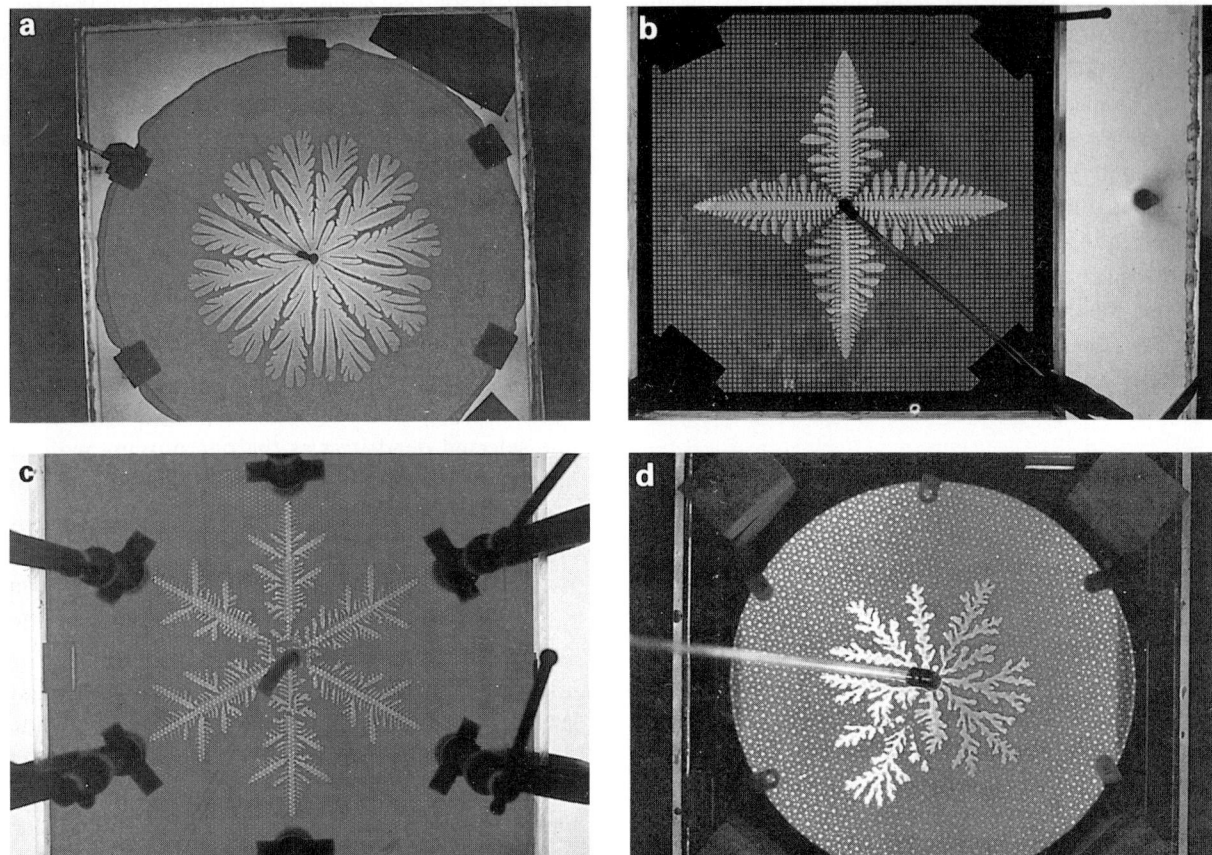

Fig. 1. The "essential shapes" in the Hele-Shaw experiment. As we explain in the text the endless array of shapes can be grouped into a small number of characteristic shapes reflecting different dominant effects. The same shapes (the geometrical characteristics) are observed in different systems and on different length scales (from meters to micrometers). (a) The dense branching morphology. The characteristic morphology in the absence of crystalline anisotropy can be characterised by the number of branches as function of the radius (the spacing between the branches and the branches' width are the characteristic length scales). Note that it has a well-defined circular envelope. (b,c) Dendrites. As we now understand anisotropy is required for dendritic growth to occur. The dendrites are characterized by a trunk with a parabolic tip moving at constant velocity [4]. The trunk is feathered with side branches which grow outward while being stationary in the laboratory frame. Under some growth conditions sidebranches are observed. We refer to the underlying trunk (the dendrite without the sidebranches) as a needle-crystal. The dendrites have characteristic length scales: the radius of the tip and the spacing between sidebranches. (d) A fractal-like shape [1,7–9]. It is characterized by the fractal dimension (related to the mass distribution as a function of the radius). This is the typical shape in the limit of small surface tension and large noise. It appears similar to the structure produced by the diffusion-limited aggregation (DLA) model [7,9]. (a) A typical example of an air DBM. The applied pressure is about 150 mbar and the spacing between the plates is about 0.4 mm. (b) A typical example of a four-fold "snowflake" in an anisotropic Hele-Shaw cell with a fourfold symmetry. Note the parabolic shape of the tip of each arm and the train of side branches shooting out from each one. (c) An example of a "snowflake" in an anisotropic Hele-Shaw cell with sixfold symmetry. We varied the applied pressure during the growth. This leads to a "decorated" structure. Similar situation occurs when real snowflakes are formed in clouds moving through regimes with different levels of supersaturation. (d) A DLA-like shape developed in the case of large noise and no surface tension: A random array of channels is engraved on the bottom with no space between the plates. The top plate was placed flat on the bottom one. This way the fluids in two adjacent channels are not connected leading to effectively no surface tension.

firmed by both experimental [15,16] and theoretical results [14,17] [#4].

New selection problems emerged with the broader understanding of pattern determination. The puzzle is no longer the dendrite's velocity and shape alone. Now the questions are as to why dendrites are selected for some parameters, and tip-splitting growth for others. To study these new selection problems we turn to an experiment in which the interfacial dynamics is expressed on the same length scale as the pattern [15,19,20]. These experiments, discussed below, permit unambiguous demonstration of morphology selection as a function of anisotropy.

3. The dense-branching morphology

To study pattern formation for isotropic interfacial dynamics, we used a modification of the Hele-Shaw cell [#5]. This simple, yet elegant, device for studying pattern formation consists of two closely spaced plexiglass plates sandwiching a layer of viscous fluid – here dyed glycerine. The top plate is circular and open to air at its edge. Through an inlet at the center of the top plate a less viscous fluid (e.g. air or water) is injected into the glycerine.

In fig. 1a an example of the dense-branching morphology (DBM) in the cell is shown. It consists of a circular envelope modulated by leading branch tips. The regular tip-splitting of the fingers distinguishes them from dendrites. Moreover, the lacunae or gaps between the fingers do not grow with the area of displaced fluid. This distinguishes the mass distribution of the DBM from that of a fractal object, such as a diffusion-limited aggregate (DLA), where the gaps grow ever larger with the overall size of the pattern.

Instead, the DBM grows as a two-dimensional object, although in some cases it may approach $d=2$ only asymptotically [20].

Experimental evidence supports the conclusion that, in the absence of anisotropy, the DBM is the generic morphology [6,20]. This is contrary to the argument that fractal growth should be the usual *macroscopic* morphological organization [25] [#6]. The DBM is observed in aggregate growth by electrochemical deposition [#7] and precipitation from supersaturated solution [29]; during solidification from undercoooled melts [30] [#8]; arising during amorphous annealing [6]; and in spherulitic growth [31].

Our present understanding of the DBM is based on analyzing its branching rate, as opposed to its mass distribution or coastline. Generally, we expect the branching rate, or surface modulation, to be the result of the interplay between the macroscopic diffusion field, which tends to make the interface irregular, and the microscopic effects of surface tension and surface kinetics. These introduce cutoff lengths and define the length scale of ordered growth. Turning again to the Hele-Shaw example, the governing equations for the pressure field p are:

$$\nabla^2 p = 0 \quad \text{(Laplace's equation)},$$

$$p_s = p_{app} - d_0 \kappa - \beta v_n^\gamma$$

(Gibbs–Thomson with kinetic terms),

$$v_n = \frac{-b^2}{12\eta} \nabla p \cdot \hat{n} \quad \text{(D'Arcy's law)}.$$

Here p_s is the pressure in the fluid at the interface, p_{app} is the pressure applied to the less viscous fluid, d_0 is the interfacial surface tension, κ is the curvature of the interface, $\hat{n}$ is the unit normal to the interface in the direction of the viscous fluid, β is the "kinetic"

[#4] For a more mathematical description of the new developments, see ref. [18].

[#5] The original Hele-Shaw cell was a long narrow channel containing a fluid sandwiched between two plates. It was used to study the flow of water around a ship's hull [21]. Variations of the Hele-Shaw cell are very much in current use. See for example, ref. [22] and the review paper by Bensimon et al. [23]. The circular geometry was first employed by Paterson [24].

[#6] Following ref. [4], first dendritic growth was expected in solidification when isotropic surface tension is included. Later it was expected that DLA-like patterns with finite width branches would develop. See for example ref. [26].

[#7] The dense branching morphology was first observed during electrochemical deposition, see for example refs. [27,28].

[#8] Fujioka observed dendritic growth in solidification of water at higher undercooling and tip-splitting at low undercooling.

coefficient, b is the spacing between the plates, and η is the fluid viscosity. The exponent γ has been computed as $\frac{2}{3}$ for a uniform wetting layer, would be different for a non-Newtonian fluid, and is taken as unity in our simple analysis below. Although the differential equation for the pressure field here is Laplace's, the physics of these equations is nearly isomorphic to problems where the governing equation is strictly the diffusion equation, e.g., precipitation from supersaturated solution, or solidification from an undercooled melt.

Linear stability analysis [32] [#9] can be used to investigate the branching rate of the DBM. Indeed, the instability of a diffusion-controlled interface to any perturbation in the absence of a Gibbs-Thomson-like stabilization, is the Mullins-Sekerka instability. Using the above equations, we can compute the relative growth rate of perturbations on a disk of radius R. Taking the perturbation to be of the form $r(\theta) = R + \delta_m \cos(m\theta)$ for small δ we find that:

$$\alpha_m(x) \equiv \frac{\dot{\delta}_m/\delta_m}{\dot{R}/R}$$

$$= -1 + m\left(x + \bar{\beta} - \frac{(m^2-1)(x \ln x^{-1} + \bar{\beta})}{\xi x - 1}\right)$$

$$\times \left[m\bar{\beta} + x\left(\frac{1-x^{2m}}{1+x^{2m}}\right)\right]^{-1},$$

where

$$\xi = \frac{p_g - p_0}{d_0/R_0},$$

$$\bar{\beta} = \beta b^2 / 12\eta R_0,$$

and

$$x = R/R_0,$$

R_0 the cell radius. Fig. 2 shows that there is a fastest growing perturbation. Experimentally, good agreement is found between this fast growing mode and the number of branches as a function of radius [6,34]. This analysis is consistent with the DLA-like dimen-

[#9] A similar linear analysis to that presented here was performed independently by Schwartz [33].

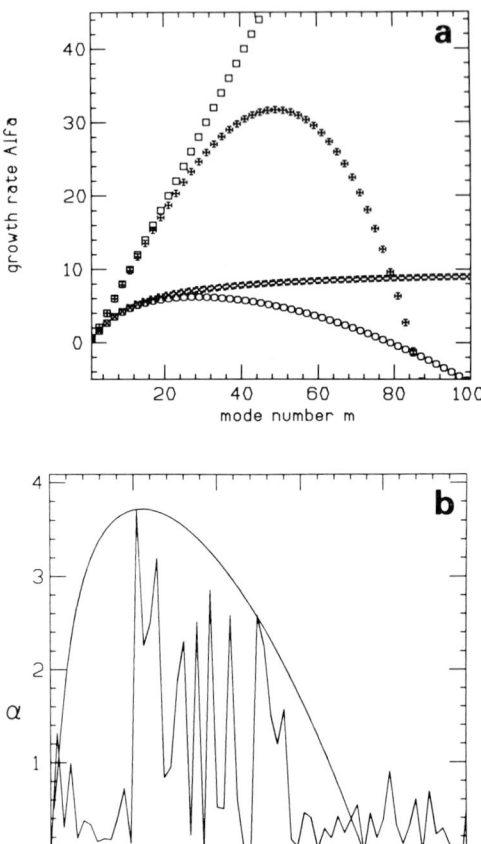

Fig. 2. The DBM and the linear stability analysis. (a) Results of the linear stability analysis, showing the initial growth rate as function of the mode number m for a sinusoidal perturbation d of a circular interface. Line (1) is for constant pressure along the interface (no surface tension and no surface kinetic), line (2) is when surface tension is included and line (3) is in the presence of surface kinetics. Note that in the latter case there is no fastest growing mode. Line (4) is in the presence of both surface tension and surface kinetic. (b) shows a comparison of the linear stability and an experimental power spectrum of the envelope of a DBM developed in the Hele-Shaw cell. See ref. [20] for more details.

sion of mass of small Hele-Shaw patterns; however, the branching rate increases with x such that past a critical value, determined by system parameters, the growth is $d=2$ as observed visually [20].

In the absence of stabilizing effects at the interface, such as surface tension or surface kinetics, the diffusive Mullins–Sekerka instability results in an unstable interface. The result is noisy, apparently fractal, growth. This is what happens in the modelling of growth by the DLA algorithm [7,9], where there is no surface tension. This is also the case when two miscible fluids are used [35], or glass beads are added [36], in fluid flow experiments. In both of these cases the net result is to reduce the interfacial stabilization effect of surface tension.

The nature of the envelope of the DBM is an integral part of the dynamics of the pattern but its understanding requires a nonlinear analysis [#10]. A naive understanding is that if one finger outgrows the others, it has more space to spread out; part of the flow goes sideways and the finger flattens and slows down. In our view, the most pressing unsolved problem is to understand the branching rates and velocity of DBM growth [19,31]. The latter is especially intriguing because there may be a selection mechanism operating with respect to the branching rate and interfacial velocity akin to the tip-radius and velocity of the dendrite selection problem [38].

4. Dendritic growth and morphology diagram in the anisotropic Hele-Shaw cell

The Hele-Shaw cell can also be used to study anisotropic growth, analogous to the solidification of a crystalline material. There is a strikingly simple way to mimic crystalline anisotropy in the fluid cell: one engraves channels on one of the plates. The channels modulate the spacing between the plates so as to create deep and shallow paths for the flow of fluid. When the grooved lattice has six-fold symmetry (three sets of parallel channels oriented at 120° to each other), the air bubble adopts beautiful snowflake-like shapes with six dendritic arms. Since snowfall on Mars is composed of CO_2 flakes with fourfold anisotropy, a fourfold lattice produces the "Martian snowflakes"

[#10] In ref. [37] it is claimed that the circularity of the envelope in electrochemical deposition can be explained solely on the basis of linear stability analysis when the aggregate has a finite resistivity.

of fig. 1b. The emergence of dendrites in the Hele-Shaw cell [15] provided the first direct experimental demonstration that anisotropy is needed for dendritic growth to occur.

But not only dendritic growth is observed in the presence of anisotropy. As we vary the applied pressure (the driving force) the air bubble assumes different shapes. Similarly, different morphologies are observed as the "microscopic" growth conditions are changed. For example, we can change the level of anisotropy simply by changing the spacing between the plates. The idea of a "morphology diagram" to organize the observations follows naturally [19, 20].

Fig. 3 depicts a morphology diagram for a cell with sixfold anisotropy. Faceted growth, the DBM, and *two types* of dendrites now occur, for different values of the applied pressure. As we will see later on, the existence of two types of dendrites play an important role in our understanding of morphology transitions. A morphology diagram is also observed in electrochemical deposition experiments [27,28], Hele-Shaw cells using liquid crystals as the viscous fluid [16], and solidification from supersaturated solutions [39].

Several questions arise now: why does anisotropy trigger dendritic growth, and if it does, why is the DBM still observed? Why do some dendrites appear when the system is driven far from equilibrium, but different ones appear close to equilibrium? Is there a general selection principle that will determine which growth will be observed for specified conditions, leading to a theoretical understanding of morphology diagrams?

5. Anisotropy and the formation of dendrites

The first understanding and clear demonstration of the singular nature of microscopic effects, and the need for anisotropy for dendritic growth, emerged from the study of two simple models of interfacial growth [14,17]. The immediate goal in the construction of these models was to pinpoint the physical effects that were essential for dendritic growth. Earlier attempts without the drastic simplifications of the lo-

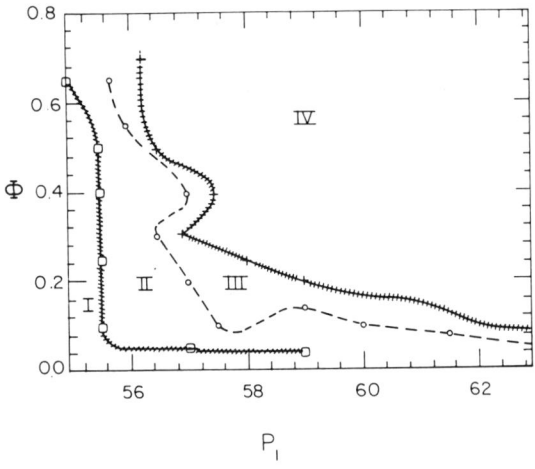 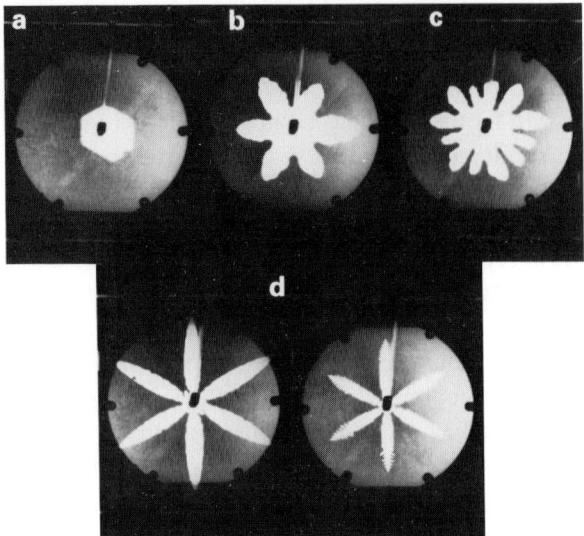

Fig. 3. Morphology diagram for a sixfold anisotropic Hele-Shaw cell (the cell is the same as described in refs. [15,19].) Here P_1 is the applied pressure measured roughly in centimeters of Hg (the actual manometer fluid was a light oil). The anisotropy of the cell is measured by the ratio $\Phi = b_1/(b_0+b_1)$, where b_1 is the depth of the grooves (0.015 in) and b_0 is the additional spacing between the top plate and the top of the grooved plate. The morphology regions are (I) faceted growth; (II) surface-tension dendritic growth (with careful inspection it is possible to observe that the dendrites point at an angle of 30° to the ruling of the grooves); (III) tip splitting growth; (IV) kinetic dendritic growth. The needle crystals grow parallel to the ruled channels. Cross hatching of curves separating labelled morphology regions indicates the possible existence of narrow regions of other morphologies, e.g. between regions I and II there is evidence for DBM growth. The dendrites pointing at 30° from the channels occur at lower pressure, for which surface tension is the crucial factor, hence the name "surface tension dendrites". In the directions of the tips of the dendrites the change in interfacial energy when moving the interface is smaller, making it easier to bend the interface. The effective surface tension is weaker and the surface tension anisotropy "prefers" these directions. At higher pressure the surface kinetics dominates, "preferring" dendrites that point along the channels [19] as the relative wetting effects are smaller and the velocity (for the same pressure gradient) is higher. Hence the name "surface kinetic dendrites". In the dividing regime the two effects (surface tension and surface kinetics anisotropy) are of comparable strength and cancel each other, leading to vanishing effective anisotropy and hence a DBM growth. There are additional details not presented in the morphology diagram. For example as the pressure is increased kinetic dendrites with different structure of sidebranches are observed. Only the main features of the morphology diagram are presented here. (a)–(d) are faceted, surface tension dendrites, DBM and surface kinetic dendrites, respectively.

cal models had failed because of the difficulty of solving even numerically the full diffusion problem. In the local models, the interface is treated as a dynamical entity (a "string" in two dimensions). Physical arguments are then used to deduce its equation of motion so as to include microscopic effects at the interface. One such model, the boundary-layer model (BLM), was inspired by solidification from an undercooled melt [14]. In this model the diffusion field is represented by a boundary layer around the string.

The great advantage in adopting such a local model is its numerical tractability. This allows a time-dependent solution on the computer explicitly showing the interfacial evolution. In this way it was first recognized that in the absence of anisotropy tip-splitting growth occurs, and that anisotropy is necessary for stable dendrite evolution. This latter observation was first considered to be an artifact of the boundary-layer model and not a reflection of universal behavior. Only after the demonstration of the anisotropic Hele-Shaw experiment was the role of anisotropy widely accepted.

A heuristic argument can be provided to explain the role of anisotropy in stabilizing and "selecting" the dendrite's tip. Consider a parabola of the form $y = -ax^2$. Let θ be the angle between a surface normal and the y-axis direction. The simplest way to introduce anisotropy is in the surface tension $d(\theta)$ where the angular dependence is presumed to arise from variations of surface tension with different crystallographic orientations. Explicitly we write the Gibbs–Thomson relationship relating the surface temperature T_s and the melting temperature T_M:

$$T_s = T_M - d(\theta) \kappa$$

and

$$d(\theta) = d_0 [1 - d_1 \cos(6\theta)]$$

for the case of sixfold symmetry. First consider the case of no anisotropy, $d_1 = 0$. Then the tip, $\theta = 0$, is the coldest point on the interface. As such it experiences the maximum temperature gradient and is the fastest growing point on the interface. The dynamic response of the system to this circumstance must be diffusion of heat along the interface toward the tip. This will cause the tip to slow down. We further reason that symmetrically placed points can develop on the interface which grow more rapidly than the tip. The result is a splitting of the tip as these other points overwhelm it. To avoid this scenario and permit a stable tip, heat flow towards the tip must be suppressed. This is exactly the effect provided by crystalline anisotropy [11]. With anisotropy the coldest point moves away from the tip to a point with a different growth direction. For large enough anisotropy (of the order of a percent or so), this will lead to a rather subtle interplay of the anisotropy and the possible needle-crystal. For a given anisotropy only the needle-crystal with the right tip velocity and tip curvature will feature the coldest point at the right temperature and the right distance from the tip (in terms of arclength) to exactly balance the original tip-splitting dynamics. The result is that instead of the original (Ivantsov) continuous family of parabolas only a discrete set of needle-crystal solutions (with close to parabolic shape) can satisfy the subtle interplay giving rise to a "solvability" criterion.

This argument motivates the role played by anisotropy in the existence of stable steady-state needle-crystal-like solutions. But, how does this lead to dendritic growth composed of a needle-crystal trunk decorated with side-branches, *and the selection of a specific dendrite*? If we perturb the needle-crystals, say by introducing a bulge near the tip, the perturbation will grow according to the diffusive instability referred to above. The bulge will grow outward at a fixed position in space, so as the tip advances, the perturbation moves backward if viewed from the tip. It can be imagined that only for the fastest needle-crystal will the bulge move backward faster than its growth rate, allowing the tip to restore its shape despite the growing perturbation. This mechanism results in decorating the *fastest* needle-crystal with side branches, and turning it into the observed dendrite, the only one that can exist.

The numerical formulation of this "microscopic solvability" criterion was discovered independently in 1984 with both the geometrical [41] and the boundary-layer models [42]. Moreover, despite the hand-waving nature of the arguments above, it has since been shown that of the discrete set of possible needle-crystals only the fastest is linearly stable. By now it is known that the same mechanism is present in the full solidification problem [43], and for Hele-Shaw in a channel geometry. This latter is known as the Saffman–Taylor problem with the channel walls providing the necessary anisotropy for selection [44]. Most recently analytic methods have been developed to compute the selected velocity in the limit of small undercooling and small anisotropy [45]. Additional support for the selection principles is provided by comparison with recent time-dependent supercomputer simulations of the full solidification problem [46]. The ultimate test, however, of the new selection hypothesis must be by comparison with experiments. Although the preliminary results are promising [47], much more study is required.

A remaining question is the nature of side-branching during dendritic growth. Much attention is given

---

[11] Other means of anisotropy can also lead to the formation of dendrites, see for example, ref. [40].

to this question now [48], with the debate focusing on the relative role of noise as opposed to deterministic dynamics in the growth of side branches. Either they emerge as a result of noise that excites the diffusive instability and linear stability is sufficient to predict their evolution, or an additional solvability principle is required. We believe the latter to be the case; however, we must leave this topic outside the scope of the present article.

6. The "fastest growing morphology" selection hypothesis and the morphology diagram

Despite the discovery of the microscopic solvability criterion, the problem of dendritic growth is not fully resolved [49]. Time-evolution studies of the interface in the boundary-layer model, and the anisotropic Hele-Shaw experiment, present a nagging problem. Both show that even with anisotropy present, dendrites are not always observed. As we decrease the driving force (pressure in the Hele-Shaw cell and undercooling in the BLM) there is a critical value below which dendritic growth is no longer observed; instead tip-splitting (the DBM) occurs. Similar behavior is also observed during freezing of water [30]. These results contradict the selection principle for dendritic growth, which suggests that as long as anisotropy is present, a specific dendrite (corresponding to the fastest needle-crystal) can exist and is linearly stable. The observation of the DBM under growth conditions suitable for dendrites as well means that with present theory the two morphologies can coexist. "Microscopic solvability" can clearly be only part of the picture. A more general principle is needed to distinguish between different morphologies and determine the one which is selected.

We have proposed [19] the more general principle that *it is the fastest growing morphology which is the dynamically selected one*. That is, if more than one morphology is possible, only the fastest one is nonlinearly stable and will be observed. Thus, one might infer that below some critical driving force the velocity of the DBM is higher than that of dendritic growth,

and so the former is selected. Motivated by our Hele-Shaw experiment we have also studied the case of competing anisotropies. Both surface tension and surface kinetic anisotropies are included and they have preferred growth directions offset by 30° as in the sixfold Hele-Shaw cell. We calculated the selected velocity both along the surface tension and the surface kinetic directions. The results (fig. 4) show that above a critical undercooling Δ_c both types of dendrites are possible, with the surface kinetic ones having the higher velocity. Time-dependent simulations of the BLM demonstrate that indeed the surface kinetic dendrites are the dynamically selected morphology in this regime.

Let us explore further the analogy between phase and morphology diagrams. For phases in equilibrium, for a given set of conditions the phase that minimizes the free energy is the selected one, independent of the prior history of the system; the concepts of a selection principle and a phase diagram go hand in hand. In contrast, nonequilibrium growth processes are time dependent, so it is not clear a priori that a morphology diagram should exist (that is, that the shape will depend only upon the growth conditions and not on the history). However, if it does exist, a selection principle must exist if a given morphology is reproducible for a given set of growth conditions. Given such a morphology selection principle, it is possible to generate a map of what shapes should be observed for what growth conditions. The existence of a morphology diagram has been confirmed experimentally in various systems, suggesting that a selection principle must exist. Is this principle the "fastest growing morphology" hypothesis that we have proposed? We believe that the latter is not the most general principle we seek, but is a step in the right direction.

When a system is driven out of equilibrium by the imposition of a gradient in one of the thermodynamic variables (e.g. the temperature or the concentration), the response of the system is described by the conjugate flux (the heat flux and particle flux, respectively). These fluxes may in general be viewed as the rate of entropy production, or the rate of ap-

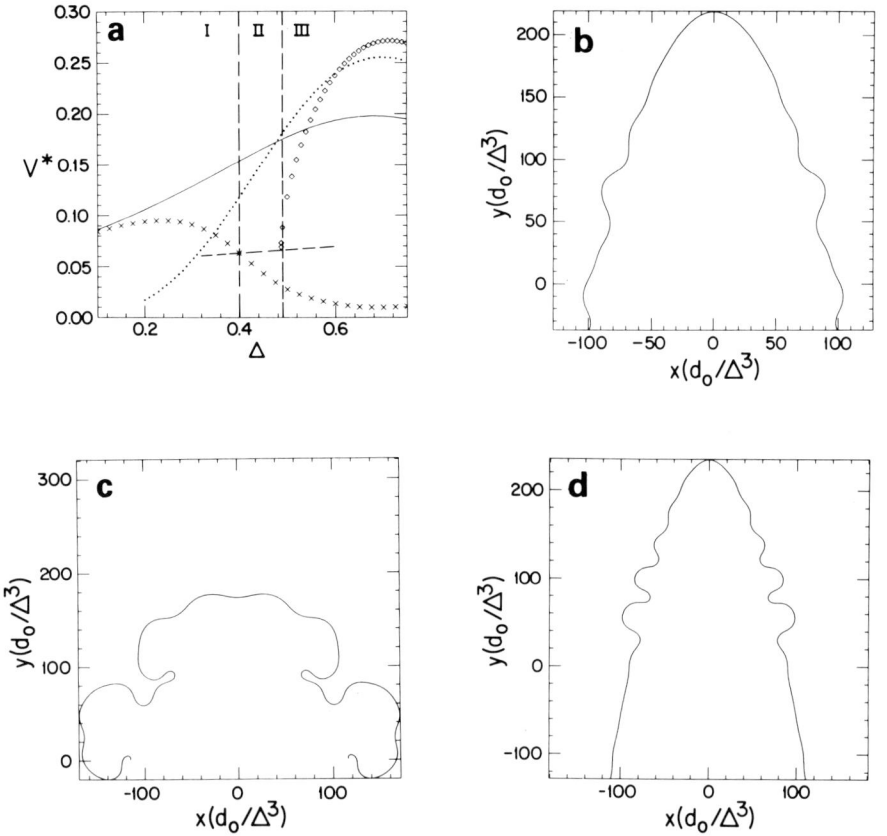

Fig. 4. (a) The morphology diagram and the needle-crystal selected velocity v^* (in the dimensionless units of refs. [19,20]), for competing anisotropies in the BLM. Both surface tension and surface kinetics anisotropies are included and they have preferred growth directions offset by 30° as in the sixfold Hele-Shaw cell. We have calculated the selected velocity of both the needle-crystals pointing in the preferred direction of the surface tension anisotropy (the crosses) and of those pointing in the surface kinetic preferred direction (diamonds): for more details see ref. [19]. The dashed lines represent our expectation of the DBM velocity. The insets show results of the time-dependent simulations in the three regimes of the morphology diagram. (I) "Surface tension" dendrites (pointing in the surface tension preferred growth directions), (II) tip-splitting, and (III) surface kinetic dendrites. In regime II the two anisotropies are close in effective strength (when acting alone, the two lead to dendrites with similar velocities as is discussed in ref. [19]). The result is a dramatic decline in the selected velocity of the surface tension dendrites, the disappearance of the surface kinetic ones and the appearance of tip-splitting. Above Δ_c the surface kinetic dendrites have the higher velocity and are the observed morphology. There is a jump in the velocity at Δ_c hence the DBM↔surface kinetic dendrites is a first-order transition. The surface tension↔DBM is a second-order (change in the slope of the velocity as function of Δ). (b), (c) and (d) are examples of time evolutions in regimes I, II and III respectively.

proach towards global equilibrium. In growth processes, specifically, the driving force (e.g. the undercooling in solidification) is the equivalent of the thermodynamic gradient. The average velocity measures the rate of approach towards equilibrium, and serves naturally as a response function. But the global rate of change of the free energy (at the interface) is given by the integral of the velocity along the interface. Thus, by the term "average velocity" we mean the velocity weighted according to the geometry of the interface, and thus take into account the global shape of the object. We expect this "average velocity" to be an important variable, but by no means the only one. It should have a counterpart (at present unknown) that will represent the equilibrium properties of the interface and the selected growing phase.

The fastest growing morphology is probably a good approximation of the general selection principle far enough from equilibrium, where the rate becomes the more dominant part in the competition. It may be viewed as a high-temperature limit of an equilibrium system coupled to a heat bath where the entropy dominates. In the same way we expect that far from equilibrium the entropy production is dominant in selecting the morphology.

The analogy with equilibrium systems may be carried even further. We have proposed the existence of two types of morphology transitions [19], as we vary the growth conditions, in analogy to phase transitions in equilibrium. The first kind shows a discontinuous jump in the velocity at the transition point (hence classified as a first order morphology transition). In the other type (characterized as second order), the velocity itself is continuous as the morphology changes, but shows discontinuity in its derivative.

In fig. 4 we show an example of both first-order and second-order morphology transitions found in the BLM. Again we wonder: is this a general phenomenon or an artifact of the BLM? Chan et al. have made a careful study of solidification from supersaturated NH_4Cl solutions [39]. In particular, their experimental data include information about the velocity of growth which fits well in the framework of morphology transitions described above. They found that, corresponding to changes in crystallographic orientation of the growing dendrites, there was either a jump discontinuity (first order) or a discontinuity in the slope (second order) of the observed dendritic velocity versus supersaturation.

Experiments in growth by electrochemical deposition also produce results in qualitative agreement with the characterization of morphology transitions advanced here. Sawada et al. [27] have plotted the interfacial velocity versus applied voltage and found sudden changes in slope when the morphology changes. In our own experiments of electrochemical

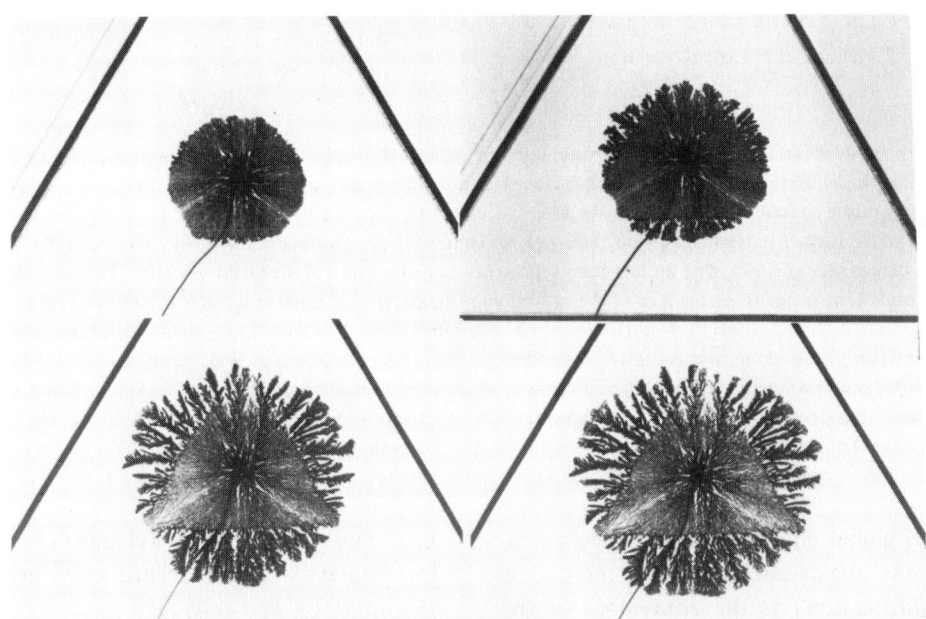

Fig. 5. Morphology transition in electrochemical deposition of Zn from 0.03 M of $ZnSO_4$ solution sandwiched between two plexiglass plates with about 0.3 mm spacing [38]. The outer anode has a triangular shape (8 cm edge). The transition is from tip-splitting growth to dendritic growth. There is also a change in color that reflects the change in the microscopic structure of the two morphologies. Similar transitions between two dense-branching morphologies with different branch densities are also observed in electrochemical deposition of copper [38,50].

deposition we have observed similar sudden changes in the interfacial velocity with associated morphology transitions [38]. An example of morphology transitions in an electrochemical deposition experiment is shown in fig. 5. It demonstrates two aspects: the sharpness of the transition and a change in the microstructure of the growing deposit (shown as a color change) corresponding to the morphology change. These observations give additional support to the use of morphology transitions nomenclature.

7. Conclusion

The field of nonequilibrium growth has made enormous strides over the past several years. However, many questions remain unanswered. The most pressing is the lack of a theory which can predict morphology selection in diffusion-controlled systems as a function of known control parameters. More general nonequilibrium principles also remain to be resolved, with the question of the nature of morphology transitions one of the most interesting.

Acknowledgements

We are very grateful to Kieran Mullen and Michal Ben-Jacob for their indispensable assistance in preparing and critiquing the manuscript. The research described here was partially supported by grants from: the US National Science Foundation, the Germany–Israel Foundation, and the Donors of the Petroleum Research Fund administered by the American Chemical Society.

References

[1] B.B. Mandelbrot, Fractals: Form, Chance and Dimension (Freeman, San Francisco, 1977); The Fractal Geometry of Nature (Freeman, San Francisco, 1982).
[2] J. Kepler, De Nive Sexangula Godfrey Tampach, Frankfurt am Main (1611).
[3] D'Arcy Wentworth Thompson, On Growth and Form (Cambridge Univ. Press, Cambridge, 1944).
[4] J.S. Langer, Rev. Mod. Phys. 52 (1980) 1.
[5] D.P. Woodruff, The Solid–Liquid Interface (Cambridge Univ. Press, Cambridge, 1973).
[6] E. Ben-Jacob, G. Deutscher, P. Garik, N.D. Goldenfeld and Y. Lareah, Phys. Rev. Lett. 57 (1986) 1903.
[7] T.A. Witten and L.M. Sander, Phys. Rev. Lett. 47 (1981) 1400; Phys. Rev. B 27 (1983) 5686;
P. Meakin, Phys. Rev. A 27 (1983) 604, 1495.
[8] M. Matsushita, M. Sano, Y. Hayakawa, H. Honjo and Y. Sawada, Phys. Rev. Lett. 53 (1984) 286.
[9] L. Pietronero and E. Tosatti, eds., Fractals in Physics (North-Holland, Amsterdam, 1985);
J. Nittmann and H.E. Stanley, Nature 321 (1986) 663;
L.M. Sander, Nature 322 (1986) 789;
H.E. Stanley and N. Ostrowsky, eds., Random Fluctuations and Pattern Growth: Experiments and Models (Kluwer, Dordrecht, 1988).
[10] G.P. Ivantsov, Dokl. Akad. Nauk. SSSR 58 (1947) 567.
[11] M.E. Glicksman, R.J. Shaefer and J.D. Ayers, Metall. Trans. A 7 (1976) 1747;
S.C. Huang and M.E. Glicksman, Acta Metall. 29 (1981) 701, 717.
[12] W. Oldfield, Mater. Sci. Eng. 11 (1973) 211.
[13] J.S. Langer and H. Müller-Krumbhaar, Acta Metall. 26 (1978) 1681, 1689, 1697.
[14] E. Ben-Jacob, N.D. Goldenfeld, J.S. Langer and G. Schön, Phys. Rev. Lett. 51 (1981) 1930; Phys. Rev. A 29 (1984) 330.
[15] E. Ben-Jacob, R. Godbey, N.D. Goldenfeld, J. Koplik, H. Levine, T. Muller and L.M. Sander, Phys. Rev. Lett. 55 (1985) 1315.
[16] A. Buka, J. Kertész and T. Vicsek, Nature 323 (1986) 424;
V. Horvath, T. Vicsek and J. Kertész, Phys. Rev. A 35 (1987) 2353.
[17] R.C. Brower, D. Kessler, J. Koplik and H. Levine, Phys. Rev. Lett. 51 (1983) 1111; Phys. Rev. A 29 (1984) 1335.
[18] D. Kessler, J. Koplik and H. Levine, Adv. Phys. 37 (1988) 255.
[19] E. Ben-Jacob, P. Garik, T. Muller and D. Grier, Phys. Rev. A 38 (1988) 1370.
[20] E. Ben-Jacob, P. Garik and D. Grier, Superlattices Microstructure 3 (1987) 599.
[21] H.S. Hele-Shaw, Nature 58 (1898) 34.
[22] J.V. Maher, Phys. Rev. Lett. 54 (1985) 1498.
[23] D. Bensimon, L.P. Kadanoff, S. Liang, B.I. Shraiman and C. Tang, Rev. Mod. Phys. 58 (1986) 977.
[24] L. Paterson, J. Fluid Mech. 113 (1981) 513; Phys. Rev. Lett. 52 (1984) 1621.
[25] S.N. Rauseo, P.D. Barnes Jr. and J.V. Maher, Phys. Rev. A 35 (1987) 1245;
A. Miller, W. Knoll and H. Mohwald, Phys. Rev. Lett. 56 (1986) 2633;

J. Nittmann and H.E. Stanley, Nature 321 (1986) 663.

[26] L.M. Sander, P. Ramanlal and E. Ben-Jacob, Phys. Rev. A 32 (1985) 3160.

[27] Y. Sawada, A. Dougherty and J.P. Gollub, Phys. Rev. Lett. 56 (1986) 1260.

[28] D. Grier, E. Ben-Jacob, R. Clarke and L.M. Sander, Phys. Rev. Lett. 56 (1986) 1264.

[29] E. Raz, E. Polturak and S. Lipson, preprint and private communications.

[30] T. Fujioka, Ph. D. Thesis, Carnegie-Mellon University (1978).

[31] N.D. Goldenfeld, J. Crystal Growth 84 (1987) 601.

[32] W.W. Mullins and R.F. Sekerka, J. Appl. Phys. 34 (1963) 323; 35 (1964) 444.

[33] L. Schwartz, Phys. Fluids 29 (1986) 3086.

[34] A. Buka and P. Palffy-Muhoray, Phys. Rev. A 36 (1987) 1527;
S. Arora, A. Buka, P. Palffy-Muhoray and Z. Racz, preprint.

[35] J. Nittman, G. Daccord and H.E. Stanley, Nature 314 (1985);
G. Daccord, J. Nittman and H.E. Stanley, Phys. Rev. Lett. 56 (1986) 336;
G. Daccord and R. Lenormand, Nature 41 (1987).

[36] K.J. Måløy, J. Feder and J. Jøssang, Phys. Rev. Lett. 55 (1985) 2681;
J. Chen and D. Wilkinson, Phys. Rev. Lett. 55 (1985) 1892.

[37] D. Grier, D.A. Kessler and L.M. Sander, Phys. Rev. Lett. 59 (1987) 2315.

[38] P. Garik, D. Barkey, E. Ben-Jacob, E. Bochner, N. Broxholm, B. Miller, B. Orr and R. Zamir, Phys. Rev. Lett. (1989), in press.

[39] S.K. Chan, H.H. Reimer and M. Kahlweit, J. Cryst. Growth 32 (1976) 303.

[40] Y. Couder, O. Cardoso, D. Depuy, P. Tavernier and W. Thom, Europhys. Lett. 2 (1986) 437;
Y. Couder, N. Gerard and M. Rabaud, Phys. Rev. A 34 (1986) 5175;
G. Zocchi, B.E. Shaw, A. Libchaber and L.P. Kadanoff, Phys. Rev. A 36 (1987) 1894.

[41] D.A. Kessler, J. Koplik and H. Levine, Phys. Rev. A 30 (1984) 3161.

[42] E. Ben-Jacob, N.D. Goldenfeld, B.G. Kotliar and J.S. Langer, Phys. Rev. Lett. 53 (1984) 2110.

[43] D. Meiron, Phys. Rev. A 33 (1986) 2704;
D. Kessler, J. Koplik and H. Levine, Phys. Rev. A 33 (1986) 3352;
M. Ben Amar and B. Moussallam, Physica D 25 (1987) 155.

[44] B.I. Shraiman, Phys. Rev. Lett. 56 (1986) 2028;
D.C. Hong and J.S. Langer, Phys. Rev. Lett. 56 (1986) 2032; Phys. Rev. A 36 (1987) 2325;
R. Combescot, T. Dombre, V. Hakim, Y. Pomeau and A. Pumir, Phys. Rev. Lett. 56 (1986) 2036; Phys. Rev. A 37 (1988) 1270;
S. Tanveer, Phys. Fluids 30 (1987) 1589;
A.T. Dorsey and O. Martin, Phys. Rev. A 35 (1987) 3989;
P. Pelce and A. Pumir, J. Cryst. Growth 73 (1985) 337.

[45] A. Barbieri, D.C. Hong and J.S. Langer, Phys. Rev. A 35 (1984) 1802;
G.A. Brener, S.V. Iordanskii and V.I. Melnikov, JETP 9 (1988);
M. Kruskal and H. Segur, Aeronautical Research Associates of Princeton, Technical Memo, 85-25 (1985), unpublished;
B. Caroli, C. Caroli, B. Roulet and J.S. Langer, Phys. Rev. A 33 (1986) 442;
P. Pelce and Y. Pomeau, Stud. Appl. Math. 74 (1986) 245.

[46] Y. Saito, G. Goldbeck-Wood and H. Muller-Krumbhaar, Phys. Rev. Lett. 58 (1987) 1541.

[47] A. Dougherty, P.D. Kaplan and J.P. Gollub, Phys. Rev. Lett. 58 (1987) 652;
A. Dougherty and J.P. Gollub, Phys. Rev. A 38 (1988) 3043.

[48] R. Pieters and J.S. Langer, Phys. Rev. Lett. 56 (1987) 1948;
M. Barber, A. Barbieri and J.S. Langer, Phys. Rev. A 36 (1987) 3340;
D.A. Kessler and H. Levine, Phys. Rev. A 36 (1987) 4123;
O. Martin and N. Goldenfeld, Phys. Rev. A 35 (1987) 1382;
R. Pieters, Phys. Rev. A 37 (1988) 3126;
J.S. Langer, Phys. Rev. A 36 (1987) 3350;
D.A. Kessler and H. Levine, Phys. Rev. A 33 (1986) 2621, 2634.

[49] J.S. Langer, Science 243 (1989) 1150.

[50] N. Hecker, Senior Thesis, University of Michigan (1988).

FALLING FRACTAL FLAKES

M.V. BERRY

H.H. Wills Physics Laboratory, Tyndall Avenue, Bristol BS8 1TL, UK

The rate at which a D-dimensional cluster of N smoke spherules falls through air is calculated. The cluster may be large or small in comparison with the mean free path of air molecules. For example, a cluster with $N=1000$ falls ten times more slowly with $D=1.8$ than when compacted into a sphere with $D=3$. Fractality is therefore important. Its effect would be to lengthen the nuclear winter, and should be taken into account in future modelling.

To celebrate Benoit Mandelbrot's birthday, here is a simple estimate of the rate at which fractal particles fall through air. I did the calculation in 1984 as part of an assessment of how the nuclear winter would be affected by the fact that smoke can be fractal [1–3]; this was not taken into account in the original studies [4–7]. The most important effect of fractality is to alter the optics of smoke [8,9]. The essential point is that coagulation of smoke spherules into fractal clusters will not reduce the absorption of light in the way that coagulation into solid spheres would. This makes the nuclear winter colder, possibly by several degrees [10]. The other effect, to be discussed here, is that fractality would prolong the nuclear winter because fractal clusters fall much more slowly than solid spheres and so would remain aloft longer.

Simons [11,12] and Hess et al. [13] have since published similar results, but my old calculation was so simple that I think it still worth presenting in its original form.

Consider a fractal cluster of dimension D, formed by the aggregation of N spherules, each of radius a and density ρ. For smoke, $D \approx 1.8$ [3] and $a \approx 20$ nm [1]. We wish to calculate the speed $v(N, D)$ with which the cluster falls under gravity (acceleration g) through air with viscosity η and density ρ_a.

For smoke, v is always small enough for the frictional force F on the cluster to be linear, i.e.

$$F = \alpha v. \tag{1}$$

Equating this to gravity minus buouancy gives

$$v = \frac{4\pi N(\rho - \rho_a) g a^3}{3\alpha}, \tag{2}$$

so the problem reduces to finding α (or, what is equivalent, the diffusion constant kT/α [14]).

There are two limiting regimes, in which the cluster radius R (defined as the rms distance between spherules) is much larger or much smaller than the mean free path L of the air molecules. R is given by the fractal relation

$$R = a N^{1/D}. \tag{3}$$

Here I have omitted a dimensionless constant whose value is close to unity. At height h in the atmosphere, L is given by

$$L = L_0 \exp(h/h_0), \tag{4}$$

where for air $L_0 \approx 60$ nm and $h_0 \approx 8$ km.

If $R \gg L$, friction is caused by air flow round the cluster, and we can estimate α by assuming that the cluster entrains the air inside it and using Stokes' law:

$$\alpha \approx 6\pi R\eta \quad (R \gg L). \tag{5}$$

If $R \ll L$, friction is caused by the impacts of individual air molecules on the spherules. These present a cross section $A \approx \pi a^2 N$ if $D < 2$ (because the cluster is geometrically transparent) and $A \approx \pi R^2 = \pi a^2 N^{2/D}$ if $D > 2$ (because the cluster is geometrically opaque). Thus

---

Essays in honour of Benoit B. Mandelbrot
Fractals in Physics – A. Aharony and J. Feder (editors)

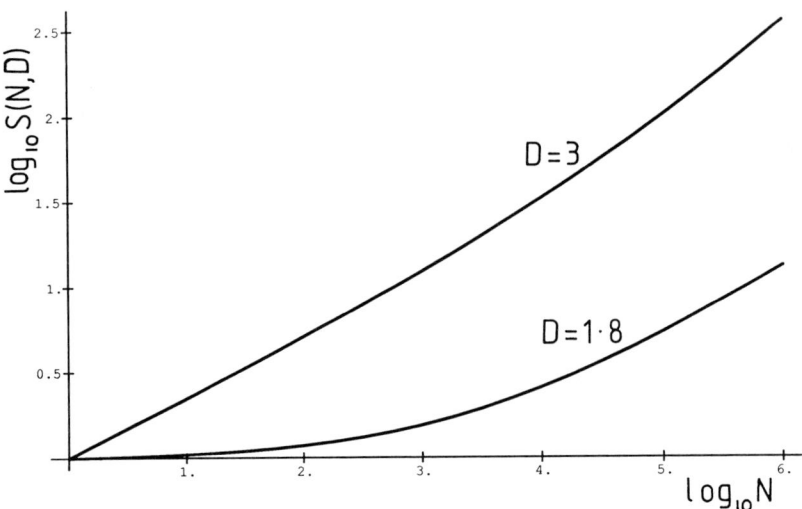

Fig. 1. Clustering speedup factor $S(N,D)$, calculated from eqs. (10) and (9).

$$A = \pi a^2 N^\beta, \quad \beta = 1 \quad \text{if } D < 2,$$
$$\beta = 2/D \quad \text{if } D > 2. \quad (6)$$

If u is the average speed of air molecules we find from momentum balance that

$$\alpha \approx 2\rho_a u A \quad (R \ll L). \quad (7)$$

We can eliminate $\rho_a u$ using $\eta = \rho_a u L/3$ [14], so that

$$\alpha \approx 6\pi a^2 N^\beta \eta / L \quad (R \ll L). \quad (8)$$

When R/L is not very small or very large, the approximations underlying (5) and (8) do not justify anything more sophisticated than the simplest interpolation, which from (2) gives the speed of fall as

$$v(N, D) = \frac{2a^2 \rho g N^{1-1/D}}{9\eta} \left(1 + \frac{L}{aN^{\beta-1/D}}\right). \quad (9)$$

For large clusters, the term involving L is negligible and $v \propto N^{1-1/D}$. When $D=3$ this gives the correct limit $N^{2/3}$. When $D=1$ it predicts v independent of N (so that, for example, the rate at which a hair falls would be unchanged if the hair were cut into pieces), which up to powers agrees with the known result $v \propto \log N$ [15]. For nuclear winter smoke, the mean free path can be large compared with R (because $h \approx 20$ km [7], cf. eq. (4)) and so its effects cannot always be neglected.

Fractality is important. For clusters with $N \approx 1000$, (9) predicts falling speeds of $v \approx 100$ m/yr for fractal clusters ($D=1.8$), as compared with $v \approx 1$ km/yr for solid clusters ($D=3$). A convenient dimensionless measure of how smoke falls faster as it coagulates is the clustering speedup factor

$$S(N, D) \equiv v(N, D)/v(1, D). \quad (10)$$

This is shown in fig. 1 for $D=1.8$ and $D=3$. Obviously the cluster speeds up much more slowly if it is fractal. For example when $N=1000$ and $D=1.8$ (corresponding to a radius $R=930$ nm) $S=1.55$, whereas the same cluster with $D=3$ (i.e. solid, with $R=200$ nm) has $S=12.4$.

Of course it would be naive to infer from (9) that the fractality of smoke implies that the nuclear winter would last ten times longer than if smoke were not fractal. One reason is that the proportion of smoke that would coagulate dry, into fractals (as opposed to wet, into spheres) is the subject of controversy. Nevertheless, fractals fall so much more slowly than non-fractals that the effect of even a small proportion on the duration of a nuclear winter seems to be severe enough to warrant it being included in future modelling.

I thank Dr. S. Simons for telling me about his and other recent contributions to this subject.

References

[1] P. Chylek, V. Ramaswamy, R. Cheng and R.G. Pinnick, Appl. Opt. 20 (1981) 2980–2985.
[2] S.R. Forrest and T.A. Witten Jr., J. Phys. A 12 (1979) L109–L117.
[3] R.D. Mountain and G.W. Mullholland, in: Kinetics of Aggregation and Gelation, Proceedings of the International Topical Conference, F. Family and D.P. Landau, eds. (North-Holland, Amsterdam, 1984).
[4] P.J. Crutzen and J.W. Birks, AMBIO 11 (1982) 114–125.
[5] R.P. Turco, O.B. Toon, T.P. Ackerman, J.B. Pollack and C. Sagan, Science 222 (1983) 1283–1292.
[6] A.B. Pittock, T.P. Ackerman, P.J. Crutzen, M.C. Maccracken and C. Sagan, in: Environmental Consequences of Nuclear War (SCOPE 28), Vol. 1. Physical and Atmospheric Effects, F. Warner, ed. (Wiley, New York, 1985).
[7] O. Greene, I.C. Percival and I. Ridge, Nuclear Winter (Polity Press, Cambridge, 1985).
[8] M.V. Berry and I.C. Percival, Optica Acta 33 (1986) 577–591.
[9] J. Nelson, J. Mod. Opt. (1989), in press.
[10] J. Nelson, Nature (1989), in press.
[11] S. Simons, J. Phys. A 19 (1986) L901–L905.
[12] S. Simons, J. Phys. D 20 (1987) 1197–1199.
[13] W. Hess, H.L. Frisch and R. Klein, Z. Phys. B 64 (1986) 65–67.
[14] F. Reif, Fundamentals of Statistical and Thermal Physics (McGraw-Hill, New York, 1965).
[15] A. Berry and L. Swain, Proc. R. Soc. (London) 101 (1923) 766–778.

THE FRACTAL STRUCTURE OF EVOLUTION

Gerd BINNIG

IBM Research Divison, Physics Group Munich, c/o Physics Section of the University,
Schellingstraße 4, D-8000 Munich 40, Fed. Rep. Germany

The claim is made that everything we know of, including our physical laws and space with its symmetries, occurred through evolution. Each evolution is a complete component of the preceding evolution. Every evolution is subject to the same mechanisms. The fact that all evolutions have self-similar structures and are encapsuled in one another calls to mind Mandelbrot's fractal geometry.

1. Introduction

Heracleitus, and other philosophers after him, came intuitively to the conclusion that in our world "everything is flux, nothing is stationary". It has become obvious in this century that intelligence, matter and life occurred through some kind of evolution. No one would ever claim that even one of these major evolutions are truly understood. There is, however, an abundance of discoveries which leaves no doubt as to their existence. The evolutions of intelligence, life and matter are described in philosophy, in the theory of evolution and in cosmology, respectively. It is possible that these are three different theoretical frameworks endeavouring to describe one and the same thing: the mechanism of evolution. It could also be called the mechanism of creation or creativity because in evolution as well as in creativity the point is the development of something new. The capability of producing new ideas and/or objects is called creativity. Unfortunately, however, this word is usually used in conjunction with the human intellect. A more general approach is to define creativity as the capacity for evolution. Matter, life and intelligence were and are obviously capable of evolution. It is clear that they have at least that much in common. After searching for further common elements one finally suspects that there are no principal differences at all. As evolutions are encapsuled in one another it is tempting to claim that they might be self-similar and can be described by Mandelbrot's theory.

2. The common elements of different evolutions

2.1. Static structure

Probably Aristotle was the first to use a picture of a fractal to describe the world. It is now known as the porphyrian tree. The entirety is the trunk that splits into tangible and intangible. The tangible splits into animate and inanimate, animate into sensitive and insensitive, and sensitive into rational and irrational representing the finest branches of the tree. Today the evolution of life is also represented by a tree. Not so long ago Reeves [1] used the model of a pyramid to describe the evolution of matter. On the lowest level known to us are quarks and leptons. These components, when combined at the next stage of the pyramid, can form more complex particles such as protons and neutrons. On the following level we have the entire range of atoms, which combine to form molecules on the next level. At the top Reeves placed the macromolecule like DNA, which is the basic foundation of life for living creatures and mankind. About 20 billion years ago this pyramid did not exist, which implies that complexity increases with time.

As already indicated by Reeves this building-block character is also obvious in the evolution of life. For

Essays in honour of Benoit B. Mandelbrot
Fractals in Physics – A. Aharony and J. Feder (editors)

this case we might structure the pyramid as follows: level n: *macromolecules* (composed of atoms and molecules); level $n+1$: *cells* (composed of macromolecules and more elementary blocks); level $n+2$; *organs* (composed of cells); level $n+3$: *living creatures* (individuals, composed of organs and more elementary parts); level $n+4$: *species* which taken together on level $n+5$ form life. Here too each level is composed of blocks from the lower levels. The same is true for the intelligence pyramid. There are very elementary thoughts like "house" or "red" or "is", and one can imagine these put together. At the next stage, having put these thoughts together, we have the statement: "The house is red". A collection of statements can portray a theory or a discovery. We would like to call the components of all pyramids and levels *interaction units*.

We also find the building-block character in all structures generated by our intelligence. Music is composed of notes. No musician would ever dream of composing a continuously running piece. One composes interaction units such as symphonies, movements, songs or melodies. Our language and literature are also assembled in this way. Our communication system is another excellent example. Let us take as basic interaction units electronic components, such as resistors, transistors, tubes, condensors, etc. With these one can build functional units like resonant circuits or simple amplifiers. Electronic equipment, such as computers, stereos, televisions and so on, is built from these units. With this electronic equipment entire networks such as world-wide television or telephone systems are set up. All networks together result in an enormous communication system (wherein is mankind's symbiotic raison d'être).

Every building block of any kind is composed of more elementary building blocks. This fractal-like structure is formed – in my opinion – as a result of optimizing the efficiency of information exchange. Every society is therefore structured in the following manner: groups are composed of subgroups. Matter might have the same good reason to form building blocks.

2.2. Structural growth

2.2.1. The reproduction of structures in space and time

Darwin certainly made the most important contribution to understanding the evolution of life. He started the ball rolling. It is generally known today that a significant – if not *the* most significant – step in the evolution of life is the self-reproduction of macromolecules. The product resulting from the application of these codes to inanimate matter is the living creature. Codes reproduce themselves; their products, living creatures, interact directly with one another. These interactions have a special structure, which we shall explain in subsection 2.2.2. Can space, matter or thoughts, reproduce themselves? The reproduction of thought is certainly one of the most important components of intelligence; our intelligence would be inconceivable without communication. Or somewhat clearer: thought patterns (not thoughts) reproduce, multiply and spread. Their products are in turn thoughts. "The house is red" is a thought which will not spread very far. The thought pattern of attributing a colour to an object is an example of a code which is carried out globally by evolution: procreation or communication equals the reproduction of thought patterns.

Does matter reproduce itself? It all depends on the conditions. If the conditions are right, reproduction can occur. Haken compares the situation in a laser to the evolution of life [2]. When the laser medium is activated one photon added to the system is rapidly reproduced by stimulated emission. Here the "primordial soup" is the excited state of the laser medium with the proper energy difference from the ground state.

It is also remarkable that classes of particles like electrons exist at all. This is sometimes called the particle zoo, and suggests a direct connection to living creatures. One could presume that the particle and the living creature zoos developed in the same way. The pattern of behaviour of certain particles is reproduced and is always the same. We do not yet know how the code of behaviour of particles is stored. A "particle DNA" might exist because in one way or another the characteristics, e.g. of an electron, must be stored. One can say very generally that the existence of the laws of nature includes reproduction: the reproduction of behaviour in space and time. As long as one cannot observe a code directly, its reproduction is only seen in the reproduction of behaviour, i.e. in reproduced interactions between the products of the code. The characteristics of thoughts or living creatures are also interaction characteristics, which can always be found newly reproduced in different places.

In the laws of nature there are, however, characteristics which are even more elementary than the laws themselves: symmetries. Our space possesses certain symmetry characteristics which can be regarded as an expression of reproduced codes. Translational invariance, for example, is nothing more than the reproduction of spatial characteristics in other parts of space. Inflationary cosmology [3] describes how space, before the conventional big bang, underwent a development – namely a colossal exponential expansion. As space also shows signs of reproduced behaviour we would like to go a step further here and suggest that space developed through evolution. This may have been an evolution similar to the evolution of life and occurred before that of matter. There is no reason to presume that that was the beginning. It is also quite possible that the universe had already experienced numerous evolutions (possibly an infinite number), each being more elementary than the subsequent one and each subsequent evolution completely contained in the preceding one. This means that one could call these structures fractals [4] as they are self-similar and encapsuled in each other. In Reeves' picture of using a pyramid to describe evolution we could see our world as a *fractal* pyramid. Every pyramid is part of a larger one. It is interesting to note that in Mandelbrot's book [4] a fractal geometrical pyramid is shown.

2.2.2. *Mutation, natural selection*

Today, the importance of the variation in the reproduction of DNA – called mutation – is quite clear to evolutionary theorists. In the context of this paper we use the word mutation in a broader sense. Mutations also occur in our daily communication. Think of the party game in which a whispered phrase is passed from one person to another. During the course of events the phrase becomes increasingly mutated. In every process of thinking, ideas, conceptions, goals or whatever mutate from one state to another. Where, however, can one find mutations in symmetries or in the laws of nature? Since the birth of quantum mechanics, one knows that the behaviour of matter is only describable statistically. The behaviour of a particle is only vaguely reproduced within a probability profile described by its wave function. One could argue here that for two identical experiments, in different places or at different times, the wave functions are reproduced precisely. On the other hand one could also specify a similar probability profile for the mutation of any gene. But this profile is generally not reproduced. As long as evolution remains very dynamic, not yet in equilibrium, one could say that the probability profiles themselves mutate. They respond to the changed conditions in the environment. In the late phases of an evolution – for example the coelacanth fish or the electron – one can only observe statistical behaviour. The probability profile shifts and changes unnoticeably. Thus individual evolutions diverge at a certain point in time in accordance with their proximity to the big bang. The statistical behaviour of particles and the vanishing of symmetries at extremely small dimensions can be seen as the mutation of matter and space, respectively. The result is an almost static probability profile – an evolutionary fossil. The older evolutions are much less dynamic than the newer ones, but one must realize that dynamics never stop as long as the pyramid grows. Simultaneous growth upwards and outwards results in changes at the lower levels of the pyramid. For example, it is generally known today that intelligence changes life (today we can intervene in DNA) and that life has changed our planet (think of the oxygen in the atmosphere) and that matter breaks the symmetries of space. If we generalize here, we have to presume that nothing is constant – not even the symmetries and the laws of nature.

Since with each synthesis new possibilities are opened up, which after just a few steps become countless, synthesis has to be followed by selections, or reductions of possibilities, which we wish to call *natural selection*. The clue in alternating mutation and selection is the evolution of chance into an increasingly structured form.

Everything we can observe in nature is bound to a certain procedure. A very basic element is dualism, which possibly developed even before space and which might have prompted Hegel to split synthesis into thesis and antithesis. The symmetries of space always display dualism. The laws of nature strictly obey symmetries. Life takes place within the laws of nature, and up to now our intelligence has moved only within the boundaries set by DNA. There is no such thing as "pure" chance. Everything must always abide by rules such as symmetries or laws of nature.

2.2.3. Goal-oriented mutation

The following hypothetical example will show how an accidental restriction of chance can come about. To start with, let us presume that a DNA molecule is "purely" statistically mutated. Let us then presume that by chance mutation blockades form which only permit mutations of certain sections of the DNA. In a separate article [5] we try to show that enzymes can, in a certain sense, actually cause such blockades. If mutation blockades are possible and hereditary, then analysis, i.e. natural selection, will result in the survival of the "fittest" or the most environmentally feasible blockades. Once they form by chance, the "right" blockades are selected by analysis. In this way chance has restricted itself. For more complex beings one can also imagine a direct feedback mechanism apart from natural selection that controls the blockades, such as via emotions. A time-limited mutation blockade of a DNA segment is equivalent to the above-mentioned unfocused goal of a thought process. It could be, "Mutate in the area responsible for the construction of the eye. Concentrate on improving the eye. Otherwise mutate as little as possible, thus

keeping down losses caused by mutation". A law is nothing but the restriction of chance. It appears to be natural that all laws had to develop by the self-restriction of chance in the same way as described here for the mutation of DNA. One could say: evolution *is* the evolution of self-limiting chance.

In the case of the mutation blockades it is obvious that for all kinds of mutations a maximum speed c_m (c_m is the maximum number of mutations per turn of the synthesis/analysis helix, τ) is reached when the blockades are reduced to zero. The blockades are controlled by the interaction between the interaction units. Before the blockades are introduced the natural speed is c_m. For the motion of particles this is the speed of light, which particles with special blockades (rest mass) cannot reach without losing their identity, the blockades being part of the genetic code. General relativity also fits qualitatively into the picture. The feedback from the higher levels of the pyramids to the lower ones has to result in a change of space by the presence of matter.

When we learn to be creative, we are in fact learning how to handle mutation blockades. What is the difference between Hamlet and a special statistical heap of words and punctuation marks? (The probability of the statistical creation is the same for both.) Hamlet has a complex impact on us, the statistical text does not. The code for writing novels or, generally speaking, for communication grew in a process of very slow evolution by means of mutation or step-by-step change. Each new mutation is followed by a selection that is based on the effect of the new mutant. In this way chance is limited step-by-step, in that accidental restrictions are found to be *effective*. For example, sentences have a certain structure; one cannot simply choose any order of words. The constant, close contact between effect, or interaction, and the evolution of chance through continuous alternation of mutation and selection results in structured chance which, with respect to effect, is optimized. No wonder therefore that Hamlet is optimal. The likelihood for the above-mentioned statistical text will, through the structure of chance, therefore be practically nil. This means that anyone who writes a text and adheres to the structured chance, i.e. journalistic rules, cannot avoid having an effect.

The self-limitation of chance is equivalent to the construction of a chance hierarchy. For the previously mentioned example of DNA we might describe the (overly simplified) process as follows: symmetries determine the laws of nature, the laws of nature determine how DNA is to be built, and moreover if and what types of mutation blockades are possible. Blockades dictate where mutations will take place, and ultimately make the accidental meeting of an actual mutation with interaction units which cause mutations possible. There is no such thing as "pure" chance, because if there were, evolution would have been too slow to have produced space or matter or something as complex as an automobile, let alone a human being.

Acknowledgement

I wish to thank all the friends, colleagues and students with whom I had so many inspiring discussions. I am especially grateful for help from H. Hörber, D. Smith and M. Niksch.

References

[1] H. Reeves, What Have We Learned about the Universe in the Last Decades?, CENS, Saclay, Talk at IBM (1987).
[2] H. Haken, Synergetics (Springer, Berlin, 1983).
[3] A. Linde, Phys. Today 40 (1987) 61.
[4] B. Mandelbrot, The Fractal Geometry of Nature (Freeman, New York, 1983).
[5] H. Hörber, M. Niksch and G. Binnig, to be published.

MULTISCALING AND MULTIFRACTALITY

Antonio CONIGLIO and Marco ZANNETTI

Dipartimento di Scienze Fisiche, Università di Napoli and Gruppo Nazionale di Struttura della Materia (CNR), Mostra d'Oltremare Pad. 19, 80125 Naples, Italy

A general scaling argument is presented which predicts for the two points correlation function for growth phenomena and critical phenomena a scaling form (multiscaling) of the type $g(r/R) = r^{-d+D(r/R)} A(r/R)$, where r is the distance from the origin, R a characteristic length and $A(x)$ is an amplitude. A connection between multifractality and multiscaling is shown. The numerical data of Plischke and Racz on diffusion-limited aggregation and a soluble model for spinodal decomposition are found in agreement with such multiscaling form.

1. Introduction

As pointed out by Mandelbrot [1] fractals play an important role in a large variety of physical phenomena. In growth phenomena [2,3] fractal patterns usually arise naturally without the necessity of tuning a parameter to reach criticality. Much studied is the diffusion-limited aggregation (DLA) model [4], which, although simple, appears to exhibit the essential features of many growth phenomena.

An interesting feature of this model was discovered by Plischke and Racz [5] by making a distinction between two regions: a frozen region characterized by practically zero growth probability and an active zone where most of the growth occurs.

The active zone exhibits an unusual and rich structure, well described by the multifractal formalism [6,7]. This formalism, which requires an infinite set of critical exponents [8] to characterize the moments of the growth probability distribution, represented a departure from standard scaling in critical phenomena, usually characterized by a single "gap" exponent. In fact an approach based on standard scaling leads inevitably to a single gap exponent for the moments of the growth probability distribution. However, a more general approach [9] based on scaling invariance, under the requirement that the transformation obeys group properties, reproduces the multifractal behavior characterized by a continuous spectrum of fractal dimension usually referred to as $f(\alpha)$ [10]. Using the same general approach we will reproduce in section 2 the scaling properties of the density profile. This new scaling form, which we will call multiscaling is characterized by a continuum set of power law decay exponents. Although we refer to DLA the approach is rather general and can be applied to any system which exhibits scaling invariance. In section 3 a connection between multiscaling and multifractality is discussed. In section 4 we give numerical evidence to show that multiscaling occurs in DLA. Finally in section 5 an exact solution is presented which shows that multiscaling holds for the pair correlation function in the spinodal decomposition process of the Ginzburg–Landau N-vector model in the limit $N \to \infty$.

2. Multiscaling

Here we investigate the scaling properties of the density profile $g(r, R)$ for DLA, which is defined as

$$g(r, R) \, d^d r = dN, \qquad (1)$$

where dN is the number of particles in the infinitesimal d-dimensional volume $d^d r$ at distance r from the origin and the dependence on the total number of particles N is expressed via the radius of gyration $R=R(N)$.

A general scaling approach [9,11] implies that, after coarse graining, the density profile, apart from a constant factor, is invariant under rescaling all lengths by a factor l,

$$\tilde{r}=r/l, \qquad (2)$$

$$\tilde{R}=R/l, \qquad (3)$$

$$g(\tilde{r},\tilde{R})=l^{d-D(r,R)}g(r,R), \qquad (4)$$

where $D(r,R)$ is a function of r and R instead of being chosen a priori to be a constant as assumed in standard scaling. As remarked above we must then require that the transformation obeys group properties. Namely, if rescaling by a factor l_1, $g(r,R)$ transforms into $g(\tilde{r}_1,\tilde{R}_1)$ and successively rescaling by a factor l_2, $g(\tilde{r}_1,\tilde{R}_1)$ transforms into $g(\tilde{r}_2,\tilde{R}_2)$, then after rescaling by a factor $l_1 l_2$, $g(r,R)$ transforms into $g(\tilde{r}_2,\tilde{R}_2)$. It is easy to verify that this requirement implies that $D(r,R)=D(\tilde{r},\tilde{R})$, which in turn implies that $D(r,R)$ is an homogeneous function of r and R, i.e. $D(r,R)=D(r/R)$. Therefore from (4), choosing $l=r$, we find

$$g(r,R)=r^{-d+D(r/R)}A(r/R), \qquad (5)$$

where $A(x)=g(1,x^{-1})$. This new scaling form, which we call multiscaling, predicts a much richer structure. In fact in the asymptotic regime of r and R large but with their ratio fixed $x=r/R$, we find a different power law decay for each value of x and consequently a different fractal dimension $D(x)$ for each shell corresponding to the ratio x. For $x=0$, i.e. well inside the frozen region or in the "static" regime corresponding to r finite and $R\to\infty$, the density profile satisfies a pure power law behavior with $D(0)\equiv D$ being the fractal dimension of the infinite aggregate. Note that standard scaling is recovered in the particular case $D(x)=$constant. We note that the argument given above is completely general and can be applied to any system which exhibits scaling invariance and is dominated by a finite characteristic length R. For example, $g(r,R)$ could be the time-dependent pair correlation function of an Ising model at the critical point, with $R=R(t)$ being the linear size of the growing domain as function of the time t.

3. Multiscaling and multifractality

We address now the question whether there is a connection between the multiscaling behavior (5) obeyed by the density profile and the multifractal behavior obeyed by the growth probability distribution. More specifically if p is the growth probability associated to a subset of the total set of perimeter sites, the subset is labeled by the singularity $\alpha=-\ln(p)/\ln(R)$. Each subset α has a fractal dimension $f(\alpha)$, namely

$$n(\alpha)\sim B(\alpha) R^{f(\alpha)}, \qquad (6)$$

where $B(\alpha)$ is an amplitude and $n(\alpha)\,\mathrm{d}\alpha$ is the number of sites in the set characterized by the interval $(\alpha, \alpha+\mathrm{d}\alpha)$. To make the connection with multiscaling, we note that the set characterized by the smallest value of α is localized at the tips of the aggregate, on the external part of it, whereas larger values of α correspond to regions closer to the origin. We assume that

$$\alpha=\alpha(r/R), \qquad (7)$$

namely that each set is localized on a shell $x=r/R$. Thus from (1), (6) and (7) we recover (5) with

$$D(x)=f(\alpha(x))$$

and

$$A(x)=B(x)\,(\mathrm{d}\alpha/\mathrm{d}x)\,x^{f(\alpha)-d}.$$

The above is only an argument that supports the idea that multiscaling is likely to occur in systems which are multifractal with respect to the measure given by the growth probability. However, multiscaling does not require necessarily multifractality; in fact it is a consequence of a general scaling invariance.

4. Numerical evidence for multiscaling in DLA

The occurrence of multiscaling on growing DLA clusters is strongly suggested by the results of Plischke and Racz [5]. These authors have carried out a numerical computation, for DLA in two dimensions, of the growth probability $P(r, N)$ defined as the probability that by adding the Nth particle it sticks at a distance r from the origin. Plischke and Racz have fitted the data with a Gaussian law

$$P(r, N) = \frac{1}{\sqrt{2\pi}\,\xi} \exp[-(r-R)^2/2\xi^2], \quad (8)$$

where $R \sim N^\nu$ is the radius of the cluster and $\xi \sim N^{\nu'}$ is the width of the interface. The numerical estimates for the exponents are $\nu \simeq 0.585$ and $\nu' \simeq 0.484$, yielding the ratio $R/\xi \sim N^{\Delta\nu}$ with $\Delta\nu \simeq 0.1$. Later Meakin and Sander [12] carried out similar computations for systems of larger and larger size, finding that $\Delta\nu$ approaches zero as the system size grows. If we conjecture that the observed behavior of $\Delta\nu$ is due to a logarithmic ratio between R and ξ of the type

$$(R/\xi)^2 \sim c \ln(R) \quad (9)$$

with c being a constant, from (8) we obtain

$$P(r, N) = \frac{\sqrt{c \ln(R)}}{\sqrt{2\pi}\,R} R^{-c(r/R-1)^2/2}. \quad (10)$$

This result is in the multiscaling form with a power law decay which is a function of r/R. Taking $c = 6.46$ and, as in ref. [5], $R = 0.693 N^{0.585}$ we find that the form (10) for the growth probability fits quite well the numerical data [13].

5. Multiscaling in spinodal decomposition

We expect this multiscaling behavior to be a general feature of many other dynamical processes, which have usually been expected to obey standard scaling. For example in spinodal decomposition, when a system initially prepared in a disordered state at high temperature is suddenly quenched below the critical point, the late stage of the ordering process is currently believed to obey standard dynamical scaling [14]. Specifically the structure function $C(\mathbf{k}, t)$ (Fourier transform of the pair correlation function), in the late stage is expected to be of the form

$$C(\mathbf{k}, t) = L^d(t) F(kL(t)), \quad (11)$$

where t is the time, d is the space dimensionality of the system and $L(t)$ is the radius of the growing domain. In a recent paper [13] we have solved analytically the time-dependent Ginzburg–Landau N-vector model [15] in the limit $N \to \infty$ at temperature $T = 0$ and we have found the following exact result for the structure function $C(\mathbf{k}, t)$ in the case of conserved order parameter,

$$C(\mathbf{k}, t) = C(\mathbf{k}, 0) \exp[k_m(t) L(t)]^4$$
$$\times \exp\{-[k^2 - k_m^2(t)]^2 L^4(t)\}, \quad (12)$$

where $L(t) \sim t^{1/4}$ is the radius of the growing domain and $k_m(t)$ is the peak wave vector. Apart from the details of the analytic forms, the qualitative behavior described by (10) and (12) are very similar. In the asymptotic time regime we have found $\xi \equiv k_m^{-1} \sim [\tfrac{1}{4} d \ln(t)/t]^{1/4}$ and from (12) we have obtained

$$C(\mathbf{k}, t) \sim [M_0 L^2(t) k_m^{2-d}(t)]^{\varphi(k/k_m)}, \quad (13)$$

where M_0 is a constant and

$$\varphi(x) = 1 - (x^2 - 1)^2.$$

The above result clearly exhibits multiscaling, in place of the standard scaling form (11). This means that in $\mathbf{k}$ space the subdomain characterized by a fixed value of $x = k/k_m$ grows according to the power law $L^{d\varphi(x)}$. In other words, according to the general multiscaling pattern, there is an infinite set of growth exponents and $d\varphi(x)$ can be identified with the fractal dimensionality of the corresponding subdomain. The interesting point is that we obtain multiscaling behavior on a quantity directly amenable to experimental observation, without having to rely on the multifractal structure of the domains with respect to the growth probability, which would be much more difficult to establish.

6. Conclusions

The two examples of multiscaling reported in this paper occur in physical contexts which are rather different: a stationary far from equilibrium dynamical process in the case of DLA and relaxation toward equilibrium in the case of spinodal decomposition. Yet they share a common feature, in both cases the dynamics is controlled by a pair of lengths which diverge according to the same power law apart from a logarithmic factor. The presence of two lengths, which diverge in a marginally different way, seems to be one of the possible mechanisms leading to multiscaling; however, we want to stress that the general analysis presented in the first part of the paper does not require such restriction. In fact a system which is scaling invariant and is characterized by a finite radius R must be described by multiscaling, characterized by a continuous set of critical exponents. As a particular case this set of critical exponents can collapse in one single exponent, in which case standard scaling is recovered. Scaling invariance cannot predict, whether one single exponent or infinitely many exponents are present. This can be decided only from experimental data or explicit calculations. If multifractality occurs with respect to the growth probability measure, it is most likely that multiscaling also occurs. However, since multiscaling is a property of the pair correlation function, it is more amenable to direct experimental observation. For example, in those cases which are believed to be in the same universality class as DLA, such as viscous fingers in a Hele Shaw cell [16], or in porous media [17], or crystal growth in aqueous solutions [18], an analysis of the density profile should be easier to perform than the growth probability distribution; consequently, it should be easier to detect the multiscaling spectrum than the multifractal spectrum. We have shown an example of a soluble model where multiscaling occurs and we have presented numerical evidence for DLA which seems to be consistent with multiscaling. However, we expect multiscaling to be a property of many other time-dependent phenomena.

References

[1] B.B Mandelbrot, The Fractal Geometry of Nature (Freeman, San Francisco, 1982).
[2] H.E. Stanley and N. Ostrowsky, eds., On Growth and Form: Fractal and Non-Fractal Patterns in Physics (Nijhoff, Dordrecht, 1985).
[3] L. Pietronero and E. Tosatti, eds., Fractals in Physics (North-Holland, Amsterdam, 1986).
[4] T.A. Witten and L.M. Sander, Phys. Rev. Lett. 47 (1981) 1400.
[5] M. Plischke and Z. Racz, Phys. Rev. Lett. 53 (1984) 415.
[6] B.B. Mandelbrot, J. Fluid Mech. 62 (1974) 331.
[7] R. Benzi, G. Paladin, G. Parisi and A. Vulpiani, J. Phys. A 17 (1984) 3521.
[8] C. Amitrano, A. Coniglio and F. di Liberto, Phys. Rev. Lett. 57 (1987) 1016.
[9] A. Coniglio, Physica A 140 (1986) 51.
[10] T.C. Halsey, M.H. Jensen, L.P. Kadanoff, I. Proccacia and B.I. Shraiman, Phys. Rev. A 33 (1986) 1141.
[11] A. Coniglio and M. Marinaro, Physica 54 (1971) 261.
[12] P. Meakin and L.M. Sanders, Phys. Rev. Lett. 54 (1985) 2053.
[13] A. Coniglio and M. Zannetti, to be published.
[14] K. Binder and D. Stauffer, Phys. Rev. Lett. 33 (1974) 1006.
[15] G.F. Mazenko and M. Zannetti, Phys. Rev. B 32 (1985) 4565.
[16] J. Nittmann, H.E. Stanley, E. Touboul and G. Daccord, Phys. Rev. Lett. 58 (1987) 619.
[17] K.J. Maloy, F. Boger, J. Feder and T. Jossang, in: Time Dependent Effects in Disordered Materials, R. Pynn and T. Riste, eds. (Plenum, New York, 1987), p. 111.
[18] S. Otha and H. Honzo, Phys. Rev. Lett. 60 (1988) 611.

FRACTONS OBSERVED

Eric COURTENS [a], René VACHER [b] and Erich STOLL [a]

[a] *IBM Research Division, Zurich Research Laboratory, CH-8803 Rüschlikon, Switzerland*
[b] *Laboratoire de Science des Matériaux Vitreux [1], Université des Sciences et Techniques du Languedoc, 34060 Montpellier, France*

Dedicated to Professor Benoit Mandelbrot on the occasion of his 65th birthday

Fractons are the strongly localized vibrational excitations of self-similar fractals. These can be visualized in simulations, and also observed in real fractal materials. It is shown that Brillouin, Raman and incoherent neutron scattering on silica aerogels, as well as low-temperature measurements, all find a consistent explanation in terms of fractons. The fractal dimension D, the spectral dimension $\bar{d}$, an internal length exponent σ and the density of vibrational states have been determined in these experiments.

1. Introduction

Fractal structures are abundant in nature [1], and the development of fractal geometry [2] led to remarkable advances in the description of many phenomena. In the solid state, fractals are commonly produced by aggregation processes, as explained at length elsewhere in these Proceedings. In the present paper, a number of experimental results concerning the *acoustical vibrations* of self-similar fractal solids are reviewed. Particularly, a series of experiments on silica aerogels, an extremely porous form of silica, have confirmed to a great extent the early scaling assumptions of the original *fracton* theory [3]. The experiments also stimulated new theoretical developments, such as those in ref. [4].

In section 2, the necessary concepts are introduced and illustrated by simulation results, while a much more complete theoretical description is presented in a companion lecture [5]. Section 3 mentions the situation in glasses and briefly reviews a number of experiments on various systems. In section 4, the structural properties of silica aerogels are explained, and we show why these materials are exceptionally well suited for the study of scaling. Section 5 presents scattering measurements performed on aerogels, investigating various aspects of their vibrational properties. These are the phonon–fracton crossover studied by Brillouin scattering, the vibrations over the full fracton regime as seen in Raman scattering, and the density of states measured by incoherent neutron scattering. Finally, in section 6, the relations of these measurements to the observed low-temperature properties of aerogels are discussed. Strong evidence is given that fractons coexist with, and are distinct from, the two-level states of fused silica. One concludes that fractons, as collective excitations of self-similar matter, have unambiguously been observed.

2. The vibrations of fractal solids

In this paper, we consider fractal solids which are self-similar over a range of lengths l, comprised between their particle (or molecular) size a and their correlation length ξ, $a < l < \xi$. Beyond ξ, the solids are homogeneous. Typically, in real solids, the ratio ξ/a can at most be $\sim 10^3$ due to limits imposed both by thermal vibrations and by gravitational stability [6]. The long-wave acoustical vibrations, of wavelength $\lambda > \xi$, are then acoustical phonons of angular frequency ω, wave vector $q = 2\pi/\lambda$, velocity $v = \omega/q$, and

[1] Associated with the CNRS, No. 1119.

density of states $N_{ph}(\omega) = \omega^{d-1}$, where d is the Euclidean dimension of the embedding space. Plane-wave excitations are scattered by the inhomogeneities of the solid. If ω is sufficiently high, this becomes the dominant broadening mechanism, leading to an elastic linewidth $\Gamma \propto \omega^4$, or a scattering length $l_{scat} \propto \omega^{-4}$. Hence, plane waves are not vibrational eigenmodes. The latter are localized states in the Anderson sense [7], with a localization length $l_{loc} > l_{scat}$.

As ω is continuously increased, one reaches a crossover frequency ω_{co1} where $q\xi = 1$. At this point one expects that the waves also become very strongly scattered, with $ql_{scat} \simeq 1$. The latter condition corresponds to the so-called *Ioffe–Regel* limit, where one enters a new regime with $\Gamma \approx \omega$ [8]. Near and beyond this point, all length scales relevant to vibrations are expected to collapse to a single one [3,4], $ql_{loc} \simeq ql_{scat} \simeq ql = 1$, where l is a measure of the spatial extent of the excitations, and the latter equality defines q in that regime. At ω sufficiently larger than ω_{co1}, l can be viewed as the size of a vibrating fractal "blob" loosely connected to other blobs [4]. These excitations have been called *fractons* [3], and their density of states was assumed to scale with ω as $N_{fr}(\omega) \propto \omega^{\bar{d}-1}$. Here $\bar{d}$ is the so-called spectral dimension, different from D, the Hausdorff dimension of the fractal, $\bar{d} < D$. The size l relates to the fracton frequency ω by $\omega \propto l^{-D/\bar{d}}$, a relation that follows quite generally from scaling arguments [4,9]. This dispersion is unusual, since $D/\bar{d} > 1$. Fractons being strongly localized eigenmodes, their lifetime is predicted to be long [10]. This should not be confused with the lifetime of plane-wave excitations, which is very short, of the order of the inverse frequency, in view of the Ioffe–Regel condition.

As the frequency is further increased, a second crossover is reached at ω_{co2}, marking the end of the fracton range, and corresponding to $l \approx a$. It follows from the fracton-dispersion relation that $\omega_{co2} \simeq \omega_{co1}(\xi/a)^{D/\bar{d}}$. The only modes of higher frequency are those of the particles. For a solid particle of near spherical shape, the lowest mode has a frequency $\omega_{min} \simeq 2\pi \times 0.83 \, v_t/2R$, where v_t is the shear-wave velocity in the bulk, and R is the radius [11]. In the Debye approximation, the density of particle modes consists of a term in ω^1 for surface contributions, plus a term in ω^2 for bulk ones [12].

The vibrational problem is easier to handle analytically, and also to simulate, in the approximation where the wave field is scalar. This assumes that scalar elasticity applies [13], as for example in rubber. The problem maps then onto that of diffusion on the fractal [14]. The latter is characterized by an exponent $\theta = d_w - 2$, where d_w is the fractal dimension of a random walk. One finds then $\bar{d} = D/(1 + \frac{1}{2}\theta)$ [3]. In that case, the conjecture that $\bar{d}$ assumes the value 4/3 for percolation networks in all dimensions $d \geq 2$ was formulated [3]. On the basis of scaling arguments in the strong scattering regime, it was even made plausible that fractons could obey this conjecture irrespective of the detailed microscopic nature of the disordered materials [8]. More recently, it was recognized that the $\bar{d}$ value, which is indeed near 4/3 for percolation clusters in scalar elasticity, depends both on the microstructure of the fractal and on the nature of the interparticle forces. In the scalar case, it is for example already different from 4/3 for a percolation backbone.

Simulations greatly help in gaining a mental picture of fractons. Examples of eigenmodes obtained by diagonalization of the dynamical matrix for a $d=2$ infinite percolation cluster at threshold are shown in fig. 1. The scalar wave equation with nearest-neighbor force constants was simulated on a 68×68 lattice with periodic boundary conditions. For that lattice, finite-size effects start being felt for $\omega \lesssim 0.4$, while the end of the fracton range is at $\omega_{co2} \simeq 2.5$. The value $\omega = 1$ is the eigenfrequency of a single spring coupled to a single mass. In spite of the small size, some important features can be made apparent, especially since all eigenfrequencies and eigenvectors are available at once in this simulation.

The first point is that fractons spanning the full range of sizes and frequencies are felt at almost every occupied site i. For example, figs. 1a–1c show three fractons selected to be in the same spatial region, and of frequencies near $\omega = 0.4$, 0.8, and 1.6, respectively. These particular eigenmodes were in fact cho-

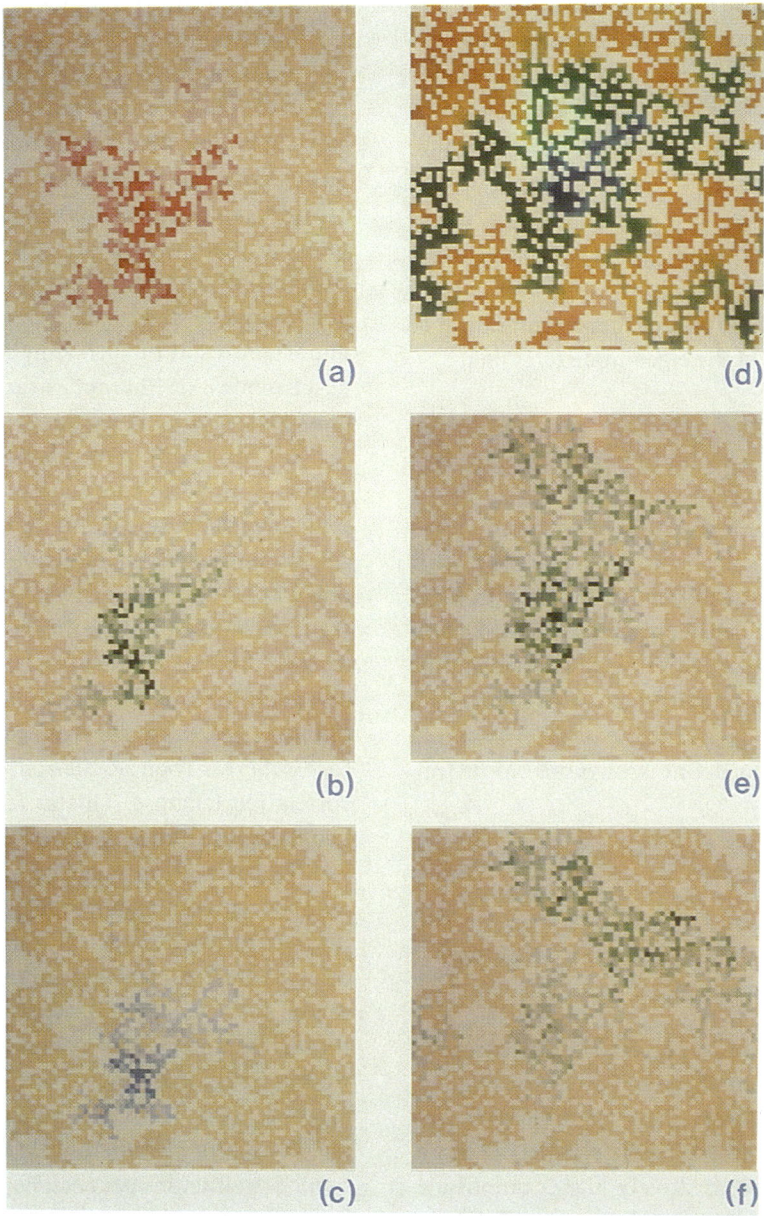

Fig. 1. Simulation results on an infinite percolation cluster at p_c; (a)–(c) present three fractons of frequency 0.397 (red), 0.814 (green) and 1.586 (blue). In each case, four shades are used to represent on a logarithmic scale the local square amplitudes. The deepest shade corresponds to the range 1 to 0.1 where 1 is the maximum, the lightest to the range 10^{-3} to 10^{-4}. The rest of the cluster is colored in yellow. The mean sizes $l(\omega_j)$ of these fractons are 9.8, 7.6, and 4.9 lattice spacings, respectively, (d) is a representation of the mean-square local frequency $\overline{\omega}_1^2$, deep blue corresponding to the highest value, and light red to the lowest one; (e) and (f) are two further fractons as in (b), with frequencies 0.792 and 0.794, and sizes 12.6 and 13.3, respectively.

sen, among many others in the same region of space and frequencies, because their sizes are typical of the mean l obtained by averaging. The size averaging was done on all fractons, in frequency intervals of 0.1, and on 14 different clusters. For this purpose, the averaging over one particular fracton of a site-dependent quantity A_i is defined by

$$\bar{A}(\omega_j) = \langle A_i \rangle = \sum_i A_i u_i^2(\omega_j) \bigg/ \sum_i u_i^2(\omega_j), \quad (1a)$$

where $u_i(\omega_j)$ is the amplitude of this eigenmode at i. The center of gravity of the eigenmode ω_j is then

$$\bar{r}(\omega_j) = \langle r_i \rangle, \quad (1b)$$

where r_i is the position vector of point i. Finally, the size $l(\omega_j)$ of the fracton at ω_j is defined by

$$l(\omega_j) = \langle |r_i - \bar{r}(\omega_j)|^2 \rangle^{1/2}. \quad (1c)$$

Figs. 1a–1c illustrate the dispersion relation. In fact, both the scaling of the density of states $N_{\mathrm{fr}}(\omega)$ and of the dispersion relation, $\omega \propto l^{-D/\bar{d}}$, were verified in this simulation and found in agreement with $D = 91/48$ and $\bar{d} \simeq 4/3$.

Conversely, one can define an average *local* frequency squared $\overline{\omega_i^2}$ by

$$\overline{\omega_i^2} = \sum_j \omega_j^2 u_i^2(\omega_j) \bigg/ \sum_j u_i^2(\omega_j)$$

$$= 2k_\mathrm{B} T \bigg/ \sum_j u_i^2(\omega_j), \quad (2)$$

where the sums are now over all modes, the local amplitudes $u_i(\omega_j)$ are normalized by equipartition, and $k_\mathrm{B} T$ is the thermal energy. The average local frequency is shown in fig. 1d, emphasizing that more connected regions have relatively higher amplitude at higher frequencies, whereas "floppy arms", like in the bottom center of the figure, oscillate most at low frequency.

A third point is that $l(\omega_j)$, defined in (1c), varies over a wide range for fractons of nearby frequencies and positions. Thus, there can be considerable spatial overlap of nearly resonant modes, and this is likely to play a decisive role in the propagation of excitations. An example is shown in figs. 1b, 1e and 1f, which present three fractons at ω very near 0.8. One sees that fracton (e) has considerable overlap both with fractons (b) and (f). However, in going from (b) to (f), the center of gravity has moved by nearly half the lattice size. This should facilitate diffusion through the fractal solid.

Much larger simulations are required to investigate the phonon–fracton crossover on percolation clusters with an occupation probability $p > p_c$, where p_c is the value of p at the percolation threshold. An important result of such simulations is that no pile-up of modes, or "hump", near the phonon–fracton crossover at ω_{col} is found [15]. The change from $N_{\mathrm{ph}}(\omega)$ to $N_{\mathrm{fr}}(\omega)$ is very smooth and progressive. This is contrary to expectations based either on simple scaling and mode-counting arguments [16], or on calculations using the effective-medium approximation [17]. We have discussed elsewhere why scaling arguments might fail on this point [18]. Essentially, these arguments assume implicitly that fractal blobs of a given size, derived from percolation systems of different $(p - p_c)$ values, are indistinguishable from each other as soon as their size is smaller than the correlation length ξ. This in fact is not the case, as known from studies of the *cyclomatic number* [19] on 2D percolation clusters [20]. The cyclomatic number represents the number of independent loops that can be trodden on a cluster. On the infinite percolation cluster, an average number of cycles per site can be defined, which is a local measure of compactness. This number increases rapidly with $(p - p_c)$, also when averaging on samples that are not large compared to ξ [20]. This dependence of the compactness on $(p - p_c)$ is presumably sufficient to modify the number of fractons per particle, invalidating the scaling argument [16] in this case.

Real systems obtained by particle aggregation are usually not percolation clusters. As explained below, there are reasons to believe that silica aerogels have in fact fewer floppy arms, or have a higher cyclomatic number per particle, or also a higher *order of ramification* [1], than infinite percolation clusters. Furthermore, in the real world, elasticity is mostly tensorial. Simulations of percolation clusters, including bond-bending elasticity, give $\bar{\bar{d}}$ equal to 0.8 or 0.9, for $d=2$ or 3, respectively [21]. Simulations on 3D models of ruptured silica have given $\bar{\bar{d}}$ values covering the range 0.5 to 2.5 [22]. Clearly $\bar{\bar{d}}$ is expected to be a strong function of the microstructure of the fractal, and a measurement of that dimension for vibrational excitations should be a goal of the experiments.

3. To be or not to be [23]

Obviously, many physical situations must exist where disordered systems are not self-similar, although very random. In such cases one should also expect localization of vibrations. Strictly speaking, these modes should not be called fractons, as the word was coined for scaling fractals [3]. Initially, however, the fracton theory was applied to glasses and cross-linked polymers [24,25]. This assumes that these systems, although homogeneous from the point of view of density down to the near atomic scale, could be *self-similar fractals* in their *connectivity*. For example, in the case of an epoxy, it was argued that the correlation length could be of the order of the molecular length ($\lesssim 30$ Å) [25]. This sparked a debate that has lasted to this day [26]. As we recently reviewed this aspect rather extensively [18], we indicate here just two important points. Firstly, the experimental observation of a hump in the density of states was repeatedly taken as evidence for fractons, e.g. in refs. [27,28]. We have seen above that such a hump is absent from simulations [15]. Furthermore, a hump is not the hallmark of amorphous systems, since it is often observed in well-ordered crystals, as e.g. in ref. [29]. Secondly, and more importantly, if the connectivity were self-similar, one should expect the elastic modulus K to scale with length as $K \propto l^{-\alpha}$. In fact, one expects rather generally that $\alpha = d - 2\sigma - D + 2D/\bar{\bar{d}}$ [4]. Here, σ is an *internal length exponent* defined in more detail in section 6. Using $d = D = 3$, and reasonable values for σ and $\bar{\bar{d}}$, one expects α of the order of 1 to 2. This would mean that the macroscopic modulus is only a small fraction of the modulus at the short cutoff length scale. In most cases, it implies then very large values of K at the latter scale, and it does not seem to agree with what we know about the relative elasticity of crystalline versus amorphous forms of dense materials.

In spite of the great interest in the rather universal low-temperature properties of glasses [30], and granting these could be fractal possibly in some non-scaling sense, it appears safer to look for genuine fractons and their properties in systems that are demonstrably scaling fractals. Besides the silica aerogels, described in detail in the following sections, the vibrations of a few other solid systems which do seem to fulfill this fractality criterium have been investigated. Most noticeable are the experiments on sintered metal powders [31,32], and those on randomly diluted antiferromagnets [33].

The former system was investigated with ultrasonic propagation, in addition to electrical conductivity and microscopic measurements. An anomalous power-law dependence of the attenuation on the ultrasonic frequency, and the onset of localization at a length which is within a factor of three of the percolation correlation length, were observed [32]. This was interpreted as a phonon–fracton crossover. The experimental crossover is smoother than predicted by an effective medium approximation (EMA) [34]. To our knowledge, these measurements of phonon propagation have been the first conclusive observation of anomalous damping in a fractal near the phonon–fracton crossover. Such experiments, by their very nature, are unfortunately not able to observe real fractons, which should be strongly localized.

On the other hand, diluted magnets form excellent realizations of percolation models. In this case, the vibrational excitations of interest are localized magnons. Among these systems, the randomly diluted antiferromagnet $Mn_xZn_{1-x}F_2$ has been extensively investigated, in particular by R.A. Cowley and collaborators. More recently, high-resolution inelastic-neutron-scattering studies of the spin dynamics in this material, with $x = 0.5$, have revealed spin waves, rather sharp at the zone center, which broaden with increasing wave vector q [33]. The signal was analyzed as the superposition of a spin-wave peak and a damped harmonic oscillator (DHO) function, the latter growing in intensity as q approaches the zone-boundary value q_{ZB}. A crossover from a dominant spin wave to a dominant DHO response was found at $q \simeq 0.3 q_{ZB}$. Remarkably, in energy space, this appears as a crossover from propagating waves to localized high-energy excitations. The crossover wave vector is in agreement with the percolation correlation ξ determined from an independent elastic diffuse magnetic scattering measurement, $\xi^{-1} \simeq 0.3 q_{ZB}$. These results provide strong evidence for a magnon-fracton crossover, as theoretically expected [35]. The broadening of the DHO should not be interpreted as the fracton lifetime, but as the lifetime of the plane-wave excitations that are observed in neutron scattering, and which are in the Ioffe–Regel regime.

Finally, among systems obtained by growth and aggregation, fumed silica powders have been found to be fractal over as many as two orders of magnitude in length [#1]. In compacted samples, a fractal dimension $D = 2.6$ was found [37]. Recently, the low-frequency density of states was measured by inelastic incoherent neutron scattering [38]. These experiments gave $N(\omega) \propto \omega^p$, with $p = 0.8$ or 1.1 at two temperatures, $T = 136$ or 265 K, respectively. This unexpected temperature dependence may, we believe, be related to adsorbed water motion. Interpreting the low-temperature value of p in terms of fractons, one finds $\bar{\bar{d}} = p + 1 = 1.8$, which is not inconsistent with the current views on the possible range of $\bar{\bar{d}}$ [22].

There are quite certainly many different systems in which the fracton concept is relevant. In some of these, such as the diluted magnets, more work can be envisaged. It is definitely worthwhile to study in sufficient detail the vibrations of systems that are well-identified scaling fractals, and to measure new dimensions. For, "if dimensions must not be multiplied beyond necessity, a multiplicity of dimensions is unavoidable" [1]. However, can one describe in a similar way the physical properties of all amorphous materials? That is the question.

4. Mutually self-similar aerogels

Aerogels are monolithic solid materials with an extremely tenuous microscopic structure [39,40]. The most thoroughly investigated aerogels are made of silica. These can be prepared with porosities higher than 99%. In consequence, aerogels exhibit unusual physical properties, making them suitable for a number of technical applications, such as Cerenkov radiators, support for catalysts, or thermal insulators. Suitably prepared materials are also excellent examples of fractal solids [41]. Their preparation and structure are now going to be explained.

4.1. Preparation of silica aerogels

The starting point for the preparation of silica aerogels is the hydrolysis of an alkoxysilane $Si(OR)_4$, where R is usually CH_3 or C_2H_5 [42]. The reaction is strongly influenced by the amount of "catalyst", either acid or base, which can be added to the water. Other important parameters are the relative concentrations of reagents, usually expressed by the ratios $[Si(OR)_4]/[H_2O]$ and $[Si(OR)_4]/[ROH]$. The former controls the degree of hydrolysis, while the amount of alcohol ROH used to dilute the reagents determines the final macroscopic density ρ of the aerogel. The reactions proceed as follows. Hydrolysis produces hydroxide groups –SiOH, which polycondense into siloxane bonds, –Si–O–Si–. Small particles start to grow in the sol. These particles bind to each other by cluster–cluster aggregation (CCA), forming more siloxane bonds, and eventually producing a disordered network filling the reaction vessel. At this point the solution gels. All reactions are of course subject to rate constants, and occur simultaneously rather than sequentially. After gelation, the so-called *alcogel* continues to evolve in time since reactions are not complete at the gel point. In order to obtain the solid porous structure, the solvent is fi-

---

[#1] For a recent review of this experimental evidence, see ref. [36].

nally removed after some "ageing" time.

If drying is done simply by evaporation in air, the capillary forces at the liquid–vapor interfaces pull on the silica network, causing considerable internal fracture and resulting in a powder of intermediate density. These are the so-called *xerogels* of modest porosities [39]. However, if the solvent is extracted above its critical point, the microscopic structure of the network is preserved, and extremely porous *aerogels* can be obtained [42]. To perform this hypercritical drying, the alcogel and additional solvent are introduced in a bomb, the temperature and pressure are raised, and the vapor is allowed to escape with the temperature maintained above the critical point. Large blocks of silica aerogels can thus be obtained whose prominent physical properties are lightness, optical translucency, solid-like elasticity, and very high internal surface.

The gel formation is clearly controlled by the chemical and physical mechanisms of reaction and aggregation, respectively. In the case of silica, the complex set of reactions is strongly influenced by the pH [43], which modifies the relative hydrolysis and polycondensation rates. These aspects are under current experimental investigation, particularly with NMR [44]. Gelation is relatively slow near pH=7, and it is generally accelerated by pH-modifiers. Acid catalysis produces many small "fluffy" particles, whereas base catalysis tends to produce fewer larger particles, possibly in part owing to redissolution of smaller ones into silicon hydroxide. To our knowledge, there has been so far no computer simulation directly applicable to this complex sequence of reactions. However, a number of relevant aggregation models has been studied in detail. They have shown that in general self-similar networks are formed. Reaction-limited cluster–monomer aggregation, the so-called Eden model, leads to a compact object ($D=3$) whose surface is fractal [45,46]. This model might relate to the early stages of particle growth, especially at $3 \leq$ pH ≤ 7. Reaction-limited CCA leads to $D \simeq 2.1$ to 2.3, depending on the amount of restructuring [47]. This might relate to the particle aggregation regime in neutral or moderately acid conditions. Finally, at the other extreme, diffusion-limited aggregation (DLA) leads to very tenuous structures with $D=1.78$ [48,49]. This may be relevant to aggregation of base-catalyzed sols.

The results of our small-angle neutron-scattering (SANS) experiments, presented in part below, demonstrate that aerogels can be prepared which are fractal over more than two orders of magnitude in length scale. All our gels were obtained from methoxysilane. Depending on catalysis conditions, D varies between $\simeq 1.8$ and $\simeq 2.4$ [41,50,51]. Furthermore, we found that the elementary particles can be smooth or rough, and confirmed that their sizes depend on catalysis. The variation of the preparation parameters is obviously sufficient to create vastly different aggregation conditions. Hence, one should always enquire about the details of preparation, and proceed to simple characterization steps, before undertaking more sophisticated experiments on these materials.

4.2. Small-angle scattering

Self-similar clusters, extending up to a correlation length ξ, can be modelled by the density–density correlation function $g(r)$,

$$g(r) - 1 \propto r^{D-3} \exp(-r/\xi). \qquad (3)$$

This reflects that the mass, $M(r)$, within a sphere centered on a particle at $r=0$, scales as $M(r) \propto r^D$. Hence, the density in the sphere scales as $\rho(r) \propto r^{D-3}$. The space Fourier transform of (3) is proportional to the structure factor $S(q)$, and to the scattered intensity in SANS [52]. The Fourier variable q is also the momentum exchange in scattering. The exponential cutoff at ξ is phenomenological, and this particular form is justified by the quality of the fits of the SANS data to the Fourier transform of (3). Two limiting regimes are of interest. At small q, $q\xi \ll 1$, a pseudo Ornstein–Zernicke behavior is observed, $S(q) \propto 1/(1+q^2\xi^2)^{(D-1)/2}$. On the other hand, when $q\xi \gg 1$, $S(q)$ is nearly proportional to q^{-D}. At even larger q, such that the neutron wavelength becomes of the order of the average particle radius R, scattering originates from the particle surfaces, and (3) must then be extended to include the appropriate particle structure factor. Fractal surfaces are characterized by a surface fractal dimension, $2 \leq D_s \leq 3$. In this regime, $S(q)$ is approximately proportional to q^{D_s-6} [53,54]. For smooth particles, $D_s = 2$, one recovers the usual Porod law, $S(q) \propto q^{-4}$ [55]. Finally, at very large q, comparable to the inverse of the bond size, it is the atomic arrangement that determines the scattering. In favorable cases, the measurement of $S(q)$ on a fractal can give the four parameters ξ, R, D, and D_s.

We performed SANS on a large number of differently prepared aerogels [41,50,51]. In particular, some measurements for a series of neutrally reacted gels [41] are summarized in fig. 2. The various curves are labelled by the macroscopic density ρ of the corresponding sample, N095, meaning "neutrally reacted" with $\rho = 95$ kg/m^3. The solid lines represent the best fits. They are extrapolated into the particle regime ($q \gtrsim 0.15$ Å^{-1}) to emphasize that the fits do not apply in that region, particularly for the denser samples. The values of D and ξ derived from these fits are shown in fig. 3. Remarkably, D is constant within experimental accuracy, $D = 2.40 \pm 0.03$. Furthermore, ξ scales with ρ, $\xi \propto \rho^{-1.67 \pm 0.05}$. We return below to the important meaning of that scaling. The departure of $S(q)$ from the q^{-D} dependence at large q indicates the presence of particles with gyration radii of a few Å. An expanded plot of the large-q range for the heaviest samples shows a region where $S(q)$ is nearly proportional to q^{-3} above $\simeq 0.15$ Å^{-1}. This is best seen in a plot of the scattered intensity times q^3 [41], which shows an almost horizontal region at high q. This plateau suggests fuzzy particles with a fractal surface. The structure at that scale can be modified by oxidation in air at 500°C, a treatment which removes remaining –CH$_3$ groups and creates new siloxane bonds. After such a treatment, one observes $S(q) q^3 \propto q^{-1}$ at large q [41]. This is the Porod law, and it demonstrates that oxidation smoothens the surface of the particles. One also notes that this oxidation does not modify the material at longer length scales. The lowest curve on fig. 2 illustrates that

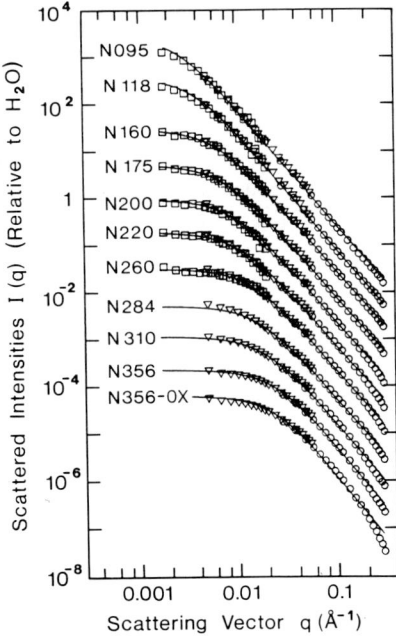

Fig. 2. SANS intensities for, from top to bottom, ten neutrally reacted samples (N), and one oxidized one. The number following N is the density in kg/m^3. The various symbols refer to different spectrometer settings [41]. For this figure, starting at the second curve from the top, each intensity is divided by 4 compared with the previous curve to separate data points. Also, points from only 20 out of 30 detectors are shown to improve visibility. From ref. [41].

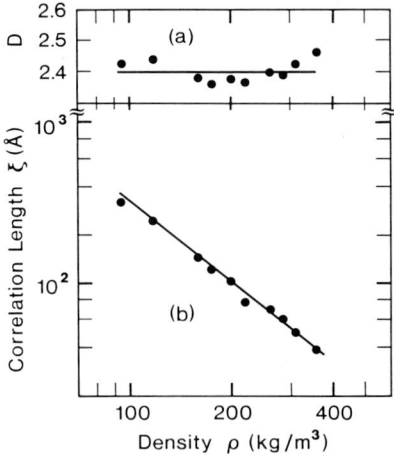

Fig. 3. Values of D and ξ ((a) and (b), respectively) obtained from the fits in fig. 2. The solid lines are from least-squares fits. From ref. [41].

N356-OX is identical to N356 except in the particle region.

4.3. Mutual self-similarity of a series

To interpret the $\xi(\rho)$ scaling law, let us consider elementary fractal clusters assembled from homogeneous particles of size a. For $d=3$, the density of such clusters at length scale $l > a$ is $\rho(l) = \rho_a (l/a)^{D-3}$. Here, ρ_a is the density of the particles. Let ξ be the final average size of the clusters, and assume that the aerogel consists of a homogeneous assembly of these clusters. The macroscopic density is $\rho \equiv \rho(\xi)$. If we now construct another aerogel in exactly the same way, only changing the value of ξ, we obtain two materials of different densities, but with identical fractal structures at length scales where they are self-similar. We have named this property *mutual self-similarity* [41]. For these materials, the scaling law obeyed by ξ as a function of ρ *for the entire series of samples*, $\xi \propto \rho^{1/(D-3)}$, is identical to the relation between l and $\rho(l)$ in the fractal region for *any one* sample. Writing $1/(D-3) = -1.67 \pm 0.05$, we find $D = 2.40 \pm 0.02$, in perfect agreement with the values of the individual samples. Since identical values of D are found from $\xi(\rho)$ and from direct measurements, the mutual self-similarity is established for that series, at least as far as the densities are concerned. Assuming that the mutual self-similarity can be extended to all relevant characteristics, e.g. to the size-dependent elasticity, it follows that a study of the dependence on ρ of a macroscopic property in such a series amounts to studying the scaling of that property as a function of the length l on a single fractal. This is a remarkable feature, as it is much easier to study a macroscopic property for a series than a microscopic property as a function of length. This makes suitably prepared aerogels particularly convenient to investigate scaling.

5. Observations of fractons in aerogels

We used several experiments to detect fractons. These have been, in order of increasing frequency, Brillouin, Raman and neutron scattering spectroscopies. The interpretation of the Brillouin results is facilitated by the availability of a mutually self-similar series. Raman scattering allows the entire fracton frequency range to be covered and gives access to a new scaling exponent. Incoherent neutron scattering can give direct $\bar{\bar{d}}$ on a single sample, and these measurements are currently being extended towards higher energy resolution. All these results will now be presented in turn.

5.1. The fracton dispersion curve

Neutrally reacted aerogels can be prepared with $2\pi\xi \approx 0.3$ μm. Thus, the crossover region can be investigated with Brillouin scattering of visible light. Indeed, this spectroscopy detects vibrational excitations of wavelengths down to $\bar{\lambda}/2n$, where $\bar{\lambda}$ is the optical wavelength and n the refractive index of the material, which is near 1. Polarized scattering was measured for several values of the momentum exchange q, and for the mutually self-similar series of samples in fig. 2, as explained in detail elsewhere [56,57]. For these aerogels, the frequency position of the peak in the scattered spectra also falls into the measurement range of Brillouin interferometry. At sufficiently small q, or sufficiently large ρ, phonons are observed with a nearly Lorentzian spectral profile. These lines broaden rapidly with increasing q. As the crossover region is reached, the line shapes become strongly non-Lorentzian, and the position of the peak becomes stationary with further increase in q.

These experimental spectra could be fitted with a theoretical profile derived from a Green-function analysis [58]. The fits used heuristic expressions for the frequency dependence of the elastic lifetime, and of the dispersion law, near crossover. A noncritical constant m fixes the sharpness of this crossover [57]. The value $m=2$ was found satisfactory for all spectra. The fits depend then on three parameters only: the crossover frequency ω_{co1}, the associated wave vector q_{co}, and the overall intensity. The fits are extremely good, with statistical χ^2 mostly around 1. Conversely, a sharp crossover (m very large) is found to be inadequate, and the EMA line shapes [58] are even worse. This emphasizes the smoothness of the crossover.

The samples of this series being mutually self-similar, the points ω_{co1} versus q_{co} for all ρ's must fall on the same fracton *dispersion curve* [59]. This is shown in fig. 4, where the different symbols correspond to different samples, while the same symbol is used for different q-values on one sample. We note that all points cluster on the curve $\omega \propto q^{D/\bar{\bar{d}}}$, covering more than one order of magnitude in ω. This is the first fracton dispersion curve to be determined experimentally. From the fit one extracts $D_{ac}/\bar{\bar{d}} \simeq 1.9$. Here, D_{ac} is an *acoustical* fractal dimension to be related to the *connectivity* rather than to the mass.

The crossover wave vector q_{co} equals $1/\xi_{ac}$, where ξ_{ac} is an acoustical correlation length. To the extent that only one length scale is relevant to determine the upper limit of fractality, one anticipates $\xi_{ac} \propto \xi$, the latter being determined by SANS. The two quantities are shown in fig. 5. As explained above, the slopes are given by $1/(D-3)$. From the Brillouin data, the acoustical value is $D_{ac} = 2.46 \pm 0.03$, in satisfactory agreement with the SANS D. One also notes that ξ_{ac}

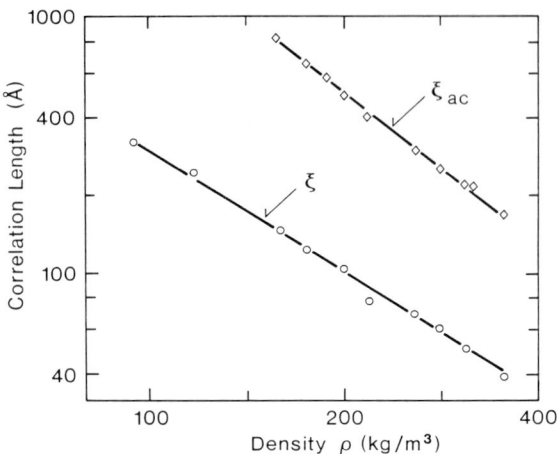

Fig. 5. The acoustically determined correlation length ξ_{ac} compared with the SANS value ξ, for the mutually self-similar series of neutrally reacted aerogels. From ref. [59].

is approximately five times larger than ξ. This numerical factor can, to a large part, be accounted for by the different definitions that are used for the two quantities. Whereas in the SANS measurements the fractality extends from the gyration radius R to ξ, in the acoustical measurement it covers the range from a to ξ_{ac}, where a is the average particle diameter, $a \simeq 2\sqrt{5/3} R$. This can account for a factor ≈ 2.5. Another factor could be due to the sensitivity of the acoustic measurement on connectivity, while SANS senses the mass. Self-similarity in the connectivity can extend further than that in the mass. Finally, the crossover ω_{co1} could actually be somewhat spread out, as discussed in the following subsection. In that case ξ_{ac} would rather correspond to the Ioffe–Regel length for phonons, which might be somewhat larger than the fractal correlation length.

From $D_{ac}/\bar{\bar{d}}$ and D_{ac}, and taking into account the uncertainty on m, one finds $\bar{\bar{d}} = 1.3 \pm 0.1$. This value happens to be close to the scalar elastic prediction $\bar{\bar{d}} = 4/3$ [3]. It is rather different from the tensorial elastic result calculated on percolation clusters in 3D, $\bar{\bar{d}} = 0.9$ [21]. However, this should be taken as an indication that the aerogels have a more connected structure than percolation clusters, rather than as evidence for scalar elasticity. Depolarized Brillouin-

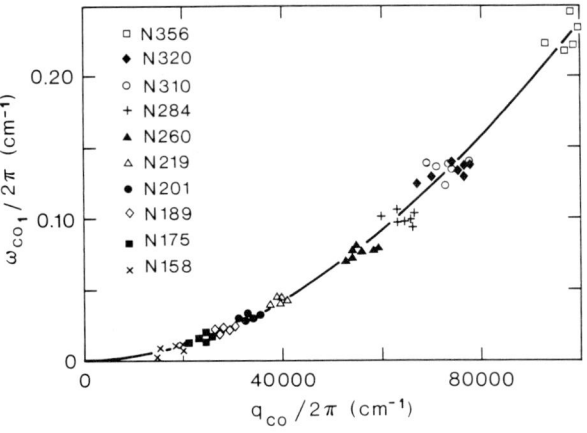

Fig. 4. A plot of the values of ω_{co1} versus q_{co} derived from various Brillouin scattering measurements performed on a series of mutually self-similar aerogels. The various symbols correspond to different sample densities, with the same notation as in fig. 2. The points for each individual sample have been obtained at various scattering angles. From ref. [59].

scattering experiments have indeed revealed shear modes on the two densest samples. Their velocities were ~0.6 times the longitudinal velocities. The scattered intensity of these modes is small, and they were not observed on the lighter samples. However, a shear velocity different from the longitudinal one rules out scalar elasticity [13]. One notes that infinite percolation clusters have numerous very floppy endings. If floppy arms should form during the aggregation process leading to the alcogels, they are likely to connect during ageing, as well as during hypercritical drying, particularly since both processes are associated with shrinkage. Whereas a few additional connections are not likely to affect appreciably the Hausdorff dimension, they can modify considerably the elasticity. An increase of rigidity can well raise $\bar{\bar{d}}$ from 0.9 to 1.3.

5.2. The internal length of fractal blobs

In addition to coherent contributions, there are incoherent contributions to light scattering in which fractons of characteristic size smaller than $\bar{\lambda}$ scatter independently from each other. Their share to the scattered intensity is the sum of their individual scattering intensities, while in the coherent case it is the scattered amplitude which is the sum of amplitudes. Incoherent contributions are not included in the Green-function formula [58]. The coherent intensity decays at large ω like $I(\omega) \propto \omega^{-3-2\bar{d}/D} \tilde{\propto} \omega^{-4}$. This is a very rapid decay. As ω is increased, the coherent signal soon becomes weaker than the incoherent one, which decays as $I(\omega) \tilde{\propto} \omega^{-1.4}$, as shown below.

The incoherent scattering should be q-independent, since fractons of sufficient frequency have localization lengths much smaller than $\bar{\lambda}$. Thus, the incoherent signal can actually be isolated in depolarized backscattering geometry, where the coherent intensity vanishes by selection rules, at least in first order. Fractons are expected up to a second crossover frequency $\omega_{co2} \simeq 10$ cm^{-1}, fairly independent of ρ for the mutually self-similar series of samples. Thus, Raman scattering from fractons in aerogels must be measured at unusually low frequencies compared with standard Raman spectrometry, a condition not previously met [60].

To obtain meaningful results, the aerogels were first oxidized in order to remove extraneous molecular groups that could otherwise contribute to the scattering. Tandem interferometry was used to achieve the high resolution required in the measurement [61]. The scattered intensities $I(\omega)$, normalized by the Bose–Einstein population factor, $n(\omega) \simeq k_B T/\hbar\omega$, are shown in fig. 6. These plots exhibit a power law in the fracton regime. The slopes do not depend significantly on ρ, and for this series one finds $I(\omega)/n(\omega) \propto \omega^{-0.37 \pm 0.02}$ [61]. The upper end of the fracton region is marked by the onset of particle modes at ω_{min}. It happens in this case that ω_{min} approximately coincides with ω_{co2} [61].

We show now that the scaling in ω of $I(\omega)$ is related to the scaling of the internal length of fractal

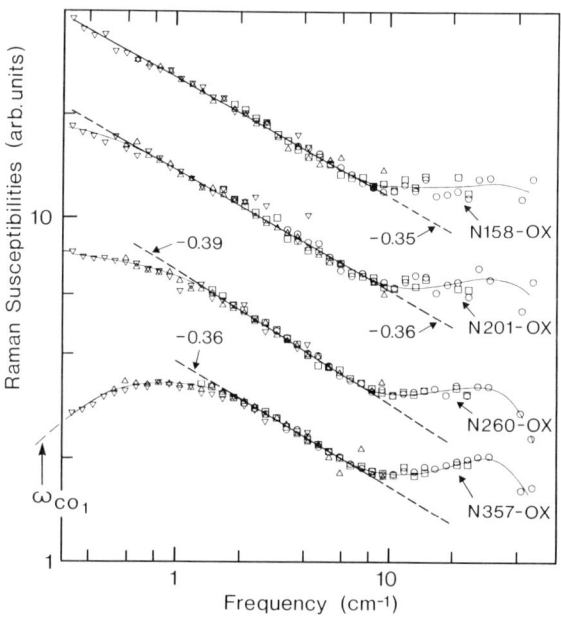

Fig. 6. The Raman susceptibilities $I(\omega)/n(\omega)$ for four oxidized samples of the mutually self-similar series. The straight lines are fits with the indicated slopes, while the thin curves are guides to the eye. The various data-point symbols indicate different spectrometer settings. From ref. [61].

blobs [4]. The intensity scattered by a single fracton of mean Euclidean length $l \propto \omega^{\bar{d}/D}$ is proportional to the square of the polarization fluctuation δP produced by the fracton strain e_l. One has $\delta P \propto l^D e_l$ since the mass M_l of the fractal blob that supports the fracton is $M_l \propto l^D \propto \omega^{-\bar{d}}$. The total intensity at ω is weighted by the fracton density of states, $N_{\text{fr}}(\omega)$. Hence, one finds [61]

$$I(\omega) \propto \omega^{\bar{d}-1} l^{2D} e_l^2. \quad (4)$$

The strain of an element of Euclidean size l is the ratio of the relative displacement of its ends, Δu_l, to its internal length, $s_l \propto l^\sigma$. The exponent $\sigma \geq 1$ characterizes the scaling of the internal length [4]. We are interested in Δu_l^2, which is related, by equipartition, to the force constant T_l of the fractal blob, $\Delta u_l^2 \propto k_B T/T_l$. Thus, one obtains $e_l^2 \propto k_B T/T_l s_l^2$. The force constant relates to the frequency and to the mass by $T_l/M_l \approx \omega^2$, hence $T_l \propto \omega^{2-\bar{d}}$. Introducing all this in (4), one finally obtains [4,61]

$$I(\omega)/n(\omega) \propto \omega^{-2+2\sigma \bar{d}/D}. \quad (5)$$

From the measured slope, -0.37 ± 0.02, we find $\sigma \simeq 1.5$, meaning that the internal length is very fractal.

On the plots of $I(\omega)/n(\omega)$ versus ω one also observes that the phonon–fracton crossover is rather extended. The arrow on the left of the lower curve in fig. 6 shows the position of ω_{co1} derived from Brillouin measurements on the same sample. Nearly one order of magnitude in ω is necessary to reach the asymptotic fracton regime above ω_{co1}. This behavior seems to scale with the position of ω_{co1} to the extent that it could be measured [61]. Hence, the extended crossover cannot be due to a progressive transition from coherent to incoherent depolarized scattering, as such an effect would depend on $\bar{\lambda}$, and thus would not scale. Further, the acoustic correlation length, which is also the largest possible fracton size, is only ≈ 150 Å on the heaviest sample in fig. 6, which is indeed much smaller than $\frac{1}{2}\bar{\lambda} \approx 2500$ Å. The extent of the crossover rather suggests that the Ioffe–Regel limit and the true phonon–fracton crossover do not exactly coincide, as discussed in ref. [8], their frequency positions being simply proportional to each other.

5.3. The fracton density of states

Incoherent neutron scattering from protons chemically bonded to the particle surfaces can be used to determine the density of states in porous media [62]. This method was applied previously to silica smoke [38]. Compared to silica smoke, the particles in our aerogels are quite a bit smaller, which should lead to higher values of ω_{co2}, a favorable feature for neutron spectroscopy. It is important in the measurement to obtain the pure incoherent contribution. We achieved this by taking the difference signal between two similar samples, one with protons and the other with deuterons attached to the particles [63]. This approach allows both samples to be dried thoroughly under identical conditions to avoid possible contributions owing to free moving water.

Measurements performed with the time-of-flight spectrometer Mibemol in Saclay revealed a nearly temperature-independent density of states, as reported in detail elsewhere [63]. These first results covered the region of particle modes, and probably the upper frequency fractons as well. It is clear that with signal coming exclusively from attached protons, the particle-mode density is not going to be reproduced very faithfully, whereas longer wave excitations such as phonons and fractons are well probed. It was nevertheless possible, using an approximate formula [12], to calibrate the particle density of states, and with additional information from Brillouin scattering [57], to obtain an overall estimate for the entire density of states [63]. This was then used to calculate the low-temperature specific heat. The result agrees remarkably well in its T-dependence, and within a factor of ~ 2 in its absolute value, with measurements by Calemczuk et al. [64] discussed in the following section.

More recently, new measurements [65] with higher resolution, on a sample prepared in a *different* manner, have confirmed an extended fracton region. Rather than aged for 15 days in the alcogel state, this particular material was hypercritically dried soon after gelation. The results at two temperatures are shown in fig. 7. The experimental points above ω_{co2} are in the region of particle modes. In the fracton domain, a power law is now observed over more than one order of magnitude. It has a slope $\bar{\bar{d}} - 1 \simeq 0.85$, rather than $\simeq 0.3$. This presumably reflects the different microstructure of the material. Soon after gelation, a significant number of small clusters remains floating in the solvent. Upon extraction of the solvent, these clusters are transported by the moving fluid, and clutter on the infinite cluster. This modification in the cluster–cluster aggregation regime can be expected to produce a more compact structure at small scales [66]. Hence, one anticipates a greater relative density of small clusters, and thus of modes at high frequencies, leading to a larger effective $\bar{\bar{d}}$. This experiment confirms that $\bar{\bar{d}}$ is not an universal dimension, but rather an additional one sensitive to the microstructure of the fractal.

6. Connection with low-temperature properties

The low-temperature specific heat, C, thermal conductivity, $\mathcal{K}$, internal friction and sound-velocity

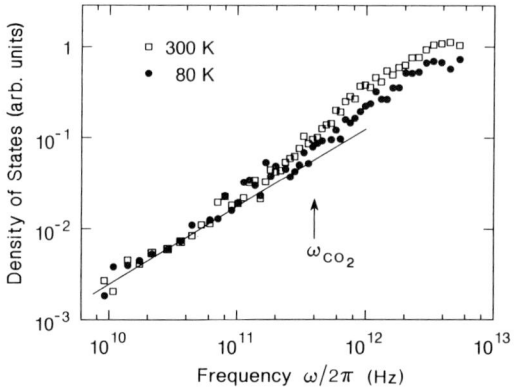

Fig. 7. Density of vibrational states derived from the difference between time-of-flight spectra on protonated and deuterated samples at 300 K and 80 K. The arrow indicates the fracton to particle-mode crossover region. The straight line is traced with a slope of 0.85. From ref. [65].

changes have been investigated for some of the same aerogels in the range of 50 mK to 10 K by the group at CEN in Grenoble [64,67]. We reproduce here the essential elements of their analysis. Plotting C/T^3 versus T logarithmically, they find for N120 and N360 samples a decay with $T^{\bar{d}-3}$ at the lowest temperatures, and up to about 5 K where the curves are flattening out. We note that this temperature corresponds fairly well to $\hbar\omega_{co2}/k_B$. From the slopes, they extract values of $\bar{\bar{d}}$ equal to 1.37 and 1.45, respectively, in fair agreement with our spectroscopic result, $\bar{\bar{d}} = 1.3 \pm 0.1$. In that region, the absolute value of the specific heat is extremely high for the aerogels, but *lower* than calculated for Debye phonons. This is in marked contrast with amorphous silica. In that case, the specific heat per mole is about two orders of magnitude smaller than in aerogels, but it is much *larger* than expected for Debye phonons. It is the common occurrence of an *excess* specific heat in amorphous materials that lead to the two-level state (TLS) models.

The difference between aerogels and amorphous silica becomes even more apparent if one considers simultaneously the thermal conductivity, $\mathcal{K}$. The conductivity of aerogels is extremely low below 2 K, and almost T-independent. In silica, the conductivity is much higher, and varies as $\mathcal{K} \propto T^2$. The latter is a typical TLS behavior, terminating in a plateau between 5 and 10 K. For the aerogels, the plateau, if any, is at much lower temperatures, and the decay of $C/T^3 \propto T^{\bar{d}-3}$ is clearly not connected with a TLS behavior in $\mathcal{K}$.

These indications that C and $\mathcal{K}$ on aerogels are not controlled by TLS are confirmed by measurements of internal friction, Q^{-1}, and of velocity changes, $\Delta v/v$, also performed by the CEN group [67]. The authors find the same order of magnitude in TLS density per unit mass in aerogels and in silica. Hence, these TLS cannot contribute appreciably to the large specific heat of the aerogels, and both the strong anomalies in C/T^3 and in $\mathcal{K}$ must relate to another, non-propagating excitation. Interestingly, fractons, which are long-lived excitations, apparently do not contribute appreciably to either Q^{-1} or to $\Delta v/v$.

The entire set of data in sections 5 and 6 allows us to conclude that fractons have been unambiguously observed based on their structure factor, their density of states and their contributions to low-temperature properties. The latter are clearly distinct from TLS contributions, which are also seen.

Acknowledgements

The experimental work on silica aerogels illustrating this review was performed in collaboration with G. Coddens, J. Pelous, J. Phalippou, Y. Tsujimi and T. Woignier. Two of us (E.C. and R.V.) thank Professors A. Aharony, S. Alexander, O. Entin-Wohlman, R. Orbach and J. Teixeira for illuminating discussions. The neutron results were obtained at the Laboratoire Léon Brillouin, a Laboratoire mixte Centre National de la Recherche Schientifique (CNRS)–Commissariat à l'Energie Atomique, Saclay, France.

References

[1] B.B. Mandelbrot, The Fractal Geometry of Nature (Freeman, San Francisco, 1982).
[2] B.B. Mandelbrot, Les Objets Fractals (Flammarion, Paris, 1975).
[3] S. Alexander and R. Orbach, J. Phys. (Paris) 43 (1982) L625.
[4] S. Alexander, Phys. Rev. B, to be published.
[5] R. Orbach, Physica D 38 (1989) 266, these Proceedings.
[6] Y. Kantor and T.A. Witten, J. Phys. (Paris) 45 (1984) L675.
[7] P.W. Anderson, Phys. Rev. 109 (1958) 1492.
[8] A. Aharony, S. Alexander, O. Entin-Wohlman and R. Orbach, Phys. Rev. Lett. 58 (1987) 132.
[9] R. Rammal and G. Toulouse, J. Phys. (Paris) 44 (1983) L13.
[10] S. Alexander, O. Entin-Wohlman and R. Orbach, Phys. Rev. B 34 (1986) 2726;
A. Jagannathan, R. Orbach and O. Entin-Wohlman, Phys. Rev. B, to be published.
[11] H. Lamb, Proc. Math. Soc. London 13 (1882) 187.
[12] H.P. Baltes and E.R. Hilf, Solid State Commun. 12 (1973) 369.
[13] S. Alexander, J. Phys. (Paris) 45 (1984) 1939.
[14] Y. Gefen, A. Aharony and S. Alexander, Phys. Rev. Lett. 50 (1983) 77.
[15] G.S. Grest and I. Webman, J. Phys. (Paris) 45 (1984) L1155;
K. Yakubo and T. Nakayama, Phys. Rev. B 36 (1987) 8933; and to be published.
[16] A. Aharony, S. Alexander, O. Entin-Wohlman and R. Orbach, Phys. Rev. B 31 (1985) 2565.
[17] B. Derrida, R. Orbach and K.-W. Wu, Phys. Rev. B 29 (1984) 6645.
[18] E. Courtens and R. Vacher, in: Fractals, L. Pietronero, ed., to be published.
[19] C. Domb and E. Stoll, J. Phys. A 10 (1977) 1141.
[20] E. Stoll and C. Domb, J. Phys. A 12 (1979) 1843.
[21] I. Webman and G. Grest, Phys. Rev. B 31 (1985) 1689.
[22] J. Kieffer and C.A. Angell, J. Non-Cryst. Solids 106 (1988) 336.
[23] W. Shakespeare, Hamlet (1600).
[24] P.F. Tua, S.J. Putterman and R. Orbach, Phys. Lett. A 98 (1983) 357.
[25] S. Alexander, C. Laermans, R. Orbach and H.M. Rosenberg, Phys. Rev. B 28 (1983) 4615.
[26] J.E. de Oliveira, J.N. Page and H.M. Rosenberg, Phys. Rev. Lett. 62 (1989) 780.
[27] H.M. Rosenberg, Phys. Rev. Lett. 54 (1985) 704.
[28] A.J. Dianoux, J.N. Page and H.M. Rosenberg, Phys. Rev. Lett. 58 (1987) 886.
[29] W.N. Lawless, Phys. Rev. B 14 (1976) 134.
[30] W.A. Phillips, ed., Amorphous Solids, Low-Temperature Properties (Springer, Berlin, 1981).
[31] D. Deptuck, J.P. Harrison and P. Zawadski, Phys. Rev. Lett. 54 (1985) 913.
[32] J.H. Page and R.D. McCulloch, Phys. Rev. Lett. 57 (1986) 1324.
[33] Y.J. Uemura and R.J. Birgeneau, Phys. Rev. Lett. 57 (1986) 1947; Phys. Rev. B 36 (1987) 7024.
[34] O. Entin-Wohlman, S. Alexander, R. Orbach and K.-W. Yu, Phys. Rev. B 29 (1984) 4588.
[35] R. Orbach and K.-W. Yu, J. Appl. Phys. 61 (1987) 3689.
[36] J.H. Page, W.J.L. Buyers, G. Dolling, P. Gerlach and J.P. Harrison, Phys. Rev. B 39 (1989) 6180.
[37] T. Freltoft, J. Kjems and S.K. Sinha, Phys. Rev. B 33 (1986) 269.
[38] T. Freltoft, J. Kjems and D. Richter, Phys. Rev. Lett. 59 (1987) 1212.
[39] J. Fricke, ed., Proceedings of the First International Conference on Aerogels (Springer, Berlin, 1985).
[40] R. Vacher, J. Phalippou, J. Pelous and T. Woignier, eds., Proceedings of the Second International Conference on Aerogels, Rev. Phys. Appl. 24-C4 (1989).
[41] R. Vacher, T. Woignier, J. Pelous and E. Courtens, Phys. Rev. B 37 (1988) 6500.

[42] S.S. Kistler, J. Phys. Chem. 36 (1932) 52.
[43] R.K. Iler, The Chemistry of Silica (Wiley, New York, 1979).
[44] R.A. Assink and B.D. Kay, J. Non-Cryst. Solids 99 (1988) 359.
[45] M. Eden, in: Proceedings of the Fourth Berkeley Symposium on Mathematical Statistics and Probability, G. Neyman, ed. (University of California Press, Berkeley, 1961).
[46] P. Meakin, Rev. B 23 (1983) 5221.
[47] P. Meakin and R. Jullien, J. Phys. (Paris) 46 (1985) 1543.
[48] P. Meakin, Phys. Rev. Lett. 51 (1983) 1119.
[49] M. Kolb, R. Botet and R. Jullien, Phys. Rev. Lett. 51 (1983) 1123.
[50] R. Vacher, T. Woignier, J. Phalippou, J. Pelous and E. Courtens, J. Non-Cryst. Solids 106 (1988) 161.
[51] R. Vacher, T. Woignier, J. Phalippou, J. Pelous and E. Courtens, Proceedings of the Second International Conference on Aerogels, Rev. Phys. Appl. 24-C4 (1989) 127.
[52] J. Teixeira, in: On Growth and Form, H.E. Stanley and N. Ostrowsky, eds. (Nijhoff, Dordrecht, 1986), p. 145.
[53] H.D. Bale and P.W. Schmidt, Phys. Rev. Lett. 53 (1984) 596.
[54] P.-Z. Wong and A. Bray, Phys. Rev. Lett. 60 (1988) 1344.
[55] A. Guinier, Théorie et Technique de la Radiocristallographie (Dunod, Paris, 1956).
[56] E. Courtens, J. Pelous, J. Phalippou, R. Vacher and T. Woignier, Phys. Rev. Lett. 58 (1987) 128.
[57] E. Courtens, R. Vacher, J. Pelous and T. Woignier, Europhys. Lett. 6 (1988) 245.
[58] G. Polatsek and O. Entin-Wohlman, Phys. Rev. B 37 (1988) 7726.
[59] E. Courtens and R. Vacher, Proc. R. Soc. London A 423 (1989) 55.
[60] A. Boukenter, B. Champagnon, E. Duval, J. Dumas, J.F. Quinson and J. Serughetti, Phys. Rev. Lett. 57 (1986) 2391.
[61] Y. Tsujimi, E. Courtens, J. Pelous and R. Vacher, Phys. Rev. Lett. 60 (1988) 2757.
[62] D. Richter and L. Passell, Phys. Rev. Lett. 44 (1980) 1593.
[63] R. Vacher, T. Woignier, J. Pelous, G. Coddens and E. Courtens, Europhys. Lett. 8 (1989) 161.
[64] R. Calemczuk, A.M. de Goër, B. Salce, R. Maynard and A. Zarembowitch, Europhys. Lett. 3 (1987) 1205.
[65] R. Vacher, E. Courtens, G. Coddens, J. Pelous and T. Woignier, Phys. Rev. B 39 (1989) 7384.
[66] P. Meakin, in: On Growth and Form, H.E. Stanley and N. Ostrowsky, eds. (Nijhoff, Dordrecht, 1986), p. 111.
[67] A.M. de Goër, R. Calemczuk, B. Salce, J. Bon, E. Bonjour and R. Maynard, Phys. Rev. B, to be published.

CANTOR SET SPECTRA AND SELF-SIMILAR CRITICAL MODES IN A 1D-QUASICRYSTAL

J.P. DESIDERI, O. LEGRAND, L. MACON [1] and D. SORNETTE

Laboratoire de Physique de la Matière Condensée, CNRS URA 190, Faculté des Sciences, Parc Valrose, 06034 Nice Cedex, France

Specific properties of the propagation of surface acoustic waves on quasiperiodically corrugated solids are reviewed. This problem is shown to correspond to the critical regime of the Anderson localization transition, characterized by critical proper modes which are neither extended nor localized and which exhibit remarkable scaling features. The spectrum is also predicted to have a Cantor-like structure. The experimental system is made of a thousand grooves engraved according to a Fibonacci sequence. For the first time, the self-similar spatial structure of the critical proper modes is observed through an optical diffraction experiment. Signatures of the fractal spectrum are also reported. These results are explained in terms of the asymptotic approximation of the quasicrystal by periodic systems of increasing periods.

1. Introduction

Self-similarity in physics usually emerges from "criticality" either in equilibrium as in phase transitions or in dynamical systems as in growth or non-equilibrium problems. A state is critical when all length scales play an equal role and produce complex long-range interwoven correlations as can be surmised by looking at the geometrical aspect of a given problem. In this respect, the concept of fractality, introduced by Mandelbrot [1] has revolutionized our vision of critical states: the simplifying mode of thinking (and imaging) given by the concept of a fractal dimension has led statistical physicists to attack, in various contexts, novel problems which were previously considered much too complicated [2].

In this contribution proposed as a tribute to Mandelbrot's achievements, we present results on a particular problem, that of the propagation of a wave in a quasiperiodic system, which exhibits several interesting and unexpected self-similar features. This problem provides an illustration that physics in random or complicated systems often leads to simple and beautiful behaviors which can be described in terms of self-similarity.

Wave propagation in an inhomogeneous medium leads to the phenomenon of Anderson localization [3] at any non-vanishing disorder for space dimension $d=1$ and $d=2$ and at sufficiently high disorder for $d=3$. The localization regime is a subtle non-perturbative effect involving coherent interferences between all the wavelets partially reflected by the quenched disordered set of scatterers, for which exist only partial theoretical scenarios [3]. Quasiperiodicity, neither true periodicity nor randomness, can be considered intermediate between these two extremes [4]. In particular, in a one-dimensional ($d=1$) quasiperiodic system, it is very interesting since a transition exists between an extended and a localized regime similarly to what occurs in a $d=3$ disordered system [5]. The existence of a transition between two regimes is always exciting because one can hope that understanding the crucial features which trigger the transition will allow to unravel the physics of the different regimes. The existence of such a critical Anderson localization transition accounts for the self-similarity of the spectrum and of the proper modes which we now describe.

---

[1] Present address: SPECTEC S.A., 14 Avenue St. Augustin, 06300 Nice, France.

Essays in honour of Benoit B. Mandelbrot
Fractals in Physics – A. Aharony and J. Feder (editors)

2. Description of the experiments

A Rayleigh surface acoustic wave (SAW) propagates at the quasiperiodically corrugated surface of a piezoelectric lithium niobate (YZ-LiNbO$_3$) substrate of total length $L=15984$ μm, with $N=10^3$ identical grooves engraved, using the well known micro-lithographic techniques [6]. The average distance a between the grooves is thus $a=L/N=15.984 \approx 16$ μm. Each groove has a width $w=5$ μm, a depth $h=0.3$ μm and a well characterized inverse plateau profile. The lateral scale of each groove (the so-called opening) is $E=2150$ μm. The groove centers are positioned on the sites of the Fibonacci sequence built recursively from successive concatenation of lower-order patterns [7]

$$S_{j+1} = \{S_{j-1}, S_j\}. \qquad (1)$$

One has $S_0=\{s\}$, $S_1=\{c\}$, $S_2=\{sc\}$, $S_3=\{csc\}$, $S_4=\{sccsc\}$ and so forth. s ($=11.6\pm0.1$ μm) and c ($=18.7\pm0.1$ μm) are the two elementary tiles of our one-dimensional quasicrystal. We have $s/c = 0.620\pm0.005$ near the inverse golden mean $t = \frac{1}{2}(\sqrt{5}-1) = 0.6180$. The ±0.05 μm precision of the position of the grooves corresponds to the limitation of the electronic etching technique.

Our typical surface acoustic wave set-up is composed of electro-mechanical transducers, laid out at the surface of the YZ-LiNbO$_3$ crystal, which surround the array of etched grooves. They perform the launching and detection of the surface acoustic waves both in reflection and transmission. The Rayleigh wave has well-known characteristics [6]. It is a mixture of longitudinal and transverse acoustic modes. It propagates along the solid–air plane boundary and is evanescent away from the solid boundary with a typical excursion of the order of the wavelength ($\lambda \approx 20$ μm at typical frequencies around 170 MHz). Its phase velocity $c_R \approx 3490$ m s^{-1} is slightly less than the transverse wave velocity c_t and its dispersion relation is linear in the absence of surface corrugation. In the presence of a single groove, the SAW is partially reflected with a reflection amplitude coefficient given by $\mu \approx 0.6(h/\lambda) \sin(2\pi w/\lambda) \approx 10^{-2}$ for our frequency range. Furthermore, a fraction $p \approx 20\mu^2$ of the SAW energy is detrapped and converted into longitudinal and shear bulk acoustic waves [6,8].

3. The spectrum

The description of the SAW propagation on the corrugated solid has been developed within a transfer matrix theory described in ref. [9]. Using the mapping approach of Kohmoto et al. [7,10] recalled and adapted to our system [9], the spectrum has been numerically calculated. In an infinite system, our results are comparable to previous ones obtained for a quasiperiodic Schrödinger equation with a *step* potential [7,10], since the transfer matrix formalism is the same in both cases. In particular, we learn from this analogy that this system exactly corresponds to the so-called "critical" regime intermediate between the extended and the localized regimes. This means that the spectrum is a Cantor set of zero measure. Thus, gaps (or stop bands) are present at all frequency scales.

These asymptotic properties are not really observed in their totality in our experiments. This comes from the finite length of the system which contains (only) $N=10^3$ grooves. Also, due to the fact that each groove is shallow ($h=0.3$ μm) compared to the typical value of the wavelength $\lambda \approx 20$ μm, most of the stop bands of the spectrum are too narrow and will not be observed.

Fig. 1 gives the dependence of the SAW reflection coefficient R obtained experimentally as a function of frequency f in the range 153–193 MHz. The large scale bell-like shape corresponds to the transfer function of the measuring transducers. The informations relevant to the study of the $d=1$ quasicrystal are the peaks which decorate this structure. We observe the existence of particular frequencies f for which the reflection coefficient is significantly increased. This can be interpreted as the largest stop bands of the system. We have checked that the pattern of peaks obtained under reflection was exactly recovered under transmission [9]. We note that the peaks in fig. 1 in the

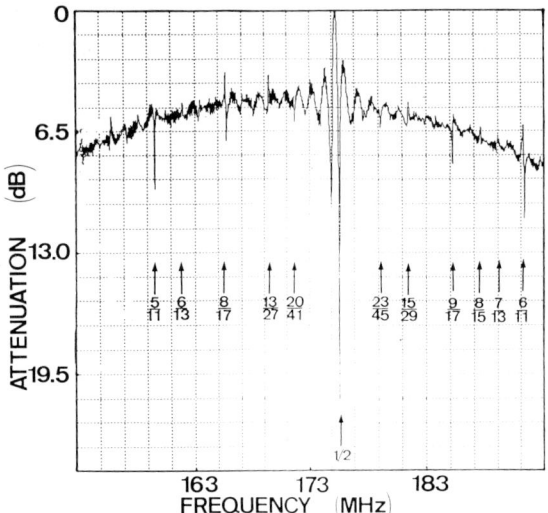

Fig. 1. Dependence of the SAW reflection modulus as a function of frequency. Each peak probes the existence of a stop band.

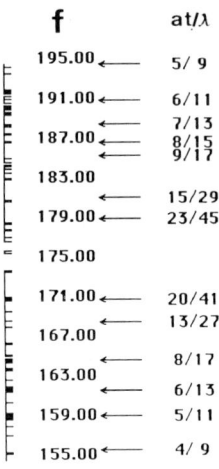

Fig. 2. Numerically determined finite size spectrum obtained from the transfer matrix formalism described in the text and in refs. [7,9,10]. Stop bands lie within vertical spaces between unconnected horizontal bars. Thick bars are due to close packing of such gaps. Values of at/λ given by eq. (2) are listed at the outmost right. The vertical f list of equidistant frequencies in the middle gives the frequency scale. The comparison between fig. 1, fig. 2 and eq. (1) is good.

neighborhood of the central frequency $f \approx 175$ MHz can be indexed by a single integer n such that the variable at/λ is of the form

$$at/\lambda = n/(2n\pm 1), \quad n=1, 2, \ldots \infty. \quad (2)$$

For instance, the central peak occurs for $0.4990 \leq at/\lambda \leq 0.5010$ ($n=1$ of the series (2) with $at/\lambda = n/(2n-1)$) corresponding to a frequency $f \approx 175$ MHz and a wavelength $\lambda \approx 19.7$ μm. In fig. 1, one can identify the different frequencies giving the smallest n for the rational expression (2), for instance, of at/λ:

$n=6$; $at/\lambda = 6/11$; $f=191.35$ MHz; $\lambda=18.12$ μm,

$n=7$; $at/\lambda = 7/13$; $f=189.35$ MHz; $\lambda=18.35$ μm,

etc.

These special frequencies, which correspond to the largest stop bands in the finite size system, are those for which the transmission is small enough or conversely the reflection is large enough in order to be detected.

Fig. 2 gives the apparent "finite size" spectrum numerically determined from the transfer matrix formalism and the mapping described in refs. [7,9,10] with the criterion that only those frequencies yielding a value of the trace of the transfer matrix for the whole system larger than 2 are selected as "effective" stop bands. We have taken explicit account of the finite size $N=10^3$ of the system and of the small value of the amplitude reflection coefficient μ in order to predict which features of the Cantor set should remain observable in our experiments. The arrows in fig. 2 show the gaps corresponding to the set (2) also observed in fig. 1. The agreement is quite good.

Qualitatively, the values of these gap frequencies can be related to the following essential property of our quasicrystal: it is the asymptotic limit of a series of periodic systems of larger and larger periods a_j, each corresponding to the successive "words" of the iterative concatenation process (1). It is well known that, for any periodic system, there should be a gap at half of any reciprocal lattice vector $2\pi/b_j$ of the crystal, and all *sums* and *differences* of the reciprocal lattice vectors of the periodic structures are reciprocal lattice vectors. As the system becomes higher- and higher-order periodic, the smallest reciprocal lattice vector of the system gets smaller and smaller. Thus, in the almost periodic limit, there should be a gap in the vicinity of every wave vector. This explains intuitively the highly fragmented Cantor-like band structure. We can go further and rationalize the existence of the observed series (2) as follows:

Consider first the smallest period a_0 corresponding to the complete pattern of letters s and c. Since s (respectively c) occur with relative frequency $t/(t+1)$ (respectively $1/(t+1)$), a_0 is given by

$$a_0 = st/(t+1) + c/(t+1). \qquad (3)$$

Thus $a_0 \equiv a$, the average period of the system. We expect a stop band at the Bragg condition $2a_0 = \lambda$, i.e. at a value of the reduced parameter $a/\lambda = 1/2$. From the self-similar structure of the infinite quasicrystal which is invariant under the following transformation in the tiling

$$c \to s, \quad sc \to c, \qquad (4)$$

a second period at^{-1} comes out. The discussion is easily generalized to the larger periods $a_j = t^{-j}a_0$, obtained by replacing the word S_j by c and S_{j-1} by s in the infinite quasicrystal. This leads to gaps at frequencies such that

$$at^{-j}/\lambda = 1/2 \quad \text{with } j \geq 0. \qquad (5)$$

Expression (5) gives only a part of the whole spectrum. Indeed, it must be generalized to take account of the interactions between the different periods in the system. Consider the two reciprocal lattice vectors $2\pi/a_0$ and $2\pi/a_1$. Their sum is $(2\pi/a_0)(1+t) = 2\pi/a_0 t$, which is again a reciprocal lattice vector with $b = a_0 t$. The Bragg condition applied to it yields a gap for $a_0 t/\lambda = 1/2$. This is the main gap observed experimentally in fig. 1. Repeating the argument for the two reciprocal lattice vectors $2\pi/a_0 t$ and $2\pi/a_0$ yields a reciprocal vector equal to $2\pi/a_0 t^2$, and so on. One thus generates gaps at all wavelengths for which eq. (5) holds but with $j < 0$. These gaps are of course a few among the infinite set of the singular continuous spectrum. However, they are the largest since they correspond to successive periodic approximation of the quasicrystal with the *smallest* periods. The smallest periods will give the largest reflection and smallest transmission since this will correspond to the largest number of periods per unit length in the effective periodic lattice.

We can now understand the existence of the full series (2) observed experimentally in the finite system. Until now, we only have considered combinations of the fundamental reciprocal wave vector $2\pi/b$. Of course, if $2\pi/b$ is a reciprocal lattice vector, $2\pi/(2n\pm 1)b$ is also a reciprocal lattice vector since it corresponds to a period $(2n\pm 1)b$, which is a multiple of the fundamental period b. Let us take $b = a$ and consider the following combination:

$$4\pi n/(2n\pm 1)at^{-1} + 4\pi n/(2n\pm 1)a$$
$$= 2\pi[2n/(2n\pm 1)]/at.$$

The Bragg condition for this reciprocal lattice vector exactly yields the series (1). Such combinations of reciprocal vectors allow to understand the peculiar role played by frequencies such that at/λ is rational which have been observed experimentally.

Up to now, the results which have been presented correspond to a fixed system. However, scaling properties of the Cantor spectrum are best seen when varying the system size. This problem has been studied previously by focusing on the "scaling index" of the spectrum (see ref. [7] for a review) (the scaling index is the exponent describing the rescaling of the gaps when increasing the size L of the system) and on the total measure of the gaps as L increases. Here, we study a complementary aspect of the problem. We fix the system size to a large value $L \approx 1.35 \times 10^6$ but change the reflection coefficient $\mu \sim h/\lambda$ per groove by varying the grooves depth h. Fig. 3 shows the gaps structure as a function of $2h$ (in μm). Fig. 3b is a magnification of the region around $\lambda_{1/2}/c = 1.0557281...$ where $\lambda_{1/2}$ is such that $at/\lambda_{1/2} = 1/2$. Note that the wavelength is given in units of c. The same fractal structure is recovered by successive magnification. $\lambda_{1/2}/c$ is thus a singular point, limit of the alternate convergent series

$$at/\lambda_m = 1/2 + (-1)^m/bm, \qquad (6)$$

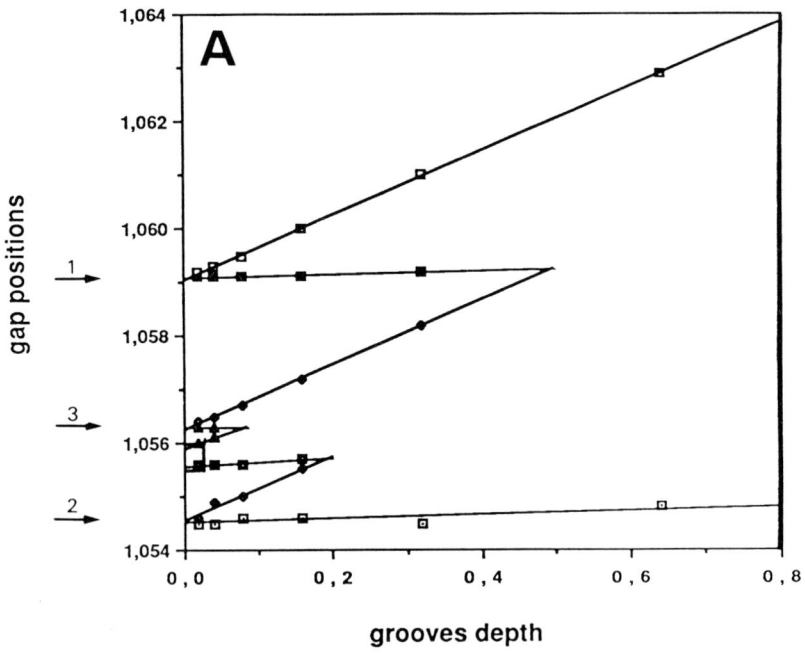

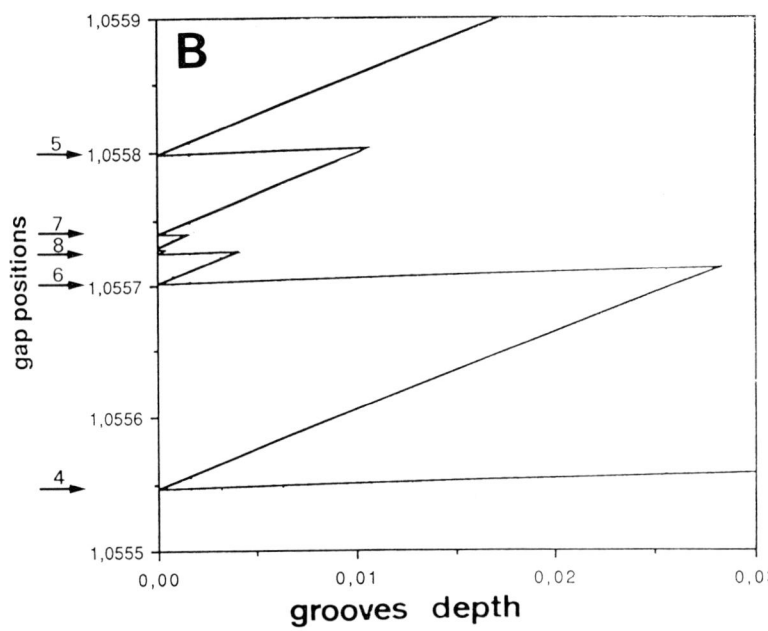

Fig. 3. Scaling of the gap structure around $at/\lambda = 1/2$ in a system of fixed length $L \approx 1.3 \times 10^6$ as a function of $2h$ (in μm) where h is the groove depth. Large gaps exist for large h which split as $h \to 0$. The particular wavelength $\lambda_{1/2}$ such that $at/\lambda_{1/2} = 1/2$ is an accumulation point of the converging set of gaps appearing as a consequence of the splitting process. Other gaps also exist between those which are presented but they have a very much smaller width. Thus, most of the measure of the gaps is taken by those which are presented.

where b is found numerically $b=220\pm20$. In the limit of small μ, it gives the number $N(r)$ of gaps whose distance to $at/\lambda_{1/2}=1/2$ is larger than r, which goes as r^{-1} thus yielding a scaling with a dimension $D=1$. This is analog to the result obtained by Thouless [11] on the scaling of the total measure of the gap as a function of system size L.

4. The spatial structure of the modes

On the basis of a transfer matrix formalism, we have shown [9] that the recursion relation obeyed by the transfer matrix for the SAW propagating on the quasicrystal is identical to the renormalization-group equation for a quasiperiodic Schrödinger equation with a *step* potential. This problem is known to correspond to a critical point just between an extended and localized regime. This criticality must be observed also on the proper modes. Along these lines, Thouless and Niu [12] have proposed that the absolute value of a wave corresponding to a proper mode should be a product of periodic functions, the period of each function being the approximate period of the almost periodic problem obtained by cutting off the continued fraction representation of the relevant incommensurate number at a particular stage.

Experimentally, our observations are reported in fig. 4 which shows the spatial structure of the SAW at particular frequencies in the close neighborhood of one of the main gaps (such that at/λ is in the close neighborhood of $8/17$). The SAW envelope intensity for different frequencies is plotted as a function of the distance x from one extremity of the lattice. These spatial proper mode structures are obtained from an optical diffraction experiment using the so-called Raman–Nath effect which has been described elsewhere [9]. The figures correspond to a "zero flux" condition, i.e. to the superposition of two counter-propagating SAWs of the same frequency, amplitude and phase launched from the two transducers on both sides of the system. We have checked on many examples [9] at different frequencies that the charac-

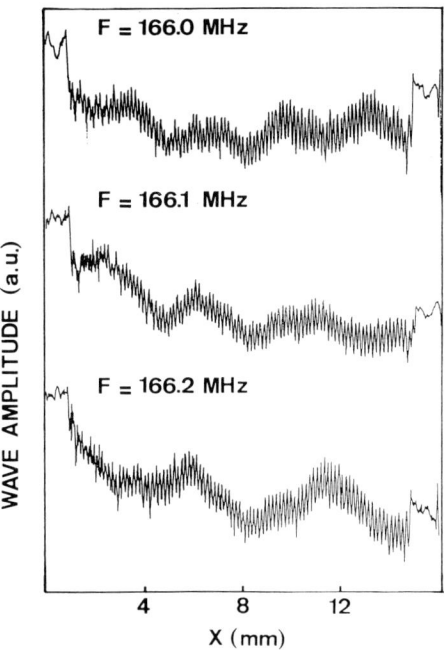

Fig. 4. SAW intensity as a function of position of the laser spot along the sample for three close frequencies in the neighborhood of one stop band shown in fig. 1 ($at/\lambda=8/17$).

teristic mode structure is not changed by using the "zero flux" condition.

In agreement with the theoretical predictions, we observe that the modes contain several sinusoidal components of widely separated spatial periods. Consider for example the mode depicted in fig. 4 whose frequency is $f\approx166.0$ MHz. Taking a value for the SAW velocity $c_R=3495$ m s^{-1} [6] gives the corresponding value for $at/\lambda=0.4697$. This value is obtained by superposition of the periods $at/\lambda=1/2$, 8, 250 since $(at/\lambda)^{-1}=2+8^{-1}+250^{-1}$ is the sum of the reciprocal wave vector $2\pi/b$ with $b=at$, $16at$, $500at$. We have used the fact that we observe experimentally the wave intensity and not its amplitude, so that the periods are divided by two. One should thus observe

(i) a period equal to a corresponding to the first period $b=at$,

(ii) a period equal to $8a$ corresponding to the second period $b=16at$,

(iii) a period equal to $250a$ corresponding to the third period $b=500at$.

Therefore, the largest scale structure exhibits $1000/250=4$ undulations, each one being decorated by $250/8 \approx 31$ smaller undulations and so on. These different periods 1, 8, 250 correspond to a total number of oscillations over the system length $L=1000a$ equal respectively to $1000/1$, $1000/8=125$ and $1000/250=4$. This is exactly what appears in fig. 4, where one observes indeed four oscillations with a very large wavelength which are decorated by a total of the order of 100 short wavelength oscillations. Due to the finite spatial resolution of the optical probing method, we do not see the shortest wavelength $\approx a$. There is thus a complete coherence between the value of at/λ corresponding to the wave frequency and its mode structure in terms of a superposition of spatial periods.

With this theory, we can understand the rapid change of the largest spatial modulation with the very small variation of SAW frequency. On going from $f=166.0$ MHz to 166.1 MHz, the large scale structure changes from having four undulations to three. Indeed, $f=166.1$ MHz corresponds to a wavelength $\lambda=21.039$ with $at/\lambda=0.4700$ and $(at/\lambda)^{-1}=2+8^{-1}+360^{-1}$. The ratio $1000/360 \approx 3$ corresponds nicely with the observation of three undulations. For $f=166.2$ MHz, we have $\lambda=21.026$ with $at/\lambda=0.4703$ and $(at/\lambda)^{-1}=2+8^{-1}+680^{-1}$. The ratio $1000/680 \approx 2$ nicely agrees with the observation of two oscillations. The existence of the large scale structure is the most typical signature of the criticality of the proper modes. The figures therefore constitute the *first direct experimental test* of the theoretical models describing wave propagation in quasi-periodic media.

We can analyse in a similar fashion the modes measured in the vicinity of $at/\lambda=1/2$ as well as all other modes [9]. In particular, by picking up the right values of at/λ, we can observe almost any spatial structure as rich or complex as desired for the critical proper modes. This complexity is completely determined from the rational expansion of $(at/\lambda)^{-1}$ as a sum of reciprocal wavevectors. In other words, by a continuous variation of the surface acoustic wavelength λ, we are able to sample the rich patterns coded by "irrational" numbers. The same discussion also applies to the family at $a/\lambda=1/2$, ..., etc.

It is also possible to predict the divergence of the largest period Λ in the mode structure as $at/\lambda \to 8/17$ (or more generally to $n/(2n\pm 1)$). Note that the modulation wavelength Λ is very large for $f=f_{8/17}$ such that $at/\lambda=8/17$ and decreases rapidly as at/λ departs from $8/17$. This is summarized in fig. 5, which shows a very rapid increase of Λ in the neighborhood of $f_{8/17}$. The log–log plot of fig. 5 is shown in fig. 6, which tells us that this divergence of Λ is well fitted by the power law

$$\Lambda \sim |f-f_{8/17}|^{-\nu} \quad \text{with } \nu=0.50\pm 0.05 . \tag{7}$$

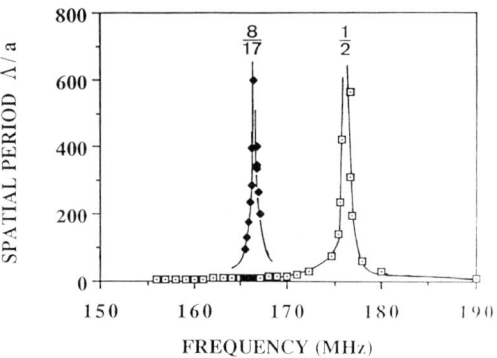

Fig. 5. Spatial period Λ of the observable proper modes versus frequency in the neighborhood of two stop bands shown in fig. 1: $at/\lambda=1/2$ (right peak) and $at/\lambda=8/17$ (left peak). Note the algebraic divergence of Λ at those particular values of the frequency.

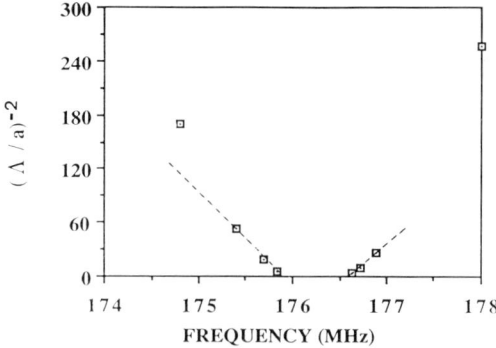

Fig. 6. Log–log representation of fig. 5 showing the scaling exponent 1/2 for the divergence of the largest spatial period in the critical modes.

The same features are observed for frequencies in the vicinity of $f_{1/2}$ for which $at/\lambda=1/2$. Again, we measure a divergence of the largest spatial wavelength of the intensity modulation which follows eq. (7). In fact, one can verify this type of behaviour for all frequencies in the neighborhood of the family $at/\lambda = n/(2n\pm1)$ which were identified in the spectrum in fig. 1.

These observations can be rationalized as follows. Consider the Bloch expression for the mode amplitude $Y_n(x)=u_n(x)\,e^{-iKx}$ with $u_n(x)$ periodic with period qa and K given by $s=e^{\pm iKqa}$. s verifies the secular equation $s^2 - (\text{Tr}\,M)s + 1 = 0$, M being the corresponding transfer matrix over a single cell. In the neighborhood of band edges, one generically has $\text{Tr}\,M \to 2$, which implies that s is close to one and K is close to zero. The sinusoidal modulation e^{-iKx} therefore develops a very long spatial period

$$\Lambda = qa(Kqa)^{-1} \approx qa(1-s)^{-1} = qa|(2-\text{Tr}\,M)/2|^{-\nu}$$

with $\nu=1/2$ in agreement with our fit. This power law is verified for all frequencies such that at/λ fulfills eq. (1).

5. Conclusion

In summary, we have given, for the first time, an experimental characterization of the critical proper modes and of the spectrum in a $d=1$ quasiperiodic system. These critical properties can be simply interpreted in terms of successive approximation of the reduced variable at/λ as a sum of reciprocal lattice vectors of decreasing norm, where a is the average lattice period, λ the surface acoustic wavelength and t the inverse golden mean. A much extended version of this work is presented in ref. [9] where the time impulse response is also discussed.

Acknowledgement

We acknowledge financial support from DRET under contract No. 86/177 for the research program "Propagation Acoustique en Milieux Aléatoires".

References

[1] B.B. Mandelbrot, The Fractal Geometry of Nature (Freeman, San Francisco, 1982).
[2] H.E. Stanley and N. Ostrowsky, Random Fluctuations and Pattern Growth: Experiments and Models (Kluwer, Dordrecht, 1988).
[3] B. Souillard, Waves and Electrons in Inhomogeneous Media, in: Chance and Matter, NATO ASI Les Houches Summer School, Session XLVI, 1986, eds. J. Souletie, J. Vannimenus and R. Stora (North-Holland, Amsterdam, 1987), and references therein;
D. Sornette, Acoustic Waves in Random Media, (I): Weak Disorder Regime, Acustica 67 (1989) 199; (II): Coherent Effects and Strong Disorder Regime, Acustica 67 (1989) 251; (III): Experimental Situations, Acustica 68 (1989) 15.
[4] D. Shechtman, I. Blech, D. Gratias and J.W. Cahn, Phys. Rev. Lett. 53 (1984) 1951.
[5] S. Aubry and C. André, Proc. Israel Phys. Soc., ed. C.G. Kuper (III) (Adam Hilger, Bristol, 1979) p. 133.
[6] E.A. Ash and E.G.S. Paige, Springer Series on Wave Phenomena. Rayleigh-Wave Theory and Application (Springer, Berlin, 1985).
[7] J.B. Sokoloff, Phys. Rep. 126 (1985) 189.
[8] D. Sornette, L. Macon and J. Coste, J. Phys. (Paris) 49 (1988) 1683.
[9] J.P. Desideri, O. Legrand, L. Macon and D. Sornette, Localization of Surface Acoustic Waves in a Quasi-Crystal, preprint.
[10] M. Kohmoto, L.P. Kadanoff and C. Tang, Phys. Rev. Lett. 50 (1983) 1870;
M. Kohmoto, B. Sutherland and C. Tang, Phys. Rev. 35 (1987) 1020.
[11] D.J. Thouless, Phys. Rev. B 28 (1983) 4272.
[12] D.J. Thouless and Q. Niu, Wavefunction Scaling in a Quasi-Periodic Potential, J. Phys. A 16 (1983) 1911.

OF MEN AND IDEAS (AFTER MANDELBROT)

Cyril DOMB

Physics Department, Bar-Ilan University, Ramat-Gan, Israel

Biographical information is provided about Hausdorff and Besicovitch, the two mathematical pioneers on whose work Mandelbrot's development of fractals is based. A number of examples are given of abstract mathematical ideas which subsequently found practical applications. It is suggested that the crisis in mathematics discussed by Mandelbrot which arose from Cantor's introductions of Mengenlehre is parallelled in the history of mathematics by other new ideas which encountered opposition but ultimately became part of the general body of mathematics.

1. Introduction

Among the most significant scientific publications that appeared in the past two decades are the essay by Benoit Mandelbrot, Fractals: Form, Chance and Dimension [1] and its revised edition The Fractal Geometry of Nature [2]. Scientific classics fall into different categories. One of the most important in the history of science is undoubtedly Willard Gibbs' essay On the Equilibrium of Heterogeneous Substances [3], which demonstrated the amazing power and wide ramifications of the second law of thermodynamics. The reader who completes this essay is overcome by a sense of awe and humility by the magnitude of Gibb's achievement. Every possible opening seems to have been explored, and wherever possible appropriate conclusions have been drawn. There are no loose ends.

Mandelbrot's essay is written in a totally different style. He tells us at the outset that he plans to deal with his subject "from a personal point of view and without attempting completeness" (ref. [1], p. 2) that it will contain "constant digressions and interruptions" (ref. [1], p. 2). When he introduces ideas he is concerned with their origin and history "an interest in the history of ideas is good for the scientist's soul" (ref. [2], p. 21), and he is excited by the personal lives of those who produce ideas and by their struggle to become accepted – particularly the mavericks who do not follow well trodden paths. Mandelbrot writes more or less as he thinks, and has produced a book in which there are fascinating unexplored paths on almost every page.

It is my aim in this note to make a few comments on the subject of people and ideas in mathematics and science, hopefully following the style of Mandelbrot (ref. [2], p. 391).

2. Felix Hausdorff (1868–1942)

Mandelbrot provides brief biographical information about Hausdorff, one of the major figures in the development of the idea of a fractional dimension. Up to the age of 35 he devoted much of his time to literature and music, and produced belles lettres under the pen name of Paul Mongré. For more details of his life we are referred to the Dictionary of Scientific Biography [4].

A few words may be relevant, however, on the role played by his treatise Grundzuge der Mengenlehre [5] in resolving the crisis in mathematics which Mandelbrot dates from 1875 to 1925. Because of World War I it took some years before reviews appeared. The following extracts are taken from the review by Henry Blumberg which appeared in the Bul-

Essays in honour of Benoit B. Mandelbrot
Fractals in Physics – A. Aharony and J. Feder (editors)

letin of the American Mathematical Society [6].

"If there are still mathematicians who hold the theory of aggregates under general suspicion, and are reluctant to grant it full recognition as a rigorous, mathematical discipline, they will find it hard to retain their doubts under fire of the logic of Hausdorff's treatise. It would be difficult to name a volume in any field of mathematics, even in the unclouded domain of number theory, that surpasses the Grundzuge in clearness and precision".

Hausdorff followed the style of Gibbs rather than that of Mandelbrot, and there are virtually no loose ends.

"As for more important criticism, one may quarrel with the author for his abstract style, for his Euclidean manner of grading the proofs, so that no difficulties remain and none but mild climaxes are reached, for his finish that may excite admiration but hardly activity on the reader's part. One may crave for a book that is built like a drama around a single idea – a more sketchy book, leaving more to the reader's imagination, a book with a less diversified and more emphatic message. But such remonstrance would be like quarrelling with Beethoven for having written symphonies instead of operas".

Hausdorff was Jewish, and one must sadly record that he committed suicide, together with his wife and her sister on January 26, 1942 to avoid internment by the Gestapo.

3. Abram Samoilovitch Besicovitch (1891–1970)

Besicovitch, a second major figure in the development of the background to fractals, did not make the Dictionary of Scientific Biography, so a more detailed biographical sketch is in order, particularly since he was such a colourful figure. I knew Besicovitch personally since I overlapped with him at Cambridge during my period as an undergraduate (1938–1941), as a graduate student (1946–1949), and as a lecturer in mathematics (1952–1954). A major source of general information is the Biographical Memoir in the Series of the Royal Society by J. C. Burkill [7]. But in addition, I have benefited greatly from discussion with Joe Gillis of the Weizmann Institute, one of his star students in the 1930's, who provided several personal anecdotes reported below; and from Dr. Ben Gross of the Jerusalem College of Technology, who was at the University of Pennsylvania when Besicovitch was a visiting Professor.

Besicovitch was descended from Karaite Jews, a breakaway sect who accepted only the written but not the oral Jewish tradition. The sect was treated more leniently by the Russian establishment than the general Jewish population, and this helped Besicovitch secure a University appointment. He was one of a talented family of six children, and graduated in 1912 at the University of St. Petersberg, where A.A. Markov was one of his teachers. The last university to be opened before the revolution in Russia, in 1916, was at Perm near the Ural mountains, as a branch of the University of St. Petersburg. In 1917 it became autonomous, and Besicovitch became a Professor in the School of Mathematics. After Soviet power was established in Perm the University developed rapidly, at the end of the summer of 1918 the Perm Physics and Mathematics Society was founded and started to publish a Journal.

During the year 1919 widespread severe hardships arose because of the Russian civil war. I.J. Schoenberg tells the following story [8] which he heard personally from Besicovitch. In 1917 the Japanese mathematician S. Kakeya had posed the following problem. Let $U = AB$ be a unit segment in the plane. We are to move U from its original position AB so as to bring it back to its original position with its end points reversed, so that the final position is BA, and during this motion U *should sweep out the least possible area.* Two obvious ways of achieving the turn over are:

(i) Turn AB around A by $180°$ to the position AB' and slide AB' along its line into the final position BA. U has then swept out a semicircle of area $\frac{1}{2}\pi$.

(ii) Rotate AB by $180°$ around its midpoint O. U sweeps out a circle of radius $\frac{1}{2}$ and area $\frac{1}{4}\pi$.

Kakeya suggested a third way which swept out a three-cusped hypocycloid of area $\frac{1}{8}\pi$. He then conjectured that $\frac{1}{8}\pi$ is the least value of the area in which U can be turned.

At the worst time of the civil war, in the winter of 1919, when there was no fuel, Besicovitch worked on, and solved, another essentially equivalent problem. He held his feet in a box filled with straw to keep warm, and came to the conclusion that the switching of the endpoints of U can be done within an arbitrarily small area. On the strength of this work he was elected in 1920 to a professorship at the University of Leningrad.

Throughout his life Besicovitch maintained a great interest in mathematical problems of all kinds. Over many years at Cambridge he ran, for the pleasure and benefit of undergraduates, a weekly feature of "contest problem". The solutions submitted were carefully read and annotated by Besicovitch and the announcement "Perfect solutions of problem 12 were sent in by M and N", spurred several young mathematicians on to develop their analytical powers.

He described himself as an expert in the "pathology" of mathematics. If someone put forward a conjecture which Besicovitch suspected to be untrue, he would keep worrying until he had produced a counter-example.

To return to academic life, in the early 1920's Besicovitch was offered a Rockefeller Fellowship to work abroad, but repeated efforts to obtain permission to accept this offer were refused by the Soviet authorities. He told Joe Gillis that on one day in 1917 when he had finished lecturing and heard the news that the Czar had abdicated he wept with tears of joy – now at long last there would be freedom. Not many months passed before he was weeping tears of sadness!

In 1924 he made plans to leave Russia illegally. Originally it was intended that three academics should be smuggled out together, A.A. Friedmann (famous for his theory of the expanding universe), J.D. Tamarkin, another mathematician, and Besicovitch. The operation was given financial support by Friedmann's father, who was wealthy (a New Economic Policy man). Regrettably Friedmann dropped out at the last minute, and died a few months later of typhoid. Tamarkin and Besicovitch went on their own, and the route to Latvia involved crossing a frozen river whose ice was not too firm. So as not to put too much pressure on a particular spot it was considered advisable to cross by lying down and rolling over, and Besicovitch complied. However, Tamarkin, whose proportions were ample, maintained that rolling was too undignified; he insisted on walking and there was considerable trepidation until he had completed the crossing successfully.

From Latvia, Besicovitch proceeded to Copenhagen, where he used the Rockefeller Fellowship to work with Harald Bohr on almost periodic functions. Whilst at Copenhagen he visited Oxford staying for several months with G.H. Hardy, who secured for him a lectureship at the University of Liverpool for 1926–1927. He moved to Cambridge in 1927 as a College and University lecturer, and in 1930 became a Fellow of Trinity College. This fellowship he retained until the end of his life. His most significant contributions to mathematics were in the area of almost periodic functions and fractional dimensions.

Besicovitch participated enthusiastically in Cambridge academic and social life. He had married the daughter of a widowed Russian mathematician, but the marriage was childless; he managed to organize his family life so that he could dine at high table (totally masculine) every evening, and drink port after dinner in the Senior Common Room with bachelors G.H. Hardy and J.E. Littlewood and other Trinity intellectuals. As College lecturer he took a great personal interest in the undergraduates, and in the early 1940's when Winchester public school (which specialized in high-grade mathematics teaching) sent to Trinity an amazing crop of undergraduates which included Freeman Dyson, James Lighthill and Michael Longuet-Higgins, who were joined by Tony Skyrme from Eton, Besicovitch was heard to boast proudly that they now had several students of the calibre of Littlewood.

A problem arose which is probably unique to Cambridge colleges. The Trinity "backs" of the river Cam contained a beautiful avenue of elms. Unfortunately, during a gale one of the elms was blown down. If this elm alone were replanted, the avenue would never regain its original beauty. It was decided to cut down and replant the whole avenue. But should it be replanted with elms or poplars? Elms took a long time to grow but were eventually more beautiful for posterity, whereas poplars might not be quite so beautiful, but grew quickly. "What is posterity" declared Besicovitch. "For me posterity begins tomorrow".

The story of Joe Gillis' Ph.D. oral examination throws an interesting light on the relaxed attitude at Cambridge in the 1930's. In addition to discussing the thesis, it was customary for the examiners to ask some general questions, and Besicovitch wanted to find out what form these might take. He asked a graduate student who had just been examined by Hardy how things had gone. "We chatted amiably for about half an hour about the thesis, then Hardy said 'I am supposed to ask you some questions. You live in Trumpington St. On your way to Trinity in the morning list in order the Colleges that you pass, starting from Fitzwilliam House'". But things did not go so easily with Gillis, and the second examiner (J.C. Burkill) began asking some quite penetrating questions about the thesis itself. Besicovitch seized an early opportunity to change the direction of the discussion to an important paper by Kolmogoroff which had appeared recently, and which he knew that Gillis had studied in detail. Gillis responded, and completed the examination impressively.

Ben Gross tells a similar story about an oral examination of a graduate student at the University of Pennsylvania in the 1960's when Besicovitch was external examiner. I.J. Schoenberg had just asked the student to prove a certain result when Besicovitch interposed "That is very hard question, Isai, how would *you* prove it?" He initiated a discussion among the examiners themselves with the student listening attentively.

Since he spoke Russian at home, Besicovitch's command of English remained stationary from his early days at Cambridge. For him the definite and indefinite articles were superfluous. Joe Gillis recalls that Besicovitch once confronted him with a moral challenge "If your friend has fault and out of kindness you ignore it, are you doing him favour?" When Joe replied negatively, Besicovitch said "Then, since you are my friend, please correct my English". But Joe did not have much success; he found that in Besicovitch's English individual words were usually correct, it was their assembly and arrangement which were unusual. In any case Besicovitch always managed to convey the intended meaning correctly.

He remained very suspicious of the Soviet Union. In the early 1930's his good friend, the distinguished physicist Peter Kapitza was working at the Mond Laboratory, and used to return to Russia every summer. Besicovitch disapproved strongly of these return visits. "My friend you go back to Russia once too often". This is exactly what happened – on one such visit the Soviet authorities refused to grant Kapitza an exit visa but provided him with a superb laboratory and excellent facilities. He spent the rest of his life very successfully in the Soviet Union.

"A mathematician's reputation rests on his bad proofs" was one of Besicovitch's surprising aphorisms. He wished to convey the idea that the originator of a result in mathematics usually establishes it by long and complicated proofs. It is later less original workers who undertake the process of simplification. About Professors at English Universities he told me "No one is elected to a professorship unless he has been respectable for at least three years. Once he has become respectable, he no longer produces creative work". Nevertheless, at the age of 59, in 1950, he succumbed to the temptation of succeeding J.E. Littlewood in the Rouse Ball Chair at Cambridge. Burkill remembered that on his thirty-sixth birthday Besicovitch, thinking that his own creative resources were drying up, proclaimed "I have had four-fifths of my life". Reminded of this on his appointment to the chair, Besicovitch sent a postcard in reply "Numerator was correct!".

4. Mathematics and practical reality

Neither Hausdorff nor Besicovitch dreamt that their ideas on the fractional dimensions of sets would be of practical use in the real world; likewise the work of Weierstrass on curves which are continuous but not differentiable at all points. It is a remarkable achievement of Mandelbrot to have identified so many practical realizations of these concepts and of the abstract mathematics which grew around them.

But, in fact, we shall draw attention to other interesting cases of abstract mathematics which subsequently had surprising practical applications. The case of Riemannian geometry and Einstein's theory of relativity is so well known that it is unnecessary to provide any further details.

In 1854 George Boole published a book entitled An Investigation into the Laws of Thought [9]. His friend Augustus De Morgan had been involved for some years in an acrimonious controversy with the Scottish philosopher Sir William Hamilton, who was critical of the role of mathematics in logic and philosophy, and Boole wished to come to De Morgan's help. He had already produced initial ideas in a slim volume in 1848 entitled The Mathematical Analysis of Logic. Now he tackled the construction in symbolic terms of logic as a doctrine-like geometry resting on a groundwork of acceptable axioms. He carried out his programme in detail in the book, reducing logic to a simple type of algebra. Logical reasoning is replaced by elementary algebraic manipulation.

For many years Boole's work received scant attention. Then at the beginning of the 20th century mathematical logic became fashionable and attracted wider attention. Russel and Whitehead's famous three-volume treatise Principia Mathematica [10] made extensive use of symbolic logic. But the discussions and applications were confined to abstract mathematics.

When digital computers were developed in the 1950's and 1960's and it was necessary to formulate logical instructions for an array of on/off switches, Boolean algebra was the natural tool for the task. Nowadays standard computer courses may ignore quantum theory, but *must* devote a few lectures to Boolean algebra.

One of the abstract topics pursued vigorously in the first few decades of this century was the Foundations of Mathematics, with three different schools initiated by Hilbert, Brouwer, and Whitehead and Russell. The Hilbert decision programme of the 1920's and 30's [11] had for its objective the discovery of a general process applicable to any mathematical theorem expressed in symbolic form, for deciding the truth or falsehood of the theorem. Gödels' incompleteness theorem [12] which made clear that the truth or falsehood of A could not be equated to probability of A or not-A in any finitely based logic, chosen once for all systems, struck a major blow at this programme. There remained the possibility of finding a mechanical process for deciding whether A or not-A, or neither, was formally provable in a given system. In 1937 Alan Turing set out to demonstrate rigorously that it was impossible. In order to do this he introduced a theoretical computing machine [13] whose operation he described in detail. This machine introduced for the purposes of abstract mathematics served as a prototype for the digital computer, whose subsequent development has changed the face of Western civilization.

After the war Turing declined an offer of a Cambridge University lectureship, and instead joined the National Physical Laboratory. I remember Besicovitch expressing his astonishment that a first rate mathematician like Turing was willing to forego an appointment at Cambridge University in favour of a government laboratory. But Turing was joining a group for the design, construction and use of a large automatic computing machine (ACE), and welcomed the opportunity to put his theoretical ideas into practice.

In his charming autobiography Enigmas of Chance [14], Mark Kac tells of his interesting encounter with Wiener measure. Wiener had developed a way of looking at Brownian motion which was remote from the physicist's picture. In fact, George Uhlenbeck (who has always been sensitive to the advantages of a rigorous mathematical approach to problems in theoretical physics) wrote the following footnote in his review paper (with Ming Chen Wong, ref. [15], p. 324) On the theory of Brownian motion II: "The authors are aware of the fact that in the mathematical literature, especially in the papers by N. Wiener, J.L. Doob and others (cf. for instance ref. [16], also for further references), the notion of a random (or stochastic process) has been defined in a much more refined way. This allows (us), for instance, to determine in certain cases the probability that the random function $y(t)$ is of bounded variation or continuous or differentiable, etc. However, it seems to us that these investigations have not helped in the solution of problems of direct physical interest and we will therefore not try to give an account of them".

Fortunately, Mark Kac relates, despite Uhlenbeck's discouragement, he was led by mathematical colleagues to pay serious attention to Wiener's theory. Having the advantage of close familiarity with probability theory, he was able to solve problems which his colleagues had found intractable. He was then led to a formula, discovered independently by Richard Feynman (and later known as the Feynman–Kac formula), which is nowadays ubiquitous throughout much of quantum physics and probability theory. Feynman had been led to the same formula by his theory of path integrals.

5. New ideas in mathematics

Reference has already been made to the crisis in mathematics associated with the introduction of Cantor's theory of infinite sets, and to the ultimate resolution of the crisis and acceptance of the new ideas. This is just one example of a general pattern which is repeated continuously in the development of new mathematical ideas. Let us look at some other examples.

The history of divergent series has been beautifully chronicled by Hardy [17]. Euler had little hesitation in manipulating and using divergent series. But after Euler opposition grew from mathematicians like d'Alembert, Laplace, and (in his later days) Lagrange, and mathematics moved slowly but steadily towards the orthodoxy imposed on it by Cauchy, Abel and their successors; divergent series were gradually banned from analysis. There were courageous analysts who continued to use them, notably Fourier and Poisson (who were more concerned with applied than with pure mathematics). Many years later Cesaro and Borel put the subject on a firm footing, and made divergent series respectable.

Oliver Heaviside (1850–1925) was a self-educated maverick who lived in conditions near to penury in Paignton, a Devon village. He was a highly individual and original thinker, and concerned himself largely with electromagnetic theory and electrical engineering. It is a tribute to his contemporaries that despite his unusual background, he was elected a Fellow of the Royal Society in 1891.

At about this time he published his operational methods for solving the differential equations of physics. His work was not systematically arranged, and in places his meaning was not clear; and he gave no proper justification of his results. It was claimed by his detractors that his methods solved no problem that could not be solved otherwise. Whilst this claim cannot easily be checked it is certainly true that in a very large class of cases the operational method gives the answer in one page when ordinary methods take five pages, and also that it gives the correct answer, when the ordinary methods, through human fallibility, are liable to give a wrong answer [18].

In dealing with small oscillations of dynamical systems with n degrees of freedom, if the actual motion arising from an initial disturbance is required, the operational method gives a solution simply and directly whilst the standard method of normal coordinates involves several ancillary stages of calculation. For continuous systems the advantage of the operational method is even greater, and for heat conduction it is particularly convenient.

For about thirty years Heaviside's work received little attention from mathematical physicists. His methods may have worked in all cases to which they had been applied, but how could one feel secure if they had not been established mathematically? Then Bromwich at Cambridge took a serious interest in Heaviside's work, and provided a justification of his results using the theory of functions of a complex variable and contour integrals. Later Carson and van der Pol related Heaviside's methods to the theory of Laplace transforms (see ref. [18] for a reference list). Nowadays the latter constitute one of the most powerful tools available to applied mathematicians.

Incidentally one of the first to publish a book on Heaviside's approach was Paul Levy, whom Benoit Mandelbrot regards as his teacher, and about whom he provides interesting and pertinent biographical information. Paul Levy had also "been kept at arm's length by the Establishment" (ref. [2], p. 398), and may well have felt spiritual kinship with Heaviside.

Heaviside in his work on electric circuits introduced a unit function whose value was 0 for $t<0$ and 1 for $t>0$. The derivative of the unit function he called "the impulse function", and many years later this function was introduced independently by Dirac as the analogue of the discrete Kronecker δ_{ij} in his development of quantum mechanics; it became widely known as the Dirac δ function.

Dirac defined the δ function by the two properties $\delta(x)=0$, $x\neq 0$, and $\int_{-\infty}^{\infty} \delta(x)\,dx=1$. Physicists could handle the function without anxiety since its physical significance was immediate – a function having a very large value for a very short interval. If any questions arose $\delta(x)$ could usually be replaced by the limit as $n\to\infty$ of a suitably chosen family of functions, say $n\exp(-nx)$ or $\sqrt{n/\pi}\exp(-nx^2)$, and the detailed behaviour traced as $n\to\infty$. But in practice they were usually guided by intuition.

Dirac used the function widely and ruthlessly and a formula like $(d/dx)(\ln x)=1/x-i\pi\delta(x)$ might be regarded by conventional mathematicians as hazardous. From 1930 to 1950 δ functions formed part of the everyday equipment of physicists, but were shunned by mathematicians. Then Laurent Schwartz (son-in-law of Paul Levy) formulated the theory of distributions, and made δ functions mathematically respectable.

A simple and cogent account of mathematically respectable δ functions is given by Lighthill [19] with the intriguing dedication "To Paul Dirac (who saw that it must be true), Laurent Schwartz (who proved it) and George Temple (who showed how simple it could be made)".

6. Conclusion

"The 'real' mathematics of the 'real' mathematicians, the mathematics of Fermat and Euler and Gauss and Abel and Riemann, is almost wholly 'useless' ". Thus wrote Hardy in 1940 in his famous book A Mathematician's Apology [20]. As usual he was being provocative and stimulating, with the aim of discouraging study for materialistic purposes.

But the counter-examples of the practical applications of the 'real' mathematics of Weierstrass, Hausdorff, Besicovitch, Boole and Hilbert must surely demolish his thesis. However, abstract and remote from reality the mathematics in which you are engaged, there is no escape from the conclusion that it may some day be put to practical use.

References

[1] B.B. Mandelbrot, Fractals: Form, Chance and Dimension (Freeman, San Francisco, 1977).
[2] B.B. Mandelbrot, The Fractal Geometry of Nature (Freeman, San Francisco, 1982).
[3] The Scientific Papers of Willard Gibbs, Vol. 1 (Longmans and Green, London, 1906), reprinted (Dover, New York, 1961), pp. 25–37.
[4] C.C. Gillespie, Dictionary of Scientific Biography, Vol. 6 (Scribner's, 1972).
[5] F. Hausdorff, Grundzuge der Mengenlehre (Leipzig, 1914), reprinted (Chelsea, New York, 1949).
[6] H. Blumberg, Bull. Am. Math. Soc. 27 (1920) 116.
[7] J.C. Burkill, Biogr. Mem. Fellows R. Soc. 17 (1971) 1.
[8] I.J. Schoenberg, Mathematical Time Exposures (Mathematical Association of America, 1982), pp. 169–171.
[9] E.T. Bell, Men of Mathematics (Simon and Schuster, New York, 1937), p. 433.
[10] B. Russell and A.N. Whitehead, Principia Mathematica (Cambridge Univ. Press, Cambridge, 1910–1913).
[11] M.H.A. Newman, Biogr. Mem. Fellows R. Soc. 1 (1955) 253.
[12] K. Gödel, Mh. Math. Phys. 38 (1931) 173.
[13] A. Turing, Proc. London Math. Soc. 42 (1937) 230.
[14] M. Kac, Enigmas of Chance (University of California Press, Berkeley, 1987).
[15] Ming Chen Wang and G. Uhlenbeck, On the theory of Brownian motion II, Rev. Mod. Phys. 17 (1945) 323–342.
[16] J.L. Doob, Ann. Math. 43 (1942) 351.
[17] G.H. Hardy, Divergent Series (Oxford Univ. Press, Oxford, 1949).
[18] H. Jeffreys, Operational Methods in Mathematical Physics, 2nd Ed. (Cambridge Univ. Press, Cambridge, 1931) preface.
[19] M.J. Lighthill, An Introduction to Fourier Analysis and Generalized Functions (Cambridge Univ. Press, Cambridge, 1958).
[20] G.H. Hardy, A Mathematician's Apology (Cambridge Univ. Press, Cambridge, 1940), p. 59.

FRACTALS IN TWO DIMENSIONS AND CONFORMAL INVARIANCE

Bertrand DUPLANTIER

Service de Physique Théorique [1], CEN de Saclay, F-91191 Gif-sur-Yvette Cedex, France

The exact fractal properties of critical geometrical systems in two dimensions are briefly reviewed. They are generically described by an infinite discrete spectrum of scaling dimensions, associated with underlying conformal operators. Applications involve polymers, percolation, Brownian motion and Ising clusters. A seemingly essential distinction between critical phenomena spectra and multifractal ones is given.

1. Introduction

As this conference celebrates Benoit Mandelbrot's 65th anniversary, it is perhaps appropriate to give a personal recollection about fractals. In our student days in Paris, a friend of mine and I, while visiting a well-known bookshop of the rue Gay-Lussac, came across a little book Les Objets Fractals: Forme, Hasard et Dimension [1] [#1]. We were definitely interested, but, being well educated buy the French mathematical system, we did not dare to buy a book with so few equations, and left it in the bookshop. Some years later, I met again the fractal dimensions, this time in polymer physics, when I realized amongst others that the correlation length critical exponent ν, with which the end-to-end distance R of a polymer of length S scales as $R \sim S^\nu$, is just the inverse of its fractal dimension D_F,

$$D_F = 1/\nu. \tag{1}$$

Nowadays, almost everybody knows about fractals [1–3] and it is well known that they provide a good mathematical description of many physical processes in condensed matter physics.

In this paper, I shall describe the case of *universal random fractals*, which arise in all critical phenomena in their geometrical formulation. In *two dimensions*, series of exact results have been obtained which provide nice illustrations of non-trivial random fractals realized in nature.

So I here review the statistical mechanics of critical fractal objects in two dimensions, such as self-avoiding walks, percolation or Ising clusters or intersection of random walks. These systems are parts of critical geometrical models, where sets of lines (e.g. polymers or random walks) or sets of points (e.g. clusters) are fluctuating and characterized by infinite series of critical exponents or (fractal) Hausdorff dimensions. More generally, infinite sets of critical exponents are hidden in the geometrical versions of the $O(n)$ and Potts models, of which polymers and percolation are particular cases. The sets of new conformal dimensions correspond to specific geometric properties of the loop or polygon representations (i.e. high-temperature expansion) of these models, and are obtained by branching and gluing together critical objects like several self-avoiding walks or several percolating clusters, etc. (fig. 1). In two dimensions (2D), in the plane, there exist conformal operators $\phi_L(X)$, sources of L such objects at a point X, with scaling dimensions x_L, depending only on the universality class of the geometrical system under consideration. The values of the scaling dimensions x_L can be obtained in general by Coulomb gas techniques, starting from the lattice realization of the models.

---

[1] Laboratoire de l'Institut de Recherche Fondamentale du Commissariat à l'Energie Atomique.

[#1] For general presentations, see also refs. [2,3].

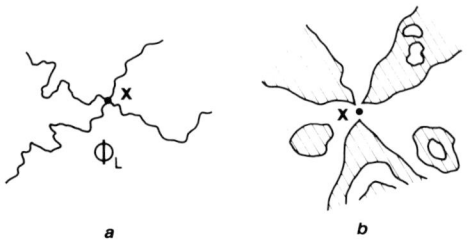

Fig. 1. (a) Branching of $L=4$ self- and mutually avoiding walks (polymers). A conformal operator ϕ_L is associated with this L-vertex. (b) Similar branching of three percolating clusters at a point X. The boundary lines form a 6-vertex.

They fit into the classification scheme of 2D conformal field theories, based on the standard Virasoro algebra.

The geometric idea behind conformal invariance is a generalization of the standard scale invariance at a critical point. For infinite critical systems, where a length scale (the correlation length) has become infinite, the rotational and scale invariances should apply *locally* and one should be able to deform by local dilations and rotations a system without changing its statistics. This is expected to be valid for critical systems with short-range interactions. The set of such local transformations is that of conformal mappings, which in two dimensions are any analytic transform. Actually this idea appeared explicitly as early as in Lévy's work on Brownian motion [4], which is perhaps the simplest model of a conformally invariant process. The idea in the more general framework of field theory appeared in an article by Polyakov [5], but the breakthrough occurred only in 1984 when it was applied to specific calculations by Belavin, Polyakov and Zamolodchikov [6] in field theory (for a review, see ref. [7]).

Here we shall concern ourselves with the geometrical scaling dimensions x_L, associated with the special vertices (point singularities), where infinite critical objects are fused together, as in fig. 1. From these scaling dimensions, one obtains all sorts of critical exponents and fractal dimensions, which provide an exact critical 2D geometry, illustrating the general field of mathematical fractals.

Let us first give a few examples showing the emergence of infinitely many scaling dimensions in critical phenomena.

2. Polymers

Consider L linear polymer chains which are self- and mutually avoiding (in a good solvent), and are trying to form a little micelle (i.e. an L-arm star polymer) (fig. 2). If the positions of the approaching heads $\{i=1, ..., L\}$ are described by vectors r_i in (2D) space, then the probability to find these heads in such positions is $P\{r_i\}$ and scales as [8–10] [#2]

$$P\{\lambda r_i\} \sim \lambda^{x_L - Lx_1} P\{r_i\}, \quad \lambda \to 0, \qquad (2)$$

where x_L is precisely the scaling dimension of the L-leg polymer vertex as described in fig. 1. The dimension x_1 is that of the 1-leg vertex, i.e. that associated with the head of a semi-infinite polymer. Since the polymers are repulsive one has necessarily [10]

$$x_L - Lx_1 > 0, \qquad (3)$$

hence gap scaling does *not* hold for such multipolymer scaling dimensions. One can prove even more inequalities by considering the contact of cores of several star polymers $i=1, ..., I$, with numbers of arms

[#2] Ref. [10] gives a review.

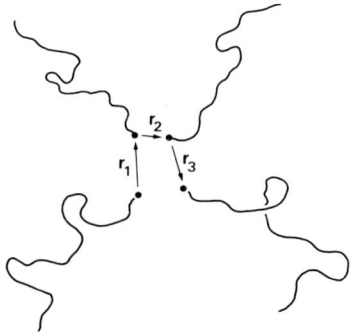

Fig. 2. L linear chain heads coming into contact to form a micelle or star with L arms. The scaling dimension before the fusion is $L \times x_1$ (gap scaling) and $x_L > Lx_1$ after the fusion (no gap scaling).

L_i, and located at positions r_i (fig. 3). There, again, the probability density of this set of positions r_i scales at short distance as

$$P\{\lambda r_i\} \sim \lambda^{\theta_{\{L_i\}}} P\{r_i\},$$

where the contact exponent [8] $\theta_{\{L_i\}}$ is expressed by the formula

$$\theta_{\{L_i\}} = x_{\Sigma_i L_i} - \sum_i x_{L_i}. \qquad (4)$$

In (4), $x_{\Sigma_i L_i}$ is the scaling dimension of the vertex obtained by *fusion* of all the L_i arms, $i=1, ..., I$, from which one has to subtract the scaling dimensions before the fusion. This strikingly simple rule (4) is, of course, the equivalent in polymer theory of the short-distance expansion (SDE) of the underlying field theory [11].

Since again the system of star cores is strongly repulsive, because of excluded volume effects, one has necessarily

$$x_{\Sigma_i L_i} - \sum_i x_{L_i} \geq 0. \qquad (5)$$

This inequality is actually *strict*, which is rather obvious from the physical origin of the problem, and ensures the stability of the universal fixed point of the field theory and is essential in the field theoretic formulation of critical phenomena. We shall see below that it also pays a fundamental role in the comparison to multifractal phenomena.

Before giving a more precise definition of the x_L as well as their values, let us mention that the first two x_1, x_2 are related to the usual critical exponents ν (1) and γ of self-avoiding walks (SAW) by

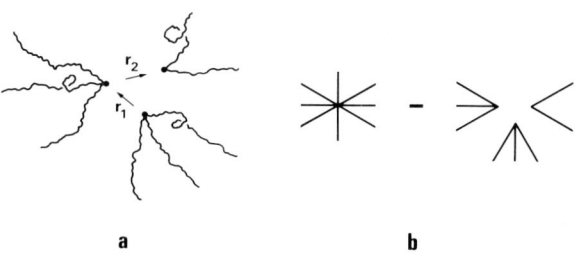

Fig. 3. Illustration of the short-distance expansion (4) for several cores coming into contact.

$$\eta = d - 2 + 2x_1,$$
$$D_F = \nu^{-1} = d - x_2, \qquad (6)$$
$$\gamma = (2 - \eta)\nu.$$

The universal configuration exponent γ governs, as usual, the number of configurations $\mathscr{L}$ of a single linear SAW of length S

$$\mathscr{L} \sim \mu^S S^{\gamma-1}, \quad S \to \infty, \qquad (7)$$

where the effective connectivity constant μ is lattice dependent, i.e. non-universal. The above scaling relations (6) are easily understood if one considers that x_1 corresponds to a sharp cut in an infinite polymer (1-leg insertion), while x_2 corresponds to inserting a point on an infinite polymer (2-leg insertion), which is like measuring the fractal dimension (fig. 4).

The scaling formulae (6) can be generalized to polymer networks of arbitrary topologies [8]. Take for instance the self-avoiding star polymer of fig. 1, made of L arms of equal lengths S. Its number of configurations $\mathscr{L}_L$ scales as

$$\mathscr{L}_L \sim \mu^{LS} S^{\gamma_L - 1}, \qquad (8)$$

where μ is the (non-universal) effective connectivity constant of the simple self-avoiding walk as in (7) and γ_L is a new universal configuration exponent. In terms of the scaling dimensions x_L of the operator $\phi_L(X)$ creating an L-vertex at point X, γ_L is written by generalized hyperscaling [8,10]

$$\gamma_L - 1 = \nu[dL - x_L - Lx_1] - L. \qquad (9)$$

We observe again the presence of one L-arm scaling dimension x_L and L times the 1-arm scaling dimensions x_1, which correspond respectively to the core of the star and to the L free extremities.

Watermelon exponents. In order to isolate only the

Fig. 4. The physical meaning of conformal operators ϕ_1 and ϕ_2 creating 1-leg and 2-leg vertices, with respective scaling dimensions x_1 and x_2 ($x_2 > 2x_1$).

dimension x_L one can consider the *watermelon* configuration of L polymers tied at their extremities X and Y in space (fig. 5). Their partition function is defined (e.g. on the lattice) as

$$\mathscr{L}(X-Y, S_1, ..., S_L) = \sum_{\Gamma_l(X,Y)} 1 ,$$

where the sum extends over all self- and mutually avoiding configurations $\Gamma_l(X, Y)$ of all paths $l=1, ..., L$ of lengths $S_1, ..., S_L$, joining X to Y. Asymptotically $\mathscr{L} \sim \mu^{\sum_{l=1}^{L} S_l}$ and one defines a correlation function by the Laplace transform (generalizing de Gennes' original idea of 1972 [12])

$$G_L(X-Y) = \sum_{\{S_l=1\}}^{\infty} K^{\sum_l S_l} \mathscr{L}(X-Y, \{S_l\}) , \quad (10a)$$

where $K \leq \mu^{-1}$ is the monomer fugacity. At the critical point $K_c = \mu^{-1}$, G_L decays as a power law

$$G_L(X-Y) = |X-Y|^{-2x_L} ,$$

where x_L is the scaling dimension of the field operator $\phi_L(X)$ generating an L-leg vertex at X such that

$$\langle \phi_L(X) \phi_L(Y) \rangle = |X-Y|^{-2x_L} . \quad (10b)$$

(In these notes we shall not be more precise about the definition of these operators and resort only onto the mathematical intuition of their existence.)

The values of the exponents x_L are known in two dimensions [9,10,13–15]

$$x_L = \tfrac{1}{48}(9L^2 - 4) , \quad (11)$$

$$x_L^D = \tfrac{1}{16}(L^2 - 4) , \quad (12)$$

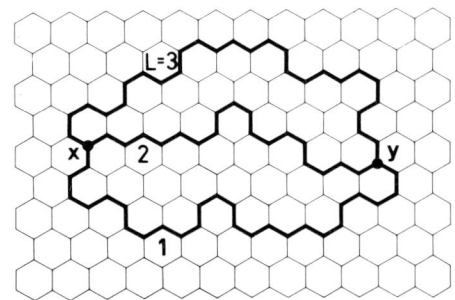

Fig. 5. $L=3$ polymer lines tied together in a watermelon configuration (here on the hexagonal lattice).

for the standard dilute polymers and for dense polymers respectively. (Dense polymers form a new universal critical phase which *fills* the 2D Euclidean space.)

In dimension $d=4-\epsilon$, $\epsilon>0$ the values of the x_L are also known [10] (for dilute SAW)

$$x_L = \tfrac{1}{2}L(d-2) + \tfrac{1}{8}\epsilon L(L-1)$$
$$+ (\tfrac{1}{8}\epsilon)^2 (\tfrac{1}{4}L)(-9L^2 + 33L - 23) . \quad (13)$$

We remark with respect to all these expressions that the inequalities (3) and (5) are fulfilled.

In particular, close to $d=4$, the *sign* of the $\mathcal{O}(\epsilon)$ quadratic $L(L-1)$ correction in (13) is crucial, and is the same as in two dimensions (eqs. (11) and (12)).

Also in 2D, from the values (11) for $L=1, 2$ we recover for (6) the Nienhuis exponents

$$D_F = \tfrac{4}{3} = \nu^{-1}, \quad \gamma = \tfrac{43}{32} , \quad (14)$$

while, using (9), we get for stars

$$\gamma_L - 1 = \tfrac{1}{64}[4 + 9L(3-L)] , \quad (15)$$

an exact formula which is very well checked numerically [10].

This then gives the solution to the number of configurations of polymer micronetworks of arbitrary topologies [8]. Take a general polymer graph $\mathscr{G}$ made of $\mathscr{N}$ chains of the same length S, connected at some prescribed vertices of arbitrary functionalities L. Let n_L be the number of vertices of type L in $\mathscr{G}$. For the star polymer for instance, $n_1 = L$, $n_L = 1$. Then the total number of self-avoiding configurations of $\mathscr{G}$, keeping the topology fixed, is asymptotically

$$\mathscr{L}_\mathscr{G} \sim \mu^{\mathscr{N} S} S^{\gamma_\mathscr{G} - 1}, \quad S \to \infty , \quad (16)$$

where $\mathscr{N}$ is the total number of chains and $\mathscr{N} S$ the total length. The exponent $\gamma_\mathscr{G}$ is a *universal* exponent, but topology dependent. Its exact expression is in terms of the dimensions x_L [8,10]

$$\gamma_\mathscr{G} - 1 = -2\nu + \sum_{L \geq 1} n_L [\nu(2-x_L) - \tfrac{1}{2}L] . \quad (17)$$

(This expression holds in 2D but is easily generalized to d dimensions [10].) We therefore find, e.g. for standard dilute SAW [8], the exact formula covering all possible topologies

$$\gamma_{\mathcal{G}} = -\tfrac{1}{2} + \tfrac{1}{64} \sum_{L \geq 1} n_L (2-L)(9L+50) . \qquad (18)$$

This formula, when compared to numerical series results by the King's College group around Gaunt [10] for various topologies (stars, H-comb, tadpoles, ...) gives excellent agreement in 2D. For instance for a network having the topology of a star, this gives $\gamma_L = \tfrac{1}{64}[68 + 9L(3-L)]$, hence $\gamma_1 = \gamma_2 \equiv \gamma = \tfrac{43}{32}$ (usual susceptibility exponent of a linear chain), $\gamma_3 = \tfrac{17}{16} = 1.00625$, $\gamma_4 = \tfrac{1}{2}$, $\gamma_5 = -\tfrac{11}{32} = -0.34374$ and $\gamma_6 = -\tfrac{47}{32} = -1.46875$. The numerical results are (see references in ref. [10]) $\gamma_3 = 1.07 \pm 0.02$, $\gamma_4 = 0.52 \pm 0.04$, $\gamma_5 = -0.29 \pm 0.04$ and $\gamma_6 = -1.33 \pm 0.05$. The γ_H exponent of a H-comb is $\gamma_H = \tfrac{25}{32} = 0.78125$, and numerically $\gamma_H = 0.79 \pm 0.02$.

Of course, the vertex decomposition formula generalizes to any space dimension [8,10]. A general proof in the realm of standard perturbative renormalization theory is still, however, lacking. It would require a careful selection of the irreducible vertex operator associated with the branching of L lines in the $O(n)$ model, among all operators mixing with φ^L in a φ^4 theory.

3. Geometrical exponents in $O(n)$ and Potts models

3.1. Watermelon configurations

For definiteness we first consider the $O(n)$ model on the honeycomb two-dimensional lattice. Its partition function is defined as [13]

$$Z_{O(n)} = \int \prod_i d\mathbf{S}_i \prod_{\langle i,j \rangle} (1 + K \mathbf{S}_i \cdot \mathbf{S}_j) , \qquad (19)$$

where the spin variables $\mathbf{S}_i$ are n-vectors with $\mathbf{S}_i^2 = n$, and where $\langle i,j \rangle$ are nearest-neighbour sites. One can show easily that on the honeycomb lattice $\mathcal{H}$ eq. (19) reads

$$Z_{O(n)} = \sum_{\text{non-intersecting loops}} K^{\mathcal{N}_B} n^{\mathcal{N}_L} , \qquad (20)$$

where the sum runs on all configurations of closed and non-intersecting loops drawn on $\mathcal{H}$, with total numbers of bonds $\mathcal{N}_B$, and of loops $\mathcal{N}_L$.

In this way, the high-temperature (i.e. small-K) expansion of the $O(n)$ model gives a *geometrical loop model*. For a fixed value of $n \in [-2, 2]$ there is a critical point $K_c = [2 + (2-n)^{1/2}]^{-1/2}$ (on the hexagonal lattice [7]) where the mean length $\langle \mathcal{N}_B \rangle \sim (K_c - K)^{-1}$ diverges. Geometrical correlation functions can be defined by considering *watermelon* configurations [14,15] on the lattice (fig. 6). One specifies to configurations $\mathscr{C}_L(X, Y)$ such that L non-intersecting lines have a source point at X and a sink at Y where they all meet together, in the presence of vacuum loops of the $O(n)$ model. The associated correlator is defined as

$$G_L(X-Y, K) = \sum_{\mathscr{C}_L(X, Y)} W(\mathscr{C}_L) / Z_{O(n)} , \qquad (21)$$

where the statistical weight of the configuration $\mathscr{C}_L$ (watermelon + vacuum loops) is

$$W(\mathscr{C}_L) = K^{\mathcal{N}_B} n^{\mathcal{N}_L} ,$$

where now $\mathcal{N}_B$ is the total length of the watermelon and loop lines. At K_c one expects a critical algebraic decay

$$G_L(X-Y, K_c) \equiv \langle \phi_L(X) \phi_L(Y) \rangle_c$$

$$\sim |X-Y|^{-2x_L} , \qquad (22)$$

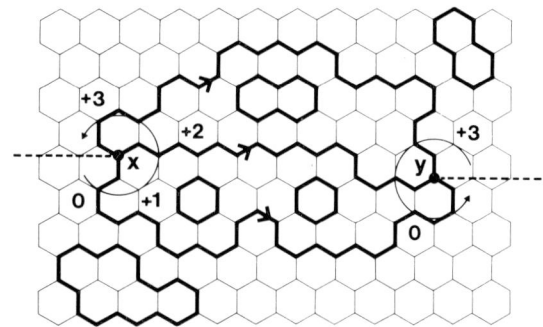

Fig. 6. Watermelon configuration of $L=3$ non-intersecting lines, in presence of vacuum loops of the $O(n)$ model. The integers indicate the heights on the equivalent SOS model, once the SAW are oriented and represent domain walls. The dotted lines represent dislocations associated with the watermelon.

where x_L is the scaling dimension of the (conformal) operator $\phi_L(X)$ assumed to represent a source of L non-intersecting *infinite* lines at a point X (fig. 1). As we shall see these x_L are not linear in L (there is no gap exponent) and are independent of each other. Hence a new geometrical critical exponent x_L appears for each star-like critical object. It is also important to notice that in the $O(n)$ model, a second critical phase appears, the *dense* phase [9,13,15], for any value $K > K_c$. Geometrically, this corresponds to taking infinite loops or lines filling the lattice with a finite (non-vanishing) density, even in the thermodynamic limit. The physical properties thus describe those of a loop melt in two dimensions [9,15,16]. In this dense phase are associated other scaling dimensions x_L^D with each source operator ϕ_L.

3.2. Coulomb gas technique

It is known that one can determine the geometrical exponents x_L by a Coulomb gas technique first devised in 1982 [17] for obtaining the usual spin and energy exponents η and ν of the $O(n)$ model. As we shall see, the x_L also belong to the conformal Kac table of the (unitary) minimal theories with central charge $c \leq 1$. Lastly, let us mention that a derivation of the same dimensions is also possible through a Bethe Ansatz on the hexagonal lattice [18]. In the standard Coulomb gas technique [13,17], the $O(n)$ model can be transformed into a SOS model by orienting the loops and the watermelon lines and in the continuum critical limit, it renormalizes onto a Gaussian field with action $A = (g/4\pi) \int (\partial \varphi)^2 \, d^2x$, where g is the Coulomb gas coupling constant such that

$$n = -2 \cos \pi g \qquad (23)$$

and $g \in [1,2]$ at K_c, while $g \in [0,1]$ in the *dense* phase [13].

Now, the watermelon correlation function generalized to the $O(n)$ model can be associated in the SOS model with the existence of a dislocation. Indeed the L lines, once oriented, are interpreted as domain walls in a height model and a circuit around X or Y gives a height discontinuity depending on L (fig. 6). As a result, the correlation function G_L (22) is simply a Coulomb gas correlator of two electromagnetic operators

$$G_L(X-Y) = \langle \mathcal{O}_{e_0,m'}(X) \mathcal{O}_{e_0,-m'}(Y) \rangle$$
$$= |X-Y|^{-gm'^2 + e_0^2/g}, \qquad (24)$$

where $e_0 = 1 - g$ is the standard floating electric charge [19] and m' the magnetic contribution coming from the L defect lines: $m' = \frac{1}{2}L$. The name "Coulomb gas" is justified when one remarks that the correlator (24) reads also

$$\exp[-(gm'^2 - e_0^2/g) \ln|X-Y|],$$

i.e. is the *Gibbs weight* of pairs of magnetic and electric charges (m', e_0) and $(-m', e_0)$ located at X and Y respectively, and interacting via the Coulomb potential in 2D ($\ln|X-Y|$). We refer the reader to refs. [13–15] for precise geometric derivations of (24). Let us simply mention that the seemingly mysterious presence of the electric charge $e_0 = 1 - g$ is necessary for correcting curvature effects along the watermelon lines, which appear in the SOS model but not in the original $O(n)$ model, when the watermelon *winds* about one of its origins.

Hence the watermelon scaling dimension is finally

$$x_L = \tfrac{1}{2}(gm'^2 - e_0^2/g) = gL^2/8 - (1-g)^2/2g \qquad (25)$$

for the $O(n)$ model, where n is parametrized by (23).

3.2.1. Conformal invariance

These scaling dimensions can be rewritten as a Kac formula [6,7,20,21]

$$h_{p,q} = \frac{[(m+1)p - mq]^2 - 1}{4m(m+1)}, \quad m \in \mathbb{N},$$

$$c = 1 - \frac{6}{m(m+1)} = 1 - \frac{6(1-g)^2}{g}, \qquad (26)$$

$$x_L = 2h_{L/2,0} \quad \text{(dilute phase, } g \in [1,2]\text{)},$$

$$x_L^D = 2h_{0,L/2} \quad \text{(dense phase, } g \in [0,1]\text{)}.$$

The values of the conformal parameter m parametrizing the universality class are $m = 1/(g-1) \in [1,$

∞[for the dilute phase and $m=g/(1-g)\in[0,\infty[$ for the dense one. The $O(n=2)$ model corresponds to the XY model at the Kosterlitz–Thouless transition point, for which $g=1$, $m=\infty$, $c=1$. The dilute and dense phases coincide there. Only this model and the $n=1$ standard Ising model, for which $m=3$, are unitary. The other $O(n)$ models (n continuous) are analytic extensions of the spin model (19) and have a purely geometrical interpretation (20) in terms of systems of non-intersecting loops. These values $h_{p,q}$ play a fundamental role in the representations of the conformal group. They are in a sense the *quantized* values allowed to the critical exponents, under the requirement that the critical system is *conformally* invariant (see ref. [7]). This is very analogous to the quantification rules of angular momentum in quantum mechanics, which this time arises from the requirement of *rotational* invariance.

3.2.2. Polymers

When $n=0$ in (20), all loops give zero weight and only the empty graph contributes, leading to a trivial partition function $Z_{O(n=0)}=1$. However, in the watermelon correlator (21), the L lines joining X to Y survive when $n\to 0$, and describe in the continuum limit the correlation function (10) of L self- and mutually avoiding polymer lines, tied at their extremities X and Y, and with *fluctuating* lengths $S_1, ..., S_L$.

The two possible phases for *polymers* ($n=0$) correspond to $g=\tfrac{3}{2}$ (dilute) or $g=\tfrac{1}{2}$ (dense). Hence (25) and (26) give the set of scaling dimensions:

$$x_L = \tfrac{1}{48}(9L^2-4), \quad c=0, \quad g=\tfrac{3}{2}, \quad m=2 \text{ (dilute)},$$

$$x_L^D = \tfrac{1}{16}(L^2-4), \quad c=-2, \quad g=\tfrac{1}{2}, \quad m=1 \text{ (dense)}. \tag{27}$$

It is interesting to remark that the usual conformal classification of "unitary" statistical systems started [7,21] with integer values $m\geq 3$ of the parameter m appearing in $h_{p,q}$ (26), $m=3$ corresponding to the Ising model, and models with higher values of m being identified later [22]. The polymer case is special, since there is no symmetric transfer matrix, or equivalently it is an $n=0$ vector model, and it is "non-unitary". It fills then precisely the two missing values $m=1,2$ of the Friedan, Qiu and Shenker classification. Notice also that the interpretation of half-integer indices in $h_{p,q}$ is not yet clear. Formulas like (26) for the $O(n)$ model are thus to be interpreted as useful analytic continuations. Notice that the thermal exponent x_2 (27) was conjectured before (from the $O(n)$ model) by Cardy and Hamber [23]. For the dense phase of polymers, we find instead of (14)

$$\nu^D = \tfrac{1}{2}, \qquad \gamma^D = \tfrac{19}{16}. \tag{28}$$

The interpretation of $\nu^D = \tfrac{1}{2}$ is immediate: the fractal dimension is $D_F = 1/\nu^D = 2$, as expected for a dense walk filling the plane. The interpretation of the γ exponent is more subtle, and requires a careful analysis [9,16] of boundary conditions for dense SAW, since these conditions are able to add non-universal terms to γ^D.

Up to now we used only the first exponents x_1, x_2. The higher-order ones x_L, $L\geq 3$ play a fundamental role, described briefly in (15) and (16) in the exponents of branched polymer networks of arbitrary topologies.

3.2.3. Percolation

Another interesting case is that of *percolation*. Let us consider the site percolation on the triangular lattice, dual of the hexagonal lattice. The critical occupation probability is $p_c = \tfrac{1}{2}$. Then the internal and external perimeter lines surrounding clusters of occupied sites are configurations of the $O(n)$-loop model for $n=1$ and $K=1$. Indeed there is an equal probability $\tfrac{1}{2}$ to occupy a site or not. One can draw freely with equal relative weight 1 any configurations of non-intersecting loops as cluster perimeters (up to the global choice empty versus occupied) and thus $n=1$, $K=1$ in eq. (20). Since the Ising-like critical point of the $O(n=1)$ model is at $K_c(n=1) = 1/\sqrt{3} < 1$, the point $K_{\text{percolation}} = 1$ lies in the (low-temperature) *dense* phase of the $O(n=1)$ model. Hence the Coulomb gas coupling constant ($g \in [0,1]$) is $g = \tfrac{2}{3}$ and the scaling dimensions (25) of L lines are

$$x_L = \tfrac{1}{12}(L^2-1). \tag{29}$$

We can now interpret them geometrically as the scaling dimensions associated with configurations of k clusters joining X to Y in the percolation problem (fig. 7), if we simply remark that the $L=2k$ external lines of these k clusters are describing all configurations $\mathscr{C}_{2k}(X, Y)$ of the $2k$-watermelon in presence of the loops of the $n=1$, $K=1$ model (i.e. the other cluster perimeters in percolation). Hence the end-to-end correlation function of k clusters at the percolation threshold decays as [24]

$$G_{2k}(X-Y) \sim |X-Y|^{-2x_{2k}},$$
$$x_{2k} = \tfrac{1}{12}(4k^2-1). \qquad (30)$$

The odd-L dimensions (29) have no obvious geometrical interpretation for percolation, but get one in terms of polymers at the tricritical Θ point [25]. As a first application, let us remark that the hull of percolation clusters corresponds to the boundary-boundary correlation exponent x_2 of a $k=1$ cluster. The hull fractal dimension is then given by scaling as [24]

$$D_\text{H} = 2 - x_2 = \tfrac{7}{4}, \qquad (31)$$

a value which is checked numerically very well [26], and was also conjectured before from an analogy to diffusion fronts [27].

3.3. Surface exponents

It is worth noting that all the previous discussion about the geometrical exponents x_L associated with the source of L lines in the $O(n)$ model can be generalized to surface geometrical phenomena [28,29]. One has simply to put the two extremities of the watermelon near the *boundary* of the half-plane. Then the correlator along the surface decays at criticality as [29]

$$G_L^\text{S}(X-Y) \sim |X-Y|^{-2x_L^\text{S}},$$

where the scaling dimension of a surface L-source operator is (fig. 8)

$$x_L^\text{S} = \tfrac{1}{4}gL^2 + \tfrac{1}{2}L(g-1). \qquad (32)$$

In terms of the Kac formula these surface dimensions are expressed, using parametrization (26), as

$$x_L^\text{S} = h_{L+1,1}, \quad \frac{m+1}{m} \qquad \text{(dilute phase)},$$
$$x_L^\text{S,D} = h_{1,L+1}, \quad \frac{m}{m+1} = g \quad \text{(dense phase)}. \qquad (33)$$

An interesting application of these formulae is the *fractal dimension of the loops of the $O(n)$ model at the boundary* (fig. 8):

$$D_\text{F}^\text{S} = 1 - x_2^\text{S} = 2(1-g).$$

These loops can be seen as hulls and for percolation

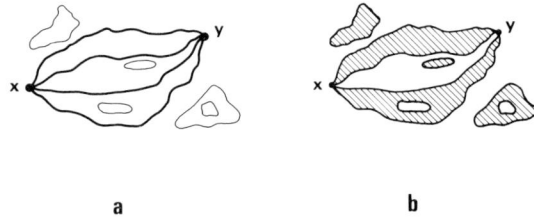

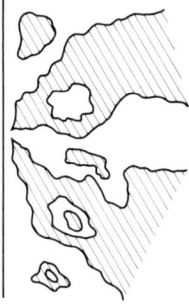

Fig. 7. The geometric correlation functions (21) and (38) in the $O(n)$ loop model and in the Potts model. The correspondence holds between L watermelon lines (heavy lines in (a)) and k bands in the Potts cluster representation (hatched areas in (b)) such that $L=2k$, and $x_L(O(n))=x'_k$ (Potts). Furthermore: $O(n)$ critical $\Leftrightarrow$ Potts tricritical, $O(n)$ low-temperature phase $\Leftrightarrow$ Potts critical.

Fig. 8. A $L=4$ line vertex near a surface line in the percolation problem (i.e. $k=2$ clusters pinched at the surface); the scaling dimension is x_4^S (eq. (32)). The hull (corresponding to x_2^S) near the boundary line has a new fractal dimension $D_\text{H}^\text{S} = 1 - x_2^\text{S} = 2/3$.

($n=1$, $g=\frac{2}{3}$) we find the *new surface hull fractal dimension*

$$D_H^S = \tfrac{2}{3}.$$

Notice that this works for the dense phase of the $O(n)$ model (or the critical Potts model, as we shall see below) such that $g \in [0, 1]$. The *dilute* loops ($g \in [1, 2]$) are repelled from the boundary and one finds formally negative fractal dimensions, as expected. Up to now, the only geometrical constraint was that all elements end at the boundary line, corresponding to free boundary conditions, i.e. Dirichlet ones in the continuum limit, or in the terminology of critical phenomena, to the *ordinary surface transition* [30]. Another case is the special surface transition where a supplementary attraction energy is given to any bond of the high-temperature loop expansion of the $O(n)$ model, which is near (or on) the surface line. At some special critical energy the loops just *adsorb*. Recent works [31,32], using numerical simulations, conformal invariance and Coulomb gas techniques indicate that the special surface dimensions for the watermelon are

$$x_L^{SP} = h_{L+1,3}$$
$$= \tfrac{1}{4} y(L+1)^2 - \tfrac{3}{2}(L+1) + 9 - (1-g)^2/4g, \quad (34)$$

valid at the dilute critical point of the $O(n)$ model, $g \in [1, 2]$. The interpretation of this special transition for $1 \leq n \leq 2$, $g \in [\frac{4}{3}, 2]$ is subtle since no special magnetic transition is expected in 2D for $n \geq 1$; the surface line and the bulk should indeed order simultaneously, and it is well known that no transition takes place in 1D for $n \geq 1$. As an application to *fractal dimensions* let us consider again that of the points of the loops adsorbed at the boundary. We have

$$D_H^{SP} = 1 - x_2^{SP} = 5 - 2g - 2/g, \quad g \in [1, 2],$$

where this formula is restricted to the standard dilute critical point of the $O(n)$ model, hence $n = -2 \cos \pi g$, $g \in [1, 2]$. For $g=1$, $n=2$ (XY model), $D_H^{SP} = 1$, the loops are filling the boundary line; for $g=2$, $n=-2$, $D_H^{SP} = 0$, and the loops are completely repelled. For polymers ($g=\frac{3}{2}$, $n=0$) we find the new fractal dimension of the *adsorbed monomers*

$$D_H^{SP} = \tfrac{2}{3}.$$

Lastly, for *Ising clusters* ($n=1$, $g=\frac{4}{3}$), the adsorbed part of their hulls has a new fractal dimension

$$D_{Hull}^{SP}(\text{Ising}) = \tfrac{5}{6}.$$

This exact value seems to have been observed numerically recently [33].

Before turning to some specific applications of the watermelon conformal spectrum, let us describe a useful geometrical equivalence to the Potts model.

3.4. *Potts model*

The Q-state Potts model is defined by the Hamiltonian $\beta H = -\beta \sum_{\langle i,j \rangle} \delta_{\sigma_i \sigma_j}$, where the Potts variables can take Q values $\sigma_i = 1, ..., Q$, and where $\langle i, j \rangle$ are nearest-neighbours on e.g. the square lattice $\mathcal{L}$. One can write the high-temperature expansion of the partition function as

$$Z_{\text{Potts}}(Q) = \sum_{\{\sigma\}} e^{-\beta H}$$
$$= \sum_{\mathcal{C}} W(\mathcal{C}) \equiv \sum_{\mathcal{C}} (e^\beta - 1)^{N_B} Q^{N_{\mathcal{C}}}, \quad (35)$$

where the configurations $\mathcal{C}$ are those of spanning graphs made of bond clusters of connected sites on $\mathcal{L}$, N_B being the total number of bonds of $\mathcal{C}$, and $N_{\mathcal{C}}$ that of the connected components of $\mathcal{C}$, including all isolated sites. This expression now defines a model for any real Q; if $Q \to 1$ one recovers bond percolation with occupancy probability $p = 1 - e^{-\beta}$. For $Q \in [0, 4]$, there is a second-order phase transition, which can be studied with use of a Coulomb-gas mapping. First, a graph $\mathcal{C}$ on the original lattice $\mathcal{L}$ in the Potts model can be associated with a polygon decomposition of the surrounding lattice $\mathcal{S}$, here another square lattice, the sites of which are the mid-points of the edges of $\mathcal{L}$. The rule is that some vertices of $\mathcal{S}$ are cut open to let the bonds of $\mathcal{L}$ go through unintersected. This also applies to the edges of the dual lattice $\mathcal{D}$ of $\mathcal{L}$. For a lattice $\mathcal{L}$ with a total number of sites $\mathcal{S}$, one has (Euler's relation) $N_L = N_B + N_{\mathcal{C}} - N_{\mathcal{S}}$, where N_L is the number of loops within the clusters of the graph. On $\mathcal{S}$, the total number of polygons one can draw around each cluster and in each loop reads $\mathcal{N}_P = \mathcal{N}_L + \mathcal{N}_{\mathscr{C}}$. Hence Z (35) can be rewritten as

$$Z_{\text{Potts}}(Q) = Q^{\mathcal{N}/2} \sum_{\mathscr{C}} [(e^{\beta}-1)Q^{-1/2}]^{\mathcal{N}_B} Q^{\mathcal{N}_P/2}. \tag{36}$$

The critical point is known by duality to be [34] $(e^{\beta_c}-1)Q^{-1/2}=1$, hence

$$Z_{\text{Potts, critical}}(Q) = Q^{\mathcal{N}/2} \sum_{\mathscr{C}} Q^{\mathcal{N}_P/2}. \tag{37}$$

In this way, the *critical* Potts model appears simply as describing a dense geometrical set of non-intersecting polygons, filling the diagonal surrounding lattice, and with a weight factor $\sqrt{Q}$ for each polygon. One guesses in this way that there should be a complete equivalence [15,24] to the *dense* phase of the $O(n)$ model, provided that $\sqrt{Q}=n$. This equivalence is checked by considering the watermelon scaling dimensions in the Potts model. One introduces correlation functions for the Potts model

$$G_k(X-Y) = \frac{1}{Z_{\text{Potts}}(Q)} \sum W(\mathscr{C}_k), \tag{38}$$

where the weight $W(\mathscr{C}_k)$ is defined in (35), and where the sum is taken over all graphs $\mathscr{C}_k$ of the surrounding lattice $\mathscr{S}$ formed by k polygons that join a neighbourhood of a point X to a neighbourhood of a point Y (the case $k=2$ is represented in fig. 7).

The polygon decomposition of the surrounding lattice $\mathscr{S}$ allows one to consider $Z_{\text{Potts}}(Q)$ in (35) and (36) as the partition function of a special kind of six-vertex model or solid-on-solid (SOS) model. We state the facts we need here and refer the reader to previous works [15,24,35] for more details. The SOS model is driven by renormalization onto a critical Coulomb gas with a coupling constant g' given by

$$Q = 2 + 2 \cos \tfrac{1}{2}\pi g', \quad g' \in [2, 4], \quad Q \in [0, 4]. \tag{39}$$

Then G_k appears as the correlation function of two combinations of vortex and spin wave with respective magnetic and electric charges $(m_X, e_X) = (\tfrac{1}{2}k, \tfrac{1}{2}g'-2)$ and $(m_Y, e_Y) = (-\tfrac{1}{2}k, \tfrac{1}{2}g'-2)$. It decays at criticality as

$$G_k(X-Y) = |X-Y|^{-2x'_k} \tag{40}$$

with a critical exponent given by the den Nijs–Nienhuis Coulomb gas formula [13,35]

$$x'_k = -\tfrac{1}{2}g' m_X m_Y - (1/2g')e_X e_Y$$
$$= \tfrac{1}{2}g'(\tfrac{1}{2}k)^2 - (4-g')^2/8g'. \tag{41}$$

This result is valid for any value of Q, $Q \in [0, 4]$, with $g' \in [2, 4]$ and thus gives the critical decay of the k-polygon correlation function (38).

Now we observe mathematically the complete equivalence between the geometrical exponents of the dense $O(n)$ model and the critical Potts model, which was anticipated from (37). The identity $\sqrt{Q}=n$, $Q \in [0, 4]$, $n \in [0, 2]$, is obtained for Coulomb gas coupling constants (23) and (39)

$$n = \sqrt{Q} \in [0, 2],$$
$$g|_{O(n),\text{low-}T} = \tfrac{1}{4}g'|_{\text{Potts, critical}} \in [\tfrac{1}{2}, 1], \tag{42}$$

which implies in (25) and (41)

$$x_{L=2k} = x'_k, \tag{43}$$

i.e. the scaling dimension of a watermelon with $L=2k$ lines in the low-temperature phase of the $O(n)$ model is the same as that of a bundle of k polygons in the critical Potts model. These two critical phases thus belong to the same *geometrical universality class*. This has some interesting consequences. For instance the low-T $n=0$ model corresponds to dense SAW, while the $Q=0$ Potts model is known to represent spanning trees [36] and Hamiltonian walks on the Manhattan lattice [15,16]. This shows in particular that Hamiltonian walks (which are SAW visiting all lattice sites) are *universal* polymer melts [9,15,16]. Another expected application is found in percolation: we have seen above that *site* percolation corresponds to the low-T phase of the $O(1)$ model. The $Q=1$ Potts model is well known to describe *bond* percolation [36], and identity (43) shows that the critical geometrical properties of the multiple hulls of site and bond percolations are the same, as it must if one believes in universality of various percolation models. It is also worth noticing a *similar geometrical equivalence between the standard critical $O(n)$ model and the tricritical Potts model, still for $n=\sqrt{Q}$*. The tricritical Potts model (which is a diluted model with percolating vacancies) is associated with the other analytic determination $g' \in [4, 6]$ in (39). The relationship (42), (43) is then extended to

$$g = \tfrac{1}{4}g' \in [1, \tfrac{3}{2}],$$

$$x_{L=2k}|_{O(n), \text{critical}} = x'_k|_{Potts, \text{tricritical}},\qquad(44)$$

$$n = \sqrt{Q},$$

and gives the same identity of watermelon exponents. An application of this is for instance the universality of SAW on the *Manhattan oriented square lattice*. This model can be transformed directly into a $Q=0$ tricritical Potts model [37], following the same method as in ref. [15]. Then (44) yields the identity to the standard $n=0$ vector model, i.e. standard SAW on the hexagonal lattice, showing the irrelevance of orientations in a 2D Manhattan lattice.

Another application can be found in fractal dimensions associated with Potts clusters. The dimension $x'_{k=1}$ (41) gives the *fractal dimension of the perimeter* of a Potts cluster

$$D_H = 2 - x'_1 = 1 + 2/g',\qquad(45)$$

where g' parametrizes $\sqrt{Q} = -2\cos\tfrac{1}{4}\pi g'$, $g' \in [2, 4]$ for the standard critical Potts model, and $g' \in [4, 6]$ for the tricritical one. For percolation $Q=1$, $g' = \tfrac{8}{3}$ and we recover the standard hull fractal dimension $D_H = \tfrac{7}{4}$. If we want the *Ising hull fractal dimension* at the Onsager critical point we can take the equivalence to the $Q=1$ Potts tricritical point and thus $g' = \tfrac{16}{3}$ giving

$$D_{\text{Hull}}(\text{Ising}) = \tfrac{11}{8}.\qquad(46)$$

The same fractal dimensions can be obtained from the standard $O(n)$ model in terms of the two-line vertex dimension x_2 (25)

$$D_H = 1 + 1/2g,\qquad(47)$$

where $n = -2\cos\pi g$, $g \in [0, 1]$ for the low-temperature $O(n)$ model, and $g \in [1, 2]$ at the critical point. Using the $O(n)$–Potts geometrical identification (42), (44) indeed gives the identity of the fractal dimensions (45) and (47) of the Potts hulls and of the $O(n)$ loops for $\sqrt{Q} = n$. Similar relationships exist for the fractal dimensions of Potts clusters. From den Nijs' work [35] one finds that the probability that two sites X and Y belong to the same Potts cluster scales as $P(X-Y) \sim |X-Y|^{-2x_M}$, where x_M is the magnetic exponent

$$x_M = 1/2g' - (4-g')^2/8g'.$$

The fractal dimension of the cluster is then given by

$$D_c = 2 - x_M = 1 + \tfrac{1}{8}g' + 3/2g'.\qquad(48)$$

For percolation ($Q=1$, $g' = \tfrac{8}{3}$) we recover the well-known result $D_c = \tfrac{91}{48}$. If we take spanning trees [17] ($Q=0$, $g'=2$) we find $D_c = 2$, as expected, since the tree fills the space by definition. If we consider the $Q=1$ tricritical Potts model, geometrically equivalent to the Ising model, we find ($g' = \tfrac{16}{3}$)

$$D_{\text{cluster}}(\text{Ising}) = \tfrac{187}{96},$$

establishing [38] a result recently conjectured [39]. A direct derivation is also possible from the critical $O(n)$ model [38], where one finds $D_c = 1 + \tfrac{1}{2}g + 3/8g$, in agreement with (48) and the identification $g' = 4g$ (42), (44).

4. Pinching points in percolation

As a second application of the watermelon formalism described in section 3, let us return to the percolation problem (i.e. the low-temperature phase of the $O(1)$ model, or the $Q=1$ Potts model) and to some specific topological questions in it [24]. One can ask: what is the typical length of a hull perimeter near the critical threshold p_c? Or, what is the probability that a point belongs to the hull of the infinite incipient cluster: or, that a point X is a *pinching point* of order $L=2k$, where k clusters come close together (fig. 1b)? Let us recall the values (30) of the percolation geometrical exponents associated with k clusters $x_L \equiv x'_k = \tfrac{1}{12}(L^2-1) = \tfrac{1}{12}(4k^2-1)$. By standard scaling analysis [26], one first shows that the hull perimeter length diverges as

$$\langle l(\mathcal{P}) \rangle \sim |p-p_c|^{-\gamma_1},\quad \gamma_1 = (2-2x'_1)\nu = 2,\qquad(49)$$

where ν is the percolation thermal exponent $\nu = \frac{4}{3}$, and x'_1 the one-cluster watermelon exponent. Accordingly the hull fractal dimension is $D_H = 2 - x'_1 = \frac{7}{4}$, as mentioned above, and the probability that a point belongs to the hull of the infinite cluster grows like $P_1 \sim (p - p_c)^{\beta_1}$, $\beta_1 = \nu x'_1 = \frac{1}{3}$. One can go to higher topologies and consider the *pinching problem* (fig. 1b). The singular part of the mean number of clusters, the external perimeter of which has the special topology of k bands pinched at their extremities (fig. 7) is

$$\langle n_k \rangle \sim |p - p_c|^{-\gamma_k},$$
$$\gamma_k = (2 - 2x'_k)\nu = \tfrac{2}{9}(13 - 4k^2). \tag{50}$$

One can also consider the probability P_k that a point belongs to a region of the perimeter of the infinite cluster, where the latter has the special topology of k bands ("peninsulas") coming close together (fig. 1b). It grows as

$$P_k \sim (p - p_c)^{\beta_k}, \tag{51}$$

where

$$\beta_k = \nu - \tfrac{1}{2}\gamma_k \equiv \nu x'_k = \tfrac{1}{9}(4k^2 - 1).$$

For $k = 2$, it is interesting to note that two touching peninsulas are equivalent to a *cutting bond*. We find $x'_2 = \frac{5}{4}$; hence we obtain a fractal dimension $D_{red} = 2 - \beta_2/\nu = \frac{3}{4} = 1/\nu$, thus giving another determination of a well-known result [41].

5. Intersections of random walks

In probability theory, and also in the representation of field theory by intersecting random walks, a simple but non-trivial problem plays an important role. Consider two random walks (RW) in $\mathbb{Z}^d$, w_1, w_2, starting at the origin. What is the probability $P_2(t)$,

$$P_2(t) \equiv P(w_1(0, t] \cap w_2(0, t] = \emptyset), \tag{52}$$

that after a given number of steps t their *paths have no mutual intersections*? (fig. 9). The notation $w(0, t]$ represents the set of points

$$w(0, t] = \{w(t'), 0 < t' \le t\}, \tag{53}$$

the origin being excluded. For large time t one expects $P_2(t)$ to decay algebraically as

$$P_2(t) \sim t^{-\zeta_2}, \quad t \to \infty, \tag{54}$$

where ζ_2 is a universal critical exponent associated with *mutually avoiding* walks. That ζ_2 is universal means that it depends only on the space dimension d and not on the lattice representation, and keeps its well-defined value in the continuum limit of Brownian motions. In this latter case, however, it is possible that there are infinitely many intersections at short distance with probability 1 in *two dimensions*, if the two paths start at the same point. So the equivalent problem to mutually avoiding walks in the continuum Brownian case is to consider Brownian paths starting at very close but different points. The nonintersection random walks constitute a critical system in a *new* universality class. The upper critical dimension is the same $d = 4$, since two Brownian paths having a fractal Hausdorff dimension $D_H = 2$, their intersection set is non-empty only if $d \le 2D_H = 4$. Hence $P_2(t) \equiv 1$ for $d > 4$ and $\zeta_2 = 0$ for $d > 4$. Several years ago, Lawler [42] proved that in four dimensions, the decay of $P_2(t)$ is logarithmic,

$$P_2(t) \sim (\ln t)^{-1/2}. \tag{55}$$

This was generalized to the non-intersection probability of L walks in $d = 4$ [43]

$$P_L(t) \sim (\ln t)^{-L(L-1)/4}. \tag{56}$$

For L walks, one expects in the same way for $d < 4$ a non-trivial exponent ζ_L generalizing ζ_2 in (54),

$$P_L(t) \sim t^{-\zeta_L}, \quad t \to \infty. \tag{57}$$

By using a direct renormalization method, the asymptotic expansion of ζ_L was obtained near four dimensions [43], $d = 4 - \epsilon$,

$$\zeta_L = \tfrac{1}{2}L(L-1)\tfrac{1}{4}\epsilon$$
$$\quad + \tfrac{1}{2}L(L-1)(5-2L)(\tfrac{1}{4}\epsilon)^2 + \mathcal{O}(\epsilon^3),$$
$$\epsilon \equiv 4 - d > 0. \tag{58}$$

For $L = 2$, this gives in particular the ϵ expansion $\zeta_2 = \tfrac{1}{4}\epsilon + (\tfrac{1}{4}\epsilon)^2 + \mathcal{O}(\epsilon^3)$.

In *two dimensions* progress has been made re-

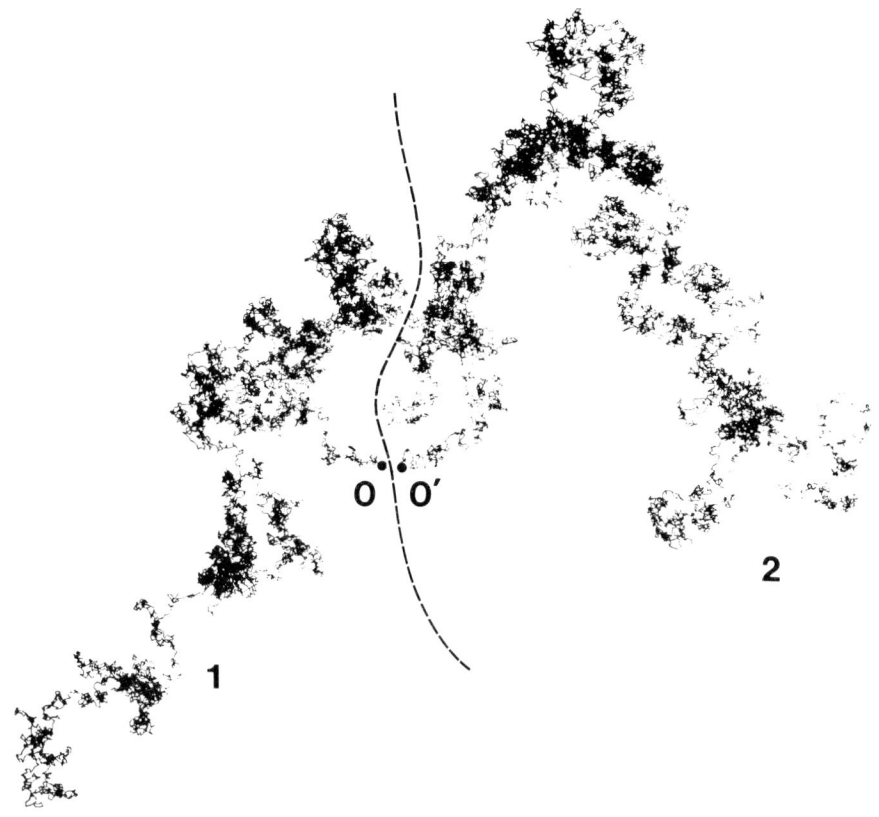

Fig. 9. Two random walks, or Brownian paths starting at close points 0 and 0', and non-intersecting.

cently. On the mathematical side [44] it was proved that (1) $\lim_{t\to\infty} \ln P_2(t)/\ln t$ exists, i.e. the exponent ζ_2 exists (and the ζ_L); (2) rigorous bounds can be given to ζ_2:

$$\tfrac{1}{2} < \zeta_2 \leq \tfrac{3}{4} \ . \tag{59}$$

On the theoretical physics side, the idea appeared [45] to use conformal invariance methods to try and determine the basic probabilistic exponents ζ_L in two dimensions.

The essential idea is that there should be a special conformal operator ϕ'_L associated with the source of L mutually avoiding walks (MAW) originating at a given point. This assumption is quite similar to that above concerning the existence of operators ϕ_L (22) in the case of the $O(n)$ model. The $n=0$ limit gave us source operators for L self- and *mutually* avoiding lines. From the scaling dimensions x'_L of the watermelon correlator in the case of MAW, one can deduce by scaling and renormalization group arguments the probability exponents ζ_L [45]:

$$\zeta_L = \nu x'_L = \tfrac{1}{2} x'_L \ ,$$

where $\nu = \tfrac{1}{2}$ is the trivial correlation length exponent of simple Brownian paths. As a result, one conjectures that the exponents ζ_L (which "act" in time t, see (57), while the scaling dimensions x'_L "act" in space, see (22)) are themselves elements of a conformal Kac table. The central charge of the underlying conformal theory is easily seen to be identically zero: $c=0$. Indeed, e.g. two mutually avoiding walks can be represented by an interacting field theory with Lagrangian $\sum_{\alpha,\beta}(\varphi_\alpha)^2(\psi_\beta)^2$, for two fields φ, ψ with components α, $\beta=1, ..., n$, and in the limit $n=0$. Hence the field partition function will be trivially $Z(\varphi,\psi)\equiv 1$, and thus $c=0$ since c can be seen as a universal finite-size scaling correction amplitude of the free energy $\ln Z \equiv 0$ (see ref. [7] and references therein). One cannot use standard transfer matrix techniques to study the MAW spectrumx'_L, since each random walk on a strip requires multiple bond occupancy, nor Coulomb gas techniques since each random walk makes infinitely many closed loops, forbidding a simple mapping onto an SOS model and then a Coulomb gas. So we resorted to Monte Carlo simulations to study directly in 2D the probabilistic exponents ζ_L. The outcome was the conformal conjecture for the spectrum ζ_L:

$$\zeta_L = h^{(c=0)}_{0,L} = \tfrac{1}{24}(4L^2-1), \quad L \geq 2, \tag{60}$$

where $h_{p,q}$ is the standard Kac formula (26) taken for $c=0$ and integer parameter $m=2$. Again we are in the same conformal table $c=0$, $m=2$ as for polymers and percolation, which extends the Friedan, Qiu and Shenker classification for unitary minimal models $c<1$, $m\geq 3$ to non-unitary geometric critical phenomena. This shows that the RW intersections in two dimensions are actually deeply related to conformal invariance, which can be used to obtain new and exact results in probability theory. Furthermore, this provides perhaps the most simple model of a conformally invariant theory. If it were solved rigorously in the future, it would be a good test for conformal invariance methods.

Two remarks: We conjecture for two walks $\zeta_2 = \tfrac{5}{8}$, which fits into the bounds (59), being even equal to their arithmetic mean. Second, the series (60) gives a non-trivial value $\zeta_1 = \tfrac{1}{8}$. This exponent cannot correspond to an intersection property of several walks and should be associated with a single-walk property. It is possible [46] that it governs the probability $P_1(t) \sim t^{-1/8}$ that a Brownian path or random walk *does not encircle* its origin, i.e. that the origin of the path is accessible from infinity without crossing the Brownian path (fig. 10).

Recently Burdzy and Lawler have also reconsi-

Fig. 10. The random walk (a) encircles, or (b) winds about its origin. In case (a) the origin cannot be reached from infinity without crossing the path.

dered the problem of the fractal dimension of the hull of a 2D (closed) Brownian motion [44]. This hull is the closed Jordan curve, which is the borderline of the infinite open set exterior to the planar Brownian motion. Mandelbrot [1] called this *Brownian perimeter* the "self-avoiding Brownian motion", since the computer simulations indicated that its Hausdorff dimension $D_{\mathrm{BP}} \approx \tfrac{4}{3}$, which is that of actual 2D SAW. Burdzy and Lawler [44]] used $D_{\mathrm{BP}} \leq 2-\zeta_2$, which can be shown by a conformal mapping, and where ζ_2 is the non-intersection exponent of two walks, together with their lower bound $\zeta_2 \geq \tfrac{1}{2} + 1/4\pi^2$, to get $D_{\mathrm{BP}} \leq \tfrac{3}{2} - 1/4\pi^2$). Using our conjecture $\zeta_2 = \tfrac{5}{8}$, we get instead the presumably exact bound

$$D_{\mathrm{Brown\,peri.}} \leq \tfrac{11}{8} = 1.375 \,.$$

This value is quite close to the numerical estimate $D_{\mathrm{BP}} \approx 1.33$, and could be the exact one.

Many other probabilistic exponents are obtained by this approach. One can take L walks all starting at a same point near the boundary line of a half-plane. In this case one obtains surface critical exponents and the probability of no intersection scales as (fig. 11)

$$P^S_L(t) \sim t^{-\zeta^S_L},$$

$$\zeta^S_L = \tfrac{1}{2}(h^{(c=0)}_{1,2L+2} - L) = \tfrac{1}{3}L(L-1)\,. \tag{61}$$

If the walks all start in the apex of a wedge angle α (fig. 11), the probability of no further encounter scales as

$$P^W_L(t,\alpha) \sim t^{-\zeta_L(\alpha)},$$

$$\zeta_L(\alpha) = (\pi/\alpha)\tfrac{1}{3}L(L-1)\,, \tag{62}$$

a formula which can be obtained from (61) by the

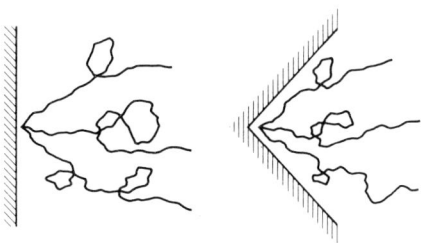

Fig. 11. Mutually avoiding walks starting at a Dirichlet boundary line, or in the apex of a wedge α, after a conformal mapping $w(z) = z^{\alpha/\pi}$. The associated exponents are ζ_L^S (61) and $\zeta_L(\alpha)$ (62).

conformal mapping $z \to z^{\alpha/\pi}$. It is interesting to remark that the case $\alpha = 2\pi$ corresponds to L mutually avoiding walks in presence of a forbidden half-line. In the many walks limit $L \to \infty$ one expects the exclusion of the half-line to be irrelevant when compared to the mutual avoidance effect for $L \to \infty$. In this case (62) gives a dominant contribution

$$\zeta_L(2\pi) \sim \tfrac{1}{6} L^2, \quad L \to \infty, \tag{63}$$

which is the same as the dominant contribution in the bulk exponent ζ_L (60) for $L \to \infty$, as expected.

As in the case of SAW, one can consider networks of mutually avoiding walks [43]. If one takes for instance $\mathcal{N}$ independent Brownian motions and ask about the probability that their paths at time t form a certain network $\mathcal{G}$, characterized by the set of numbers $\{n_L\}$ of L-walk vertices. In the absence of mutual avoidance, the probability scales as

$$P_\mathcal{G}^{RW}(t) \sim t^{-d\mathcal{L}/2}, \tag{64}$$

where $\mathcal{L}$ is the number of independent loops in $\mathcal{G}$,

$$\mathcal{L} = \tfrac{1}{2} \sum_{L \geq 1} n_L(L-2) + 1. $$

Now, one can implement mutual avoidance between all the $\mathcal{N}$ walks building up $\mathcal{G}$. Then the probability will scale as [43]

$$P_\mathcal{G}(t) \sim t^{-d\mathcal{L}/2 - \Sigma_{L \geq 2} n_L \zeta_L}, \tag{65}$$

where ζ_L are the MAW exponents defined above for L walks starting at a same point. This formula describes the factorization over the vertices and is entirely similar for SAW [8]. The relative probability that, knowing that the $\mathcal{N}$ RW form a given graph $\mathcal{G}$, they further have no mutual intersection, is simply the ratio of (64) and (65)

$$P'_\mathcal{G} = P_\mathcal{G}/P_\mathcal{G}^{RW} \sim t^{-\Sigma_{L \geq 2} n_L \zeta_L}$$

$$= t^{-\Sigma_{L \geq 2} n_L(4L^2-1)/24}. \tag{66}$$

Notice that if one further introduces the constraint that the free extremities of the walks (corresponding to $L=1$) are *not encircled*, this gives $n_{L=1}$ new constraints which are simply taken into account by a supplementary factor $t^{-n_1 \zeta_1} = t^{-n_1/8}$ in (65) or (66), completing the summation over L down to $L=1$.

As an application, we can, for instance, calculate the probability $P_{\mathcal{W}_L}$ that L independent random walks starting together meet again anywhere at time t without having any mutual crossing point in between (watermelon $\mathcal{W}_L$). It scales like $P_{\mathcal{W}_L}(t) \sim t^{\gamma_{\mathcal{W}_L}-1}$, where the scaling theory of (65) gives

$$\gamma_{\mathcal{W}_L} - 1 = -(L-1) - 2\zeta_L$$

$$= -\tfrac{1}{12}(4L^2 + 12L - 13). \tag{67}$$

It follows from (57) and (67) that the (conditional) probability that L mutually avoiding RW starting at the same point enjoy a reunion at time t at any point scales like

$$P_{\mathcal{W}_L}/P_L \sim t^{\gamma_{\mathcal{W}_L}-1+\zeta_L} \sim t^{-(4L^2+24L-25)/24}. \tag{68}$$

Direct walks. Note that the solution for (symmetric) *directed* walks [47] gives different exponents for probabilities (57), (67), and (68), $\tilde{\zeta}_L = \tfrac{1}{4}L(L-1)$, $\tilde{\gamma}_{\mathcal{W}_L} - 1 = -\tfrac{1}{2}(L^2-1)$, $\tilde{\gamma}_{\mathcal{W}_L} - 1 + \tilde{\zeta}_L = -\tfrac{1}{4}(L-1)(L+2)$, which are also quadratic in L, but of course differ from ours. These directed MAWs are extremely sensitive to non-symmetric diffusion constants, which lead to non-universal and continuously varying exponents [48]. There the space anisotropy forbids conformal invariance, in contrast to our work.

6. Comparison to multifractals

In the previous sections, we have presented infinite (though discrete) spectra of conformal scaling dimensions associated with geometrical critical phenomena. The question then arises of the possible analogy or distinctness from multifractal spectra [#3]. This question has also arisen in other contexts [52]. Here I shall only give a brief description of the subtle distinction between the multiple exponents given above and multifractal ones. Details can be found in a recent work written in collaboration with Ludwig [53].

In section 2, we showed how the polymer scaling dimensions x_L (11) obey the inequality (5). We now observe that the scaling dimensions of $O(n)$ or Potts model (25), (41), including the multiple spectra (29) of percolation pinching points, and (60) of multiple mutually avoiding walks, *all* obey the same inequality (5). We state [53] that this short-distance expansion inequality is fundamental in all these *critical phenomena spectra*:

$$x\left(\sum_i L_i\right) \geq \sum_i x(L_i) \quad \text{(critical phenomena)}.$$

(69)

Gap scaling would mean equality. In all our examples the *inequality* is strict, showing that gap scaling is not at all characteristic of critical phenomena (see also ref. [52]). It nevertheless shows that since one usually has to pick up the *lowest* scaling dimension in order to get the *most* relevant critical behaviour, one is led to retain $x = \sum_i x(L_i)$ instead of $x(\sum_i L_i) > x$, as an *overall* dimension for correlation functions. So one *superficially observes* gap scaling.

Now, for multifractals, we state [53] that the situation is just the *opposite*. One can show that the analogous scaling dimensions x_L for multifractals (properly defined from correlation functions of the multifractal measure) obey the reverse inequality

---

[#3] For Mandelbrot's views on multifractal measures, see ref. [49]. See also refs. [50,51].

$$x\left(\sum_i L_i\right) \leq \sum_i x(L_i) \quad \text{(multifractal spectra)}.$$

(70)

This inequality is usually strict, and is related to the decrease of the $D(q)$ generalized dimensions [54] and to the existence of the $f(\alpha)$ Legendre transform [51]. Indeed the latter requires a convexity property, which appears as a particular case of inequality (70) in our scaling dimension formalism. The relation to the usual formulation in terms of the function $\tau(q) = (q-1)D(q)$ is obtained via the identity [53]

$$\tau(q) = x_q - D - q(x_1 - D),$$

(71)

where $\{x_q\}$ are the analog of the scaling dimensions x_L described above, and where D is the fractal dimension of the support of the multifractal measure. The inequality (70) explains why one sees directly the absence of gap scaling in multifractals, and *immediately the whole spectrum*: one has to pick up the smallest scaling dimension and because of (70) it is $x(\sum_i L_i)$ instead of $\sum_i x(L_i)$. We refer the reader to ref. [53] for a derivation of these basic inequalities and illustrative examples.

References

[1] B.B. Mandelbrot, Les Objets Fractals: Forme, Hasard et Dimension (Flammarion, Paris, 1975); The Fractal Geometry of Nature (Freeman, San Francisco, 1983).
[2] L. Pietronero and E. Tosatti, eds., Fractals in Physics, Proceedings of the 6th Trieste International Symposium (North-Holland, Amsterdam, 1986).
[3] J. Feder, Fractals (Plenum, New York, 1988).
[4] P. Lévy, Processus Stochastiques et Mouvement Brownien (Gauthier-Villars, Paris, 1965).
[5] A.M. Polyakov, JETP Lett. 12 (1970) 381.
[6] A.A. Belavin, A.M. Polyakov and A.B. Zamolodchikov, Nucl. Phys. B 241 (1984) 333.
[7] J.L. Cardy, in: Phase Transitions and Critical Phenomena, Vol. 11, C. Domb and J.L. Lebowitz, eds. (Academic Press, New York, 1987).
[8] B. Duplantier, Phys. Rev. Lett. 57 (1986) 941; Phys. Rev. B 37 (1987) 5290.
[9] B. Duplantier and H. Saleur, Nucl. Phys. B 290 (1987) 291.
[10] B. Duplantier, J. Stat. Phys. 54 (1989) 581.

[11] L.P. Kadanoff, Phys. Rev. Lett. 29 (1969) 1430;
K.G. Wilson, Phys. Rev. 179 (1969) 1499.
[12] P.G. de Gennes, Phys. Lett. A 38 (1972) 339.
[13] B. Nienhuis, in: Phase Transitions and Critical Phenomena, Vol. 11, C. Domb and J.L. Lebowitz, eds. (Academic Press, New York, 1987).
[14] H. Saleur, J. Phys. A 19 (1986) L807; A 20 (1987) 455.
[15] B. Duplantier, J. Stat. Phys. 49 (1987) 411.
[16] B. Duplantier and F. David, J. Stat. Phys. 51 (1988) 327.
[17] B. Nienhuis, Phys. Rev. Lett. 49 (1982) 1062.
[18] M.T. Batchelor and H.W.J. Blöte, Phys. Rev. Lett. 61 (1988) 138.
[19] Vl.S. Dotsenko and V.A. Fateev, Nucl. Phys. B 240 [FS12] (1984) 312.
[20] V.G. Kac, in: Lecture Notes in Physics, Vol. 94 (Springer, Berlin, 1979) p. 441.
[21] D. Friedan, Z. Qiu and S. Shenker, Phys. Rev. Lett. 52 (1984) 1575.
[22] D.A. Huse, Phys. Rev. B 30 (1984) 3908.
[23] J. Cardy and H.W. Hamber, Phys. Rev. Lett. 45 (1980) 499.
[24] H. Saleur and B. Duplantier, Phys. Rev. Lett. 58 (1987) 2325.
[25] B. Duplantier and H. Salier, Phys. Rev. Lett. 59 (1987) 539;
A. Coniglio, N. Jan, I. Majid and H.E. Stanley, Phys. Rev. B 35 (1987) 3617;
T. Grossman and A. Aharony, J. Phys. A 19 (1986) L745.
[26] R.M. Ziff, Phys. Rev. Lett. 56 (1986) 545.
[27] B. Sapoval, M. Rosso and J.F. Gouyet, J. Phys. (Paris) 46 (1985) L149.
[28] J.L. Cardy, Nucl. Phys. B 240 (1984) 514.
[29] B. Duplantier and H. Saleur, Phys. Rev. Lett. 57 (1986) 3179.
[30] K. Binder, in: Phase Transitions and Critical Phenomena, Vol. 8, C. Domb and J.L. Lebowitz, eds. (Academic Press, New York, 1983).
[31] I. Guim and T.W. Burkhardt, Temple University preprint (1988).
[32] B. Duplantier and H. Saleur, in preparation.
[33] C. Vanderzande and A. Stella, J. Phys. A 22 (1989) L445.
[34] F.Y. Wu, Rev. Mod. Phys. 54 (1982) 235.
[35] M. den Nijs, Phys. Rev. B 27 (1983) 1674; J. Phys. A 12 (1979) 1857.
[36] P.W. Kasteleyn and C.M. Fortuin, J. Phys. Soc. Japan 26 (Suppl.) (1969) 11.
[37] B. Duplantier, unpublished;
B. Nienhuis, unpublished.
[38] B. Duplantier and H. Saleur, to be published.
[39] A.L. Stella and C. Vanderzande, Phys. Rev. Lett. 62 (1989) 1067.
[40] M. den Nijs, J. Phys. A 12 (1979) 1857.
[41] A. Coniglio, J. Phys. A 15 (1982) 3829.
[42] G.F. Lawler, Commun. Math. Phys. 86 (1982) 539.
[43] B. Duplantier, Commun. Math. Phys. 117 (1988) 279.
[44] K. Burdzy, G.F. Lawler and T. Polaski, preprint (1988);
K. Burdzy and G.F. Lawler, preprint (1989).
[45] B. Duplantier and K.H. Kwon, Phys. Rev. Lett. 61 (1988) 2514.
[46] B. Duplantier and K. Kwon, unpublished.
[47] D.A. Huse and M.E. Fisher, Phys. Rev. B 29 (1984) 239;
M.E. Fisher, J. Stat. Phys. 34 (1984) 667.
[48] M.E. Fisher and M.P. Gelfand, J. Stat. Phys. 53 (1988) 175.
[49] C.H. Schulz and B.B. Mandelbrot, eds., Springer Issue of Fractals in Geophysics, Pure Appl. Geophys. 131, Nos. 1/2 (1989);
B.B. Mandelbrot, J. Fluid Mech. 62 (1974) 331.
[50] U. Frisch and G. Parisi, in: Turbulence and Predictability in Geophysical Fluid Dynamics and Climate Dynamics, International School of Physics "Enrico Fermi" 1988, M. Ghil, ed. (North-Holland, Amsterdam, 1985).
[51] T.C. Halsey, M.H. Jensen, L.P. Kadanoff, I. Procaccia and B.I. Shraiman, Phys. Rev. A 33 (1986) 1141.
[52] B. Fourcade and A.M.S. Tremblay, Phys. Rev. B 39 (1989). 6819.
[53] B. Duplantier and A.A. Ludwig, to be published (1989).
[54] H.G.E. Hentschel and I. Procaccia, Physica D 8 (1983) 435.

THE TRANSMISSION OF STRESS IN AN AGGREGATE

S.F. EDWARDS and R.B.S. OAKESHOTT

Cavendish Laboratory, Cambridge CB3 0HE, UK

We consider the question of how stress is transmitted in aggregates. The macroscopic state of a powder is summarized by its compactivity $X \equiv (\partial V/\partial S)$. In aggregates where the particles have stuck together (e.g. lipid crystals in margarines, or in flocculated colloids) the material has a fractal structure on which stress is carried. In a non-cohesive powder we consider the possibility that the stress will be carried by domes of particles, giving a dimensionality for the set of stressed particles between 2 and 3. Finally we observe that there is a coupling between the structure (described by its compactivity X) and the distribution of stress.

1. Introduction

There is an increasing interest in theoretical physics in the problems of assemblages of particles which are sufficiently numerous and are assembled under well-defined circumstances, so that it is reasonable to expect the physical properties of the assembly to be predictable (and interesting). Whereas in problems dominated by thermal behaviour one can expect ergodicity, an aggregate can be perverse, i.e. a Maxwell Demon can put an aggregate together in a way which has no particular relationships to a physical law. The stones in the Great Pyramid follow the instructions of the Pharaoh, but if they were poured out of a gigantic hopper we could expect to be able to predict the shape of the resulting heap. It will be assumed in this paper that systems which are formed by extensive operations, i.e. by shaking or stirring, or aggregation according to some explicit rules, will be predictable and capable of having their specification solved. This is the anologue of the ergodicity of thermal systems.

So suppose we have a powder and apply stress to it, e.g. suppose a powder fills a vertical cylinder and it is compressed. How does it rearrange itself internally to sustain the stress, and how is that stress transmitted? A variant of this is to consider a horizontal cylinder filled with powder with a piston at one end but free to move at the other, remote, end. If the piston moves to compress the powder, how is stress transmitted through it?

The reason that this topic is offered as part of the celebrations of Professor Mandelbrot's 65th birthday is that the dimensionality of the transmission is not clear. We conjecture that the pattern is fractal; but has extra complications in that the fractal dimension may depend on the extent of the compression consequent on the stress. This paper does little more than state the problem and discuss the framework for a solution. However, it is of such technological importance as well as being a challenge just as physics, it is certain to gain more attention in future.

2. The description of an aggregate

In the presence of an appropriate background, or with particles forming strong inter-particle bonds, aggregates can take up configurations ranging from the low density found in DLA clusters, to that of a space filling powder. Although there will be reference to DLA-type clusters, more attention will be paid to space-filling systems and it is here that a simple description should be possible and will be attempted.

The key fact is that a powder can be formed with a variety of densities. In the best known case of hard

Essays in honour of Benoit B. Mandelbrot
Fractals in Physics – A. Aharony and J. Feder (editors)

spheres there are two limiting densities of random, loose and close packing (0.6 and 0.6366). Making the assumption that the density is sufficient to characterize the powder completely, one may make a fruitful analogy with normal statistical mechanics. There the energy has a value E, so that if H is the Hamiltonian

$$P = e^{S/k} \delta(E-H) \tag{1}$$

gives the microcanonical probability that the system is found with E, and S is the normalization

$$\int P = 1, \tag{2}$$

$$e^{S/k} = \int \delta(E-H) \, d(\text{all}), \tag{3}$$

k serving to give the entropy S the dimension of energy. Further one defines temperature $T = (\partial E/\partial S)$ and can transfer to the Canonical ensemble with

$$P = e^{(F-H)/kT}, \tag{4}$$

$$e^{-F/kT} = \int e^{-H/kT} \, d(\text{all}), \tag{5}$$

where

$$E = F - T \frac{\partial F}{\partial T}, \qquad S = -\frac{\partial F}{\partial T}.$$

Suppose now that there is a function W which plays the role of H in the powder in that the volume V is the value taken by W for the particular configurations of the grains. Then if we define λ to be a constant which gives entropy the dimension of volume

$$P = e^{-S/\lambda} \delta(V-W), \tag{6}$$

$$e^{S/\lambda} = \int \delta(V-W) \, d(\text{all}). \tag{7}$$

The interesting step is to go to the canonical ensemble, defining a compactivity X by

$$X = \frac{\partial V}{\partial S}, \tag{8}$$

$$P = e^{(Y-W)/\lambda X}, \tag{9}$$

$$V = Y - X \frac{\partial Y}{\partial X}, \tag{10}$$

where we call Y the effective volume; it has the role of free energy and

$$S = -\frac{\partial Y}{\partial X}. \tag{11}$$

Thus, just as it is easier to think of a temperature in a thermal system rather than its energy, we propose a compactivity X, linked to an effective volume Y, which is easier to handle than the density of the powder. Just as in a thermal system one can then go over to a temperature gradient, and a heat source of a point or plane giving rise to such a temperature gradient. We can now think of an injection of stress into a system at a point or over a plane, which leads to a gradient of X emanating from that point or plane.

The effect of X is most apparent in the coordination distribution it entails. For example, a simple calculation of the distribution of coordination number n_c shows that this reaches a maximum at $X=0$ and is most uniformly distributed at $X=\infty$. Thus the most highly close-packed system has $X=0$ and just as one cannot be more packed than that, X cannot be negative. The lowest density corresponds to $X=\infty$, where all configurations are equally probable, subject, however, to coordinates which imply a solid, i.e. not having so slow a coordination as to have unconnected or unstable material. A typical formula at the simplest level of approximation applied to a model of only two possible coordinations is

$$W = n_0 v_0 + n_1 v_1, \qquad n_0 + n_1 = N, \tag{12}$$

leading to

$$V = \tfrac{1}{2} N(v_0 + v_1) + N(v_0 - v_1) \tanh\left(\frac{v_0 - v_1}{\lambda X}\right), \tag{13}$$

where $X=0$ gives Nv_0 and $X=\infty$ $\tfrac{1}{2}N(v_0+v_1)$. Although this is a gross oversimplification, it gives one a flavour of a true description.

3. The emergence of structures

Section 2 considered a three-dimensional space-filling aggregate and we return to this in section 4.

Stress can be carried by precursor structures, and indeed such structures appear central in understanding systems which are just emerging as solids. For example a margarine has a balance of oils and lipid crystals, the latter forming chains which when of sufficient density make the margarine a solid. For a low density of lipid crystals the material is not a solid. The lipid crystals have some fractal structure in space which goes over into the usual material for a high enough content. This range of materials is very similar mathematically to a flocculating colloid, and the same transition of

particles → aggregate into fern-like fractal
→ orthodox amorphous solid

appears in a host of everyday materials. Brown and Ball [1] have given a theory for this process. They consider a typical diffusion-limited cluster which has a fractal dimension d_f. In addition another exponent d_{chem} can be defined in terms of the scaling behaviour of the electrical conductivity of the cluster. Brown and Ball are then able to deduce an elastic spring constant for the aggregate provided they use the longest and shortest length scales R and a of the aggregate. The constant is given by

$$k \approx a\left(\frac{R}{a}\right)^{-d_{chem}}. \tag{14}$$

For a mono-disperse system, by numerical simulation

$$d_{chem} = 1.066 \pm 0.07,$$

whereas a poly-disperse system has

$$d_{chem} = 0.96 \pm 0.033.$$

When an assembly of these clusters is formed by packing them together with a density ρ, the resulting modulus is given by

$$G \propto \rho^{(3+d_{chem})/(d-d_f)}, \tag{15}$$

which agrees well with the experiments of Buscall, and of Buscall, Stewart and Sutton [2], who were able to measure G and control ρ by ultra-centrifuging (see fig. 1).

It is clear that stress is transmitted in these mate-

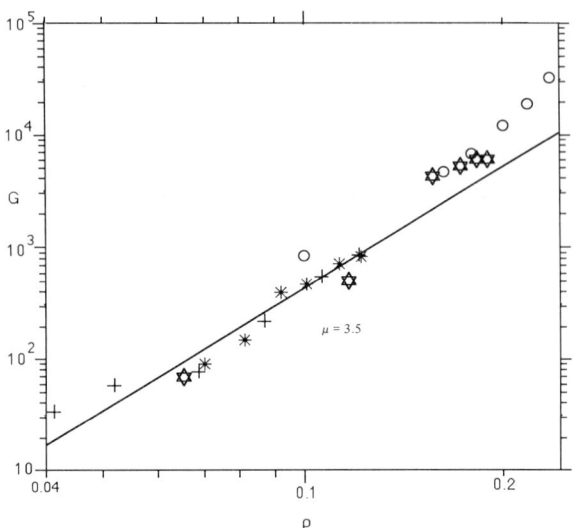

Fig. 1. Elastic shear modulus G versus density ρ for fully flocculated 0.3 μm acrylic spheres (crosses), 0.33 μm polystyrene spheres (asterisks), 0.5 μm polystyrene spheres (stars), 0.96 μm polystyrene spheres (circles). (From the Ph.D. thesis of W.D. Brown [3].)

rials in one-dimensional paths which are branched and are characterised by a fractal dimension. The crystals or flocs stick together and can sustain an arbitrary stress tensor. For a powder the situation is not so clear for one has to have stress spread amongst several neighbours to maintain stability. Nevertheless one can imagine a locus of stress in this way. Suppose a powder is stressed by the application of a force in some way (e.g. at a point or plane). Mark all particles where the stress exceeds a certain level. What locus do the marked particles take up? This is effectively possible using photoelasticity as in the work of Dantu [4]. One can imagine a picture as in fig. 2, which is uncovered by removing overlying particles.

In this picture the stress pathway resembles that of the packed floc aggregates, or the lines of lipid crystals in the margarine. Although a fractal index is involved, the picture is made of lines related by some topology. There are, however, other possibilities.

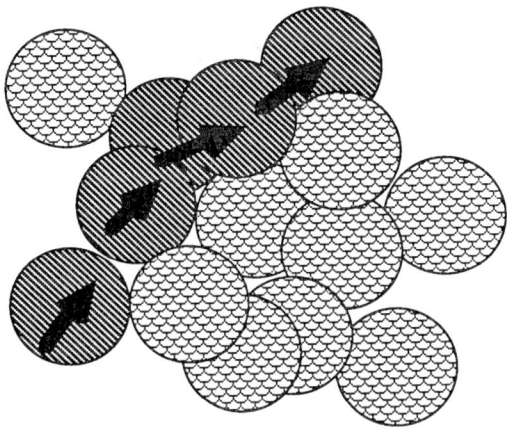

Fig. 2. Possible line of stress transmission in the highlighted section of the powder.

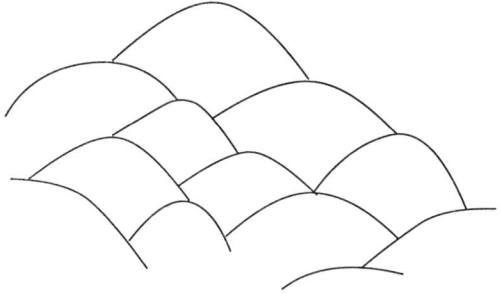

Fig. 3. Stress-forming domes in powder.

4. The dimensionality of stress transmission

At this point we reach the most difficult problem of our discussion. We argued above that if we were to mark those grains sustaining a stress above a given level, one might expect a series of branched lines in space resembling the physical lines of lipid crystals in a margarine with high oil content or one of the numerous other forms of percolation diagrams such as the path of an electric current in dielectric breakdown. These all have the familiar form of starting at a point and progressing in a directed manner in a tree-like growth.

There are very familiar examples of a quite different structure however. Consider for example a funnel, i.e. a body of diminishing cross section. Everyone is familiar with the problem that a powder flowing through such a body is liable to jam, i.e. form a dome which is able to sustain the stress caused by the weight of the powder above it, and transmits the stress to the container. The dome of powder grains holds just like the dome of a building or arch of a bridge. This process could be a central one in the transmission of stress through a powder, i.e. a point injection of stress produces a dome where the injection point has the role of a capstone to the dome. The picture would then be a series of domes terminating at a dome below them, i.e. in a section like fig. 3.

This can be regarded as a two-dimensional transmission of stress in contrast to the one-dimensional model above. We consider some consequences of these models below, but must also make the obvious comment that when the powder is at its maximum close packing and even more so if bonded, the transmission of stress must be adequately described by a three-dimensional transmission of stress as in a continuum theory of elasticity, although the powder problem remains non-linear since the material can support only compressive and not tensile stresses.

Thus in addition to the kind of fingering in a directed DLA assembly, the fingers themselves have dimensionality which can be between 1 and 3. It is reasonable to suppose that this dimensionality will depend on the compactivity so that there will be an X-dependent index associated with the transmission with $X=0$ associated with high dimensionality and $X=\infty$ with low. The precise form of this law has yet to be resolved and offers an important problem. We can, however, suggest some crude consequences of the two-dimensional stress flow as against the one-dimensional form discussed in the last section.

To do this, consider the pressure exerted by a conical heap of powder as in the diagram in fig. 4. The lines drawn in this diagram are at an angle steeper than that of the angle at repose of the powder. If the powder becomes filled with domes or, in the two-dimension section shown, arches, then one can expect the stress to be transmitted down in such a way, from arch to arch, that dx on the base supports the mass of powder formed in the section inclined at the 'arch'

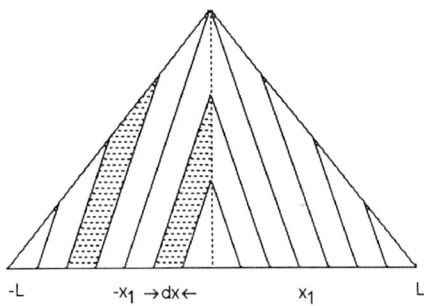

Fig. 4. Lines of stress in fully arched pyramid.

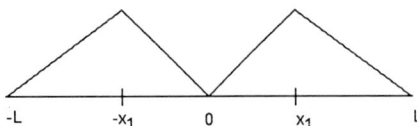

Fig. 5. Pressure distribution under 2D pyramid.

angle to the vertical. Two examples are shaded in.

Thus the pressure distribution can be expected to be as in fig. 5 with the total pressure equating to the gravitational force of the powder. The maxima occur at $\pm x_1$, where the angle which is postulated to arise from the arching defines a line from apex to base. A similar analysis is found in Trollope [5]. Something like this is found by Briscoe, Pope and Adams [6] but as one would expect is much smoother than the crude diagram above.

Rather complex experiments on the dynamics of powders forced through their angle of repose are being reported (Jaeger, Liu and Nagel [7], Evesque and Rajchenbach [8]) and offer a fascinating challenge, but even simple experiments still require explanation and the purely static situation offers a fractal problem which has yet to be resolved.

The problem above is one of many which we can characterise this way. There is a compactivity X in a powder which has a role like temperature in a thermal system, i.e. an inhomogeneous system which is extensively created will have a single X throughout, just as a gas of particles of different mass will have inhomogeneous distribution in space under gravity even though there is a single temperature. This compactivity can then have a weak and slow variation $X(r,t)$. When a point source of stress is introduced, a variation in X is induced, but unlike $T(r, t)$ which will satisfy a differential equation, X may fluctuate as in a discharge phenomenon. This fractal form can of course give rise to a simple average behaviour described by a differential equation, but need not, and particular studies so far have not resulted in simple solutions of equations such as Fick's equation. However, quantities like $\langle X \rangle$, $\langle XX \rangle$ can be studied and it is hoped to define a proper calculus of this problem in due course.

References

[1] W.D. Brown and R.C. Ball, J. Phys. A 18 (1985) L517–L521.
[2] R. Buscall, Colloids Surf. (1982) 269–283;
    R. Buscall, R.F. Stewart and D. Sutton, Filtr. Sep. 21 (1984) 183–186.
[3] W.D. Brown, Ph.D. thesis, Cambridge (1986).
[4] P. Dantu, Ann. Ponts Chausées 4 (1967) 193.
[5] D.H. Trollope, in: Rock Mechanics in Engineering Practice, K.G. Stagg and O.C. Zienkiewicz, eds. (Wiley, London, 1969).
[6] S.J. Briscoe, L. Pope and M.J. Adams, Powder Technol. 37 (1984) 169.
[7] H.M. Jaeger, Ch. Liu and S. R. Nagel, Phys. Rev. Lett. 62 (1989) 40–43.
[8] P. Evesque and J. Rajchenbach, Phys. Rev. Lett. 62 (1989) 44.

DYNAMIC STRUCTURE FACTOR OF FRACTALS

O. ENTIN-WOHLMAN [a], U. SIVAN [b], R. BLUMENFELD [a] and Y. MEIR [c]

[a] *School of Physics and Astronomy, Raymond and Beverly Sackler Faculty of Exact Sciences, Tel Aviv University, Tel Aviv 69978, Israel*
[b] *IBM Thomas J. Watson Research Center, Yorktown Heights, NY 10598, USA*
[c] *Department of Physics, Weizmann Institute of Science, Rehovot 76100, Israel*

The dynamic structure factor $S(q, \omega)$ of the vibrational modes of a deterministic fractal is analyzed as a function of both frequency (ω) and momentum (q) transfer. It is found that $S(q, \omega)$ is peaked at $q_{max} \approx \omega^{2/(2+\theta)}$, where θ is the anomalous diffusion exponent, and is a scaling function of $q^{(2+\theta)/2}/\omega$. The results are obtained by a novel recursion method for the calculation of the vibration Green's function of a deterministic fractal. They confirm predictions based upon the fracton scaling model.

1. Introduction

The dynamics of disordered systems is a subject of considerable interest. In particular, much work has been devoted to studies of geometrically disordered systems (e.g. percolation clusters) by scaling considerations [1,2], by numerical simulations [3] and by scattering experiments [4,5]. These structures appear to be homogeneous at length scales longer than the connectivity length ξ, and exhibit fractal characteristics at shorter length scales. The dynamic excitations of the latter regime are termed fractons [1,2]. Fracton modes are localized; their localization length decreases rapidly with increasing frequency [2] and is smaller than ξ. By scaling considerations it was argued [1] that the fracton density of states $N(\omega)$ varies as $\omega^{\tilde{d}-1}$, where ω is the frequency and $\tilde{d}$ is the fracton dimensionality [1]. The excitations pertaining to the homogeneous regime are phonons, with sound velocity that depends upon ξ. Thus, within the scaling picture, the low-frequency portion of the spectrum consists of extended phonons and the higher-frequency part of localized fractons. The crossover frequency ω_c tends to zero as ξ increases (e.g. as the percolation threshold is approached), $\omega_c \approx \xi^{-(2+\theta)/2}$, where θ is the exponent characterizing the diffusion on a fractal [6]. The crossover from fractons to phonons has recently been established by light scattering experiments [4]. Fracton modes were also invoked to interpret Raman scattering data of silica aerogels [7].

An excitation spectrum is commonly probed by inelastic neutron or light scattering, which measures the dynamic structure factor $S(q, \omega)$ as a function of frequency and momentum transfer of the scattered particles. This method has recently been applied to silica aerogels [4] and also to diluted antiferromagnets [5], in which the short-length-scale spin modes are expected to have fracton features [2].

The detailed form of the dynamic structure factor of tenuous systems is yet unknown. Some general remarks can be made using scaling arguments [8] and an explicit analysis was carried out within the effective medium approximation [9]. Here we present an exact calculation of the dynamic structure factor of a fractal object [10].

The concept of fractal geometry, first introduced by Mandelbrot [11], has turned out to be a most useful tool in studies of tenuous materials. For example, silica aerogels at short length scales are characterized by their fractal dimensions [4]. The same is true for a percolating system close enough to the percolation threshold [12]. It is therefore of interest to investigate the dynamic properties of a fractal structure. In section 2 we compute $S(q, \omega)$ for a modified version of the Sierpinski gasket [11] (see fig. 1). We present there a novel recursion method for calculating the vibration Green's function of a deterministic fractal. We find that $S(q, \omega)$ has a well-defined maximum as a function of q or ω and exhibits single-variable scaling. The analysis of the peak position as a function of frequency confirms the fracton model prediction [1,2,8]. Section 3 includes some conclusions.

2. The structure factor

The determination of $S(q, \omega)$ necessitates the knowledge of the full Green's function of the vibrating structure. We present here a new algorithm for a successive construction of the Green's function, and use it to derive a recursion relation for $S(q, \omega)$. The algorithm is extremely efficient and can be exploited in computations of other properties related to vibration dynamics, e.g. the heat diffusion.

The model considered (see fig. 1) is a version of the two-dimensional Sierpinski gasket [11], on which we solve the scalar elasticity equations

$$-m\omega_+^2 u_i = \sum_{j=\text{nn}} t(u_{i+j} - u_i), \quad \omega_+ = \omega + i\eta, \quad (1)$$

where u_i is the displacement of the ith mass from its equilibrium position, nn denotes nearest-neighbours and t is the spring constant. The distance between two vertices of different triangles is set to zero in order to make the structure geometrically self-similar. This structure is more amenable to efficient calculations than the usual Sierpinski gasket. Though it is not self-similar, the low-frequency spectrum rapidly converges (as a function of iteration order) to that of the common Sierpinski gasket [11]. It simulates a structure with holes on all length scales – a feature exhibited by some of the systems [4] studied in scattering experiments.

The solution to eq. (1) may be written in terms of the retarded Green's function, whose i,j- matrix element is denoted $G_{ij}(\omega_+)$. Here i and j are site indices. The structure factor is then given by

$$S(q, \omega) = -\frac{1}{\pi} \operatorname{Im} \sum_{i,j} \exp[i\mathbf{q}\cdot(\mathbf{R}_i - \mathbf{R}_j)] G_{ij}(\omega_+), \quad (2)$$

in which ω is the real part of ω_+ and $\mathbf{R}_i$ denotes the coordinate of the ith site.

We now outline the derivation of the recursion relation for the Green's function matrix elements which are used in eq. (2) to obtain the recursion formulae

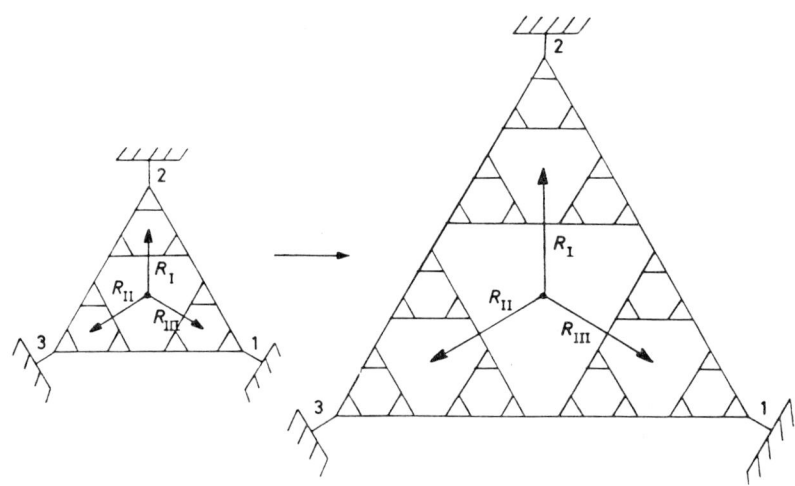

Fig. 1. Two stages in the construction of the gasket.

of $S(q, \omega)$. Suppose we know the full Green's function of the $(n-1)$th stage, $G^{(n-1)}$. The Green's function of the nth stage, $G^{(n)}$, can be written, in matrix notation, as

$$\mathbf{G}^{(n)} = [\mathbf{g}^{-1} + \mathbf{\Gamma}]^{-1}, \qquad (3)$$

where

$$\mathbf{g} = \begin{pmatrix} \mathbf{G}^{(n-1)} & 0 & 0 \\ 0 & \mathbf{G}^{(n-1)} & 0 \\ 0 & 0 & \mathbf{G}^{(n-1)} \end{pmatrix} \qquad (4)$$

and $\mathbf{\Gamma}$ is a coupling matrix, connecting the three $(n-1)$th gaskets to form the nth stage gasket. Note that $\mathbf{G}^{(n)}$ is an N matrix, where N is the number of sites ($N=3^n$) in the nth stage gasket. Referring to fig. 1, one notes that there are only six nonzero matrix elements in $\mathbf{\Gamma}$, the ones connecting together the three $(n-1)$th stage gaskets. Expanding eq. (3), it is easily seen that each term of the expansion includes those matrix elements of $\mathbf{\Gamma}$, and the diagonal (α) and nondiagonal (β) matrix elements of $\mathbf{G}^{(n-1)}$ with respect to the external sites of the $(n-1)$th stage. (For example, G_{11} and G_{12} in fig. 1.) It therefore follows that

$$\mathbf{G}^{(n)} = \mathbf{g} - \mathbf{gTg}, \quad \mathbf{T} = \mathbf{\Gamma}[\mathbf{I} + \mathbf{g\Gamma}]^{-1}, \qquad (5)$$

and to obtain $\mathbf{T}$ explicitly, one has to invert a 6 by 6 matrix. The elements of the symmetric matrix $\mathbf{T}$ are given in terms of α and β, which, in turn, are calculated recursively. The evaluation of $S(q, \omega)$ involves the site coordinates. To incorporate those into the recursive procedure, we have adopted an hierarchical scheme in which the coordinates of each site are defined with respect to the center of the smallest gasket to which it belongs, the latter being measured with respect to the second smallest gasket, and so on.

The results are presented for a gasket of 3^{10} sites and wavevectors q at an angle $\pi/4$ relative to one of the edges (this direction is incommensurate with the gasket symmetry axes). $S(q, \omega)$ has a well-defined maximum as a function of q or ω (see fig. 2). The analysis of the peak position, q_{max}, as a function of frequency yields

$$q_{\text{max}} \approx \omega^{2/(2+\theta)}, \qquad (6)$$

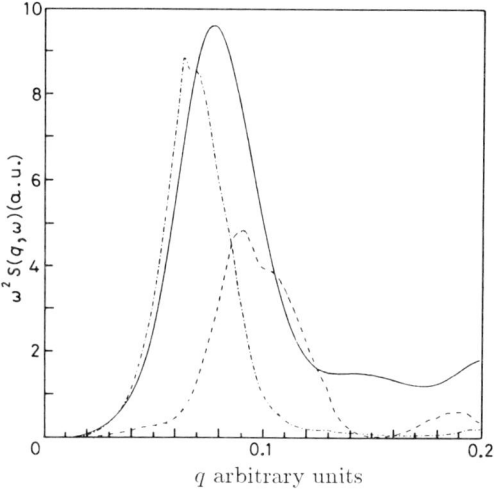

Fig. 2. Typical curves of $\omega^2 S(q, \omega)$ versus q for three closeby frequencies belonging to different subsets.

which confirms the fracton model prediction [1,2]. Here $2+\theta = \ln(5)/\ln(2)$. Indeed, eq. (6) agrees with the scaling assumption [1,2] for the mass m and the coupling constant t, in conjunction with the basic relation $\omega^2 = t/m$. Thus, excitations of frequency ω correspond to modes of spatial variation q_{max}^{-1}. The spectrum of the gasket includes a small number of nondegenerate eigenvalues characteristic to its specific construction procedure [13]. These were neglected in the analysis of the overall shape and scaling properties and we concentrated on eigenstates characteristic to the fractal nature of the structure. Most of these eigenvalues are highly degenerate. They separate into subsets of frequencies [13], within each the frequencies relate by $\omega_i^2 = 2^{2+\theta}\omega_{i-1}^2$ (or $q_{\text{max},i} = 2q_{\text{max},i-1}$), where i enumerates members of the same subset. The degeneracy of the ω_i^2 eigenvalue is approximately proportional to 3^i.

The fracton model assumes the existence of a single, frequency-dependent, length that sets up the scale for the dynamics [1,2]; this implies scaling of the structure factor with a single length. We now construct this scaling.

In terms of the (normalized) eigenmodes ϕ_α and eigenfrequencies ω_α the structure factor takes the form

$$S(q, \omega) = \sum_\alpha \phi_\alpha(q) \phi_\alpha^+(q) \delta(\omega^2 - \omega_\alpha^2),$$

where $\phi_\alpha(q)$ is the spatial Fourier transform of $\phi_\alpha(\mathbf{R})$. As $S(q, \omega)$ satisfies the sum rule $\int d\omega\, \omega S(q, \omega) = 1$, the assumption of single-length scaling implies

$$S(q, \omega) = \omega^2 F(\omega/q^{(2+\theta)/2}), \tag{7}$$

where F is a scaling function. To test this form, we have calculated $S(q, \omega)$ for successive eigenfrequencies belonging to the same subset. For numerical reasons, the δ function was eliminated, by adding an arbitrarily small imaginary term $i\eta$ to ω^2, calculating $S(q, \omega)$, and then multiplying the result by η. For $|\omega^2 - \omega_\alpha^2| < \eta$, the result is independent of η. Effectively, this procedure is equivalent to the calculation of $\sum' \phi_\alpha^+(q) \phi_\alpha(q)$, where the sum runs over all degenerate states of eigenfrequency ω_α. The corresponding curves as a function of q, fall one onto the other when the q axis is rescaled by a factor of 2 for each successive frequency (see fig. 3). Hence, though the shape of $S(q, \omega)$ of two frequencies belonging to different subsets is not the same, the structure factor of frequencies of the same subset obey the scaling form (7), which again confirms the fracton model predictions [8].

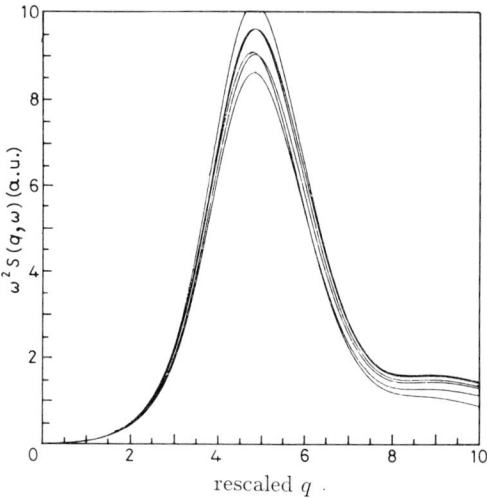

Fig. 3. Rescaled curves of $\omega^2 S(q, \omega)$ for eigenvalues belonging to a given subset.

A technical point should be noted. The structure factor and hence the scaling function F are expected to obey scaling in the $\eta \to 0$ limit. The deviation from exact scaling observed in fig. 3 is due to numerical iterative approach towards that limit: one chooses a value for η, calculates the approximate eigenvalue, reduces η, then recalculates the eigenvalue until the desired accuracy is achieved. The nonsystematic deviations from scaling can be removed by further iterations. The scaling of the structure factor is, however, clearly demonstrated by fig. 3.

3. Summary

We have shown that the excitation spectrum pertaining to the equation of motion (1) obeys single length scaling for a fractal structure. This holds not only for the peak position of the structure factor, but for its shape as well.

This computation cannot describe the phonon–fracton crossover, as it is valid only in the fracton regime. To consider the entire frequency range, the effective medium approximation (EMA) was exploited [9]. The spirit of this model is to replace a random system by an effective homogeneous (and periodic) medium, which one can solve for the dynamics. The parameters of the effective medium depend, in a self-consistent way, upon the parameters of the random system. The line shape of the structure factor, within EMA, can be written in terms of an effective sound velocity and linewidth, τ^{-1}, both functions of frequency [9]. One can then analyze their limiting behaviours, in the phonon ($\omega < \omega_c$) and in the fracton ($\omega > \omega_c$) regimes. By interpolating between these two limits and using the EMA line shape, Courtens et al. [4] were able to fit beautifully their Brillouin scattering data.

Within EMA, the sound velocity, up to ω_c, was found [9] to be independent of ω. Above ω_c it increased with ω. The scattering width τ^{-1} followed the Rayleigh law in the phonon regime, $\tau^{-1} \approx \omega^{d+1}$, and became proportional to the frequency in the fracton regime, thus confirming the Ioffe–Regel limit for fractons, derived from scaling arguments [8]. However, within EMA, the linewidth did not scale with ω_c. It is conceivable that this failure of scaling is connected with the approximations involved in the effective medium approach. The calculation presented here indicates that in the fracton regime the dynamics does obey single variable scaling.

We finally speculate upon the spatial extension of the eigenfunctions corresponding to a given frequency. The 3^i degeneracy of the ω_i^2 mode suggests that there is a representation in which each eigenfunction extends over 3^{n-i} sites. This implies that the spatial extension is approximately $2^{n-i} > 1/q_{\max}$, in a complete agreement with the width of the computed structure factor. The spatial extension of the wavefunctions is most effectively probed by the heat diffusivity; it would be of interest to explore this quantity.

Acknowledgements

Useful discussions with A. Aharony, R. Orbach and G. Polatsek are gratefully acknowledged. The work was supported by the fund for Basic Research administered by the Israel Academy of Sciences and Humanities.

References

[1] S. Alexander and R. Orbach, J. Phys. (Paris) 43 (1982) L625.

[2] S. Alexander, Ann. Israel. Phys. Soc. 5 (1983) 144;
R. Orbach, Science 231 (1986) 814;
O. Entin-Wohlman, in: The T.D. Holstein Symposium, Condensed Matter Physics, R. Orbach, ed. (Springer, Berlin, 1986), p. 160.

[3] K. Yakubo and T. Nakayama, Phys. Rev. B 36 (1987) 8933.

[4] E. Courtens, J. Pelous, J. Phalippou, R. Vacher and T. Woignier, Phys. Rev. Lett. 58 (1987) 128;
R. Vacher, T. Woignier, J. Pelous and E. Courtens, Phys. Rev. B 37 (1988) 6500;
E. Courtens, R. Vacher, J. Pelous and T. Woignier, Europhys. Lett. 6 (1988) 245;
Y. Tsujimi, E. Courtens, J. Pelous and R. Vacher, Phys. Rev. Lett. 60 (1988) 2757.

[5] Y.J. Uemura and R.J. Birgeneau, Phys. Rev. B 36 (1987) 7024.

[6] Y. Gefen, A. Aharony and S. Alexander, Phys. Rev. Lett. 50 (1983) 77.

[7] A. Boukenter, B. Champagnon, E. Duval, J. Quinson and J. Serughetti, Phys. Rev. Lett. 57 (1986) 2391;
A. Boukenter, E. Duval and J. Serughetti, Phys. Rev. Lett. 59 (1987) 604.

[8] A. Aharony, S. Alexander, O. Entin-Wohlman and R. Orbach, Phys. Rev. Lett. 58 (1987) 132;
A. Aharony, O. Entin-Wohlman and R. Orbach, in: Time Dependent Effects in Disordered Materials, T. Riste and R. Pynn, eds. (Plenum, New York, 1988), p. 233.

[9] G. Polatsek and O. Entin-Wohlman, Phys. Rev. B 37 (1988) 7726;
G. Polatsek, O. Entin-Wohlman and R. Orbach, Phys. Rev. (1989), in press.

[10] U. Sivan, R. Blumenfeld, Y. Meir and O. Entin-Wohlman, Europhys. Lett. 7 (1988) 249.

[11] B.B. Mandelbrot, The Fractal Geometry of Nature (Freeman, San Fransisco, 1982).

[12] A. Aharony, in: Scaling Phenomena in Disordered Systems, R. Pynn and A. Skjeltorp, eds. (Plenum, New York, 1985), p. 335.

[13] E. Domany, S. Alexander, D. Bensimon and L.P. Kadanoff, Phys. Rev. B 28 (1983) 3110.

FRACTAL PATTERN FORMATION IN HUMAN RETINAL VESSELS

Fereydoon FAMILY
Department of Physics, Emory University, Atlanta, GA 30322, USA

Barry R. MASTERS
Department of Ophthalmology, Emory University, Atlanta, GA 30322, USA

Daniel E. PLATT [1]
Department of Physics, Emory University, Atlanta, GA 30322, USA

Dedicated to B.B. Mandelbrot on the occasion of his 65th birthday

The mechanism for the formation of retinal vessel patterns in the developing human eye is an unresolved question of considerable importance. The current hypothesis is based on the existence of a variable oxygen gradient across the developing photoreceptors which stimulates the release of angiogenic factors which diffuse in the plane of the retina and result in the growth of retinal vessels. This implies that the limiting step in the formation of retinal blood vessels is a diffusion process. To test this hypothesis we have performed a fractal analysis of the human retinal vessels using two different methods. Within the limited range of length scales available in the red-free fundus photographs, we find that the human retinal blood vessels have a self-similar structure with a fractal dimension $D \approx 1.7$. Since this value of D is the same as the value found for a diffusion limited growth process, our result supports the hypothesis that diffusion is the fundamental process in the formation of human retinal vessel patterns.

1. Introduction

In a paper published in 1951 Benoit Mandelbrot pioneered the application of fractal concepts to describe complex natural shapes and structures as well as mathematical sets and functions having intricately irregular form. The full impact of this remarkable discovery was not recognized until the publication of Mandelbrot's classic French monograph Les Objets Fractals: Forme, Hasard et Dimension [1], and the following two books, Fractals: Form, Chance and Dimension [2], and The Fractal Geometry of Nature [3]. Since the publication of these books, there has been an explosion of activity in applying the powerful mathematical language of fractals to many unresolved problems at the frontiers of almost all scientific disciplines. Beginning with the study of the kinetics and structure of disordered materials [4], such as polymers, colloids, aerosols and gels, fractal concepts have found applications in numerous other areas, including transport phenomena in disordered systems and dynamics of random materials [5,6], the growth and form of complex patterns in flow through porous media and hydrodynamic instabilities such as viscous fingering [7,8], chaotic vibrations [9] and chaos in dynamical systems [10], data compression of images [11] and pattern formation in far from equilibrium growth phenomena [12,13].

Biology has immense potential for the application of fractal concepts. Biological growth almost invariably leads to the formation of complex shapes, forms and patterns [14]. From the bifurcating structure of

---

[1] Present address: Thomas J. Watson Research Center, I.B.M., Yorktown Heights, NY 10598, USA.

Essays in honour of Benoit B. Mandelbrot
Fractals in Physics – A. Aharony and J. Feder (editors)

trees to such diverse patterns as the bronchial tree, and the network of nerves and blood vessels, there are many biological patterns that appear to be self-similar [2,3,15]. This has been illustrated by studies of blood vessel pattern in the chick embryo [16], neuronal dendritic arborizations [17], vascular heterogeneity in the heart [18,19], the structure of the bronchial tree [20–22] and the cerebral surface of the normal human brain [23].

In this paper we present the first quantitative analysis of the geometry of blood vessels in the normal human retina, using fractal concepts. Preliminary results have been presented by Masters et al. [24,25]. On the basis of two methods for determining the fractal dimension we conclude that the network of blood vessels in the inner human retina is fractal. Within the limited number of bifurcations which appear in the photographs obtained with a retinal camera (see fig. 1), we find that the fractal dimension of the vascular network is about 1.7. The current model for the development of retinal vessels is based on the diffusion of angiogenic factors in the plane of the retina [26–30], which implies that the formation of blood vessel patterns is diffusion limited. Since the retina blood vessels have a similar pattern, and the same fractal dimension as two dimensional diffusion-limited aggregation [31,32] and Laplacian fractal [33,34] patterns, our result provides direct quantitative support for the model [26–30].

2. Development of the retinal circulation system

The retina has the highest oxygen requirement of any tissue in the body, and any alteration in circulation may result in functional impairment and tissue damage. Diseases of the retinal circulation which can lead to blindness if untreated include the following: diabetic retinopathy, retinopathy of prematurity, and hypertensive vascular disease. The retina is supplied by two major systems of blood vessels. The inner layer of nerves and glial cells are supplied by the retinal circulation. In humans this is supplied by the central retinal artery and has one main collecting trunk, the central retinal vein. The veins and the arteries do not cross themselves, but occasionally a vein and an artery do overlap. There are smaller branches of these major vessels; the arterioles, the venules, and the smallest vessels which are the capillaries. The second circulatory system of the retina is the choroidal circulation which supplies the outer layer of the cells of the neural retina (photoreceptors) and the retinal pigment epithelium. This system consists of three layers of choroidal vessels. The retinal circulation is visible clinically; however, the choroidal circulation is not visible except in depigmented areas of the retina. We have analyzed the patterns of the retinal circulatory system.

Endothelial cells form a single layer that lines all blood vessels. Vessels develop from the walls of existing small vessels by the outgrowth of these endothelial cells. The endothelial cells are formed by division of existing endothelial cells. New capillaries form by sprouting from existing small vessels and develop into new vessels. This process is called angiogenesis [35–37]. The growth of the capillary network is controlled by angiogenic factors released by the surrounding tissues [35–37].

Kretzer, Hittner and coworkers [26–30] have developed a model for the development of the inner retinal vasculature. The primary components of this hypothesis are the following: there is a relationship between inner retinal blood vessel development and the maturation of the photoreceptors. During this developmental stage, the maturing photoreceptors consume progressively more oxygen, decreasing the oxygen available to the inner retina. The migrating spindle cells in the avascular inner retina sense this diminished concentration of oxygen and release angiogenic factors. The angiogenic factors diffuse in the plane of the retina and stimulate the growth of new retinal blood vessels. The decrease in the transretinal flux of oxygen from the choroidal vasculature is compensated by a new vascular source on the inner retina. Therefore, diffusion of angiogenic factors is the physical process responsible for the development of retinal vessel patterns.

3. Fractal analysis of retinal blood vessels

We analyzed five red-free (a green 540 nm cutoff filter is used to enhance the blood vessels) photographs (30 cm × 30 cm) and one fluorescein angiogram of the human retina in order to determine if the retinal vessels have a fractal structure. A typical example of retinal blood vessels is shown in fig. 1. We first made tracings of the retinal arteries and veins comprising only the retinal circulation. Although vessel width varies with distance from the optic nerve head, we ignored this variation and measured the vessel length. We obtained the mid-line coordinates and the lengths of the vessels in the images using a digitizing pad.

We used two methods to determine the fractal dimension of the retina circulation patterns. One method, called the mass–radius method [13,32], is based on finding the relation between the mass $M(r)$, within circles of radius r whose origin is placed at a point on the object, and the distance r. The fractal dimension D is then determined from the relation

$$M(r) \sim r^D. \tag{1}$$

In order to apply this method, we assumed that $M(r)$ is proportional to the length of the traced vessels within a circle of radius r. The fractal dimension D was obtained from the slope of the log–log plots of $M(r)$ versus r, and the standard error was calculated using a linear regression method. A typical mass–radius plot is shown in fig. 2, and the values of D are given in table 1.

A random fractal is self-similar only statistically and within a finite range of length scales. An effective method for determining the fractal dimension of a

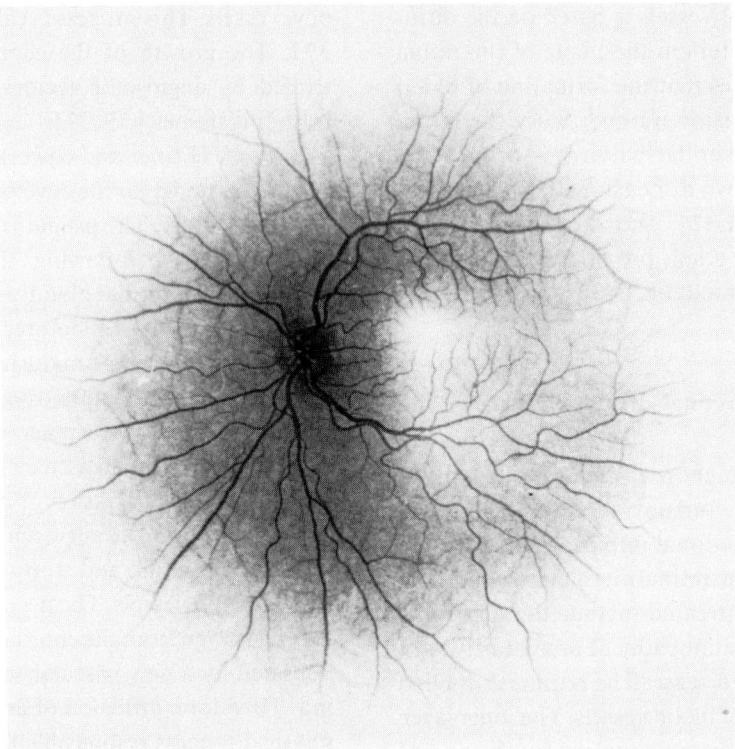

Fig. 1. Fluorescein angiogram of patient E made with a 140° Pomerantzeff fundus camera. The figure is the negative of the fluorescein angiogram with the blood vessels appearing black on a light background.

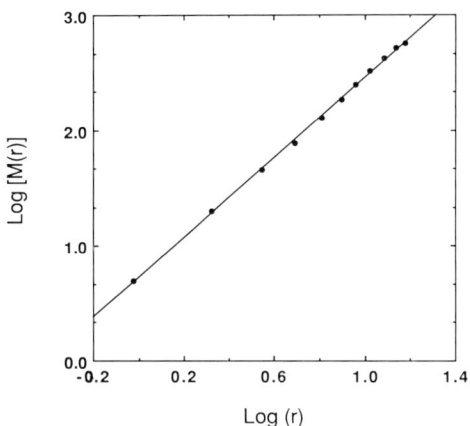

Fig. 2. A typical plot of the logarithm of the length of the vessels, $M(r)$, versus the logarithm of the radius r. The slope of the line gives $D = 1.72 \pm 0.03$. The data are from a fundus photograph of patient C.

random structure is based on the scaling of the two-point correlation function [13,31,32]

$$C(r) = \frac{1}{N} \sum_{r'} \rho(r+r')\rho(r'), \qquad (2)$$

where $\rho(r)$ is the local density, i.e. $\rho(r) = 1$ if there is a particle at point r, otherwise it is zero. The normalization factor N gives $C(r)$ the interpretation of an averaged density of the cluster. Self-similarity implies that $C(r)$ decays with r as

$$C(r) \sim r^{-\alpha}, \qquad (3)$$

where

$$\alpha = d - D \qquad (4)$$

and d is the Euclidean dimension, which is 2 for a plane.

We determined the two-point correlation function using the coordinates of the points belonging to the retina patterns obtained from the digitizer. A typical plot of $C(r)$ is shown in fig. 3. As can be seen in the figure, $C(r)$ has three different types of behavior. For small r, $C(r)$ decays rapidly, indicating that the individual branches have a highly tortuous path with a small fractal dimension. In the intermediate regime, the cluster is self-similar and $C(r)$ decays with the slope α. The correlations vanish rapidly at length scales larger than the cluster radius. We determined the values of the slopes α and the fractal dimension D for all six images. The results are shown in table 1. Due to the small range over which the patterns are self-similar, we find slightly different values of D using the mass–radius and the two-point correlation function methods. The mass–radius method probes the mass within a given length scale, whereas the two-point correlation function is an *average* over the entire cluster [13,31,32]. Thus, in small scale simulations, or in natural patterns with a limited range of length scales, these two methods give slightly different values for D [32].

Table 1
Fractal dimension of human retinal vessels determined by the mass–radius relation (method I) and the two-point correlation function (method II) techniques. The images were obtained from photomontages of nine fields in the eye using a 60° Cannon fundus camera (PM), 30° Zeiss fundus camera and 140° Pomerantzeff fundus camera. The figure for patient E (see fig. 1) was a fluorescein angiogram.

ID	Age	Sex	Diagnosis	Image	Fractal dimension	
					method I	method II
A	25	M	vasculitis	PM	1.64 ± 0.03	1.71 ± 0.02
B	15	F	angioid streaks	PM	1.66 ± 0.03	1.71 ± 0.03
C	14	F	normal	30°	1.72 ± 0.03	1.75 ± 0.03
D	27	F	normal	140°	1.75 ± 0.02	1.82 ± 0.04
E	41	M	normal	140°	1.69 ± 0.03	1.88 ± 0.05
F	30	F	normal	140°	1.73 ± 0.04	1.82 ± 0.04

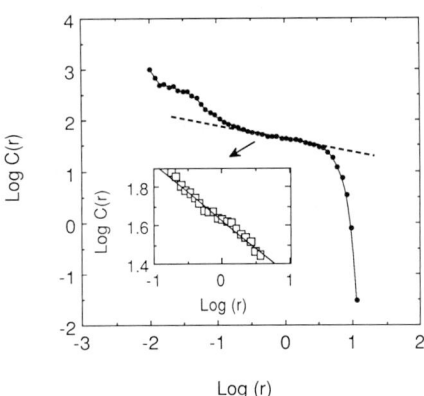

Fig. 3. The logarithm of the two-point correlation function $C(r)$ versus the logarithm of the distance r. The linear region of the plot in the range $-1 < \log(r) < 1$ indicates that the patterns are self-similar in this range of length scales. An enlarged view of this range is shown as an insert. The slope, $\alpha = -0.29$, indicates that $D = 1.71 \pm 0.03$. The data shown are from a fundus photograph of patient B.

4. Discussion and conclusions

The geometry of the eye could affect the measurements of the fractal dimension. A projection from a two-dimensional curved surface to a two-dimensional flat surface was used in producing a photograph of the retinal vessels. This projection involves the introduction of a fixed length scale – namely the radius of curvature of the eyeball. Asymptotically, the measurement of the fractal dimension should not be sensitive to such an effect. In addition, the projection of a fractal embedded in three dimensions to a plane does not change D as long as $D < d$ [38]. However, since our measurements only encompass about two decades in length scale, the projection might have a small, but noticeable, effect.

In conclusion, we analyzed six different photographs of human retina images and determined that, within the limited range of length scales available in the photographs, the retinal circulation is fractal with $D \approx 1.7$. The current model for the development of inner retinal vessels is based on the diffusion of angiogenic factors in the plane of the retina [26–30], which implies that diffusion is the fundamental physical process. Since the retina blood vessels have a similar pattern, and the same fractal dimension as two-dimensional diffusion-limited aggregation [31,32] and Laplacian fractals [33,34], our results provide direct quantitative support for the model [26–30].

There is a great need in ophthalmology for noninvasive methods to diagnose retinal diseases and to quantitate its severity and progression. This is even more important if the retinal disease can be detected prior to cellular damage. The analysis presented could have the potential of being used as a clinical diagnostic tool for detection of vascular diseases. For example, if it can be demonstrated that the fractal dimension of retinal blood vessel patterns differs between normals and diabetics, then this technique can be used as a discriminant in diagnostic screening. Another potential application is to identify those infants who are at risk from retinopathy of prematurity and require surgical intervention. Studies with matched populations of normals and patients with vascular ocular disease are in progress and the results will be presented elsewhere.

Acknowledgements

We would like to thank J. Gilman, M. Pankratov and D. Bartlett for providing us with the retinal photographs. We would also like to thank Tamás Vicsek, Carl Evertsz and Tom Witten for helpful comments and discussions. This research was supported by NIH grant EY-06958 (BRM). FF would like to thank the Office of Naval Research and the donors of the Petroleum Research Fund, administered by the American Chemical Society, for partial support of this work.

References

[1] B.B. Mandelbrot, Les Objets Fractals: Forme, Hasard et Dimension (Flammarion, Paris, 1975).
[2] B.B. Mandelbrot, Fractals: Form, Chance and Dimension (Freeman, San Francisco, 1977).

[3] B.B. Mandelbrot, The Fractal Geometry of Nature (Freeman, San Francisco, 1982).
[4] F. Family and D.P. Landau, eds., Kinetics of Aggregation and Gelation (North-Holland, Amsterdam, 1984).
[5] M.F. Shlesinger, B.B. Mandelbrot and R.J. Rubin, eds., Proceedings of a Symposium on Fractals in the Physical Sciences, National Bureau of Standards, Gaithersburg, MD, J. Stat. Phys. 36, Nos. 5/6 (1984).
[6] N. Boccara and M. Daoud, eds., Physics of Finely Divided Matter (Springer, Berlin, 1985).
[7] J. Feder, Fractals (Plenum, New York, 1988).
[8] H.E. Stanley and N. Ostrowsky, eds., On Growth and Form: Fractal and Non-Fractal Patterns in Physics (Nijhoff, Dordrecht, 1986).
[9] F.C. Moon, Chaotic Vibrations (Wiley, New York, 1987).
[10] M.F. Barnsley and S.G. Demko, eds., Chaotic Dynamics and Fractals (Academic Press, Orlando, Florida, 1986).
[11] M.F. Barnsley, Fractals Everywhere (Academic Press, San Diego, 1988).
[12] H.E. Stanley and N. Ostrowsky, eds., Random Fluctuations and Pattern Growth (Kluwer, Dordrecht, 1988).
[13] T. Vicsek, Fractal Growth Phenomena (World Scientific, Singapore, 1989).
[14] H. Meinhardt, Models of Biological Pattern Formation (Academic Press, New York, 1982).
[15] P. Meakin, J. Theo. Biol. 118 (1986) 101.
[16] A.A. Tsonis and P.A. Tsonis, Perspect. Biol. Med. 30 (1987) 355.
[17] H.E. Stanley, Bull. Am. Phys. Soc. 34 (1989) 716.
[18] J.H.G.M. van Beek, J.B. Bassingthwaighte and R.B. King, Biophys. J. 53 (Suppl.) (1988) 401.
[19] J.B. Bassingthwaighte, Bull. Am. Phys. Soc. 34 (1989) 715.
[20] B.J. West, Bull. Am. Phys. Soc. 34 (1989) 716.
[21] B.J. West and A.L. Goldberger, J. Appl. Phys. 60 (1986) 189.
[22] B.J. West and A.L. Goldberger, Am. Sci. 75 (1987) 354.
[23] S. Majmudar and R.R. Prasad, Comp. Phys. 2 (1988) 69.
[24] B.R. Masters, F. Family and D.E. Platt, Biophys. J. 55 (Suppl.) (1989) 575.
[25] B.R. Masters and D.E. Platt, Invest. Ophthalmol. Vis. Sci. 30 (Suppl.) (1989) 391.
[26] F.L. Kretzer, R.S. Mehta, A.T. Johnson, D.G. Hunter, E.S. Brown and H.M. Hittner, Nature 309 (1984) 793.
[27] F.L. Kretzer and H.M. Hittner, in: Retinopathy of Prematurity, W.A. Silverman and J.T. Flynn, eds. (Blackwell Scientific Publications, London, 1985).
[28] F.L. Kretzer, A.R. McPherson and H.M. Hittner, Graefe's Arch. Clin. Exp. Ophthalmol. 224 (1986) 205.
[29] F.L. Kretzer and H.M. Hittner, Arch. Dis. Child. 63 (1988) 1151.
[30] F.L. Kretzer and H.M. Hittner, in: Retinopathy of Prematurity: Problem and Challenge, J.T. Flynn and D.L. Phelps, eds. (March of Dimes Birth Defects Foundation, 1988), p. 147.
[31] T.A. Witten and L.M. Sander, Phys. Rev. Lett. 47 (1981) 1400.
[32] P. Meakin, in: Phase Transitions and Critical Phenomena, C. Domb and J.L. Lebowitz, eds. (Academic Press, New York, 1988).
[33] C.J.G. Evertsz, Laplacian Fractals, Ph.D. Thesis, University of Groningen (1989), to be published.
[34] F. Family, in: Computer Simulation Studies in Condensed Matter Physics, D.P. Landau, K.K. Mon and H.-B. Schüttler, eds. (Springer, Berlin, 1988), p. 65.
[35] J. Folkman, E. Merler, C. Abernathy and G. Williams, J. Exp. Med. 133 (1971) 275.
[36] J. Folkman, Sci. Am. 234(5) (1976) 58.
[37] J. Folkman and C. Haudenschild, Nature 288 (1980) 551.
[38] D.A. Weitz and M. Oliveria, Phys. Rev. Lett. 52 (1984) 1433.

GEOMETRICAL CROSSOVER AND SELF-SIMILARITY OF DLA AND VISCOUS FINGERING CLUSTERS

Jens FEDER, Einar L. HINRICHSEN, Knut Jørgen MÅLØY and Torstein JØSSANG

Department of Physics, University of Oslo, Box 1048, Blindern, 0316 Oslo 3, Norway

Diffusion-limited aggregation (DLA) clusters and the structures observed in viscous fingering experiments at high capillary numbers are tree-like fractals. The different branches may be assigned a *branch order* in a way that exhibits scaling, and permits a self-similar characterization by the bifurcation ratio r_N and the length ratio r_L of branches of different orders. The fractal dimension is given by $D = \log(r_N)/\log(1/r_L)$. Good agreement between experiments and simulations is found. A crossover function characterizes the branch orders, and we conclude that DLA is in a state of geometrical crossover: branches are linear up to their length, L, but fractally distributed on length scales much larger than L. The effective fractal dimension of DLA depends on how D is measured and over what part of the cluster.

1. Introduction

The diffusion-limited aggregation (DLA) model [1] generates fractal [2–4] structures and is the prototype of "Laplacian" [5] growth models that lead to ramified tree-like structures. In simulations of the DLA model one finds that the radius of gyration R_g of the growing cluster scales with the mass M of the cluster as [6] $M \sim R_g^{D_g}$, where D_g, the radius-of-gyration dimension or growth dimension, is 1.71 in two and 2.5 in three dimensions. Many two-dimensional growth processes exhibit regimes where the observed structures resemble DLA clusters with $D \simeq 1.7$ [7,8]. The analogy between DLA and viscous fingering (VF) in porous media was first pointed out by Paterson [9] and a modified DLA model [10,11] simulates the observed VF dynamics accurately. For a discussion of irreversible growth models see refs. [3,4,8].

DLA clusters have different scaling properties in the radial and azimuthal directions [12,13]. This gives rise to the question: Is DLA self-similar? In spite of the recent progress of a renormalization group approach [14–16] in estimating the fractal dimension of DLA, much remains to be done, particularly in reconciling the many *different* fractal dimensions that arise when DLA-like structures are analyzed, see table 1 below.

We introduced [17] a *new* way to characterize DLA clusters and VF patterns by a hierarchy of *branch orders*. The concept of branch orders has previously been used by Horton [18] in the description of river systems. Horton found that the bifurcation ratio r_N between the number of steams of two subsequent orders was constant for many river systems. He also found that the length ratio r_L between two subsequent orders was constant.

We have discovered that both DLA clusters and VF patterns have a branching ratio r_N and a length ratio r_L independent of branch order n. The self-similarity of these statistical tree-like structures is therefore analogous to well-known deterministic recursive structures such as the triadic Koch curve [2]. We also find for DLA and VF that the longest ($n=0$) branches are one-dimensional. The higher-order branches are (for a given n) fractally distributed on scales above their average length L_n and one-dimensional on scales below L_n. This behavior is characterized by a *crossover function* intrinsic to DLA and VF. Numerically we find good agreement between simulations and experiments, lending further support to the hypothesis that VF and DLA are in the same universality class.

Essays in honour of Benoit B. Mandelbrot
Fractals in Physics – A. Aharony and J. Feder (editors)

Recently Vannimenus [19,20] studied tree structures using "Strahler numbers" (originally used to analyze river systems [21]) to analyze DLA and other tree structures in terms of a *ramification matrix*. The Strahler classification of branch orders is related to the Horton scheme, and we consider the two approaches to be complementary.

2. Definition of branch orders

For DLA and other branching structures without loops it is possible to assign branch orders in the following way. Each branch defines a continuous line, starting at a tip, and ending on another branch of lower order if it is not the "trunk" (zeroth-order branch). More than two branches may meet at a single point. The highest-order branches are those that have no side branches. The next to highest order branches have side branches of the highest order, and so on. If branches of the same order n meet, the longest branch is relabeled and assigned order $n-1$. The minimal side branch order of a given point on a branch is the lowest side branch order found when going from the tip to this point. A unique assignment of branch orders is obtained if the minimal side branch order does not decrease in steps larger than one when going from the tip to root point.

2.1. A simple model

The hierarchical ordering of self-similar fractal trees is best illustrated by discussing a deterministic algorithm as defined in fig. 1. The *initiator* [2] shown in (a) is a line segment. The *generator* (b) replaces this line segment by $\mathcal{N}=5$ new line segments each scaled down by the ratio $r=1/3$. In (c) the generator is applied again giving the second generation *prefractal* [3]. After an infinite number of applications of this algorithm one arrives at a fractal set. This set is self-similar and has a similarity dimension [2,3] $D_s = \log(\mathcal{N})/\log(1/r) = 1.46...$, equal to the fractal dimension of the set. Each piece of the fractal is a scaled-down version of the whole set, as is required for self-similar fractals.

The branches of this structure (and of the DLA structure shown in fig. 2) may be characterized as follows: In (a) we have a tree-like structure containing only the trunk, which we identify as a branch of order 0 and length 1. In (b) we have added two side branches of order 1 and length $1/3$. In (c) the structure contains one branch of order 0, 2 branches of order 1, and 10 new branches of order 2. These new branches have a length of $1/9$. If we continue this process, we will at each stage have a branching prefractal with branches of order $0, 1, \dots$. The number N_n and length L_n of branches of order n are given by

$$L_n = L_m r_L^{n-m} \quad \text{and} \quad N_n = N_m r_N^{n-m}, \qquad (1)$$

where in this example $r_L = 1/3$, $r_N = 5$, and $m = 1$ is a lower cutoff for the validity of the scaling relations. In general the *similarity dimension* is given by

$$D_s = \ln(r_N)/\ln(1/r_L), \qquad (2)$$

for tree structures with fixed r_N and r_L.

2.2. Self-similarity and box counting

This result is also obtained by "box counting". Cover the structure with small "boxes" having shapes chosen for optimal coverage but with the same length $\delta = L_n$ and width $\sim 2L_{n-1} \simeq 2r_L \delta$. Using (1) we find that the number of boxes needed to cover the fractal is given by

$$N(\delta) = \sum_{i=0}^{n} \frac{N_i L_i}{\delta} = A\left(\frac{\delta}{L_m}\right)^{-D} + B\left(\frac{\delta}{L_m}\right)^{-1}, \qquad (3)$$

where $D = \ln(r_N)/\ln(1/r_L)$ is the fractal dimension. A and B are constants. The branches of order $n=0$ to $n=m-1$, which do not follow the scaling relation in (1), contribute only to the last term in (3). This method of finding the fractal dimension using optimally shaped pieces to cover the fractal is closer to the Hausdorff–Besicovitch definition than the box-counting method [2]. Neglecting the corrections to scaling, i.e., the last term, in (3), we find that the

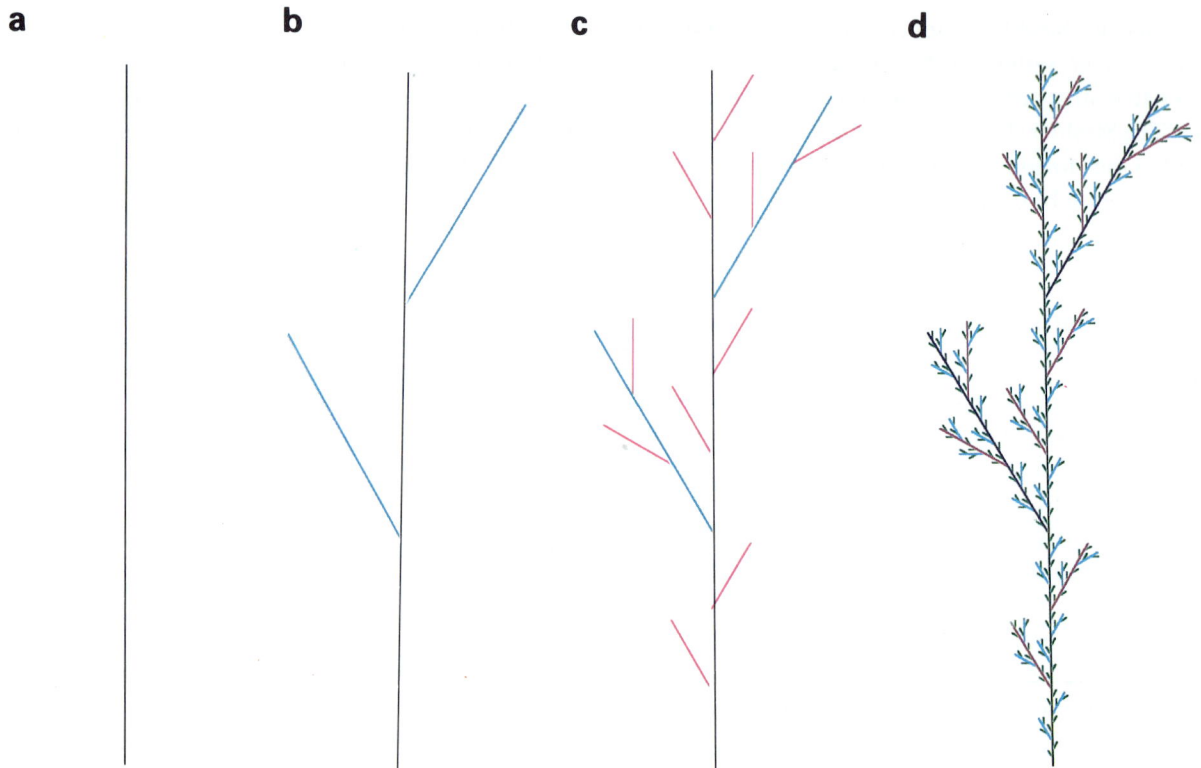

Fig. 1. A recursively defined tree structure. The colors indicate the branch orders; black: a zero-order branch, dark blue: first-order branches, red: second-order branches, blue: third-order branches and green: fourth-order branches. See text for a full explanation.

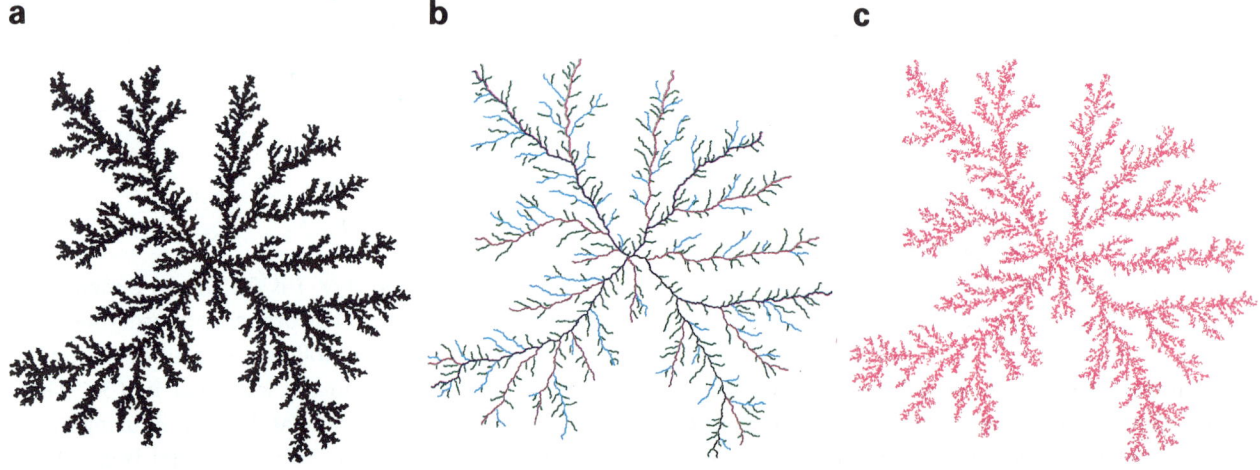

Fig. 2. Branch structure of an off-lattice DLA cluster consisting of 250 000 particles. (a) The full cluster. (b) The low-order (long) branches. The color scheme is as in fig. 1. (c) The highest-order (shortest) branches. This figure is indistinguishable from the full cluster.

number of pieces $N(\delta)$ of "diameter" δ needed to cover the fractal is given by

$$N(\delta) \sim \left(\frac{\delta}{L_m}\right)^{-D}, \quad \text{with } D = D_s. \qquad (4)$$

This result shows that the fractal dimension for scaling trees is given by the similarity dimension. If one instead covers the fractal with square boxes of side δ, one finds that the number of boxes needed to cover the fractal is $N(\delta) \sim \delta^{-D_b}$, where D_b is the box dimension [2].

3. Branch structure of DLA

We have analyzed viscous fingering patterns obtained from experiments in a two-dimensional porous medium [10,11] and also computer-simulated off-lattice DLA clusters using these ideas. These structures have almost no loops, and may therefore be described as trees.

3.1. Simulations

We generated 200 DLA clusters of size 10^3, 50 of size 5×10^3, 30 of size 2.5×10^4, 20 of size 5×10^4, and 10 of size 2.5×10^5. Each particle added to the cluster was assigned a pointer to the particle it attached to. The tree structure was identified after the DLA seed particle in the center was removed. To identify one branch, we started at the last particle and used the pointer to find the particle to which it attached, and then to the particle to which this one was attached, and so on all the way to the DLA seed particle. The other branches were found in a similar way, keeping track of the branch order.

The branching ratio r_N and length ratio r_L were obtained by fitting the number and length of order n to the power laws $N_n \sim r_N^n$ and $L_n \sim r_L^n$. In the simulations, the average length of a branch was defined as either the average mass of the branch or the average tip-to-root distance of a branch. Both of these definitions gave about the same value for r_L. In figs. 3 and 4 we show the experimental result together with

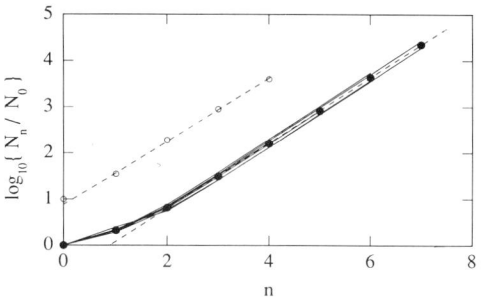

Fig. 3. The number of branches N_n at a given order as a function of branch order n for VF experiments ($\circ$), and DLA simulations ($\bullet$). The lines represent fits of $N_n \sim r_N^n$ to the data, with $r_N = 4.8$ for VF and 5.2 for DLA. The experimental points have been shifted up one decade for clarity. Figure from ref. [17].

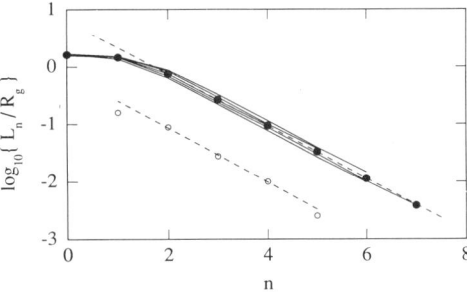

Fig. 4. The scaled average length L_n/R_g plotted against branch order n. L_n is given by the tip-to-root distance in DLA ($\bullet$) and the average mass in the VF experiments ($\circ$). Lines are fits of $L_n \sim r_L^n$ to the data, with $r_L = 0.34$ for VF and 0.36 for DLA. The experimental points have been shifted down for clarity. Figure from ref. [17].

results obtained from simulations for all the different cluster sizes. The points shown are the average values for these different cluster sizes. For a given size, the values fall on a straight line either above or below the average curve. The size of the cluster is, however, not correlated with this deviation from the average value. This scatter is related to the fact that averaging is done with fixed mass and not for a fixed maximal order. Note also that the scaling of the different cluster sizes in fig. 4, show that $R_g \sim L_n$ for all orders n, including orders $n < m$ ($\simeq 2$ for DLA) that do not follow (1). Using this result in eq. (4) gives the identification $D_g = D_s$ since the total mass $M \sim N(\delta) \sim R_g^{D_g}$. The values of r_L and r_N do not change if the analysis is limited to the part of the cluster within R_g. The scaling relations are therefore not changed in the crossover regime, see section 5.2.

The two ratios characterizing the branching structure of the tree-like fractals give $D=1.6$ from (4) for DLA clusters, consistent with other dimensions quoted in table 1 (see below). The viscous fingering result $D=1.5$ is also consistent with other values quoted in table 1.

4. Geometrical crossover

For the DLA clusters, we have analyzed each branch order separately in terms of the box-counting algorithm. The insert in fig. 5 shows the number $N_n(\delta)$ of filled boxes of order n as function of the box size δ. These curves have a linear regime, $N_n(\delta) \sim \delta$, for box sizes less than L_n and a fractal regime, $N_n(\delta) \sim \delta^{-D}$, for δ larger than L_n. We therefore expect $N_n(\delta)$ to have the scaling form

$$N_n(\delta) = M\delta^{-D}g(\delta/L_n) . \qquad (5)$$

The mass M is the total cluster mass. The crossover function $g(x)$ is constant for $\delta > L_n$ and tends to x^{D-1} for $\delta < L_n$. This is indeed demonstrated in fig. 5, where

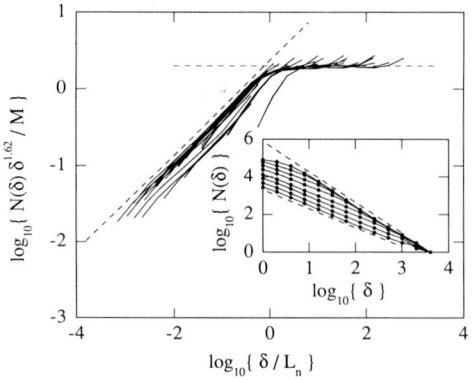

Fig. 5. The insert shows the result of box counting individual orders for clusters of sizes 2.5×10^5 particles. Slopes of dashed lines are 1.0 (lower line) and 1.62. The main figure shows the scaling function $g(\delta/L_n)$ obtained by scaling similar curves for six different cluster sizes. Slope of dashed line is 0.62. Figure from ref. [17].

all branch orders from clusters of six different sizes are scaled onto one single curve. Note that $g(\delta/L_n)$ depends on n through the ratio δ/L_n. The length of the lowest orders is not fully grown because it is calculated by taking the mean value of different clusters of the same size and not at a given order. For that reason we cannot expect the lowest order to fit well in the scaling plot, explaining some of the scatter seen. Also, as the insert in fig. 5 shows, the highest-order branches cross over at small δ to a slope close to 0 instead of 1. These branches consist of 1–2 particles and are therefore points. The highest-order branches from all the different cluster sizes collapse without scatter onto one curve in this regime, which is the curve that deviates most from the scaling function seen in fig. 5. It would be desirable of course to increase the size of the simulated clusters in order to obtain a better data collapse for the geometrical crossover. Unfortunately one needs to increase the mass of the cluster by a factor of $10^D \simeq 40$ in order to gain just one order of magnitude in cluster size or δ/L_n.

5. Fractal dimension

The fractal dimensions obtained for these clusters are summarized in table 1. There has been some dis-

Table 1
Fractal dimensions for DLA and VF

		DLA	VF
		branch order ratios	
$r_N = N_n/N_{n-1}$	number	5.2 ± 0.2	4.8 ± 0.5
$r_L = L_n/L_{n-1}$	length	0.35 ± 0.01	0.34 ± 0.04
		dimensions	
$D_s = \dfrac{\log(r_N)}{\log(1/r_L)}$	self-sim.	1.6 ± 0.1	1.5 ± 0.1
$N(\delta) \sim \delta^{-D_b}$	box	1.62 ± 0.02 [a]	1.51 ± 0.06
$N(\delta) \sim \delta^{-D_b}$	box [b]	1.67 ± 0.03	–
$M(R_g) \sim R_g^{D_g}$	growth	1.710 ± 0.005	–
$M(r) \sim r^{D_c}$	cluster	1.69 ± 0.01 [a]	1.62 ± 0.05 [a]

[a] Estimated by scaling different cluster sizes onto the same curve.
[b] Box counting, only points with $r < R_g$.

cussion whether D_b is equal to D_g [13,22]. Also we want to compare the traditional ways of estimating fractal dimensions for DLA clusters with the similarity dimension introduced here.

5.1. The radius of gyration dimension D_g

Numerically the scaling $M \sim R_g^{D_g}$ appears to be the most robust of the scaling relations we have tested, showing no sign of correction terms after the initial growth. Therefore we plot in fig. 6 the deviation from the scaling form and find that $\log(M)$ as function of $\log(R_g)$ has small oscillations around its mean value.

We have fitted these oscillations with a damped cosine function as explained in the caption to fig. 6. We see that a single component gives a reasonable fit and that the high-frequency noise at smaller R_g damps out more quickly than the main component. We have no explanation for these small oscillations. However, since the oscillations are small and damped we conclude that $M(R_g) \sim R_g^{D_g}$ and that D_g is accurately estimated from our data. We are somewhat surprised that we find no finite size effects in relation.

5.2. The cluster dimension D_c

The cumulative mass within radius r,

$$M(r) = M \cdot (r/R_g)^{D_c} f(r/R_g), \qquad (6)$$

is easy to measure experimentally [10] and gives a robust measure of the scaling structure of a given cluster. Here M is the total mass or equivalently the total number of particles of a *given* cluster, and f is a crossover function accounting for the finite cluster size. D_c is called the cluster dimension. In fig. 7 we have plotted $M(r)/M \cdot (r/R_g)^{D_c}$ as a function of r/R_g and find a data collapse for all the data obtained from the simulation. Such a data collapse was previously obtained for viscous fingering clusters of different sizes and in different fluids [10].

The scaling function $f(r/R_g)$ shown in fig. 7 shows clearly the crossover due to finite-size effects for $r/R_g > 1$. Measuring D_c of the cumulative mass $M(r)$ for $r < R_g$, where f is constant, gives $D_c = 1.69$.

The behavior of the total mass M as a function of R_g necessarily includes the crossover regime. However, since D_c is consistent with D_g this is an indication that the mass and width of the growth zone scale in the same way.

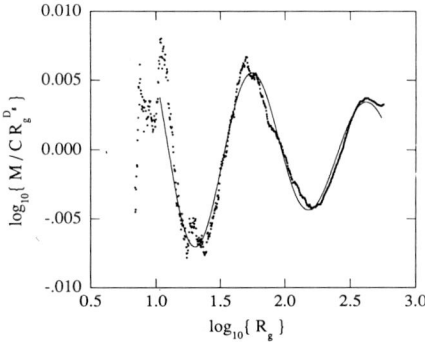

Fig. 6. The deviation of the cluster mass from the scaling form: $M(R_g)/CR_g^{D_g}$, as a function of the radius of gyration R_g. The full drawn line is a fit to the function $B_4 \exp(B_3 \log R_g) \cos(B_1 \times \log R_g + B_2)$. The values of the fitted parameters are: $D_g = 1.71010(7)$, $\log C = 0.5583(2)$, $B_1 = 7.15(2)$, $B_2 = 6.31(4)$, $B_3 = 0.55(2)$ and $B_4 = 0.0144(5)$. The numbers in parentheses are statistical errors in the last digit.

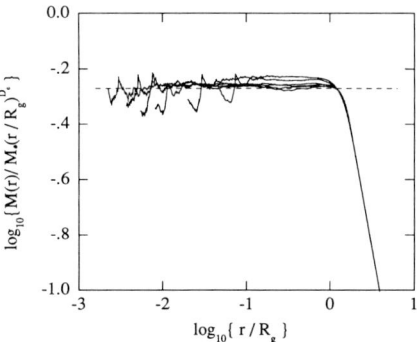

Fig. 7. The crossover function $f(r/R_g)$ for the cumulative mass $M(r)$ within a radius r as a function of r/R_g. The value $D_c = 1.69$ ensures that the scaling function is independent of its argument for $r/R_g < 1$.

5.3. The box dimension D_b

In fig. 8 we show the number of boxes $N(\delta)/M$ needed to cover a cluster of mass M as a function of the box size δ for all our clusters. Finite-size effects are clearly visible in fig. 8 when δ is larger than the radius of gyration, R_g, of the cluster. Note, however, that the family of curves have a well-defined lower bound that is a straight line representing the asymptotic behavior. The slope of the dotted line corresponds to $D = 1.62$. Having the asymptotic slope we can estimate where the finite particle size affects the box counting. Fractal structures have always a lower and an upper cutoff in practice. We feel that unless they are visible in the data one has little guidance in choosing the region where the data should be fitted to obtain an estimate for D. Averaging over box positions and several other methods suggested [22] do not help in this regard.

Naively box counting the whole cluster at a given stage of growth gives values of D_b in the range 1.59 to 1.63, depending on the range of δ used in the fit. If the box counting is limited to the part of the cluster inside a box of size R_g, we find values of D_b in the range 1.63 to 1.69. Even though this leads to an estimate of D_b lower than D_g, they may be equal within errors.

6. Viscous fingering experiments

The experiments were carried out in a two-dimensional porous model 40 cm in diameter, consisting of a single random layer of 1 mm diameter glass beads glued between two glass plates. The experimental arrangement has been described before [10,11]. The pore space was initially filled with glycerol. Air at a constant pressure was then injected in the center. The structures observed at high capillary numbers were the typical DLA like viscous fingering structures [11] shown in fig. 9. The fractal dimensions given in table 1 were obtained by digitizing photographs of the VF structures at a resolution of 2000×2000 pixels. The branch orders were, however, identified manually from the pictures. In the experiments, the average length of a branch was identified as its average mass.

7. Conclusions

We conclude that DLA and VF patterns are tree-like structures having fixed bifurcation and length ra-

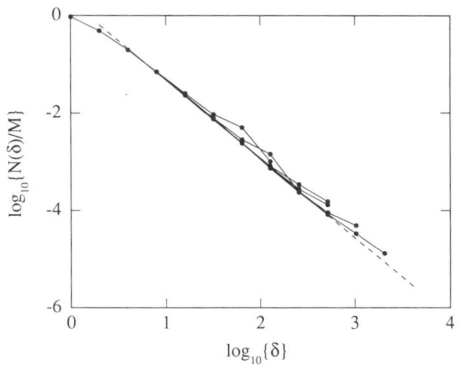

Fig. 8. Box-counting clusters of size 1000 to 250 000 particles. The number of filled boxes $N(\delta)$ of a given size δ needed to cover the entire cluster divided by the total cluster mass M as a function of box size δ. The dashed line has slope 1.62.

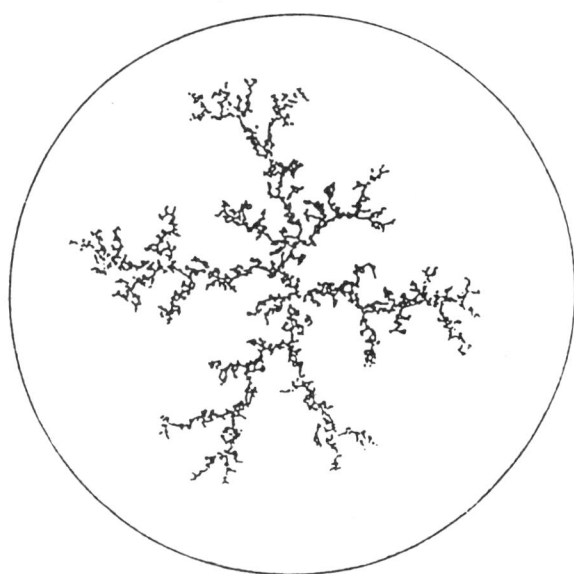

Fig. 9. Fractal viscous fingering with $D_c = 1.64 \pm 0.04$ in a two-dimensional porous medium. Air (black) injected at the center displaces glycerol at a capillary number Ca = 0.15. Figure from ref. [11].

tios independent of branch order. From these two ratios, a fractal dimension may be defined analogous to the self-similarity dimension of deterministic self-similar fractals such as the deterministic tree in fig. 1. This dimension is consistent with values obtained from other methods. We have shown that DLA and VF clusters are self-similar in terms of branch order. The branches of DLA are linear on length scales less than the average branch length, and fractally distributed on larger scales. This feature is characterized by the scaling function $g(\delta/L_n)$, valid for all but the largest and smallest branches. We have also shown that the lengths L_n all scale with R_g.

Comparing the results obtained both from the simulations of the DLA clusters, and the viscous fingering experiments at high capillary numbers, we see a close agreement in both the bifurcation ratio and the length ratio. This lends further support to the hypothesis that these two processes are in the same universality class.

The geometrical crossover of DLA remains for arbitrarily large clusters, and the estimate of the fractal dimension is sensitive to the method used. We believe that D_g equals D_c and the box dimension, if D_b is obtained by analyzing the central region $r < R_g$ of the cluster.

Our definition of a tree structure on the DLA cluster may be used for other fractal aggregates as well. Preliminary studies show that Eden clusters do not obey branch order scaling, whereas ballistic growth and screened growth models [23] appear to exhibit branch order scaling.

Acknowledgements

We thank Benoit Mandelbrot for many exciting and stimulating discussions and for his inspiring lectures in Oslo. We thank A. Aharony and P. Meakin for helpful and stimulating discussions. This work has been supported by VISTA, a research cooperation between the Norwegian Academy of Science and Letters and Den Norske Stats Oljeselskap A.S. (STATOIL).

References

[1] T.A. Witten and L.M. Sander, Phys. Rev. Lett. 47 (1981) 1400.
[2] B.B. Mandelbrot, The Fractal Geometry of Nature (Freeman, San Francisco, 1982).
[3] J. Feder, Fractal (Plenum, New York, 1988).
[4] T. Vicsek, Fractal Growth Phenomena (World Scientific, Singapore, 1989).
[5] L. Niemeyer, L. Pietronero and H.J. Wiesmann, Phys. Rev. Lett. 52 (1984) 1033–1036.
[6] P. Meakin, Phys. Rev. A 27 (1983) 1495–1507.
[7] H.J. Herrmann, Phys. Rep. 136 (1986) 154–227.
[8] P. Meakin, Fractal aggregates and their fractal measures, in: Phase Transitions and Critical Phenomena, C. Domb and J.L. Lebowitz, eds. (Academic Press, New York, 1987) pp. 336–489.
[9] L. Paterson, Phys. Rev. Lett. 52 (1984) 1621–1624.
[10] K.J. Måløy, J. Feder and T. Jøssang, Phys. Rev. Lett. 55(1985) 2688–2691.
[11] K.J. Måløy, F. Boger, J. Feder, T. Jøssang and P. Meakin, Phys. Rev. A 36 (1987) 318–324.
[12] P. Meakin and T. Vicsek, in: Fractals in Physics, L. Pietronero and E. Tosatti, eds. (North-Holland, Amsterdam, 1986), pp. 213–216.
[13] P. Meakin, in: Fractals in Physics, L. Pietronero, ed., Proceedings from the Erice meeting on Fractals, 1988 (Plenum, New York, 1989).
[14] M. Kolb, J. Phys. A 20 (1987) L285.
[15] T. Nagatani, J. Phys. A 20 (1987) 6603.
[16] L. Pietronero, A. Erzan and C. Evertsz, Phys. Rev. Lett. 61 (1988) 861.
[17] E.L. Hinrichsen, K.J. Måløy, J. Feder and T. Jøssang, J. Phys. A 22 (1989) L271–L227.
[18] R.E. Horton, Geol. Soc. Am. Bull. 56 (1945) 275–370.
[19] J. Vannimenus, in: Universalities in Condensed Matters, R. Jullien, L. Peliti, R. Rammal and N. Boccara, eds. (Springer, Berlin, 1988).
[20] J. Vannimenus and X.G. Viennot, J. Stat. Phys. (1989), in press.
[21] A.N. Strahler, Bull. Geol. Soc. America 63 (1952) 275.
[22] F. Argoul, A. Arneodo, G. Grasseau and H.L. Swinney, Phys. Rev. Lett. 61 (1988) 2558.
[23] P. Meakin, Physica A 155 (1989) 37–51.

FRACTAL AND NONFRACTAL SHAPES IN TWO-DIMENSIONAL VESICLES

Michael E. FISHER

Institute for Physical Science and Technology, University of Maryland, College Park, MD 20742, USA

Ongoing collaborative work on the statistical mechanics of two-dimensional vesicles is reviewed with emphasis on the range of fractal and nonfractal shapes exhibited by the bead-and-tether model of a closed membrane which also embodies an osmotic pressure difference, Δp, and a bending rigidity modulus, κ; Monte Carlo simulations and scaling analyses are described. Flaccid vesicles, with $\Delta p = \kappa = 0$, represent closed self-avoiding rings; their mean area, $\langle A \rangle$, scales with the mean square size, $\langle R_G^2 \rangle$; their mean shape is close to elliptical. Inflated, $\Delta p > 0$ vesicles become circular; deflated, $\Delta p < 0$ vesicles collapse to form *branched polymers*. The vesicle shapes appear to vary continuously with the scaling combination $\Delta p N^{2\nu}$, where N is the number of monomers/beads and $\nu = 3/4$. For $\Delta p < 0$ and $\kappa/k_B T$ large relative to N^3 a range of characteristic nonfractal shapes or *cytotypes* appears.

1. Introduction

Benoit Mandelbrot has taught us that many shapes seen in the physical and biological worlds, or arising in graphical analyses of data obtained from real life in its manifold aspects, belong *not* to the realm of classical geometry but rather display aspects of self-similarity or self-affinity, typically of a statistical nature, and so lie in the domain of *fractal geometry*. Indeed, under the ubiquitous macroscopic power laws, which describe the variation of physical observables as external parameters or ranges of observation are changed, frequently lie microscopic configurations of a fractal nature. The shapes formed by these configurations, and their details, may be of intrinsic mathematical, scientific, engineering, or purely artistic interest! At other times, the shapes, or particular aspects of them not always obvious a priori, may directly reflect the underlying physical process driving the phenomena in question. In other circumstances the shapes may act as diagnostics of anomalous behavior. Sometimes knowledge of the shapes may provide a starting point for speculation and the building of new theories.

In this talk I will describe some ongoing numerical investigations of two-dimensional vesicles [1–4]. The original work, performed in collaboration with Leibler and Singh [1], was inspired by the problem of understanding the statistical mechanics of *membranes*, that is ($D=2$)-dimensional surfaces with *intrinsic structure* embedded in ($d=3$)-dimensional space. An example of biophysical importance is provided by *bilipid membranes* [5–7] which form the basis of all biological cell walls. A membrane closed on itself constitutes a vesicle. Bilipid vesicles (and red blood cells) display a range of more-or-less well defined but fluctuating shapes, discocytes, stomatocytes, etc., as the solution pH, the osmotic pressure, the temperature, etc. are changed. One would like to characterize such shapes, understand the physical parameters controlling their properties and elucidate the transitions between various shapes. Since thermal fluctuations are always present, and may even play a crucial role, a statistical mechanical treatment is appropriate.

The statistical mechanics of ($D=2$)-dimensional membranes in three-dimensional space poses a variety of subtle problems, including different universality classes, crumpling transitions, etc. [6,7] [#1], even when one restricts attention only to *open* membranes. Accordingly, it is reasonable to start an inves-

---

[#1] The chapter by S. Leibler in ref. [7] describes many applications.

Essays in honour of Benoit B. Mandelbrot
Fractals in Physics – A. Aharony and J. Feder (editors)

tigation of vesicles by examining closed ($D=1$)-dimensional surfaces, i.e., loops or polygons, in a plane, i.e. in $d=2$ dimensions. Certainly the codimension, $\bar{D}=d-D$, is the same in both cases. Furthermore, the numerical simulation of open surfaces in three dimensions is sufficiently challenging on present day computers that the effective study of closed surfaces seems barely feasible.

2. A statistical model

A reasonable model of a fluctuating membrane must include at least three features: (a) a flexible *connectivity* of the basic units (or monomers); (b) *self-avoidance* between parts of the membrane, however distant along the surface; and (c) a representation of the *rigidity* arising from the structure of the membrane. In $d=2$ dimensions the "pearl-necklace" model used traditionally for polymers, embodies (a) and (b): N hard *beads* (discs) of diameter a are linked into a closed loop by loose bonds or "tethers" [8] of maximum center-to-center length $l_0 > a$. No two beads are allowed to overlap. Self-crossing of the chain is prevented by taking $l_0 < 2a$ and eschewing large Monte Carlo steps in simulations; in the work described here $l_0/a = 9/5$ was adopted [1-3].

If r_i is the position of the center of the ith bead and θ_i is the angle between the vectors $s_i = r_i - r_{i-1}$ and s_{i+1}, a total bending energy

$$E_b = \frac{\kappa}{a} \sum_{i=1}^{N} (1 - \cos \theta_i) \qquad (1)$$

is assigned to the configuration $\{r_i\}$; κ is the *rigidity modulus*. This formula is a precise analog of the continuum elastic energy expression $\frac{1}{2}\kappa \int H^2 \, ds$, where H is the curvature and ds an element of surface/perimeter [6,7].

A closed membrane divides space into an interior and exterior region which, in a real vesicle, will normally be filled with fluids of distinct, but controllable composition. Accordingly a model for vesicles must allow (d) for a variable osmotic *pressure difference*

$$\Delta p \equiv p_{\text{int}} - p_{\text{ext}}, \qquad (2)$$

which may be positive *or* negative. Indeed, it transpires that the interplay of a *negative* Δp with a fairly large rigidity, κ, is responsible for the characteristic cell-like shapes or *cytotypes* discovered in the model [1]. In a stress ensemble, which will be assumed here, Δp is fixed and the area (or two-dimensional volume) of a vesicle, A, fluctuates subject to a Boltzmann weighting factor [#2] $\exp(\Delta p A / k_B T)$. (The area may be defined as that of the polygon specified by the $\{r_i\}$.)

This model has been studied [1-4] on the Cornell supercomputer by simulations using simple, single-bead, off-lattice dynamics and a standard Metropolis algorithm. The number of beads, N, has not exceeded 100 since it proves impossible, in available times, to achieve an adequate sampling of the full range of equilibrium configurations for larger N. For the simplest properties $(1-50) \times 10^5$ Monte Carlo steps per bead were satisfactory but some interesting properties in certain regimes appeared to equilibrate more slowly.

It is worth remarking that the model has an obvious lattice analog, say on a triangular lattice. The interesting "universal" shape and size properties should be the same for such models. As we will see, a surprising range of behavior is encompassed in these primitive vesicle models.

3. Flaccid vesicles

Consider first the case $\Delta p = 0$ and $\kappa = 0$: we then have a *flaccid vesicle*, which is equivalent to a closed self-avoiding walk or polygon. Typical samples generated by the simulations [1] are shown in part (i) of fig. 1. The "size" of such a vesicle may be measured by the mean square radius of gyration, $\langle R_G^2 \rangle$, or by the mean area, $\langle A \rangle$. Open self-avoiding chains of polymers are characterized by

$$\langle R_G^2 \rangle \approx R_0^2 N^{2\nu} \quad \text{as } N \to \infty, \qquad (3)$$

with an exponent ν now believed to be exactly 3/4. One must expect the same to hold for a polymeric

---

[#2] Note that ref. [1] incorrectly specified $-\Delta p$ in place of $+\Delta p$.

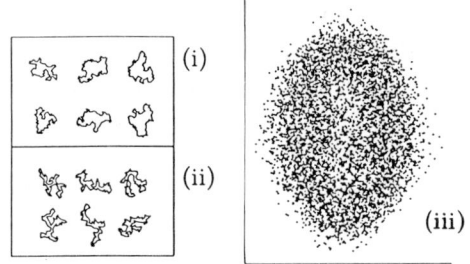

Fig. 1. Sample shapes for vesicles of $N=60$ beads/monomers with zero rigidity ($\kappa=0$). In (i), for $\Delta p=0$, and (ii) for $\bar{p}=\Delta p\, a^2/k_B T = -1.25$, the centers of the beads have been joined by straight lines. (After ref. [1].) In (iii) about 50 statistically independent vesicle configurations for $\Delta p = 0$ have been superimposed with common center of mass and aligned principal axes of inertia: only the bead centers are plotted [2].

ring. This is confirmed by the simulations which yield [1] $R_0 \simeq 0.34 a$ and

$$\nu = 0.755 \pm 0.018 \, . \tag{4}$$

(See also, e.g., Privman and Rudnick [9], and Bishop and Saltiel [10].) Since $\nu < 1$ the vesicle is obviously fractal: see fig. 1. Does this fractal polygon fill out its interior so that $\langle A \rangle / \langle R_G^2 \rangle \to 0$? Or does it have an area proportional to the square of its linear dimension so that if

$$\langle A \rangle \approx A_0 N^{2\nu_A} \, , \tag{5}$$

one has $\nu_A = \nu$? The latter surmise seems to be correct since we estimate $\nu_A/\nu = 1.007 \pm 0.013$ [1]. Furthermore, many years ago Hiley and Sykes [11] studied the areas of polygons on the square and triangular lattices by exact enumeration and concluded $2\nu_a = 1.50 \pm 0.04$. Evidently lattice and continuum models do, indeed, display the same fractal behavior!

How large is the relative area of a flaccid vesicle? If we accept $\nu = \nu_A$ one estimates

$$\langle A \rangle / \langle R_G^2 \rangle \approx A_0 / R_0^2 = 2.5 \pm 0.3 \, . \tag{6}$$

For $N \to \infty$ this value should be universal, independent of the details of the model. It clearly reflects on the *shape* of a flaccid vesicle since for a circle one has $\langle A \rangle / \langle R_G^2 \rangle = \pi$. In fact flaccid vesicles are, on average, *an*isotropic in shape. Perhaps the simplest measure of this anisotropy is

$$S = \langle R_{G1}^2 \rangle / \langle R_{G2}^2 \rangle \, , \tag{7}$$

where $\langle R_{G1}^2 \rangle$ and $\langle R_{G2}^2 \rangle$ denote the smallest and largest eigenvalues of the moment of inertia tensor. (Various other measures of anisotropy or "asphericity" have featured in the recent literature: e.g. refs. [10, 12, 13].) The simulation data yield the universal value [4]

$$S \approx S_0 \equiv S(\Delta p = 0) \simeq 0.41_2 \, , \tag{8}$$

revealing a strong degree of anisotropy. Indeed, if one studies the mean shapes of the vesicles by superposing appropriately oriented plots of many flaccid vesicles, as in part (iii) of fig. 1, one sees a roughly elliptical "cloud" which is less dense near the center. The semiaxes of this mean cloud have a ratio $\bar{a}/\bar{b} \simeq 0.64$; since $(0.64)^2 \simeq 0.41$ this is in surprising concordance with (8)!

4. Inflated and deflated vesicles

The introduction of a nonzero pressure difference, Δp, obviously changes the size of a vesicle. Guided by the theory of critical phenomena and the renormalization group, we anticipate that (3) and (5) should be replaced by the scaling forms [1]

$$\langle R_G^2 \rangle \approx R_0^2 N^{2\nu} X(x), \qquad \langle A \rangle \approx A_0 N^{2\nu} Y(x) \, , \tag{9}$$

with scaled pressure

$$x = D \bar{p} N^{\varphi \nu}, \quad \text{where } \bar{p} = \Delta p \, a^2 / k_B T \, , \tag{10}$$

while D is a metrical factor (discussed below).

Now the *crossover exponent*, φ, may be estimated numerically by examining derivatives with respect to $\bar{p}$. Thus one has for example

$$\left. \frac{\partial \langle A \rangle}{\partial \bar{p}} \right|_0 \propto \langle \Delta A^2 \rangle_0 \equiv \langle A^2 \rangle_0 - \langle A \rangle_0^2 \sim N^{2\nu + \varphi \nu} \, , \tag{11}$$

where the subscript 0 denotes $\bar{p} = 0$. If all large lengths in flaccid vesicles ($\bar{p}=0$) scale with exponent ν this implies $\varphi = 2$. Fitting to data for $\langle \Delta A^2 \rangle_0$, etc. yields

[1,2] $\varphi = 2.13 \pm 0.17$; we regard this as supporting the conclusion $\varphi = 2$ (although possibly suggesting fairly large finite-N corrections to the asymptotic behavior).

Plots of $\langle R_G^2 \rangle / N^{2\nu}$ and $\langle A \rangle / N^{2\nu}$ versus $\bar{p}N^{\varphi\nu}$ for a range of $\bar{p}$ and N values now serve to crosscheck the scaling forms (9) and to evaluate the scaling functions: see fig. 1 in ref. [1]. One can expand these as

$$X(x) = \sum_{j=0} X_j x^j, \qquad Y(x) = \sum_{k=0} Y_k x^k, \qquad (12)$$

with, by (3), (5) and (9), $X_0 = Y_0 = 1$. One may also impose a further normalization, say, $Y_1 = 1$; then the nonuniversal "metrical factor" D, in (10), is determined and all the remaining expansion coefficients $X_1, X_2, Y_2, \ldots$ should be universal. A tentative analysis [4] indicates $X_1 \simeq 0.6_7$.

Highly *inflated* vesicles with $\Delta p > 0$ and $x = \bar{p}N^{2\nu} \gg 1$ should approach a near-circular but somewhat fuzzy shape with $\langle R_G^2 \rangle, \langle A \rangle \sim N^2$ [1]. This implies $X(x), Y(x) \sim x^\omega$ as $x \to \infty$, with $\omega = (1-\nu)/\nu = 1/3$. However, owing to computer limitations it has not yet proved possible to check this surmise.

On the other hand, for *deflated* vesicles with $\Delta p < 0$ one observes, for $x \ll -1$, the power laws

$$X(x) \approx X_{-}/|x|^\sigma \quad \text{and} \quad Y(x) \approx Y_{-}/|x|^\tau, \qquad (13)$$

with $\sigma = 0.13 \pm 0.05$ and $\tau = 0.25 \pm 0.04$. What does this mean? Notice, first, that the smallest area a vesicle can display is proportional to N; it is then in a "fully collapsed" state and, indeed, the corresponding simulations reveal vesicles looking like seaweed with assorted narrow branches: see part (ii) of fig. 1. Now (13) implies $\langle A \rangle \sim N^{2\nu_A^-}$ with $2\nu_A^- = (2-\varphi\tau)\nu = 1.10 \pm 0.10$, which, allowing for the uncertainties, is consistent with the collapsed, branched picture. But what about the mean square radius of gyration? The form (13) implies $\langle R_G^2 \rangle \sim N^{2\nu^-}$ with $\nu^- = \nu(1-\sigma\varphi) = 0.65 \pm 0.04$. But our seaweedy vesicles resemble, in form, freely *branched polymers* or trees which, as regards their universal properties, behave in the same fashion as *lattice animals*, i.e. general clusters made of N connected sites or bonds: see e.g. refs. [12–15]. For these fractal objects, precise numerical studies [16] indicate $\nu^- \simeq 0.640_8$, which

agrees perfectly with the vesicle data. Thus we conclude that deflated vesicles reduce to *branched polymers* (with, essentially, each arm formed of a doubled chain).

Now branched polymers and, hence, deflated vesicles certainly have a different shape than flaccid vesicles (or self-avoiding polygons): see fig. 1. Indeed, a tentative estimate [4] of the ratio S, defined in (7), for deflated vesicles gives $S(\Delta p = -\infty) \simeq 0.23$ in place of (8). This value is consistent with previous studies of lattice animals per se, bearing in mind that the mean ratio, $\langle R_{G1}^2/R_{G2}^2 \rangle$, was studied rather than S [12,17]. Conversely, for fully inflated, circular vesicles one must have $S(\Delta p = +\infty) = 1$. But what, we may ask, are the *shapes*, as against the sizes, of vesicles for *intermediate* pressures, positive or negative?

Now it is not hard to see that if $N \to \infty$ at *fixed* Δp, the vesicles inflate to form circles whenever $\Delta p > 0$ or deflate to branched polymers for any negative Δp. This suggests that S might behave discontinuously with Δp as indicated by the dotted lines in fig. 2, which depicts schematically a plot versus the scaled variable $y = x/(c+|x|)$ (for a convenient value of c: one has $y = \pm \frac{1}{2}$ for $x = \pm c$ and $-1 \leq y \leq 1$). Conversely, one might speculate that the shapes for all intermediate values of x (or y) are the same as for flaccid vesicles, corresponding to the dashed lines in fig. 2. Finally, it is perhaps most plausible to guess that the

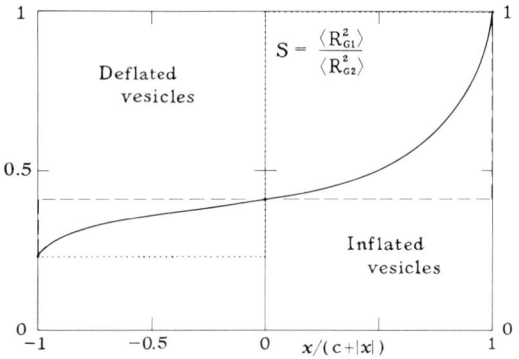

Fig. 2. Sketch illustrating the variation of vesicle shapes, as measured by the anisotropy parameter $S(N, \Delta p) = \langle R_{G1}^2 \rangle / \langle R_{G2}^2 \rangle$, with the scaled pressure variable $x \propto \Delta p\, N^{2\nu}$, where N is the number of beads. (The plot corresponds roughly to $c = 3$.)

shapes *vary in a continuous fashion* according to the scaling form

$$S(N, \Delta p) \approx S_0 W(x) = S_0 [1 + W_1 x + \ldots] . \quad (14)$$

A preliminary study [4] of the data for a range of Δp and $N \lesssim 90$ do, in fact, support this scaling ansatz and yield a plot like the solid curve sketched in fig. 1. The expansion coefficient W_1, which should be universal, can be estimated, following (11), by studying the scaled fluctuation quantity

$$N^{-2\nu}[\langle R_{G1}^2 A \rangle_0 / \langle R_{G1}^2 \rangle_0 - \langle R_{G2}^2 A \rangle_0 / \langle R_{G2}^2 \rangle_0] , \quad (15)$$

in the flaccid, $\Delta p = 0$, limit. Initial data indicate $W_1 \simeq 0.8$. The finiteness of this value represents, of course, supporting evidence for the scaling expression (14).

Note that the behavior of the scaling function $W(x)$ as $x \to -\infty$, and hence the shape of the plot near $y = -1$ in fig. 1, is expected to be determined by the leading correction exponent for lattice animals. This is $\theta \simeq 0.87$ (see e.g. Privman [18]) which suggests $W(x) - W(-\infty) \sim x^{-\psi}$ with $\psi = \theta/2\nu \simeq 0.58$; the data provide some evidence to support this. Similar considerations should apply for $x \to +\infty$ but have not been pursued.

5. The effects of rigidity

Consider now the effects of nonzero rigidity, κ, in the bending energy (1), first for the case $\Delta p = 0$. For a sufficiently long open chain the only *effect* of κ is to change the effective tether size or the Kuhn length, say $\tilde{a}(\kappa)$, which enters through $\langle R_G^2 \rangle^{1/2} \approx \tilde{a}(\kappa) N^\nu$. One anticipates

$$\tilde{a}(\kappa) = a_0 + a_1 \bar{\kappa} + \ldots, \quad \bar{\kappa} = \kappa / a k_B T , \quad (16)$$

with, in our simulations, $a_0 \simeq 2.5a$ [2]. The same behavior should apply to rings, and is, indeed, observed [2].

When, however, κ becomes large one enters a *stiff regime* where *new* scaling laws apply. The nature of these laws can be learned by analyzing stiff chains with self-avoidance neglected; this can be done exactly as

an example of the rod-to-coil transition [2]. The appropriate new scaling combination is found to be

$$y = L/l_\kappa, \quad L = Na, \quad l_\kappa = \kappa / k_B T . \quad (17)$$

Here l_κ is a *rigidity length* while L may be regarded as the *contour length* of the chain or vesicle. (One might prefer to take $L = N l_0$ but that changes no orders of magnitude.) It is plausible that this same scaling should apply when self-avoidance is incorporated. Thus in the stiff regime for vesicles we anticipate, in contrast to (9), the scaling laws

$$\langle R_G^2 \rangle \approx \tilde{R}_0^2 N^2 U(y) \quad \text{and} \quad \langle A \rangle \approx \tilde{A}_0 N^2 V(y) , \quad (18)$$

with $U(0) = V(0) = 1$.

In fact the simulations confirm (18) quite well for y less than N. In this regime the vesicle shape is basically nonfractal and, in fact, near-circular for small y, with a weakly fluctuating perimeter: see the plot labelled $\bar{p} = 0$ in fig. 3. As y increases, however, the perimeter becomes increasingly convoluted and, eventually, fractal in character. Correspondingly, the scaling functions for $y \gtrsim 10$ approach the power laws $U(y), V(y) \sim 1/y^{2(1-\nu)}$; these describe the ultimate crossover to the original, flaccid, small-κ behavior with $\langle R_G^2 \rangle, \langle A \rangle \sim N^{2\nu}$. For $y \gtrsim N$ the stiff scaling laws lose validity since $\tilde{a}(\kappa)$ then becomes of order a_0.

6. The deflated stiff regime: cytotypes

As just seen, stiffening the membrane by increasing κ at fixed N results, even for $\Delta p = 0$, in vesicles

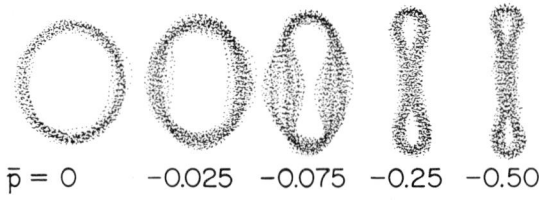

Fig. 3. Vesicle cytotypes for $N = 60$, $\bar{\kappa} \equiv \kappa / a k_B T \equiv l_\kappa / a = 50$, and increasing pressure differences $\Delta p \propto \bar{p}$ [1]. (The construction of the plots is the same as in part (iii) of fig. 1.)

which become increasingly circular. Inflating the vesicles by imposing a positive Δp can only strengthen this tendency. On the other hand, if one *deflates* a vesicle in the stiff regime strikingly new phenomena arise. Fig. 3 shows the effect of increasing $|\Delta p|$ at fixed $N=60$ and reduced rigidity $\bar{\kappa} \equiv l_\kappa/a = 50$. The initial, near-circular, fuzzy walled vesicle, or *ellipsocyte*, elongates somewhat and the sides start fluctuating increasingly strongly. In real vesicle systems these fluctuations correspond to low frequency observed "flickering". For $\bar{p} \simeq 0.06$–0.09 the fluctuations in the model become anomalously large, perhaps even suggesting a fractal character, as evident in fig. 3. This behavior appears to correspond to a shape *transition* since a completely new type of shape – a dumbbell or bi-lobocyte – appears when $\bar{p} \gtrsim 0.2$: see fig. 3. This new shape is strongly reminiscent of the cross-sections of discocytes seen in red blood cells and artificial lipid vesicles.

When the transition region around $\bar{p} = 0.08$ is examined by following the Monte Carlo time sequences, one finds that the flickering is strongly *nonlinear*: for a period of time the shape resembles a fairly broad, but well-defined, bi-lobocyte; then, fairly rapidly, it expands to a more elliptical shape; afterwards it contracts again, and so forth. In a similar regime observed [1] for $\bar{\kappa} = 10$ and $\bar{p} = -0.075$ the bi-lobocyte rather abruptly curves one way or the other to form a shallow "cup", or stomatocyte, which breaks reflection invariance; a while later the vesicle switches its shape again to the straight form or to the oppositely oriented stomatocyte; and so on! Free energy barriers of a few $k_B T$ seem to separate the different configurations. It might be possible to observe such nonlinear flickering in real systems but, as regards actual time-dependence, the artificiality of the Monte Carlo dynamics used for the model must be kept strongly in mind. It is also possible that simulations at larger N would resolve this region into regimes with distinct, stable shapes.

To identify the domains in parameter space where the distinct cytotypes arise, let us idealize a bi-lobocyte as consisting of two circular lobes of radius ρ joined by a double-sided bar of length some fraction of $L = Na$ or less. If $A_\rho = \pi \rho^2$ is the area of a lobe and $N_\rho \approx 2\pi\rho/a$ is the number of beads/monomers in a lobe, the energy for $\Delta p < 0$ can be estimated from (1) and (2) as

$$E_{\text{lobe}} \approx |\Delta p| A_\rho + \tfrac{1}{2} N_\rho \kappa a/\rho^2 . \tag{19}$$

Minimization on ρ yields the characteristic lobe radius [1]

$$\rho^* = (\kappa/2|\Delta p|)^{1/3} \equiv (\tfrac{1}{2} l_\kappa l_p^2)^{1/3} , \tag{20}$$

where we have introduced a *pressure* or *baric length*, $l_p = (k_B T/|\Delta p|)^{1/2}$. The total circumference of the lobes, $4\pi\rho$, cannot exceed the contour length. Hence, as a criterion for the presence of cytotypes we find $y = L/l_\kappa \lesssim 1$ and

$$z \equiv l_\kappa l_p^2/L^3 \equiv \kappa/N^3 |\Delta p| a^3 \lesssim 1 . \tag{21}$$

Inasfar as this analysis captures the essence of the phenomenon, all the cytotypes would be essentially nonfractal with the relative fluctuations of the vesicle walls decreasing as N increases. Furthermore, similar shapes should be realized for the same values of z as κ, Δp and N are varied, at least when all these parameters are large. This surmise is currently being tested [4]. The detailed characterization of the cytotypes and the possibility of new forms also awaits further investigation, as does the degree to which the transformations of one shape into another might be regarded as true phase transitions in the limit $N \to \infty$. It is clear, however, that in some real sense the existence of definite, nonfractal cytotypes represents only a *finite-size phenomenon* since when N becomes large at *fixed* κ and $\Delta p < 0$ the vesicles will always collapse to the form of highly fractal, branched polymers: fractality is the rule!

Acknowledgements

The material reported here has relied on collaborative work with Dr. Stanislas Leibler, Dr. Rajiv R.P. Singh and Carlos J. Camacho. The interest and comments of Dr. M.F. Sykes and of Professors J.L. Cardy, V. Privman and S.G. Whittington have been much appreciated. The support of the National Science

Foundation through the Condensed Matter Theory Program [#3] is gratefully acknowledged. The computations reported on here were performed on the National Supercomputer Facility at Cornell University which is supported by the National Science Foundation, New York State, and the IBM Corporation.

[#3] Under Grant No. DMR 87-01223/96299.

References

[1] S. Leibler, R.P.P. Singh and M.E. Fisher, Phys. Rev. Lett. 59 (1987) 1989.
[2] R.R.P. Singh, M.E. Fisher and S. Leibler, to be published.
[3] M.E. Fisher, Nucl. Phys. B (Proc. Suppl.) 5A (1988) 165; J. Math. Chem. (1989), in press.
[4] C.J. Camacho and M.E. Fisher, to be published.
[5] V. Degiorgio and M. Corti, eds., Physics of Amphiphiles: Micelles, Vesicles and Microemulsions (North-Holland, Amsterdam, 1985).
[6] J. Meunier, D. Langevin and N. Boccara, eds., Physics of Amphiphilic Layers (Springer, Berlin, 1987).
[7] D.R. Nelson, T. Piran and S. Weinberg, eds., Statistical Mechanics of Membranes and Surfaces (World Scientific, Singapore, 1989).
[8] Y. Kantor, M. Kardar and D.R. Nelson, Phys. Rev. Lett. 57 (1986) 791.
[9] V. Privman and J. Rudnick, J. Phys. A 18 (1985) L789.
[10] M. Bishop and C.J. Saltiel, J. Chem. Phys. 85 (1986) 6728; 88 (1988) 3976.
[11] B.J. Hiley and M.F. Sykes, J. Chem. Phys. 34 (1961) 1531.
[12] F. Family, T. Vicsek and P. Meakin, Phys. Rev. Lett. 55 (1985) 641.
[13] J. Rudnick and G. Gaspari, J. Phys. A 19 (1986) L191.
[14] T.C. Lubensky and J. Isaacson, Phys. Rev. A 20 (1979) 2130.
[15] G. Parisi and N. Sourlas, Phys. Rev. Lett. 46 (1981) 871.
[16] B. Derrida and D. Stauffer, J. Phys. (Paris) 46 (1985) 1623.
[17] J.P. Straley and M.J. Stephen, J. Phys. A 20 (1987) 6501.
[18] V. Privman, Physica A 123 (1984) 428.

THE BUILDING BLOCKS OF RANDOM WALKS

Yuval GEFEN

Department of Nuclear Physics, Weizmann Institute of Science, Rehovot 76100, Israel

Isaac GOLDHIRSCH

Department of Fluid Mechanics & Heat Transfer, Faculty of Engineering, Tel Aviv University, Ramat Aviv, Tel Aviv 69978, Israel

It is shown that certain generating functions, related to random walk processes on networks, constitute a basic set; other generating functions are merely algebraic expressions in the basic ones. Moreover, the basic generating functions corresponding to a network can be expressed in terms of elementary generating functions for the basic constituents of the network, such as "single lines". The basic generating functions can be directly related to physical properties of such a network, e.g., conductance. Using such relations, a generalized Einstein relation for inhomogeneous systems is derived. Results for some elementary structures such as loops and systems with dangling branches are presented. Some analytic properties are discussed.

1. Introduction

Long ago it has been recognized by Benoit Mandelbrot that the importance of fractals goes beyond characterization of static irregular geometries, and that a large variety of dynamical (often random) processes could be assigned fractal dimensionalities [1]. It later became evident that many kinds of irregular dynamical processes (e.g. random walks) occur on objects having fractal geometries [2].

Ever since its inception as a means of modelling of physical phenomena, at the beginning of the century [3], the concept of random walk has been an unfailing source of applications, methods and consequently of physical understanding [4–6]. The gambler's ruin and the drunkman's walk have been generalized to model transport phenomena in solids and in liquids, energy transitions in molecules, and conformation of polymers to mention a few applications. Theory has gone a long way from analyzing simple random walks in simple environments to transport in random media and to quantum effects. Present research in this field seems to be still thriving, in spite of its apparent coming of age. The pioneering simple analytic methods devised by Polya have been generalized; additional methods such as effective medium theory and the renormalization group have been involved, and numerical methods, exploiting the permanently increasing power of computing, have been an important source of information, new questions and new theories. Qualitative methods and applications of ideas of rather general applicability, such as the Einstein relation, have been an additional source of progress.

When dealing with irregular networks, several questions are in order: (a) What part of the dynamics is due to the nature of the elementary (say, hopping) process, and which part can be attributed to the geometry of the system? (b) Can one find "combination rules", such as the Kirchhoff laws for electrical networks, that would quantitatively relate the property of a network to those of its constituents? In particular, can one devise combination rules for "black boxes" of which a network is made, and what is the information that has to be known about the black boxes? (c) Are any general theorems, such as the Einstein relation, valid for inhomogeneous networks?

Surely, many more questions can be asked. Here we shall try to concentrate on the abovementioned questions.

The structure of this paper is as follows: section 2 presents the fundamentals of the method, section 3 is devoted to its applications to simple networks, section 4 provides a generalization of the Einstein relation, and section 5 provides a brief discussion.

2. Basic generating functions

In the following it is assumed, for simplicity, that a network consists of N points: 1, 2, ..., N. The random walk is discrete in time, all time units being equal. Let $p_{i,j}$ be the probability to move in one time unit from point i to point j. The probability to stay at point i for n time steps is thus p_{ii}^n. It is convenient to define the following elementary probabilities [7]: (a) $\hat{T}_{i,j}(n)$ is the probability to leave point i on the first step and reach point j, after n time units, for the first time, without having returned to i in the process. (b) $\hat{Q}_{ij}(n)$ is the probability of leaving point i on the first step and returning to it, for the first time, after n time units, without having visited point j on the way. The quantities $\hat{T}$, $\hat{Q}$ and the staying probability p_{ii}^n are the basic building blocks of any walk. A quantity of interest, which is a function of the building blocks, is the first passage time probability, $\hat{G}_{ij}(n)$. It is defined as the probability of starting at point i and reaching point j, for the first time, following n steps. It is convenient to associate a generating function $P(z)$, to each probability distribution $\hat{P}(n)$, by defining

$$P(z) = \sum_{n=0}^{\infty} z^n \hat{P}(n),$$

z being a (possibly) complex variable. Generating functions can be shown to add and multiply like their corresponding probabilities [7], except that one does not have to keep track of the number of steps. For example, if $\hat{P}(n) = \sum_{n_1=0}^{n} \hat{P}_1(N-n_1) \hat{P}_2(N_1)$, then $P(z) = P_1(z) P_2(z)$. It follows that the generating function for moving from point i to point j in one step is merely $zp_{i,j}$. The generating functions for staying at point i, denoted as $X(z)$, equals

$$X_i(z) = \frac{1}{1 - zp_{i,i}}. \tag{1}$$

It is also easy to see that

$$G_{i,j}(z) = \frac{X_i(z)}{1 - X_i(z) + Q_{ij}(z)} T_{i,j}(z). \tag{2}$$

As a first application of these definitions, we demonstrate a method for combining black boxes [8]. Consider the case depicted in fig. 1, of two black boxes having two exits each. Let $\alpha = 1, 2$ be a superscript differentiating between the boxes. The elementary generating functions are: X_A, X_B, X_C, T_{AB}^1, T_{BA}^1, T_{BC}^2, Q_{BA}^1 and Q_{BC}^2. Notice that Q_{BA}^1 assumes that all "trajectories" are contained in box 1. It now follows that

$$T_{AC} = T_{AB}^1 (X_B + X_B(Q_{BA}^1 + Q_{BC}^2) X_B + X_B(Q_{BA}^1 + Q_{BC}^2) X_B (Q_{BA}^1 + Q_{BC}^2) X_B + ...) T_{BC}^2. \tag{3}$$

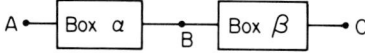

Fig. 1. Two coupled black boxes.

Eq. (3) means as follows: in order to reach C, the walker leaves A and reaches B for the first time (with probability T^1_{AB}), stays at B for some time (with probability X_B), then wanders into box 1 or box 2 and returns to B for the first time (with probability $Q^1_{BA} + Q^2_{BC}$), and repeats this process until he leaves B for good and reaches C (with probability T^2_{BC}). Summing eq. (3) we find

$$T_{AC} = \frac{T^1_{AB} X_B T^2_{BC}}{1 - X_B(Q^1_{BA} + Q^2_{BC})}. \tag{4}$$

Similarly

$$G_{AC} = \frac{X_A T_{AC}}{1 - X_A \left(Q_{AB} + \frac{T^1_{AB} X_B T^1_{BA}}{1 - X_B(Q^1_{BA} + Q^2_{BC})} \right)}. \tag{5}$$

Next, we use the same type of reasoning to compute generating functions pertaining to a one-dimensional finite lattice of points, $0 \leq i \leq N$. Assume that the nonzero hopping probabilities are: $p_{i,i+1} = p$ and $p_{i,i-1} = q$; it then follows that

$$Q_{0,N}(z) = \frac{pqz^2 X_{1-p-q}}{1 - X_{1-p-q} Q_{1,N}(z)}, \tag{6}$$

where $X_{1-p-q} \equiv 1/[1 - z(1-p-q)]$ is the generating function for staying at any point $0 < i < N$. Obviously $Q_{j,k}(z)$ depends only on $k - j$ (where $k > j$ is assumed). Since the number of steps leading from j back to j is always even, it also follows that $Q_{j,k} = Q_{k,j}$ (because $Q_{j,k}$ is obtained from $Q_{k,j}$ by changing p into q). Thus we can denote $Q_{j,k} = Q_{k,j} = \tilde{Q}_{|j-k|}$. It follows that (6) is a recursion relation, in r, for $\tilde{Q}_r$. Since $\tilde{Q}_1 = 0$ (in a lattice of unit length one cannot leave a point without reaching the other end point), the solution for $\tilde{Q}_N$ is

$$Q_{0,N} = \tilde{Q}_N = pqz^2 X_{1-p-q} \frac{\lambda_1^{N-1} - \lambda_2^{N-1}}{\lambda_1^N - \lambda_2^N}, \tag{7}$$

where $\lambda_{1,2} = (1 \pm \sqrt{1-4a})/2$ and $a = X^2_{1-p-q} pqz^2$. Next, by considering separately the set of all trajectories leading from site 0 back to site 0, which reach site $N-1$, and those which do not, we obtain

$$\tilde{Q}_N = \tilde{Q}_{N-1} + T_{0,N-1} T_{N-1,0} \frac{X_{1-p-q}}{1 - X_{1-p-q} \tilde{Q}_{N-1}}. \tag{8}$$

Each path leading from site 0 to site N corresponds to a path going from N to 0. The ratio of their respective probabilities is $(p/q)^N$, for all paths. Consequently $T_{0,N-1}(z) = (p/q)^N T_{N-1,0}(z)$. Substituting this relation into eq. (8) and using eq. (7), one obtains a closed expression for $T_{0,N}$. Also, the generating function for the first passage to N (starting at 0) is

$$G_{0,N}(z) = \frac{X_{1-p} T_{0,N}}{1 - X_{1,p} \tilde{Q}_N}, \tag{9}$$

where $X_{1-p} = 1/[1 - z(1-p)]$.

These expressions can be used to compute various moments of arrival time, as well as their full distribution. The mean first passage time, τ, is

$$\tau = \left. \frac{d \ln G}{dz} \right|_{z=1} = \frac{N}{p-q} - \frac{q(p^N - q^N)}{p^N(p-q)^2}. \tag{10}$$

As $p \to q$ (no bias), $\tau \to N(N+1)/2q$. As $N \to \infty$ for $p > q$, one obtains $\tau \propto N$, i.e. a net drift velocity, with an effective "resistance" of N/ρ_0, ρ_0 being the density of "carriers".

3. Application to simple networks

Once the basic generating functions for simple segments are known, it is a matter of straightforward, though sometimes tedious, algebra to compute properties related to complex networks. In this section we present results pertaining to some simple composite networks.

Consider the case of a one-dimensional lattice with a single dangling branch, composed of n_3 points, attached to it (cf. fig. 2), so that there are n_1 points to the left of the intersection and n_2 point to the right of it. Assuming a homogeneous nearest neighbour hopping probability of $\frac{1}{4}$ everywhere, we obtain for the mean first passage time (MFT) between the endpoints

$$\tau = 2(n_1 + n_2)^2 + 2(n_1 + n_2) + 4n_2 n_3. \tag{11}$$

Note the asymmetry between n_1 and n_2 in eq. (11). It follows from the fact that when one moves from endpoint A to endpoint B (cf. fig. 1), one encounters the dangling branch at a different "time" than when moving from B to A. Another fact worth noticing is the linearity in n_3, indicating a uniform probability distribution inside the dangling branch. When a case of many dangling branches (cf. fig. 3) with uniform nearest neighbour hopping probability is considered, one obtains, using the same technique, for the MFT between endpoints

$$\tau = 2N^2 n(n+m) + 2Nn(m+1), \tag{12}$$

where m is the length of a dangling branch, n is the distance between intersections, and N is the number of branches. Eq. (12) is invariant under the replacement $N \to N/k$, $m \to km$, $n \to kn$ for large n, which shows that by diluting the density of dangling branches, while keeping their total length fixed, the MFT remains unchanged. As long as m is not too large, the effective diffusion constant $\tilde{D} = N^2 n^2 / 2\tau$ is asymptotically fixed.

Another elementary structure of interest is a loop (fig. 4). The expression for the corresponding MFT is rather

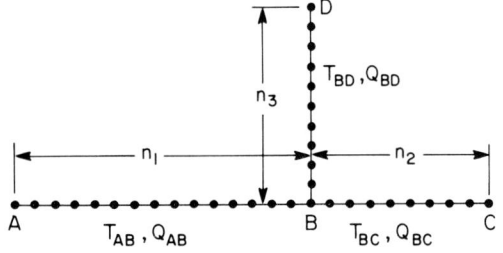

Fig. 2. A segment with a single dangling bond.

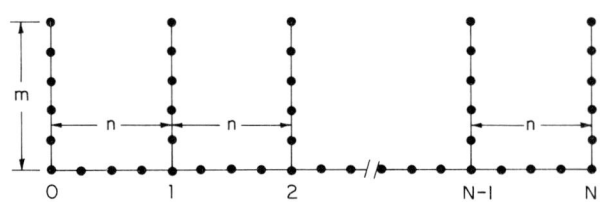

Fig. 3. A segment with many dangling bonds.

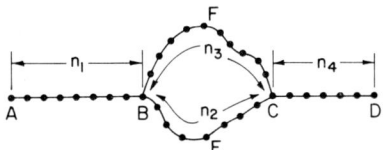

Fig. 4. Loop configuration.

lengthy and not very transparent [7]. Here we quote the result for the MFT when $n_1=n_2=n_4=n$ (see fig. 4): $\tau=20n^2$. Also, when $n_1=n_2=n_4=n$ and $n_3=n+n\delta$, $|\delta|\ll 1$, one obtains: $\tau=20n^2+7n^2\delta+\mathcal{O}(\delta^2)$. When $n_1=n_4=0$ (two segments in parallel) one obtains: $\tau=2n_2n_3$.

These results become rather involved in the presence of bias [8,9]. The gist of the result is that when moving in the direction of the bias, the MFT is proportional to the resistance of an equivalent system of resistors, each segment being equivalent to a resistance proportional to its length and inversely proportional to the "bias" (i.e., p_1-p_2, p_1 being the larger probability of hopping and p_2 the smaller one). When the motion is opposed to the bias the MFT is proportional to $(p_1/p_2)^N$, for a segment, N being its length. The MFT for a complex network in the presence of bias depends nontrivially on the structure of the network. In the case of a single dangling branch with a bias at 45° to the intersection [9] (and thus tending to move "walkers" to the end of the dangling branch), the MFT is given by (cf. fig. 2)

$$\tau \sim \frac{p_1 r^{-n_3}}{(p_1-p_2)^2}(1-r^{n_2}) + \frac{n_1+n_2}{p_1-p_2} + \mathcal{O}(nr^n). \tag{13}$$

If one attaches (noninteracting) particle reservoirs at the ends (A, C in fig. 2) of the segment, with equal chemical potentials, it is easy to see that the total, steady current through the system is given by

$$I_{AB} = \rho[T_{AB}(z=1) - T_{BA}(z=1)]/\Delta t, \tag{14}$$

where Δt is the size of a time step. Both T_{BA} and T_{AB} are independent of the existence of side branches (as in the case of resistors). When a field E is applied along the system of length N (as the source of bias), the potential drop being $V/N=E$, we find the net drift velocity to be

$$\vartheta_D = \frac{1}{n\,\Delta t}\,\mathrm{tgh}\left(\frac{V}{nk_BT}\right), \tag{15}$$

where the hopping rates $p_1 \propto \exp(V/nk_BT)$ and $p_2 \propto \exp(-V/nk_BT)$ in the direction of the bias and against it, respectively, are assumed.

4. Generalization of the Einstein relation

Consider a network of one-dimensional segments of variable lengths and an equivalent system of resistors (of the same topology), such that the resistance of each resistor is proportional to the length of the corresponding segments [10]. All (discrete) hopping rates are to nearest neighbours and are assumed to equal p. The probability to stay at a point j, is $1-r_jp$, where r_j is the number of nearest neighbours. Let p_A be the probability to leave point A per unit time τ, and define similarly p_B for a point B. Attach A to a reservoir so that the number of walkers there is fixed and equal to N_A; similarly keep the number of walkers at B equal to N_B. In a steady state situation, the current from A to B is

$$I = T_{AB}(z=1)\frac{N_A-N_B}{\tau}. \tag{16}$$

Regarding N_A/τ and N_B/τ as potentials, it follows that $T_{AB}(z=1)$ can be identified as the conductance between A and B.

In the equivalent resistor network, let A be at a potential $N_A\vartheta_0$ and B at a potential $N_B\vartheta_0$. We claim that the potential at any node i, equals $N_i\vartheta_0$, where N_i is the average number of particles in the equivalent network of segments. The proof is straightforward, since the particle current in each node sums up to zero in a steady state situation, and the current in each segment is proportional to the difference of densities at its ends (thus satisfying the Kirchhoff rules). We thus find that the conductance of the network of resistors is proportionate to $T_{AB}(z=1)$ in the equivalent network of segments. The generalization of this idea is shown below to lead to a generalized Einstein relation for networks. The original Einstein relation was proven for a homogeneous medium, whereas here we derive a similar relation for an inhomogeneous network. We consider general time-dependent particle reservoirs.

Consider a time-dependent problem. Let $N_A(n)$ and $N_B(n)$ be the number of particles at points A and B, respectively, at time $n\tau$. These numbers are assumed to be *predetermined* for all n by (time-dependent) external agencies (particle baths). We wish to find $I_{in}(n)$, the particle current leaving A at time n, and $I_{out}(n)$, the particle current entering B at time n. Since we deal with a time-dependent problem, $I_A(n)$ needs not be equal to $I_B(n)$; particles can be delayed in the network. The current leaving A is composed of three contributions: (1) the current leaving A at time $n\tau$: $(1/\tau)p_A N_A(n)$; (2) the current composed of all particles that left A in the "past" and return to A at time $n\tau$, without having reached B in the process; and (3) the current composed of all particles that left B and reached A for the first time, without having returned to B. Altogether, the average particle current leaving A at time n is

$$I_{in}(n) = \frac{1}{\tau} p_A N_A(n) - \frac{1}{\tau} \sum_{m=0}^{\infty} Q_A(m) N_A(n-m) - \frac{1}{\tau} \sum_{m=0}^{\infty} T_{BA}(m) N_B(n-m). \tag{17}$$

Similarly, the current entering B at time n is

$$I_{out}(n) = \frac{1}{\tau} \sum_{m=0}^{\infty} T_{AB}(m) N_A(n-m) + \frac{1}{\tau} \sum_{m=0}^{\infty} Q_B(m) N_B(n-m) - \frac{1}{\tau} p_B N_B(n). \tag{18}$$

Defining the discrete Fourier transform of a function $f(n)$ to be $\hat{f}(\omega) = \sum_{n=-\infty}^{\infty} f(n) e^{i\omega n}$, we find from (17) and (18)

$$\hat{I}_{in}(\omega) = \frac{1}{\tau} p_A \hat{N}_A(\omega) - \frac{1}{\tau} \tilde{Q}_{AB}(\omega) \hat{N}_A(\omega) - \frac{1}{\tau} \tilde{T}_{BA}(\omega) \hat{N}_B(\omega), \tag{19}$$

$$\hat{I}_{out}(\omega) = \frac{1}{\tau} \tilde{T}_{AB}(\omega) \tilde{N}_A(\omega) + \frac{1}{\tau} \tilde{Q}_{BA}(\omega) \hat{N}_B(\omega) - \frac{1}{\tau} p_B \tilde{N}_B(\omega), \tag{20}$$

where $\tilde{T}_{AB}(\omega) \equiv T_{AB}(z=e^{i\omega})$ and similarly for T_{AB}, $\tilde{Q}_{AB}$ and $\tilde{Q}_{BA}$. Hence

$$\hat{I}_{in}(\omega) - \hat{I}_{out}(\omega) = \frac{1}{\tau} \tilde{N}_A(\omega) [p_A - \tilde{Q}_{AB}(\omega) - \tilde{T}_{AB}(\omega)] + \frac{1}{\tau} \hat{N}_B(\omega) [p_B - \tilde{Q}_{BA}(\omega) - \tilde{T}_{BA}(\omega)]. \tag{21}$$

The terms in the square brackets of eq. (21) vanish due to particle number conservation. Hence

$$\frac{d}{d\,i\omega} [\hat{I}_{in}(\omega) - \hat{I}_{out}]|_{\omega=0} = -\frac{1}{\tau} \hat{N}_A(0) \frac{d}{d\,i\omega} [\tilde{Q}_{AB}(\omega) + \tilde{T}_{AB}(\omega)]|_{\omega=0}$$

$$- \frac{1}{\tau} \hat{N}_B(0) \frac{d}{d\,i\omega} [\tilde{Q}_{BA}(\omega) + \tilde{T}_{AB}(\omega)]|_{\omega=0}. \tag{22}$$

The value of the derivatives in eq. (22) is found as follows. Recall that

$$G_{AB}(z=e^{i\omega}) = \frac{X_A(z=e^{i\omega}) \tilde{T}_{AB}(\omega)}{1 - X_A(Z=e^{i\omega}) \tilde{Q}_{AB}(\omega)}, \tag{23}$$

where G_{AB} is the generating function for first passage from A to B. Also $G_{AB}(z=0) = 1$. Defining

$\tilde{G}_{AB}(\omega) = G_{AB}(z=e^{i\omega})$, we find for the mean first passage time t_{AB}, from A to B, that: $t_{AB} = [1/\tilde{G}_{AB}(\omega)] \, d\tilde{G}/d\, i\omega|_{\omega=0}$ or

$$t_{AB} = \frac{1}{1-X_A(0)\tilde{Q}_{AB}(0)} \frac{1}{X_A(0)} \frac{d\tilde{X}_A(i\omega)}{d\, i\omega}\bigg|_{\omega=0} + \frac{1}{\tilde{T}_{AB}(0)} \frac{d\tilde{T}_{AB}(\omega)}{d\, i\omega}\bigg|_{\omega=0}$$
$$+ \frac{\tilde{X}_A(0)}{1-\tilde{X}_A(0)Q_{AB}(0)} \frac{d\tilde{Q}_{AB}(\omega)}{d\, i\omega}\bigg|_{\omega=0}. \tag{24}$$

Using (22) and (23) we find

$$t_{AB} = \frac{1}{\tilde{T}_{AB}(0)} \frac{1}{\tilde{X}_A^2} \frac{d\tilde{X}_A(\omega)}{d\, i\omega}\bigg|_{\omega=0} + \frac{1}{\tilde{T}_{AB}(0)} \frac{d\tilde{I}_{AB}}{d\, i\omega} + \frac{l}{\tilde{I}_{AB}(0)} \frac{d\tilde{Q}_A}{d\, i\omega}\bigg|_{\omega=0}. \tag{25}$$

Thus

$$t_{AB}\tilde{T}_{AB}(0) = \frac{d}{d\, i\omega}[\tilde{T}_{AB}(\omega) + \tilde{Q}_{AB}(\omega)]|_\omega - \frac{d}{d\, i\omega}\left(\frac{l}{\tilde{X}_A}\right)\bigg|_{\omega=0}. \tag{26}$$

Recalling the definition of X_A, $\tilde{X}_A(\omega) = 1/[1-(1-p_A)e^{i\omega}]$, we obtain

$$\frac{d}{d\, i\omega}[\tilde{T}_{AB}(\omega) + \tilde{Q}_{AB}(\omega)]|_{\omega=0} = t_{AB}\tilde{T}_{BA}(0) - (1-p_A). \tag{27}$$

Similarly, it is straightforward to show that

$$\frac{d}{d\, i\omega}[\tilde{T}_{BA}(\omega) + \tilde{Q}_{BA}(\omega)]|_{\omega=0} = t_{BA}\tilde{T}_{BA}(0) - (1-p_B). \tag{28}$$

Substituting (27) and (28) into (21) and using $\tilde{T}_{AB} = \tilde{T}_{BA}$,

$$\frac{d}{d\, i\omega}[I_{\text{out}}(\omega) - I_{\text{in}}(\omega)] = \frac{\tilde{T}_{BA}}{\tau}[\hat{N}_A(0)t_{AB} + \hat{N}_B(0)t_{BA}] + \frac{1}{\tau}[\hat{N}_A(0)(p_B-1) + \hat{N}_B(0)(p_A-1)]. \tag{29}$$

Defining a weighted average of the mean first passage time between A and B ($\hat{N}_A(0)$ and $\hat{N}_B(0)$ are the dc components of the occupations at A and B, respectively) as

$$\bar{t}_{AB} = \frac{\hat{N}_A(0)t_{AB} + \hat{N}_B(0)t_{BA}}{\hat{N}_A(0) + \hat{N}_B(0)}, \tag{30}$$

and a weighted staying probability q_{AB} as

$$q_{AB} = \frac{\hat{N}_A(0)(1-p_A) + \hat{N}_B(0)(1-p_B)}{\hat{N}_A(0) + \hat{N}_B(0)}, \tag{31}$$

we obtain

$$\tau\frac{d}{d\, i\omega}[I_{\text{out}}(\omega)]|_{\omega=0} = [\hat{N}_A(0) + \hat{N}_B(0)](T_{AB}\bar{t}_{AB} - q_{AB}). \tag{32}$$

We next wish to reformulate the left-hand side of (32). To this end, consider any node j and the nodes connected to it by single bonds. Let $N_j(n)$ be the average number of particles at j at time n. The average number of particles entering j at time n is obviously $N_j(n+1) - N_j(n)$. The average current entering j at time n is $I_j(n) = (1/\tau)[N_j(n+1) - N_j(n)]$.

This relation does not hold for $j = A, B$, since $N_A(n)$ and $N_B(n)$ are externally controlled. Since the current entering one mode through a given bond is the negative of the current entering the other mode through that bond, the sum of $I_j(n)$ over all nodes $j \neq A, B$ is the sum of the current leaving A (and entering other bonds), $I_{\text{in}}(n)$, and the current leaving B, $-I_{\text{out}}(n)$,

$$I_{\text{in}}(n) - I_{\text{out}}(n) = \frac{1}{\tau} \sum_{j (\neq A, B)} [N_j(n+1) - N_j(n)]. \tag{33}$$

Fourier transforming this equation we find

$$\hat{I}_{\text{in}}(\omega) - \hat{I}_{\text{out}}(\omega) = \frac{1}{\tau} \sum_{j (\neq A, B)} (e^{-i\omega} - 1) \hat{N}_j(\omega). \tag{34}$$

Denoting by $\hat{N}$ the total number of particles in the system (excluding A and B), we find

$$\hat{I}_{\text{in}}(\omega) - \hat{I}_{\text{out}}(\omega) = \frac{1}{\tau} (e^{-i\omega} - 1) \hat{N}(\omega). \tag{35}$$

Hence,

$$\frac{d}{d\, i\omega} [I_{\text{out}}(\omega) - I_{\text{in}}(\omega)] |_{\omega = 0} = \frac{1}{\tau} \hat{N}(0). \tag{36}$$

Comparing (36) and (32) we find

$$\hat{N}(0) = [\hat{N}_A(0) + \hat{N}_B(0)](T_{AB} \bar{t}_{AB} - q_{AB}). \tag{37}$$

Let us embed our network in a box of length L and cross section A (say, L is the distance between nodes A and B). We define an effective diffusion coefficient, $D = L^2 / \bar{t}_{AB}$; an effective dc density of particles, $\rho = \hat{N}(0)/AL$, and an effective conductivity σ, $\sigma_{AB} = L\Sigma_{AB}/A$ (Σ is the conductance). It follows that

$$\frac{\sigma_{AB}}{D} = \frac{\Sigma_0}{T_0} \left(\frac{\rho}{\hat{N}_A(0) - \hat{N}_B(0)} + \frac{q_{AB}}{AL} \right).$$

Here Σ_0, T_0 are the conductance and the operator T respectively, associated with a single bond. Denote $\alpha \equiv (\Sigma_0/T_0)\{1/[\hat{N}_A(0) + \hat{N}_B(0)]\}$. α is a constant, depending on the properties of a single bond and the endpoint dc occupation numbers. The term q_{AB}/AL vanishes in the limit of large volumes and then we obtain

$$\sigma_{AB}/D = \alpha \rho. \tag{38}$$

Eq. (38) is an Einstein relation valid for networks, be they homogeneous or inhomogeneous.

Notice that the Einstein relation obtained in this section can be understood either as a connection between the conductivity σ_{AB} of an equivalent resistor network and the diffusion constant of the network, or as a property of the hopping current through the network. We have both a physical statement regarding hopping conductivity and a formal analogy to a resistor network.

5. Discussion

We have discussed here the principles of our method and demonstrated its applications. In principle it is applicable to general networks. The underlying idea can be explained considering a 3-site one-dimensional segment, which is a part of a complex network. The path 1–2–3 corresponds to an infinite set of trajectories that start at 1 and arrive in 3, e.g., 1-2-3, 1-2-1-2-3, 1-2-2-1-2-3 etc. (the numbers denote the position of the walker at a given discrete time step). All these trajectories (or subsets of these trajectories) are summed up to yield the basic "probabilities", in terms of which physical quantities may be calculated. Summing over trajectories that begin and end at site 2 in the above example, one may obtain a renormalized probability for staying at 2 (for an unspecified period of time). One may also use the concept of renormalization, decimate out site 2, and obtain an effective transmission probability for going from site 1 to site 3.

The ideas outlined here were further extended to full distribution functions, and were also employed to predict $1/f$ noise in calculations of the certain inhomogeneous systems [11-13]. This method is amenable to further generalizations. One may consider hopping beyond nearest neighbors. The effect of external magnetic fields can be accounted for. Finally, it is possible to consider cases in which the hopping probability itself is a random function of space [14].

Acknowledgements

We have benefited from numerous discussions with many of our colleagues. In particular we would like to acknowledge useful comments made by Professors A. Aharony, R. Landauer, B.B. Mandelbrot, D. Stauffer and D.J. Thouless. Y. Gefen acknowledges the hospitality of the Sackler Institute of Solid State Physics at Tel Aviv University, where part of this work was done.

References

[1] B.B. Mandelbrot, The Fractal Geometry of Nature (Freeman, San Francisco, 1982), and references therein.
[2] Y. Gefen, A. Aharony and S. Alexander, Phys. Rev. Lett. 50 (1983) 77;
 S. Alexander and R. Orbach, J. Phys. Lett. 43 (1982) L625;
 H. Nakanishi, Y. Meir, Y. Gefen, A. Aharony and P. Schofield, J. Phys. A 20 (1987) L153.
[3] A. Einstein, Ann. Physik 17 (1905) 549; 19 (1906) 37.
[4] M. Lax, Rev. Mod. Phys. 32 (1960) 26, and references therein.
[5] R. Kubo, J. Phys. Soc. Japan 12 (1957) 570;
 H. Scher and M. Lax, Phys. Rev. B 7 (1973) 4491;
 T. Odagaki and M. Lax, Phys. Rev. B 24 (1981) 5284.
[6] M.N. Barber and B.W. Ninham, Random and Restricted Walks (Gordon and Breach, New York, 1970).
[7] I. Goldhirsch and Y. Gefen, Phys. Rev. A 33 (1986) 2583.
[8] I. Goldhirsch and Y. Gefen, Phys. Rev. A 35 (1987) 1371.
[9] Y. Gefen and I. Goldhirsch, J. Phys. A 18 (1985) L1037.
[10] Y. Gefen and I. Goldhirsch, Phys. Rev. B 35 (1987) 8639.
[11] Y. Gefen, I. Goldhirsch and R. Laibowitz, Fractal Aspects of Materials (extended abstracts), Proceedings of Symposium N, 1985 Fall Meeting of the Materials Research Society, Boston 1985, R.B. Laibowitz, B.B. Mandelbrot and D.C. Passoja, eds. (Materials Research Society, Pittsburgh, 1985), p. 95.
[12] I. Goldhirsch and S.H. Noskowicz, J. Stat. Phys. 48 (1987) 291.
[13] S.A. Noskowicz and I. Goldhirsch, J. Stat. Phys. 48 (1987) 255.
[14] S.A. Noskowicz and I. Goldhirsch, Phys. Rev. Lett. 61 (1988) 500.

FRACTAL MOTION OF MAMMALIAN CELLS

Ivar GIAEVER and C.R. KEESE

School of Science, Rensselaer Polytechnic Institute, Troy, NY 12180-3590, USA

Hurst's rescaled range analysis has been applied to electrical signals that reflect the motion of fibroblastic human cells. According to the analysis the cells have a fractional Brownian motion with a Hurst exponent of about $H=0.65$. The normal WI-38 cell line has a Hurst exponent that is slightly smaller than the transformed WI-38 VA13 cell line.

1. Introduction

Since the publication of Benoit B. Mandelbrot's book The Fractal Geometry of Nature [1], the way we perceive and model our surroundings have changed. Euclidean geometry is often ill suited to describe everyday experiences, and Mandelbrot states: "Responding to this challenge, I conceived and developed a new geometry of nature and implemented its use in a number of diverse fields". This paper deals with fractal records in time produced by the mechanical motions of cells growing in tissue culture. We deal with two different cell lines purchased from the American Type Culture Collection – a normal human embryonic lung fibroblast called WI-38 and its transformed (cancer) counterpart, the WI-38 VA13 cell. The data from these cells have been analyzed making use of Hurst's rescaled range analysis (R/S analysis), a statistical method devised by Hurst et al. [2] and later discussed extensively by Mandelbrot and Wallis [3,4].

2. R/S analysis

A detailed description of R/S analysis can be found in many places [1,2,5], but since the method is not generally known we will give a brief outline here. In this work we have obtained a discrete record in time of electrical signals that reflect the motion of mammalian cells. By choosing a time period (referred to as a lag) shorter than the total record, we first calculate the increments I, the change in the signal between two adjacent times, for this lag and then the average A and the standard deviation S of the increments. By constructing a set of new increments $(I-A)$ and sequentially adding them, we reconstruct an image of the original curve in this lag along a horizontal axis. The range R is defined as the maximum value minus the minimum value of this curve, and by dividing R with S we obtain a dimensionless number R/S for the lag in question. If the lag was chosen say $1/100$ of the total period, we have one hundred independent estimates of the R/S for this lag, and the final value is the average of all of them.

For many natural phenomena Hurst and later Mandelbrot found that $R/S \sim (\text{lag})^H$ where H is now referred to as the Hurst exponent. It is easy to show that for a random record in time the Hurst exponent is 0.5 [5]. It is therefore very surprising that for the natural phenomena studied, H is always significantly larger than 0.5. Mandelbrot has described data when $H > 0.5$ as persistent and when $H < 0.5$ the data are said to be antipersistent.

3. The experimental system

It is possible to culture mammalian cells in vitro where the cells can be regarded as independent but interacting organisms [6]. Normally cells are studied using an optical microscope, sometimes in connection with time lapse photography. We have recently introduced a new method where the cells can be monitored electrically in real time [7,8]. The cells are cultured on small gold electrodes evaporated on the bottom of standard tissue culture dishes. Normally we apply a constant ac current at 4000 Hz between a large counter electrode and a small active electrode while we monitor the voltage (see fig. 1). Since this is a two-probe measurement, it is important that the active electrode is small, such that the impedance associated with it dominates the system. In practice, this requires an electrode with an area less than 10^{-4} cm^2. In the work reported here, the human fibroblastic cells were grown to confluence, and then the fluctuations in the in- and out-of-phase voltage were followed as a function of time. It is known that cells in culture attach to surfaces with small foot-like adhesion plaques but that the ventral surface of a cell is maybe 50 nm from the substrate on the average. By model calculations, it is possible to show that the impedance is a measure of the average distance the cells are away from the substrate, i.e. the impedance fluctuations for confluent layers of cells are a measure of changes in the size of this space due to cell motion [9].

The in- and out-of-phase voltage is measured with the aid of a lock-in amplifier that is interfaced with an IBM AT computer. The time constant used is 0.3 s, and data are sampled at every 0.1 s for a total time of approximately 20 h. Fig. 2 shows the change in the in-phase voltage recorded for the two cell lines used. By killing the cells, the fluctuations are essentially eliminated as is shown using formalin with the normal cell line (fig. 2). This clearly demonstrates that these fluctuations are due to the living state.

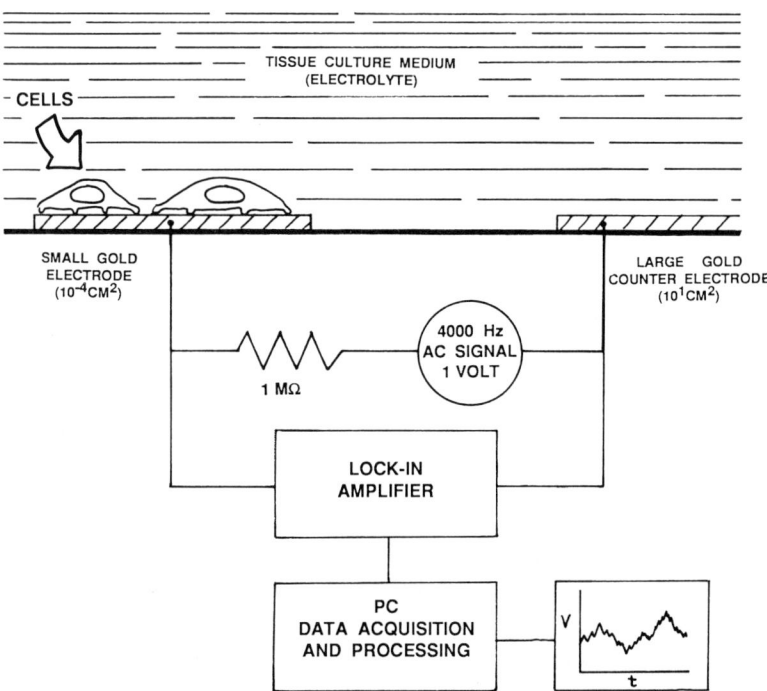

Fig. 1. A schematic of the experimental arrangement. Note the fact that the active electrode must be small. The 1 MΩ resistance provides an essential constant current through the cell culture device.

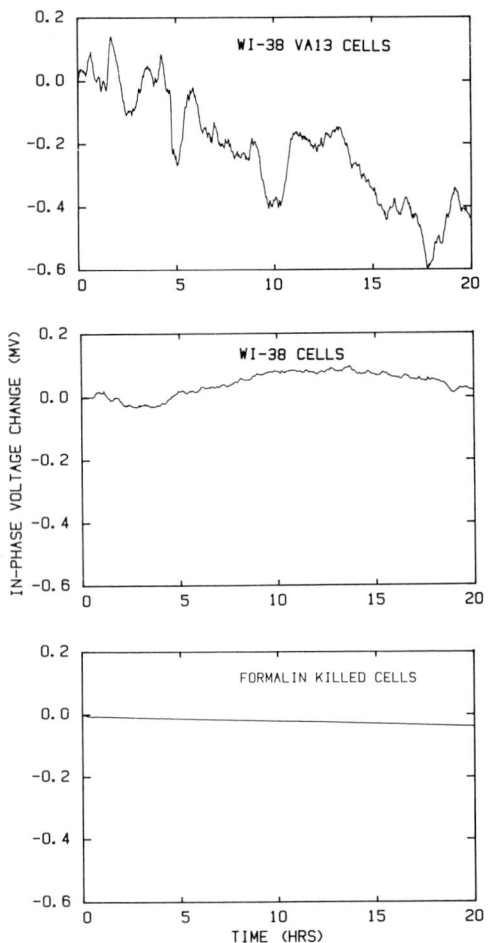

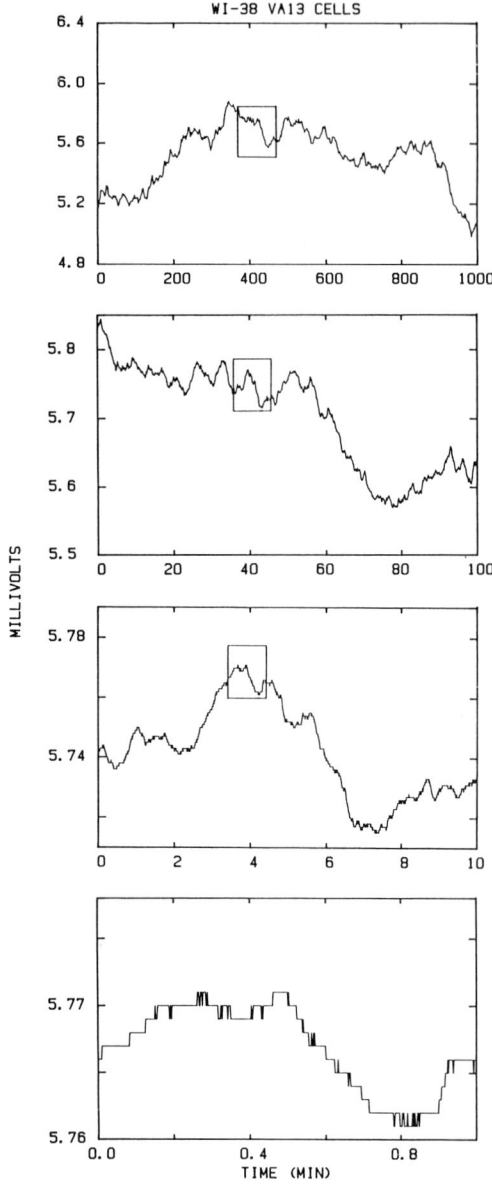

Fig. 2. The change in the in-phase voltage for WI-38 VA13 cells (top), the WI-38 cells (middle) and WI-38 cells killed with 10% formalin (bottom). This type of data is the basis for the R/S analysis.

4. Results

Fig. 3 shows the in-phase voltage for various recorded time periods for the transformed cell line. The steps visible in the curve for shorter times are due to the digital output of the amplifier, and thus they have no physical significance. In this case, each step corresponds to a 1.0 μV change in the sampled voltage. The similarities in the curves at various times suggest that the data are of a fractal nature, and indeed an extensive search for periodic behavior has been negative.

Fig. 3. The in-phase voltage for WI-38 VA13 cells for four different time intervals. The steps visible in some curves are due to the digital amplifier and have no physical significance.

In fig. 4 is shown the result of an R/S analysis of this data set. The data are collected at 0.1 s intervals, but since they are stored in the computer we can also calculate the R/S values for different sampling times. In the figure, the results are shown for sampling times

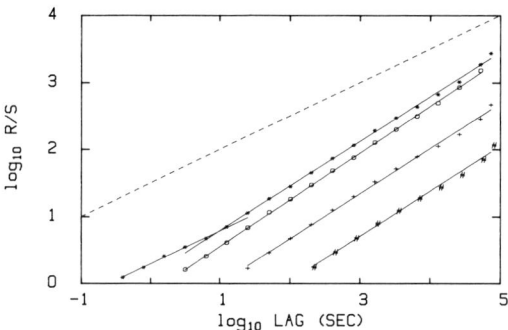

Fig. 4. The R/S analysis for four different sampling times 0.1 (∗), 0.8 (○), 6.4 (+) and 51.2 (♯) s. Since the slope does not change significantly, the data are truly fractal. The dotted line has a slope of 0.5 and is drawn for comparison.

of 0.1, 0.8, 6.4, and 51.2 s, and as seen, the Hurst exponent is essentially the same. The distribution of the voltage changes or increments for the various sampling times and the corresponding Hurst exponents are shown in table 1.

The autocorrelation functions for the same sampling times have been calculated and are displayed in fig. 5. For the sampling time of 0.1 s the autocorrelation function show antipersistence. This is due to "flipping noise" arising from the digital amplifier when the actual output value is between two adjacent amplifier values. This noise is clearly evident in the short-time record of fig. 3. For longer sampling times the autocorrelation functions demonstrate that the data are persistent and, since these functions do not change significantly with the sampling time, fractal.

A comparison between the two cell types using R/S analysis is shown in fig. 6. Each point in the figure is an average over eight separate experiments for the transformed cell type and six separate experiments for the normal cells. As seen, fluctuations of the transformed cell result in a larger Hurst exponent, however, the difference between the values for the two cell lines is not large. Since this method is robust with regards to the details of the experiment it is attractive; however, because there are overlaps in the individual experiments between the two cell lines, it is not highly sensitive to the difference between these cells.

5. Discussion

As seen from fig. 6, for short lag times the Hurst exponent is close to 0.5 while for larger lags it shows persistence. The amplifier noise is probably responsible for the Brownian character of the data obtained at short lags. From fig. 5 it is clear that for short times the data are antipersistent, thus it is at first sight surprising that this is not evident in fig. 6. The R/S analysis is not very sensitive for short times when the data set is digital, and a better way of analyzing the data is to use the fact that for fractional Brownian motion, the average of the standard deviation $\langle S(\Delta t) \rangle$ for a sampling time of Δt is a function of the Hurst exponent $\langle S(\Delta t) \rangle \sim (\Delta t)^H$. This relation is very good for short times, while the Hurst R/S analysis is better for

Table 1
The distribution of the voltage increments for various sampling times. Each number represents the number of increments with a given microvolt value; the center bold numbers represent increments of 0 μV, and the values change by +1 and −1 μV going right and left respectively.

Sampling time (s)	Hurst exponent	Distribution of increments
0.1	0.67	0 58431 **620877** 57972 0
0.8	0.70	0 218 16008 **60264** 15360 305 5 0
6.4	0.67	0 9 23 63 209 445 773 1415 1942 **2040** 1862 1267 751 402 182 83 37 13 0 4 0
51.2	0.67	17 18 26 17 37 34 38 48 55 45 50 44 59 64 57 **60** 55 69 61 57 62 48 46 38 36 32 22 27 21 11 13 0 1 2 0 2 3 5 8 3 7 11 5 15 16 14 11 10 8 8 5 8 6 5 2 1 2 1 2 1 0 1 0

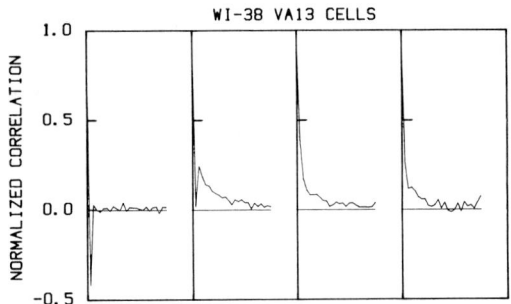

Fig. 5. Twenty-five points of the autocorrelation functions for sampling times of 0.1, 0.8, 6.4 and 51.2 s are displayed from left to right. The function for 0.1 s sampling time reflects the amplifier noise and shows antipersistence. The other functions show persistence and, since they do not change their appearance significantly with sampling time, illustrate that the data are fractal.

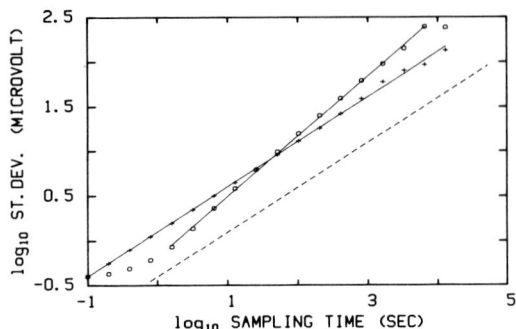

Fig. 7. The standard deviation at different sampling times, as a function of the sampling time, for the same experiment as in fig. 4 (○) and for the data set scrambled (+). The slope is equal to the Hurst exponent and is 0.67 and 0.503 for the cell data and the scrambled data respectively.

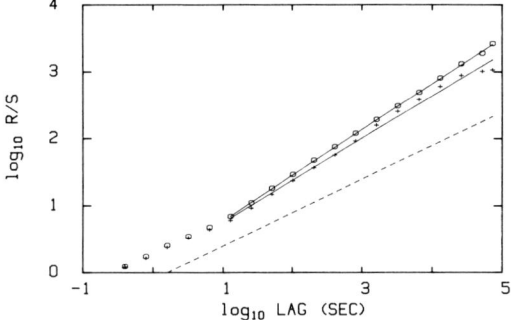

Fig. 6. A comparison between the WI-38 VA13 cells (○) with a Hurst exponent $H=0.68$ and the WI-38 cell line (+) with $H=0.63$. The points are an average of several experiments (see text).

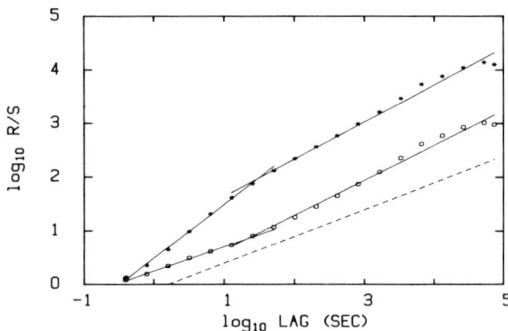

Fig. 8. R/S curve for a specific experiment done with (∗) and without (○) a 32-point (3.2 s) "running average".

longer times. An example is shown in fig. 7, where it is clear that for short times the data are antipersistent and the Hurst exponent is less than 0.5, while for longer times $H=0.67$.

If we could have extended the R/S measurements in fig. 6 to significantly longer times, the slope of the curve would approach zero. The reason is that the standard deviation of the increments becomes essentially independent of the lag time for lags greater than about 100 s, and now the value of the range determines the value of R/S, i.e. $R/S \sim R \sim (\text{lag})^H$. Because this is a physical system the in-phase voltage and hence the range is bounded; the lower bound is set by the voltage measurement for an empty electrode and the upper bound is at most 10 times this value; in practice the limits are smaller. In fig. 6, there is a hint that the normal cell line data are approaching the limitations where R/S becomes a constant.

Sometimes when lock-in amplifiers are used, a higher accuracy can be obtained by doing a "running average", i.e. the value of each measurement is the average of a fixed number of previous points. This has a drastic effect on an R/S calculation as shown in fig. 8. Here a running average of 32 points is used. At low lags the curve is highly persistent and the persistence lingers for almost an order of magnitude higher in lag time than the length of the averaging time (3.2 s). The Hurst exponent for longer lags is the same as before, but the absolute values of R/S are much greater. The reason is that the increments and therefore the standard deviations decrease by this procedure, while the range is unaffected. If a larger time constant had been used while collecting the data instead of a "running average", a similar result would have emerged.

The different values of the Hurst exponents for the two cell types are probably significant, but the difference is small. It is, however, easy to distinguish these two cell types by other electrical measurements with this system. One obvious result is that the transformed cell increases the in-phase voltage of an empty electrode by a factor of two more than the normal cell and exhibits much larger increments in the electrical signal. This effect is clearly seen for the living cells in fig. 2.

It is hard to understand why natural phenomena generally have a Hurst exponent greater than 0.5, and more specifically, why the fluctuations in the measured voltage that reflect the cell motion show such a dependence. That the phenomenon is real and associated with living cells can be further demonstrated by estimating the Hurst's exponent for a data set that has been scrambled (fig. 7). When this is done, as would be expected, the Hurst exponent is close to 0.5, but surprisingly almost always slightly larger than 0.5. Persistence implies that the cells have a "memory" of past behavior. It is not clear where this "memory" resides; however, the circadian rhythm is one example of such behavior that in all probability resides in the complex biochemical systems of living organisms. It is known that the generation time for mammalian cells in tissue culture is about 20 h, but it is difficult to see why this should lead to persistence from a few seconds to at least 20 h. Mandelbrot introduced fractional Brownian motion to change random data to persistent or antipersistent data, and at least mathematically this requires an infinite "memory". Hurst on the other hand simulated persistent behavior using simple models reminiscent of the "running average" described here. In that case the persistence lingers for maybe an order of magnitude longer than the built-in "memory" effects. In these experiments, we deal with the average behavior of 20–50 cells that interact with each other and with the active electrode; it may be that the persistence is somehow associated with a coherent behavior in this population. Since it is now possible to repeat these experiments with single cells, it will be interesting to discover whether the persistent nature remains the same.

Acknowledgement

This work was performed pursuant to a contract with the National Foundation for Cancer Research. Special thanks to Dr. Jens Feder for introducing us to the mysteries of the R/S analysis.

References

[1] B.B. Mandelbrot, The Fractal Geometry of Nature (Freeman, San Francisco, 1983).
[2] H.E. Hurst, R.P. Black and Y.M. Simaika, Long-Term Storage: An Experimental Study (Constable, London, 1965).
[3] B.B. Mandelbrot and J.R. Wallis, Water Resource Res. 4 (1965) 909.
[4] B.B. Mandelbrot and J.R. Wallis, Water Resource Res. 5 (1969) 321.
[5] J. Feder, Fractals (Plenum Press, New York, 1988).
[6] R.I. Freshney, Culture of Animal Cells (Liss, New York, 1983).
[7] I. Giaever and C.R. Keese, Proc. Natl. Acad. Sci. US 81 (1984) 3761.
[8] I. Giaever and C.R. Keese, IEEE Trans. Biomed. Eng. 33 (1986) 242.
[9] I. Giaever and C.R. Keese, to be published.

A LIGHT SCATTERING STUDY OF TURBULENCE

W.I. GOLDBURG, P. TONG[1] and H.K. PAK

Department of Physics and Astronomy, University of Pittsburgh, Pittsburgh, PA 15260, USA

By scattering light from a turbulent fluid seeded with small particles, one obtains information about turbulent velocity fluctuations over varying spatial scales, R. The measured intensity autocorrelation function, $g(t)$, is related to the probability density $P(V(R))$ of finding velocity fluctuations of magnitude $V(R)$ associated with eddies of size R. The measurements described here strongly suggest that the energy-containing eddies occupy a fractal region whose dimension (or spectrum of dimensions) increases with the Reynolds number Re when Re exceeds some threshold value.

1. Introduction

In his seminal book, The Fractal Geometry of Nature, Benoit Mandelbrot [1] makes clear his deep interest in the geometrical nature of turbulence. As he points out, the description of the visual appearance of a turbulent fluid, such as smoke curling up from a cigarette, taxes our powers of description. It seems that present-day speech is not well suited to evoking the image of self-similar structures. After all, it takes a series of images, one magnified with respect to the other, to identify fractal structures. And turbulence is, by all evidence, a fractal thing at its roots[2].

There are many ways of revealing the fractal or spotty nature of a turbulent fluid. One technique is to measure the time variation of the square of the velocity at a point in the fluid [3]. Another is to add a small amount of long-chain molecules to the fluid and observe it through crossed polaroids [4]. The molecules are locally aligned by turbulent shear forces. These molecules, being anisotropic scatterers, depolarize the light in regions where the shear is large, making the local structure of the strong vorticity directly visible.

Herein we describe experiments, carried out at the University of Pittsburgh, which provide a new approach to the study of the small-scale structure of turbulence. The method involves a measurement of the autocorrelation function of the light intensity scattered by small particles suspended in the turbulent fluid. For this technique there is no need to invoke the "frozen turbulence assumption" to translate temporal information to spatial information. According to this assumption, small-scale eddies (the ones of interest), are transported past a velocity measuring device with the mean velocity U of the flow. If these small-scale eddies remain intact for a long enough time, a time record of the velocity $v(t)$ at a point will reveal spatial features of the flow through the equation $v(t) = v(x/U)$. The frozen turbulence assumption fails unless the velocity fluctuations $V(R)$ associated with eddies of size R are uncoupled from the larger-scale eddies.

The technique of photon correlation homodyne spectroscopy (HS) [5], which we have used in our experiments, is that of recording the beating of scattered light waves that have been Doppler shifted by pairs of particles seeded in the turbulent fluid. The technique was introduced many years ago by Bourke et al. [6], but seems largely to have been ignored. Being an optical technique, it permits non-invasive observation of velocity fluctuations at very small scales.

The homodyne scheme is readily understood from

---

[1] Present address: Exxon Research and Engineering Co., Annandale, NJ, 08801, USA.

fig. 1, which shows two moving particles at a particular instant of time when their separation is R and their velocities are v_1 and v_2. The seed particles are small enough that they scatter light isotropically. A photodetector (PMT), located at an angle θ with respect to the incident beam, receives the light from both particles. The scattered light from each particle is Doppler shifted by an amount $\mathbf{k} \cdot \mathbf{v}_1$ and $\mathbf{k} \cdot \mathbf{v}_2$ respectively, where $\mathbf{k}$ is the scattering vector, of magnitude $k = (4\pi n/\lambda) \sin(\tfrac{1}{2}\theta)$. Here λ is the vacuum wavelength of the light ($\lambda = 488$ nm in our experiments), and n is the refractive index of the turbulent fluid; in our case the fluid was water. The photomultiplier, which receives the light from the particle pair, is a square-law detector, so that its output current, $I(t)$, contains a beating term proportional to $\cos[kV_k(R)t]$, where V_k is the projection of the velocity difference $v_1 - v_2$ along the direction of $\mathbf{k}$. Henceforth the subscript on $V_k(R)$ will be dropped, but its R dependence will be retained.

The essential aspect of turbulence is that the velocity difference between two points in the fluid depends on the separation R of these two points. According to the theory of Kolmogorov [6], the moments of the velocity fluctuations $V(R)$ obey a scaling law

$$\langle V(R)^n \rangle \sim u(R)^n \sim R^{\xi_n}, \qquad (1)$$

with $\xi_n = \tfrac{1}{3}n$. The homodyne technique is well suited to measure the lower moments of $V(R)$, but not the higher moments. On the other hand, the method yields information about the functional form of the probability density $P(V(R))$, that two points in the fluid, separated by a distance R, have velocity difference lying within $V(R)$ and $V(R) + \mathrm{d}V(R)$. Our central finding is that $P(V(R))$ is well represented by a Lorentzian function,

$$P(V(R)) \propto \{1 + [V(R)/u(R)]^2\}^{-1}, \qquad (2)$$

for relatively small values of $V(R)$. We also find that the scaling velocity $u(R) \sim R^\zeta$, where ζ is a function of Reynolds number, Re. The measurements were made at very modest values of the Reynolds number. In fact the turbulence was so weak that one might not have thought the flow would exhibit the self-similarity which was indeed observed. Throughout this paper the Reynolds number is defined as $\mathrm{Re} = Ul_0/\nu$, where U is the mean velocity of the flow, l_0 is the outer scale of the turbulence, and ν is the kinematic viscosity of the fluid.

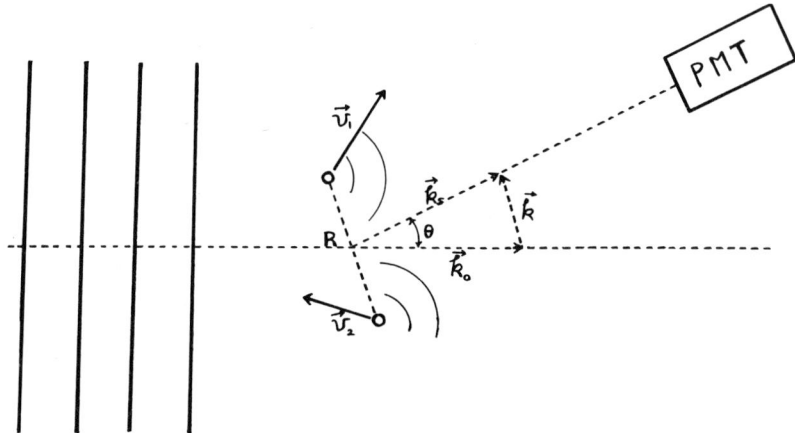

Fig. 1. A schematic diagram showing scattering geometry. The scattering vector $\mathbf{k} = \mathbf{k}_s - \mathbf{k}_0$, where $\mathbf{k}_s$ and $\mathbf{k}_0$ are the scattered and incident wave vectors respectively.

2. Experimental

The detailed experimental setup can be found in ref. [8]. The fluid flow was generated in a closed water tunnel comprised of a cylindrical pipe and a pump of variable speed. The turbulence is generated by a grid within the pipe. The grid can be removed to permit study of wall-generated turbulence (pipe flow). A baffle section placed in the high-pressure side of the grid, suppresses the turbulence generated by the pump and by those sections of pipe on the high-pressure side of the grid. With this arrangement all of the turbulence is generated by the grid only. In most of the experiments discussed here the diameter of the pipe was 4.4 cm, and the aperture size of the grid was 3.1 mm. These parameters are taken to be l_0 in calculating the Reynolds number. The measurements were made 28 cm downstream from the grid. The water which flowed through the pipe was seeded with polystyrene spheres 60 nm in diameter. These particles were small enough to scatter light isotropically and in sufficient concentration that their mean separation was much less than the Kolmogorov dissipation length l_d, which was estimated to be a fraction of a millimeter.

On the downstream side of the grid there is an optically transparent section of piping to admit the incident laser beam and observe the scattering. Because the flow is seeded, a thin column of the scattered light is produced in the water and that light is imaged with a lens, on a slit of variable width, L. By varying L, the homodyne scheme permits the probing of velocity fluctuations $V(R)$ from the smallest scale l_d to that of the width of the slit, L.

Using a standard light scattering apparatus and a digital correlator, we measure the intensity autocorrelation function, $g(t) = \langle I(t') I(t'+t) \rangle / \langle I(t') \rangle^2$, where $I(t)$ is the scattered light intensity measured at scattering angle θ, and the angle brackets represent a time average over t'. One can show [8] that the correlation function $g(t)$ has the following form:

$$g(t) = 1 + f(A) G(t). \tag{3}$$

The geometrical factor $f(A)$ is of order unity if the photodetector receives light from only one coherence area [5]. All of the interesting physics is contained in $G(t)$, which is proportional to a sum of the time-averaged phase factors $\cos(ktV)$ coming from the Doppler shift of all particle pairs in the scattering volume. The function $G(t)$ can be written as [8]

$$G(t) = \int_0^L dR\, h(R) \int_{-\infty}^{\infty} dV(R)\, P(V(R))$$
$$\times \cos[kV(R)t], \tag{4}$$

where $h(R)$ is the probability of finding a particle pair, separated by R, in the columnar region of length L.

If the image on the slit is taken to be quasi-one-dimensional, which is valid when the slit width remains large compared to the diameter of the laser beam, $h(R) = 2(1-R/L)/L$. Note that the inner integral in eq. (4) is the Fourier cosine transform of $P(V(R))$, and the $G(t)$ may be thought of as a transform of the characteristic function. If the probability density $P(V(R))$ has the scaling form $P(V(R)) = Q[V(R)/u(R)]/u(R)$, eq. (4) becomes

$$G(t) = \int_0^L dR\, h(R) F(ktu(R)), \tag{5}$$

where F is the Fourier cosine transform of $Q[V(R)/u(R)]$. It is easy to show that the scaling law in eq. (1) follows if the probability density function $P(V(R))$ has the above mentioned scaling form.

The above equations for $g(t)$ have quite general validity. They hold, for example, even if the fluid is stationary, and the seed particles are undergoing Brownian motion only. In that case, $V(R)$ is independent of R and $P(V(R))$ is a Gaussian function. Then the function $G(t)$ is an exponentially decaying function [5], $G(t) = \exp(-2Dk^2 t)$, where D is the diffusivity of the Brownian particles, and is given by Stokes' law. This contribution to the decay of $G(t)$ will be present, even when the fluid is turbulent. However, in a turbulent fluid, the decay time T of $G(t)$ is much shorter than the diffusive decay time, $T_d = 1/2Dk^2$, so that the latter contribution can be safely ignored. From eq. (5), it follows that the turbulent decay time should be of the order of $T= 1/ku(L)$, because the fastest decay rate is associated with the largest eddies of size L.

3. Results

Over a wide range of slit widths and Reynolds numbers, we find that the function $G(t)$ exhibits the scaling form,

$$G(kt, L, \text{Re}) = G(\kappa), \qquad (6)$$

with $\kappa \sim k^\mu L^\zeta t$. This scaling behavior of $G(t)$ is observed only when the Reynolds number exceeded a certain value Re_c. In the case of the grid flow described above, Re_c was roughly 500, which corresponds to much weaker turbulence than that one normally associates with scaling behavior. It is quite possible that the scaling behavior is seen at such small values of Re because the simultaneous velocity difference $V(R)$ is measured and no frozen turbulence assumption is needed in the data analysis.

The exponent μ in eq. (6) was measured as follows. For a fixed slit width L and a fixed Re, $G(t)$ was measured at several scattering angles and hence several values of k. All of the plots of $\log[G(t)]$ versus $\log(t)$ could be superimposed by horizontal translation of one graph with respect to another. The amount of translation, $\delta(k)$, is found to be roughly proportional to k, i.e. $\mu = 1$, when Re exceeded the critical value Re_c. However, in the absence of flow, $\mu = 2$, as expected for Brownian motion of the seed particles. Similar measurements were made in which k and Re were held fixed, and L was varied. Again all the plots of $\log[G(t)]$ versus $\log(t)$ could be superimposed, yielding the result $\kappa \sim kL^\zeta t$, as long as Re exceeded Re_c. In these experiments, ζ is found to be Re dependent. We return to this important observation below.

Fig. 2, a log-log plot of $G(\kappa)$ versus κ, shows the scaling behavior of $G(t)$ discussed above. The measurements correspond to several values of scattering angle, or k-value, several slit widths and at various

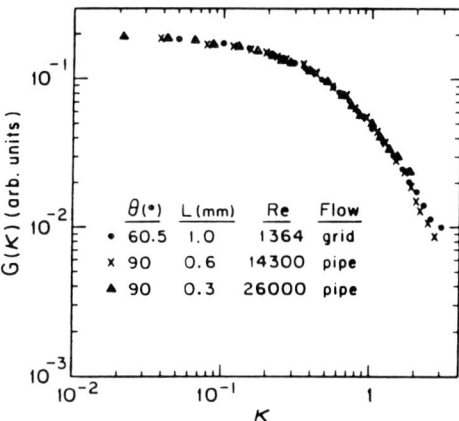

Fig. 2. The scaling function $G(\kappa)$ versus $\kappa = qu(L)t$ in pipe flow and grid flow.

Reynolds numbers. In one set of measurements (closed circles), the grid was present; in the other two sets (crosses and triangles), the grid was removed (pipe flow). The correlation functions $G(t)$ have been horizontally (and vertically) translated so that they coincide. In the pipe flow measurements, the Reynolds number is based on the pipe diameter, making it an order of magnitude larger than that for the grid flow, even when the mean flow velocities U are comparable in both cases.

An alternative way to determine the exponent ζ was to plot, on a double logarithmic scale, the slit-width dependence of the decay time, T, of $G(t)$, keeping Re and k fixed. As is shown in fig. 3, linear variation of $\log(T)$ with $\log(L)$ was seen at intermediate values of L. The data in fig. 3 were obtained in the grid flow at three different Reynolds numbers Re = 460, 1400, and 2200. Since $T \approx 1/ku(L)$ and $u(L) \sim L^\zeta$, the slope of this line yields the exponent ζ, which is $1/3$ in the Kolmogorov theory. We have verified that the power law behavior at large L was limited by the outer scale, l_0, of the turbulence. At small values of L the beam diameter was no longer negligibly small, which could account for the decrease in T at small values of L. Imperfections in the optical system may also be responsible for the decrease in T at small slit widths.

The behavior of $T(L)$ at small L has more recently been reexamined in a water tunnel of much superior

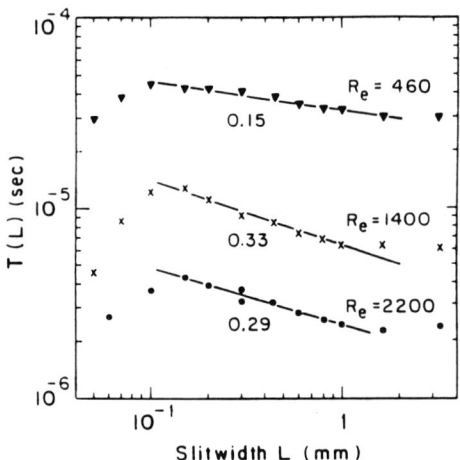

Fig. 3. The decay time $T(L)$ versus slit width L in grid flow. The number below a line is the slope of that line.

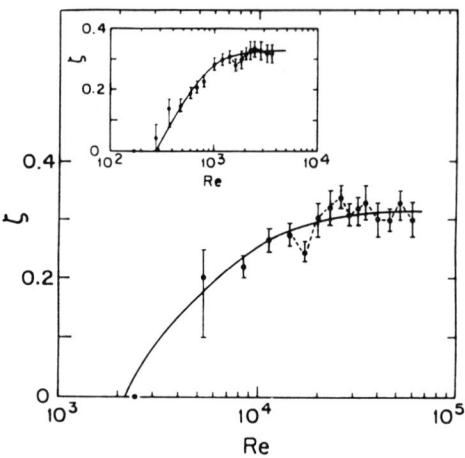

Fig. 4. The exponent ζ as a function of Re in pipe flow. The solid line is drawn by eye through the data points, and the dashed curve shows the oscillatory behavior of ζ. The inset shows ζ versus Re in grid flow.

design to that used in the studies reported above. In this experiment, the optically transparent pipe, where $g(t)$ was measured, was square in cross section, rather than cylindrical, so that the laser beam was undistorted in passing through it. In this square pipe, the beam diameter was less than 0.1 mm, which is smaller than the smallest value of L at which $g(t)$ was measured. Using laser Doppler velocimetry and invoking the frozen turbulence assumption one can determine the smallest eddy size l_d. At Re=850, we obtained $l_d=0.4$ mm. At values of L between 0.4 mm ($=l_d$) and 0.1 mm, the decay time of $G(t)$ became independent of L, i.e. $\zeta \approx 0$ when $L < l_d$. This result is very different from the Kolmogorov prediction, $\zeta = 1$ when $L < l_d$.

From the straight-line segment (solid line in fig. 3) we can extract the slope ζ which shows a Re-dependent feature. Fig. 4 shows ζ as a function of Re for both pipe flow and grid flow (insert). The exponent ζ is seen to increase from 0 to $\approx 1/3$ (the Kolmogorov value) as Re is increased. When the Reynolds number is below Re_c, $G(t)$ fails to exhibit scaling behaviour. The measured $Re_c \approx 300–400$ in the grid flow and $Re_c \approx 3000–4000$ when the grid was removed. Measurements in the improved water tunnel give similar results. These observations are consistent with

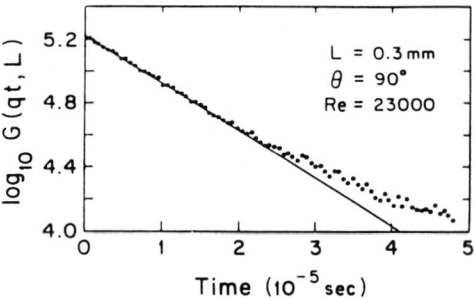

Fig. 5. A plot of $\log[G(qt, L)]$ versus t in pipe flow at indicated parameters.

the notion that the turbulence becomes increasingly three-dimensional as Re is increased above Re_c and that in the vicinity of Re_c, the turbulence is two-dimensional [9].

We now turn to the discussion about the functional form of $G(t)$. Fig. 5 is a semilog plot of $G(t)$ versus t in pipe flow at the indicated values of L, θ, and Re. The straight line is a linear fit to the data points at small t. It is seen that only at large time does the curve start to deviate from the linear behavior. If we assume that the characteristic function $F(ku(R)t)$ in eq. (5) has the form $F \sim \exp[-ku(R)t]$, $G(t)$ then becomes an incomplete gamma function with $ku(L)t$ as its argument [8]. This equation is well fitted to our measurement of $G(t)$. An example of this good fit is shown in fig. 6. Note that the assumption of $F(x)$ being a single exponential decaying function implies that $P(V(R))$ is of Lorentzian form as shown in eq. (2). This function has a diverging second moment, to which the energy density in the fluid is proportional. Therefore $G(t)$ cannot have this form for large values of $V(R)$. We indeed observed departures from this Lorentzian form for $P(V(R))$ with very large values of $V(R)$ (corresponding to very small t for $G(t)$) [10,11]. However, these observations will not be discussed further here. Most theories of turbulence concentrate on the scaling behavior of the moments of $V(R)$, rather than in $P(V(R))$ itself. Quite often, $P(V(R))$ is assumed to be of Gaussian form, $P(V(R)) \sim \exp\{-[V(R)/u(R)]^2\}$, but this form of $P(V(R))$ is clearly contrary to our findings.

How can one understand that the exponent ζ increases from 0 to $\approx 1/3$ as the fluid becomes increasingly turbulent? A fundamental understanding of this result is lacking, but it can be said that the observation is consistent with the notion that the turbulent active region is a fractal [12]. Let the fractal dimension of the turbulent region be D. Since the turbulent energy is confined to active regions of dimension $D<3$, the concentration of the turbulent energy is increased to smaller regions, relative to the case of volume-filling turbulence. Modifying the Kolmogorov theory to take this effect into account [13], one has $u(R) \sim R^\zeta$, with $\zeta = \frac{1}{3}(1+D-3)$. According to this model, the increase of ζ from 0 to $\approx \frac{1}{3}$ corresponds to an increase of D from 2 to 3.

It should be stressed that our measurements of $g(t)$ described above, do not directly give information about the fractal dimension of the energy-containing eddies; it can only be said that the data invite such an interpretation. The above interpretation of the data in fig. 4 is supported by the recent work of Shreenivasan et al. [3]. They measured the fractal dimension of the interface of two counter-flowing fluids, one of which has been dyed. Such measurements, made in the vicinity of Re_c, support the conclusion that increasing Re above Re_c increases the dimensionality of the turbulent active region. With one adjustable parameter, Re_c, the data of Shreenivasan et al. can be directly superimposed on the measurements in fig. 4 [3].

Even if the energy-containing eddies in a turbulent field occupy regions with dimensionality less than 3, it is not necessary that the entire turbulent region be characterized by a homogeneous fractal. Benzi et al. [14] have proposed a model that the turbulent region is a multifractal object. In their model there is a probability, x, that the turbulent region is space filling ($D=3$) and a probability, $1-x$, that $D=2$. Our measurements are consistent with this model, provided one makes an additional assumption that x is a function of the Reynolds number. The details of the model have been worked out for a general function of $x(Re)$ and fitted to experiment [9]. At present, however, our measurements are not precise enough to confirm that a multifractal model is required to explain the observations.

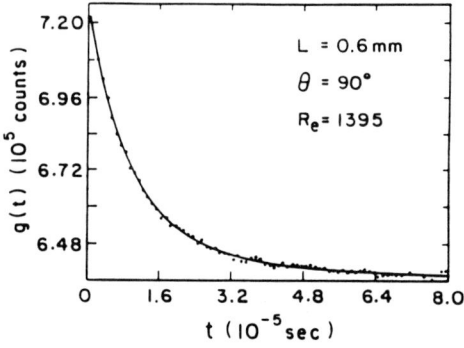

Fig. 6. A typical autocorrelation function $g(t)$ versus t in grid flow. The solid line is a fit to the incomplete gamma function.

4. Concluding remarks

What started out as a study carried out in the time domain (the measurement of $g(t)$), has ended up by yielding spatial information about turbulent flow. The homodyne experiments provide further confirmation of the notion of Mandelbrot that the energy-containing eddies are fractal in their geometrical structure. This finding is not new. What seems to have gone unnoticed before, is that the fractal dimension of the turbulence changes with changing Reynolds number, when some critical value of this parameter is exceeded. The interpretation of these experiments makes no appeal to the frozen turbulence assumption. By using the technique of photon correlation homodyne spectroscopy, we have been able to observe the self-similar behavior of turbulent flows at moderate Reynolds number that were heretofore regarded as too weak to exhibit universal features.

Acknowledgements

We have benefited from illuminating discussions and correspondence with K.R. Shreenivasan, M. Nelkin, and J. Stavans and have enjoyed a continuing fruitful interaction with A. Onuki. We are grateful for the collaboration of our colleague, A. Sirivat and to C.K. Chan for his essential contributions in the early stages of this work. This research was supported by the National Science Foundation under Grant No. DMR-8611666.

References

[1] B.B. Mandelbrot, The Fractal Geometry of Nature (Freeman, San Francisco, 1982).
[2] B. Mandelbrot, Phys. Fluids Suppl. 10 (1967) S302.
[3] K.R. Sreenivasan, R. Ramshankar and C. Meneveau, Proc. Roy. Soc. (London) A 421 (1989) 79.
[4] E.R. Lindgren, Arch. Fys. (1959) 97.
[5] B.J. Berne and R. Pecora, Dynamic Light Scattering (Wiley, New York, 1976).
[6] P.J. Bourke et al., J. Phys. A 3 (1970) 216.
[7] A.N. Kolmogorov, C.R. (Dokl.) Acad. Sci. URSS 30 (1941) 301; 31 (1941) 538.
[8] P. Tong et al., Phys. Rev. A 37 (1988) 2125;
P. Tong and W.I. Goldburg, Phys. Fluids 31 (1988) 2841.
[9] P. Tong and W.I. Goldburg, Phys. Fluid 31 (1988) 3253.
[10] A. Onuki, Phys. Lett. A 127 (1988) 143.
[11] P. Tong and W.I. Goldburg, Phys. Lett. A 127 (1988) 147.
[12] B.B. Mandelbrot, J. Fluid Mech. 62 (1974) 331; in: Lecture Notes in Physics, Vol. 12. Statistical Models and Turbulence, M. Rosenbblatt and C.W. Van Atta, eds. (Springer, Berlin, 1972), p. 333.
[13] U. Frisch, P. Sulem and M. Nelkin, J. Fluid Mech. 87 (1978) 719.
[14] R. Benzi et al., J. Phys. A 17 (1984) 3521.

ASYMMETRIC RANDOM WALK ON A RANDOM THUE–MORSE LATTICE [★]

S. GOLDSTEIN, K. KELLY, J.L. LEBOWITZ [1] and D. SZASZ [2]

Department of Mathematics, Rutgers University, New Brunswick, NJ 08903, USA

Dedicated to Benoit Mandelbrot on the occasion of his 65th birthday

We study the behavior of an asymmetric random walk in a one-dimensional environment whose nonuniformity is in between that of quasi-periodic and random. We construct the environment from arithmetic subsequences of the Thue–Morse sequence. The construction induces in a natural way a measure μ on the space of environments which is invariant and ergodic with respect to translations but is not mixing and has zero entropy. The behavior of the random walk is rather similar to that found by Sinai for the Bernoulli case, when μ is a product measure for which the entropy has its maximum value; i.e. the particle motion is subdiffusive, the displacement growing in time as $(\log t)^{1/\beta}$, $\beta = \log 3/\log 4$. The nature of the dramatic Sinai–Golosov "localization" is however quite different, exhibiting an interesting fractal structure whose nature depends upon the time scale of observation.

1. Introduction

We study the behavior of an asymmetric random walk in a one-dimensional environment whose nonuniformity is in between that of quasi-periodic and random. We will specify the environment by a "spin" configuration $\xi = \{\xi_j\}$, $\xi_j = \pm 1$, $j \in \mathbb{Z}$. Given ξ and some $0 < \epsilon < 1$, the random walk has a transition probability at site j to the right, p_j (and to the left, $1 - p_j$), of the simple form $p_j = \frac{1}{2}(1 + \epsilon \xi_j)$.

We will consider environments $\xi = \{\xi_j\}$ which are obtained from arithmetic subsequences of the Thue–Morse substitutional sequence [1,2], in a manner to be described in section 3. The construction will induce in a natural way a unique measure μ on the space of configurations (possible environments) $\{-1, 1\}^{\mathbb{Z}}$ [2]. This measure μ is invariant and ergodic with respect to translations but is not mixing, and has zero entropy.

We shall later see that despite this lack of randomness in the environment, the behavior of the random walk is rather similar to that found by Sinai [3] for the Bernoulli case, when μ is a product measure for which the entropy has its maximum value; i.e. the particle motion is subdiffusive, the displacement growing as a power of a logarithm in time. To see how this comes about we now discuss briefly the general setting of the Sinai theorem, while retaining the simple relation $p_j = \frac{1}{2}(1 + \epsilon \xi_j)$.

In one dimension it is always possible to define a potential energy function $U(j)$ so that the transition probabilities satisfy the detailed balance condition with respect to the (non-normalized) measure $\exp[-U(j)]$, i.e.

$$p_j / (1 - p_{j+1}) = \exp[U(j) - U(j+1)]. \tag{1}$$

Hence the measure $\exp[-U(j)]$ is stationary. Also, it follows that for a translation-invariant ergodic measure μ on the ξ_j, the condition of no drift is simply $\langle \xi_j \rangle_\mu = 0$.

[★] Research supported in part by NSF Grants DMR-86-12369 and DMS-89-03047.
[1] Also at Department of Physics, Rutgers University, New Brunswick, NJ 08903, USA.
[2] Permanent address: Mathematical Institute of the Hungarian Academy of Sciences, Budapest, Hungary. Also at Princeton University and the Institute for Advanced Study, Princeton, NJ, USA. Research partially supported by the Hungarian Science Research Foundation No. OTKA-819/I.

Essays in honour of Benoit B. Mandelbrot
Fractals in Physics – A. Aharony and J. Feder (editors)

Let us now consider the position X_t of the random walk at integer times t, in an ergodic statistically translation-invariant random environment $\boldsymbol{\xi}$, starting at $X_0=0$. Sinai [3] considers the case in which the ξ_j's are independent random variables. As we shall see in section 2, the potential $U(n)$ grows in proportion to $\sum_{j=1}^{n}\xi_j$; consequently, the fluctuations in $U(n)$ will grow like $\sqrt{n}$ for a typical environment. Now the time necessary for a particle to diffuse over a potential barrier of height H grows (asymptotically) like $\exp(H)$. Therefore the time for the random walk to get to n should grow like $\exp(c\sqrt{n})$ (for some constant c) and the displacement $X_t(\boldsymbol{\xi})$ of the random walk in time t should grow like $(\log t)^2$,

$$X_t(\boldsymbol{\xi}) \sim (\log t)^2 \tag{2}$$

(and $\langle X_t^2(\boldsymbol{\xi}) \rangle \sim (\log t)^4$), for almost all $\boldsymbol{\xi}$ with respect to the product measure [3].

It is clear from the heuristic discussion given here that the essential feature of the environment leading to the behavior (2) is the fluctuation behavior of $\sum \xi_j$ in (2). When this sum goes like n^β then the power of $(\log t)$ on the right-hand side of (2) will be $1/\beta$. In the model we shall consider $\beta = \log 3/\log 4 > \frac{1}{2}$, so the random walk will be even more confined than in the random case.

In addition to the "confinement" expressed by (2), Sinai also showed that for the Bernoulli environment the particle exhibits a more striking form of localization: Sinai proves that the environment $\boldsymbol{\xi}$ determines a function $m(t) = m(t, \boldsymbol{\xi})$ which is a good indicator of the position X_t of the particle in the sense that as $t \to \infty$, $X_t - m(t) = \mathscr{O}[(\log t)^2]$. More striking still, Golosov [4] has shown that $X_t - m(t) = \mathscr{O}(1)$, i.e., that up to an error of order unity, $m(t)$ gives the particle's position at time t, *for arbitrarily large t*. In other words, for any $\epsilon > 0$, almost all of the randomness in X_t arises from the environment, for t sufficiently large.

The localization of our random walk exhibits a richer structure than in the case of a Bernoulli environment. For a typical environment and large t there are three possibilities for the position X_t:

(i) X_t is near the center $m(t)$ of a "V-shaped valley". In this case $X_t - m(t) = \mathscr{O}(1)$.

(ii) X_t is near the center $m(t)$ of a "W-shaped valley". In this case $X_t - m(t) = \mathscr{O}(\log t)^{1/\beta}$ and has an asymptotically singular distribution on a Cantor set.

(iii) X_t is near the center of a random valley.

Here $m(t) = m(t, \boldsymbol{\xi})$ is determined by the environment $\boldsymbol{\xi}$; the environment also determines which of the three cases occurs at time t; for a typical environment all three cases occur repeatedly.

We also analyze the behavior of our random walk on "macroscopic" time scales, again finding several possibilities. In particular, on "critical" time scales, we find a jump process of transitions between valleys. We also analyze more refined asymptotic distributions.

In section 2 we describe some relevant generalities concerning one-dimensional random walks: the potential function and escape times from valleys. In section 3 we define our random environment, and analyze its hierarchical structure. Section 4 contains the results on the asymptotics of the random walk in our random environment.

2. Potentials, valleys and escape times for one-dimensional random walks

The potential function $U(n)$ which satisfies (1) is most conveniently written as a linear interpolation of a function U defined at the midpoints between sites. Specifically, for $n \in \mathbb{Z}$,

$$U(n) = \tfrac{1}{2}[U(n-\tfrac{1}{2}) + U(n+\tfrac{1}{2})], \tag{3}$$

where

$$U(n+\tfrac{1}{2}) = -\log(1+\epsilon\xi_0) - \alpha \sum_{j=1}^{n} \xi_j, \quad \text{for } n \geq 0,$$
$$= -\log(1-\epsilon\xi_0) + \alpha \sum_{j=n+1}^{-1} \xi_j, \quad \text{for } n < 0, \tag{4}$$

where $\alpha = \log[(1+\epsilon)/(1-\epsilon)]$ and where $\log[(1+\epsilon\xi_j)/(1-\epsilon\xi_j)] = \alpha\xi_j$ has been used. This corresponds to the normalization $U(0) = -\tfrac{1}{2}\log(1-\epsilon^2)$.

The potential U defines a stationary (in fact, reversible) measure for the random walk X_t in the random environment ξ. All the basic probabilistic characteristics of this process, such as exit probabilities, expected exit times from an interval, and stationary probabilities for an interval with reflecting endpoints (which can be explicitly obtained by solving the corresponding Dirichlet problem) can be naturally expressed in terms of U. Moreover, the structural features of the graph of U, i.e. the nature of the "valleys", play the critical role in the asymptotic analysis of the process X_t. We say that $[a, b]$ (for $a, b \in \mathbb{Z} + \tfrac{1}{2}$) is a valley for the potential energy function $U(n)$ if there exists a $d \in \mathbb{Z} + \tfrac{1}{2}$ such that $\min_{n \in [a,b]} U(n) = U(d)$, $\max_{n \in [a,d]} U(n) = U(a)$, and $\max_{n \in [d,b]} U(n) = U(b)$ [3]. The height of such a valley is $H = \min\{U(a), U(b)\} - U(d)$.

Let T_H denote the time required to escape from the bottom of a valley of height H. Now as $H \to \infty$, T_H becomes exponential with mean

$$\bar{t} \equiv \langle T_H \rangle \sim e^H. \tag{5}$$

Hence

$$\text{Prob}[T_H < t] \sim 1 - e^{-t/\bar{t}} \approx t e^{-H}, \quad \text{for } t \ll \bar{t},$$
$$\approx 1, \quad \text{for } t \gg \bar{t}. \tag{6}$$

Therefore the probability of escaping by time t from a valley of size $c \log t$ is, for $c > 1$, proportional to $t e^{-c \log t} = t^{1-c} \to 0$ as $t \to \infty$, while, for $c < 1$, this probability approaches 1 as $t \to \infty$. Thus we have for the "critical potential barrier for time t", $H_{cr} \sim \log t$. Hence if the fluctuations in $U(m) \sim m^\beta$, we have $H_{cr} \sim m_{cr}^\beta$, and X_t should be near $m_{cr} \sim (\log t)^{1/\beta}$.

3. A random environment developed from the Thue–Morse sequence

For any non-negative integer n, let $b(n)$ denote the sum of the binary digits of n, and let $a_n = (-1)^{b(n)}$. The sequence $a_0, a_1, a_2, \ldots$, is the well-known Thue–Morse sequence [1]. Equivalently, this sequence may be generated by the substitution rule γ: $+1 \mapsto +1, -1$; $-1 \mapsto -1, +1$. γ maps any finite ± 1 sequence to another; we observe that $\gamma^k(+1)$ generates the first 2^k terms of the Thue–Morse sequence. The dynamical systems resulting from substitutions of this form are studied extensively by Queffélec [2].

The Thue–Morse sequence possesses an important block structure. For $k \geq 0$, let A_k denote the block $a_0, a_1, \ldots, a_{4^k-1}$, and let $\bar{A}_k$ denote $-a_0, -a_1, \ldots, -a_{4^k-1}$. Then, due to the substitution rule, the block A_{k+1} of size 4^{k+1} may be formed by concatenating blocks of size 4^k:

$$A_{k+1} = A_k \bar{A}_k \bar{A}_k A_k. \tag{7}$$

We refer to the blocks A_k and $\bar{A}_k$ as the k-blocks, since any block of size 4^k (starting from a multiple of 4^k) in the sequence will be either A_k or $\bar{A}_k$.

Define $\Omega = \{\text{limit points of } S^n(\gamma^\infty(+1)) \mid n \to \infty\}$, where S is the left shift on $\{+1, -1\}^\mathbb{Z}$, i.e. Ω is the set of doubly infinite sequences obtained by shifting the semi-infinite Thue–Morse sequence to the left by n steps and then taking limits $n \to \infty$ along subsequences. Let μ be the unique S-invariant probability measure on Ω. (μ, S) is ergodic, but not even weakly mixing (in fact, S^2 is not ergodic), and it has entropy 0 [2].

The block structure of the Thue–Morse sequence allows a more "constructive" definition of Ω and μ. Let ω_0, η_1, η_2, ... be a sequence of independent random variables, with ω_0 taking on values $+1$ and -1 with equal probability, and each η_j taking on values 1, 2, 3, and 4 with equal probability. This sequence determines a (random) element $\omega = \{\omega_n | n \in \mathbb{Z}\} \in \Omega$ as follows: For every $k \geq 1$, ω is built out of a doubly infinite sequence of k-blocks. η_1 gives the position of the origin in the 1-block containing the origin: $\eta_1 = i$ if the origin is the ith site in this block. Similarly, η_k gives the position of the $(k-1)$-block containing the origin in the k-block containing the origin. Finally ω_0 is of course the value of ω at the origin. Note that ω_0 and η_1 determine whether the 1-block containing the origin is A_1 or $\bar{A}_1$; these together with η_2 determine whether the 2-block containing the origin is A_2 or $\bar{A}_2$; and so on. Moreover, the product measure distribution on ω_0, η_1, η_2, ... gives rise to the previously mentioned distribution μ on Ω. (Sequences with $\eta_k = 1$ for all $k > N$ or with $\eta_k = 4$ for all $k > N$, for some $N > 0$, do not correspond to elements of Ω because they define only a semi-infinite sequence. But there are exactly two elements of Ω to which such a sequence can be extended. The set of all such sequences (which of course has probability zero) is just $\pm$ the Thue–Morse sequence, preceded by $\pm$ its reflection, together with all of its translates. Some of these sequences will play an important role in the asymptotics.)

We now describe the random environment we shall consider for the rest of this paper. Since the behavior of a random walk depends heavily upon the fluctuation behavior of $\Sigma \xi_j$, we base our environment upon its arithmetic subsequences of difference 3, i.e. we define the set of all possible environments to be $\Xi = \{\xi : \xi_j = \omega_{3j}, \omega \in \Omega\}$. Now, since every $\omega \in \Omega$ has the block structure described in the previous paragraph, it suffices to consider the fluctuations associated with the rarefactions $(a_i, a_{i+3}, a_{i+6}, \ldots)$, $i \in \{0, 1, 2\}$, of the Thue–Morse sequence itself. Unlike the original sequence, these rarefied sequences exhibit significant fluctuations, cf. Newman [5] and Coquet [6].

More precisely, to determine the fluctuations, we calculate rarefied sums of the form

$$S_i(n) = \sum_{0 \leq j < n} a_j, \quad j \equiv i \pmod{3}. \tag{8}$$

The following relations are easily derived from the block structure of the Thue–Morse sequence, and the fact that $4^k \equiv 1 \pmod{3}$: First, the block A_k (of size 4^k) is symmetric, so for $k \geq 0$

$$S_1(4^k) = S_2(4^k). \tag{9}$$

Next, since $A_{k+1} = A_k \bar{A}_k \bar{A}_k A_k$,

$$S_i(4^{k+1}) = S_i(4^k) - S_{i+2}(4^k) - S_{i+1}(4^k) + S_i(4^k). \tag{10}$$

Applying these relations, and the fact that $S_0(4^1) = 2$ and $S_1(4^1) = -1$, we see that for $k > 0$

$$S_0(4^k) = 2 \cdot 3^{k-1}, \quad S_1(4^k) = S_2(4^k) = -3^{k-1}, \tag{11}$$

corresponding to fluctuations in $\sum_{j=0}^{n} \xi_j$ on the order of n^β, where $\beta = \log 3 / \log 4$.

We will now analyze the structure of the potential $U(n)$ by examining the detailed graphs of the $S_i(n)$. We claim that (11), together with the hierarchical structure of our environment, completely determines U. As suggested by (3) and (4), it is convenient to focus on the values of U on $\mathbb{Z} + \frac{1}{2}$, for which we define a sequence of piecewise linear approximations. These approximations will be built up from pieces $\Sigma \xi_j$ over basic blocks of the environment. We begin by defining, for each $k > 0$, linear functions $R_0^{(k)}$ on $[0, \frac{1}{3}(4^k + 2)]$, and $R_1^{(k)}$ and $R_2^{(k)}$ on $[0, \frac{1}{3}(4^k - 1)]$ by specifying their values at their endpoints: $R_i^{(k)}(0) = S_i(0) = 0$ (for $i = 0, 1, 2$); $R_0^{(k)}(\frac{1}{3}(4^k + 2)) = S_0(4^k) = 2 \cdot 3^{k-1}$; and $R_i^{(k)}(\frac{1}{3}(4^k - 1)) = S_i(4^k) = -3^{k-1}$ (for $i = 1, 2$). Clearly, $R_i^{(k)}(n)$ is a linear approximation to $S_i(3n)$ on the interval of definition; we may obtain a better approximation on the same interval by "concatenating" several $R_i^{(k-1)}$ functions:

The "concatenation" of two continuous functions is the operation of putting the graphs of the functions end to end. Suppose $f_1: [0, x_1] \to \mathbb{R}$ and $f_2: [0, x_2] \to \mathbb{R}$. We define a function $f_1 \square f_2$ (the "concatenation" of f_1 and f_2) on $[0, x_1 + x_2]$ as follows:

$$(f_1 \square f_2)(x) = f_1(x), \qquad \text{for } 0 \leq x < x_1, \qquad (12)$$
$$= f_1(x_1) + f_2(x - x_1), \quad \text{for } x_1 \leq x \leq x_2.$$

We now recall that, for $k > 0$, $A_k = A_{k-1} \bar{A}_{k-1} \bar{A}_{k-1} A_{k-1}$, and $4^{k-1} \equiv 1 \pmod 3$. Hence we obtain the following "refined" approximation of $S_i(3n)$:

$$S_i(3n) \approx [R_i^{(k-1)} \square (-R_{i+1}^{(k-1)}) \square (-R_{i+1}^{(k-1)}) \square R_i^{(k-1)}](n), \qquad (13)$$

with equality holding at all endpoints and concatenation points. We call this approximation a "refinement" of $R_i^{(k)}$. Now, each of the terms of (13) may be further refined, yielding successively better approximations, until after k iterations, we obtain a function which coincides with $S_i(3n)$ for all integers n in the interval of definition.

Figs. 1a–1c show the first refinements of $R_i^{(k)}$, for $i \in \{0, 1, 2\}$. These three graphs, together with their opposites ($-R_i^{(k)}$, for $i \in \{0, 1, 2\}$), form the "building blocks" for successive refinements of each $R_i^{(k)}$. Fig. 2 illustrates this process through the construction of the next refinement of $R_1^{(k)}$.

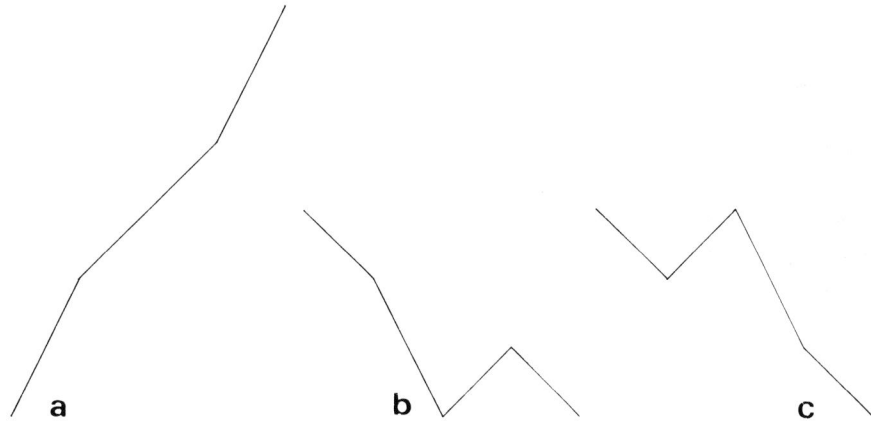

Fig. 1. (a)–(c) Refinements of the graphs of $R_i^{(k)}$ (for $i = 0, 1, 2$ respectively), as given by (13).

Fig. 2. Second refinement of the graph of $R_1^{(k)}$.

Since every sequence $\xi \in \Xi$ possesses the same block structure as the rarefied Thue–Morse sequence, we can describe the fluctuations of U corresponding to any ξ by using the linear pieces $R_i^{(k)}$ defined above. Given ξ, let $\{b_j^{(k)} \equiv b_j^{(k)}(\xi)\}_{j=-\infty}^{\infty} \subset \mathbb{Z} + \frac{1}{2}$, $b_j^{(k)} < b_{j+1}^{(k)}$, denote the set of points in between the rarefied k-blocks of ξ. (It is not hard to see that this set is uniquely determined by ξ.) We shall call the graph of the piecewise linear function which interpolates U on $\{b_j^{(k)}\}$ the *k-graph* of U. This coincides (up to a translation and multiplication by α) with a doubly infinite concatenation of $\pm R_i^{(k)}$'s. In fact, this concatenation will be of the form

$$\ldots \Box \pm R_i^{(k)} \Box \pm R_{i+2}^{(k)} \Box \pm R_{i+1}^{(k)} \Box \pm R_i^{(k)} \Box \ldots, \tag{14}$$

with the signs determined by the corresponding $\omega \in \Omega$. Inspection of figs. 1a–1c demonstrates that absolute extrema of U on $[b_j^{(k)}, b_{j+1}^{(k)}]$ occur at the endpoints. Thus, in order to find the valleys of U of height, say, 3^k, it suffices to check the k-graph of U, and hence to check this concatenation for local minima at the concatenation points. We find that three of the possible concatenations lead to local minima at the concatenation point:

$$\ldots \Box R_2^{(k)} \Box - R_1^{(k)} \Box \ldots, \tag{15}$$

$$\ldots \Box R_1^{(k)} \Box R_0^{(k)} \Box \ldots, \tag{16}$$

$$\ldots \Box - R_0^{(k)} \Box - R_2^{(k)} \Box \ldots. \tag{17}$$

The type of minimum seen in (15) behaves completely differently from the other two during subsequent refinements. Figs. 1b and 1c reveal that the first type of local minimum is conserved, always remaining the unique deepest point in a symmetric, roughly V-shaped valley.

On the other hand, one may conclude from fig. 1 and (13) (see also fig. 2) that the minima which arise from (16) and (17) are doubled at each successive refinement, thus leading to approximate Cantor-set-like minima. We will refer to these valleys as W-shaped valleys.

Note that further refinement of a valley of height 3^k in the k-graph of U will produce subvalleys of height at most 3^{k-1}.

We require some terminology for describing the valleys in our structure. Since all V- and W-shaped valleys will have height 3^k for some $k \in \mathbb{Z}$, we refer to such valleys as "valleys of order k", or simply, "k-valleys". Also, when we specify the endpoints $[a, b]$ (where $a, b \in \mathbb{Z} + \frac{1}{2}$) of a k-valley, we will require that a and b be the endpoints of rarefied k-blocks in the environment. (This distinction will become important in our discussion of escape times from k-valleys.)

4. Results

We present here some results concerning the long time behavior of the random walk in our Thue–Morse environment ξ chosen according to the measure μ. Most of our results deal with the motion of a particle in a fixed environment ξ which is typical for the measure μ: We let P_ξ denote the probability distribution on random walk trajectories in the environment ξ. Our results are consequences of the recursive nature of our construction, and of the structure of valleys in our model.

We begin with the most basic result about the growth behavior of the random walk.

Proposition 1. There exists a positive constant C such that for μ-almost every ξ

$$\lim_{t \to \infty} {}^{\sup}_{\inf} X_t / (\log t)^{1/\beta} = \pm C \tag{18}$$

P_ξ almost surely, where $\beta = \log 3/\log 4$.

This proposition is a consequence of several of our later results, as well as the ergodic theorem, applied to the renormalization group transformation $\varphi: \Xi \to \Xi$ defined by $\varphi((\lambda, \omega_0, \eta_1, \eta_2, ...)) = (\lambda', \pm\omega_0, \eta_2, \eta_3, ...)$ where the sign of $\pm\omega_0$ is $+$ if $\eta_1 = 1$ or 4 and $-$ if $\eta_1 = 2$ or 3, where $\lambda' = \lambda + \eta_1 - 1$ and the "λ" coordinate is the integer (mod 3) indicating the "type" of rarefied block containing the origin ($\pm R_0$, $\pm R_1$, or $\pm R_2$) on the relevant scale, with initially, on the microscopic scale, $\lambda = 0$, and where $(\omega_0, \eta_1, \eta_2, ...)$ encodes the environment ω as in section 3.

Our next proposition deals with our ability to predict the location of a particle after a long time. Proposition 2 will imply that for most choices of time t and environment ξ, we can predict the location and type of valley in which the particle will be trapped. The deterministic functions $m(t)$ and $u(t)$, which depend on ξ, and which we are about to define, will give respectively the center of the valley and the type of valley (V- or W-shaped) in which the particle will be trapped, for most choices of ξ and t.

The positions of our particle at large times will typically be near the "centers" $m_k = m_k(\xi) \in \mathbb{Z} + \frac{1}{2}$ of certain k-valleys, where $k = k(t)$ and m_k depend on t (and ξ) in a manner which we will now specify. Consider the k-graph of U. Starting at the point $(0, U(0))$ on this graph, proceed "downhill" to the first minimum; this point will be a minimum of the k-valley we wish to specify. Suppose this point is $(s, U(s))$; then we define m_k according to the type of concatenation at that point. If the concatenation is of type (15), s is the unique minimum in a V-shaped valley, so we set $m_k = s$. If the concatenation is of type (16) or (17), s is a minimum of a W-shaped valley, but this minimum is not unique. Since we want m_k to represent the center of the W-shaped valley, we set $m_k = s - \frac{1}{3}(4^{k-1} - 1)$ for type (16) valleys, and $m_k = s + \frac{1}{3}(4^{k-1} - 1)$ for type (17) valleys. Furthermore, we denote by $u_k \equiv u_k(\xi)$ the type of k-valley at m_k; i.e., $u_k = $ V or W depending upon whether m_k is the center of a V- or W-shaped k-valley.

Next, we define a sequence of "critical" times $t_k \equiv t_k(\xi)$ for our process. First, we define t_k^V and t_k^W, the "escape times" from V- and W-shaped k-valleys, in the following manner. Let $[a, b]$ be a V-shaped k-valley (with endpoints specified according to the remark at the end of section 3). Then $t_k^V = 2E_{(a+b)/2 \pm 1/2}(T_{\{a,b\}})$, i.e. twice the expected time of arrival at either endpoint of the valley, having started from the bottom. (We multiply by 2 here because we want $(t_k^V)^{-1}$ to represent the rate of "one-sided escape".) Taking into account the detailed, fractal-like structure near the top of the V-valley, we find that $t_k^V = c_k^V e^{\alpha 3^k}$, where c_k^V is of order 2^k, and $\alpha = \log[(1+\epsilon)/(1-\epsilon)]$.

Now, let $[a, b]$ be a W-shaped k-valley. To be definite, suppose $[a, b]$ is a k-valley formed by a concatenation of type (17). Then $t_k^W = 2E_{(a+b)/2 \pm 1/2}^{(r)}(T_{\{b\}})$, twice the expected time of arrival at the right (i.e. "lower") endpoint, assuming that we impose, say, a reflecting boundary condition on the left end of the interval. We multiply by 2 to account for the fact that once the particle reaches point b, it is equally likely (by symmetry) to proceed down from the peak at b in either direction. (The reader should convince himself that b is at a concatenation point of the form $...\square - R_2^{(k)} \square R_1^{(k)} \square ...$, which gives the aforementioned symmetry.) Now, $t_k^W = c_k^W e^{\alpha 3^k}$, where c_k^W is of order unity.

For k large, the one-sided escape times T_k^V and T_k^W from (the bottom of) V- and W-shaped k-valleys are approximately exponential with means t_k^V and t_k^W. More precisely, as $k \to \infty$, T_k^V/t_k^V and T_k^W/t_k^W converge in distribution to exponential random variables of mean 1.

Finally, we define the critical time $t_k \equiv t_k(\xi)$ by $t_k = t_k^{u_k}$.

We may now define $k \equiv k(t)$, the order of the valley in which our particle is likely to be trapped at time t, by setting $k(t) = k$ whenever $t_{k-1} < t \leq t_k$. Using $k(t)$, we define (with a slight abuse of notation) $m(t)$ and $u(t)$ by $m(t) = m_{k(t)}$ and $u(t) = u_{k(t)}$. For "most" times t, our particle will be near $m(t)$, the center of a $k(t)$-valley of type $u(t)$. (In this regard, the reader should convince himself that when a W-shaped k-valley "destabilizes" into a V-shaped k-valley, the latter valley in fact lies at the center of a V-shaped $(k+1)$-valley.)

We note, however, that if the origin of our random walk is in a V-shaped $(k-1)$-valley resulting from a single further refinement of either $...\square R_0^{(k)} \square R_2^{(k)} \square ...$ or $... \square - R_1^{(k)} \square - R_0^{(k)} \square ...$, then m_k is not a very good indicator of the position of the particle, even at times t for which $k(t) = k$, since the particle could have descended from the V-shaped $(k-1)$-valley in either direction with probability of order unity. Let $\Xi_k \subset \Xi$ be the complement (in Ξ) of the set of environments of the type just described. A simple computation shows that $\mu(\Xi_k) = \frac{17}{18} + o(1)$ (as $k \to \infty$). Again abusing notation, we define $\Xi(t)$ by setting $\Xi(t) = \Xi_k$ if $t_{k-1}^V < t \le t_k^V$. (If $\xi \notin \Xi_k$, then, typically, $t_{k-1} = t_{k-1}^V$.) For environments $\xi \in \Xi(t)$, the position X_t of the random walk should be close to $m(t)$.

In stating the following propositions, we require the function $[\log](t) \equiv 3^{k(t)}$. (Note that this depends, through $k(t)$, on ξ.) $[\log](t)$ is a left-continuous step function with jumps at t_k, $k = 1, 2, ...$, and for $t = t_k$, $[\log](t) = \alpha^{-1} \log(t/c_k^{u_k})$.

Proposition 2. For every $\xi \in \Xi$, as "$t \to \infty$" (in a sense explained after this proposition),

$$P_\xi\{[X_t - m(t)]/B_{u(t)}(t) < x\} - F_{u(t)}(x) \to 0, \tag{19}$$

where

$$F_u = \mathcal{V} \quad \text{if } u = V,$$
$$= \mathcal{W} \quad \text{if } u = W,$$

with $\mathcal{V}$ and $\mathcal{W}$ denoting specific probability distribution functions which will be described later, and

$$B_u(t) = 1 \quad \text{if } u = V,$$
$$= 4^{k(t)} [= ([\log](t))^{1/\beta}] \quad \text{if } u = W.$$

In the statement of the proposition, by "$t \to \infty$" we mean that $t \to \infty$ in such a way that $\xi \in \Xi(t)$, and t stays away from the times t_k, in the sense that $\text{dist}(\log t, \{\log t^k\}_{k=1}^\infty) \to \infty$.

For V-shaped valleys, the random walk exhibits a strong, Golosov-type localization, while for W-shaped valleys the random walk spreads out around $m(t)$ on a scale of order $([\log]t)^{1/\beta}$. Figs. 3 and 4 illustrate the distribution of a particle during random walks in V- and W-shaped valleys respectively.

We remark that while for $\xi \notin \Xi(t)$, $m(t)$ is not a good indicator of X_t, it is nonetheless possible to give a complete description of the situation. (For brevity, we omit the details.)

Proposition 3. As $t \to \infty$,

$$\mu(\{m(k)/4^k < x | \xi \in \Xi_k\}) \to \mathcal{M}(x), \tag{20}$$

where $\mathcal{M}$ is a probability distribution function.

The following proposition describes the behavior of our process X_t when observed on "macroscopic" length and time scales; after suitable rescaling we obtain a process $X_t^{(T)}$ where T defines the time scale and $t \in \mathbb{R}^+$ is the time variable on time scale T. Parts (ii) and (iii) follow from proposition 2 and the fact that if $k(T) = k$, then on time scale T our particle quickly escapes from valleys of order $k-1$ or smaller. Part (i) is an immediate consequence of part (iii).

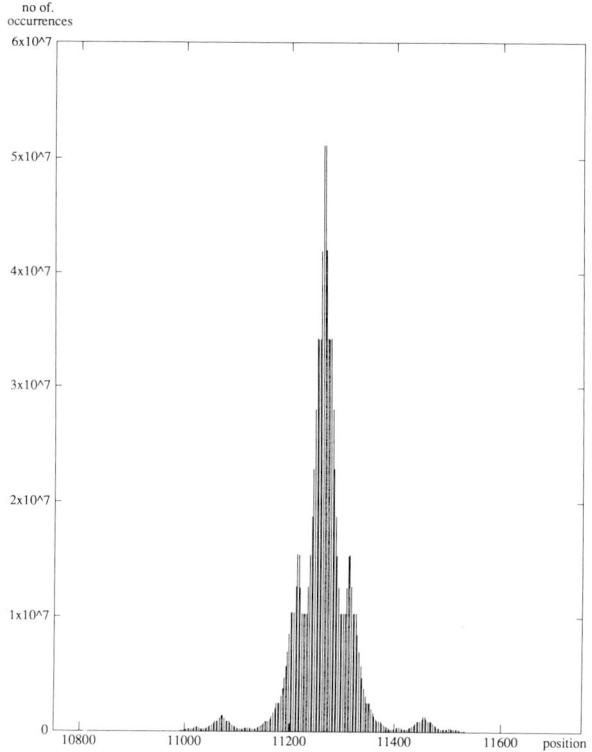

Fig. 3. Distribution of a random walk of 10^9 steps in a V-shaped valley ($\epsilon=0.1$).

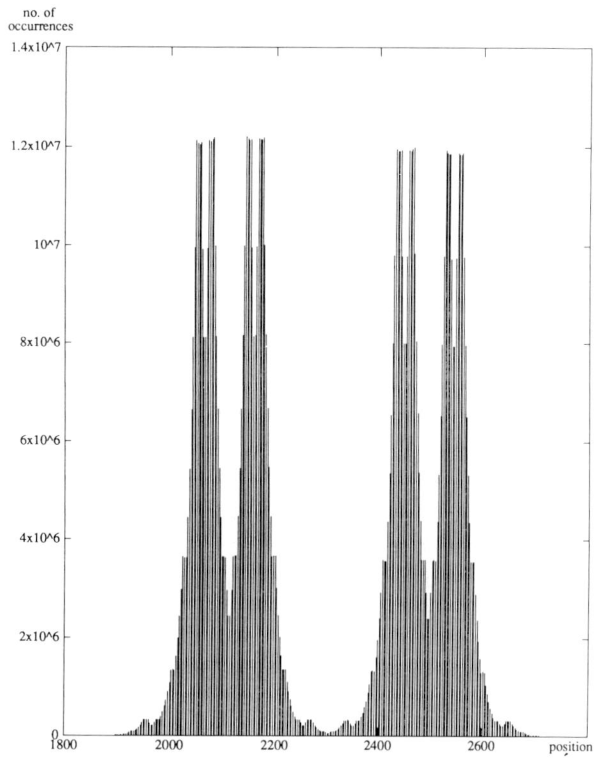

Fig. 4. Distribution of a random walk of 10^9 steps in a W-shaped valley ($\epsilon=0.1$).

Proposition 4. Consider the rescaled process (for $t \geq 0$, $T \in \mathbb{Z}^+$)

$$X_t^{(T)} := [X_{Tt} - m(T)]/([\log]T)^{1/\beta}. \tag{21}$$

Then for every $\xi \in \Xi$ the following assertions are valid:

(i) As "$T \to \infty$" (in the same sense as in proposition 2) in such a way that $u(T) = V$, we have $X_t^{(T)} \to 0$ in P_ξ probability.

(ii) As "$T \to \infty$" in such a way that $u(T) = W$, we have $\{X_t^{(T)}\}_{t>0} \to \{Y_t\}_{t>0}$, where the convergence is to be understood in the sense of finite-dimensional distributions, and where $\{Y_t\}_{t>0}$ is a process with independent, identically distributed values whose common distribution is $\mathscr{W}$ (see proposition 2).

(iii) As "$T \to \infty$" in such a way that $u(T) = V$, we have $\{X_{Tt} - m(T)\}_{t>0} \to \{Z_t\}_{t>0}$, where the sense of convergence is as in (ii), and where $\{Z_t\}_{t>0}$ is a process with independent, identically distributed values whose common distribution is $\mathscr{V}$.

Proposition 4 describes the behavior of the rescaled process $X_t^{(T)}$ for values of T which are not near any of the critical times t_k. According to the proposition, on such time scales T, our particle stays near $m(T)$ on a microscopic scale if $u(T) = V$, or, if $u(T) = W$, jumps very rapidly among the sites in a W-valley of a "macroscopic" size ($[\log]T)^{1/\beta}$. In the latter case, this motion would be visible only as a glob (a "Cantor glob", see proposition 6). Thus when viewed on macroscopic scales our process defines either point-like or extended structures (see the initial part of fig. 5). These apparent structures, their sizes and their locations depend upon the time scale of observation.

The transition between structures visible on time scales $T' < t_k < T''$ separated by a critical time t_k is best observed on a critical time scale $T = ct_k$, where c is a constant of order unity. On such a time scale, the k-valley at m_k "destabilizes", as does the structure defined by X_t^T, resulting in a Markov process of jumps between k-valleys which ends, after at most a few jumps, either in a V-shaped k-valley (which lies at the bottom of a V-shaped $(k+1)$-valley), or in a W-shaped $(k+1)$-valley, with associated "Cantor glob" (recall that $t_k^V \sim 2^k t_k^W$), or in a state of oscillation between two W-shaped k-valleys (which together form a W-shaped $(k+1)$-valley), corresponding to oscillation between the associated "Cantor globs" (see the initial part of fig. 5). In either case, the essential characteristic of the rescaled process on a critical time scale is that the particle makes a small number of transitions between valleys of height 3^k, and then reaches a valley of height 3^{k+1}, where of course it will remain trapped until times on the next critical time scale. (It is not difficult to give a complete description of all the jump Markov processes which arise from the process $X_t^{(T)}$, for T a critical time scale. There are only a small number of possibilities; we leave them as an exercise for the reader.)

We remark that for the Sinai (Bernoulli) environment a similar picture should hold, the main differences being that in the Bernoulli case there will be no extended structures – only Golosov (microscopic) localization – and no oscillations; in fact, the process of jumps between valleys will involve only a single jump, almost surely.

The remaining propositions concern asymptotic distributions. First consider the distribution function $\mathscr{V}$, first mentioned in proposition 2. As $k \to \infty$, the environment for which a V-shaped k-valley is centered near the origin,

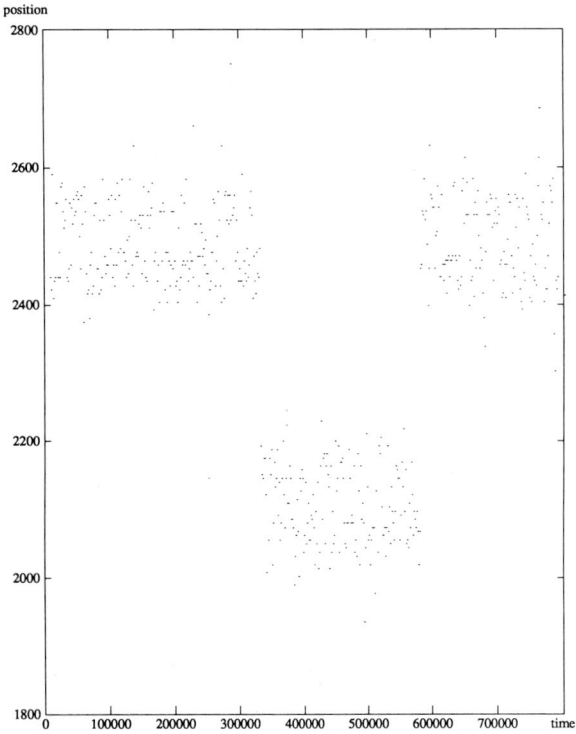

Fig. 5. Path of a typical random walk in a W-shaped valley ($\epsilon = 0.1$; position recorded every 1000 steps).

say at $-\frac{1}{2}$, converges to the environment consisting of the rarefied negative Thue–Morse sequence $\{-a_{3n+1}\}_{n=0}^{\infty}$, preceded by the opposite of the reflection of this sequence about $-\frac{1}{2}$. Since $m(t) \in \mathbb{Z} + \frac{1}{2}$, we easily obtain the following proposition.

Proposition 5. $\mathscr{V}$ is the probability distribution function for the probability measure ρ_V on $\mathbb{Z} + \frac{1}{2}$ given by

$$\rho_V(n+\tfrac{1}{2}) = \frac{e^{-U(n)}}{\sum e^{-U(m)}}, \quad \text{for } n \in \mathbb{Z}, \tag{22}$$

where U is the potential for the environment

$$\xi_j = -a_{3j+1}, \quad j = 0, 1, 2, \ldots$$
$$= a_{-3j-2}, \quad j = -1, -2, \ldots.$$

The distribution function $\mathscr{W}$, describing the asymptotic distribution in a W-valley, is somewhat more intricate. Note that the limit of the environments which give rise to the centered W-shaped k-valleys is of little interest. This limit is, in fact, the negative of the V-type environment of proposition 5. This inverted V-type environment is of little interest because our particle will asymptotically be infinitely far from the center, so this environment has little to do with the environment seen by the particle. (The denominator of (22) for this environment would be infinite.)

Proposition 6. $\mathscr{W}$ is the probability distribution function for the Cantor measure ρ_W on the interval $I = [-\frac{1}{12}, \frac{1}{12}]$ based on the sequence of removals of middle halves, starting with (Lebesgue measure on) I.

The measure ρ_W is the limit of the sequence of probability measures $\rho^{(k)}$ describing a uniform distribution over the 2^k intervals which remain after k interactions of the removal of middle halves. In order to understand proposition 6, it is useful to note that ρ_W may be characterized as follows: ρ_W is the (unique) probability measure supported on the (middle-half-removal) Cantor set which gives equal weight to each of the 2^k intervals $I_j^{(k)}$ remaining after k iterations. Moreover, if ρ_n is a sequence of probability measures which, for each fixed k, asymptotically (as $n \to \infty$) (i) gives equal weight to each interval $I_j^{(k)}$, and (ii) is supported by $\bigcup_j I_j^{(k)}$, then $\rho_n \to \rho_W$. Such a sequence is provided by the distribution ρ_n of $(X-m)/4^n$, where m is the center of a W-shaped n-valley $[a, b]$ with corresponding potential U_W, and X is distributed on $[a, b] \cap \mathbb{Z}$ according to $\exp[-U_W(j)]/\sum \exp[-U_W(i)]$. Moreover, for $n = n(t)$, $u(t) = W$, and t large, ρ_n is a good approximation to the distribution of $[X_t - m(t)]/4^n$.

Our last two propositions concern more refined asymptotic distributions, namely, the distribution of the "environment seen by the particle" and the joint distribution of this environment and the position of the particle, at times for which the particle is in a W-valley. (These distributions are trivial when $u(t) = V$.)

Proposition 7. Let $S: \Xi \to \Xi$ be a translation by one unit to the left, $S\{\xi_j\} = \{\xi_j'\}$ where $\xi_j' = \xi_{j+1}$. Then for every $\xi \in \Xi$ the pair $([X_t - m(t)]/([\log]t)^{1/\beta}, S^{-X_t}\xi)$ converges in distribution to $(Y, \tilde{\xi})$ as "$t \to \infty$" in such a way that $u(t) = W$, where Y has distribution ρ_W (Cantor measure, with distribution function $\mathscr{W}$), $\tilde{\xi}$ is a random environment with distribution ν (to be specified in proposition 8), and Y and $\tilde{\xi}$ are independent.

The explanation of proposition 7 is as follows: For k large, the (rescaled) position of our particle when in a W-shaped k-valley may be well approximated by the location of the nearest of the 2^n absolute minima of the

$(k-n)$-graph of U over this W-valley. Each of these minima, upon further refinement, yield precisely the same (Cantor-like) local detailed structure for the potential. The convergence in distribution of the proposition refers to the usual weak convergence of measures (i.e. convergence on bounded continuous functions), which for the convergence of the environment distribution amounts to convergence on (microscopically) local functions of the environment. Thus, since each approximate graph of U around an absolute minimum splits upon refinement into 2 pieces which are mirror images of each other, the proposition follows, with ν as described in the next proposition.

Note that independence is achieved only in the limit; in fact, for any finite t, the environment $S^{-X_t}\xi$ seen by the particle is completely determined by X_t (for fixed ξ). Independence is achieved in the limit because the position on the macroscopic scale is well approximated in terms of structures related to refinements from the macroscopic scale down, while the environment seen by the particle is well approximated by structures built up from the microscopic scale.

Proposition 8. For every $\boldsymbol{\sigma}=\{\sigma_k\}_{k=2}^\infty$, $\sigma_k=\mathrm{L},\mathrm{R}$, let $\boldsymbol{\xi}^{(\sigma)}\in\varXi$ be the (unique) environment such that for every $k\geq 2$ the origin is in a W-valley of order k, with an absolute minimum at $-\tfrac{1}{2}$, which is on the left (right) side of the W-valley if $\sigma_k=\mathrm{L}(\mathrm{R})$. (If $\boldsymbol{\sigma}$ ends with all L's or all R's, ξ^σ is "one-sided" and is not actually an element of $\varXi$.)

Let $\tilde{\nu}$, on $\tilde{\varXi}_\mathrm{W}$, be the image of the $(\tfrac{1}{2},\tfrac{1}{2})$ Bernoulli measure on $\{\boldsymbol{\sigma}\}$ under the map $\boldsymbol{\sigma}\to\boldsymbol{\xi}^{(\sigma)}$. Let $\varXi_\mathrm{W}=\bigcup_{j=-\infty}^\infty S^j(\tilde{\varXi}_\mathrm{W})$, and let ν' be the translation invariant measure on $\varXi$ which agrees with $\tilde{\nu}$ on $\tilde{\varXi}_\mathrm{W}$. Then the probability measure ν of proposition 7 satisfies

$$\nu(\mathrm{d}\xi)\equiv\mathrm{d}\nu=\frac{\mathrm{e}^{-U(0)}\tilde{\nu}(\mathrm{d}\xi)}{\int\mathrm{d}\tilde{\nu}\,\mathrm{e}^{-U(0)}}=\frac{\mathrm{e}^{-U(0)}\,\mathrm{d}\tilde{\nu}}{\int\mathrm{d}\tilde{\nu}\,\mathrm{e}^{-U(0)}}, \qquad (23)$$

where the potential $U(n)\equiv U(n,\xi)$ is now normalized, so that U is, say, 0 at the absolute minima.

We remark that in terms of the coding of section 3, $\tilde{\varXi}_\mathrm{W}\subset\varXi$ consists of environments coded by $(\omega_0,\eta_1,\eta_2,...)$ for which

$$\begin{aligned}\eta_{k+1}&=1\quad\text{or}\quad 3,\quad\text{for }\eta_k=1,2,\\ &=2\quad\text{or}\quad 4,\quad\text{for }\eta_k=3,4,\end{aligned} \qquad (24)$$

for $k\geq 2$ and $\eta_1=2$, $\omega_0=+1$ when η_2 is even, $\eta_1=3$, $\omega_0=-1$ when η_2 is odd. Note that $\mu(\varXi_\mathrm{W})=0$. The environments of $\varXi_\mathrm{W}$ are precisely those with potential $U(n)$ bounded below, apart from the translates of the V environment of proposition 5. It would be natural to regard the measure ν' on $\varXi_\mathrm{W}$ as $\mu(\cdot|\varXi_\mathrm{W})$, the conditional distribution given $\varXi_\mathrm{W}$ arising from μ. However $\nu'(\varXi_\mathrm{W})=\infty$, and it is only when ν' is adjusted to reflect the dynamical properties of our random walk that a normalizable measure is obtained.

We conclude by mentioning one aspect of the asymptotics which we have not analyzed in detail, namely, "the motion on a log–log time scale". More precisely, by this we mean the process m_k, $k=1,2,...$, or the process (m_k,u_k), $k=1,2,...$, on the probability space $\{\varXi,\mu\}$. The basic structure of these processes should follow from the hierarchical structure of $\varXi$. Note that the time 1 map $(k\to k+1)$ for these processes is induced by the renormalization group transformation φ described after proposition 1. We leave the computation to the reader.

Acknowledgements

We thank J. Peyrière for useful discussions regarding the Thue–Morse sequence. This work was completed during D. Szasz' visit at the Departments of Mathematics of Rutgers University and Princeton University and in the School of Mathematics of the Institute for Advanced Study, in Spring 1988 and 1989. D. Szasz expresses his sincere gratitude to Joel Lebowitz, John Mather, Tom Spencer and Arthur Wightman for their kind hospitality.

References

[1] M. Morse, Trans. AMS 22 (1921) 84–100.
[2] M. Queffélec, Substitution Dynamical Systems – Spectral Analysis, Lecture Notes in Mathematics, vol. 1294 (Springer, Berlin, 1987).
[3] Ya.G. Sinai, Theory Prob. Its Appl. 27 (1982) 256–268.
[4] A.O. Golosov, Russian Mathematical Surveys 39 (1984) 157–158.
[5] D.J. Newman, Proc. AMS 21 (1969) 719–721.
[6] J. Coquet, Inventiones Mathematicae 73 (1983) 107–115.

AGGREGATES, BROCCOLI AND CAULIFLOWER

Francois GREY and Jørgen K. KJEMS

Risø National Laboratory, DK-4000 Roskilde, Denmark

Naturally grown structures with fractal characters like broccoli and cauliflower are discussed and compared with DLA-type aggregates. It is suggested that the branching density can be used to characterize the growth process and an experimental method to determine this parameter is proposed.

1. Introduction

Numerous structures that result from natural growth processes have been analyzed in terms of fractal geometry since Mandelbrot pointed out the power of this approach [1]. In this paper we want to discuss the relation between examples of typical organic tree-grown structures like broccoli and cauliflower and the inorganic aggregate structures that are found in colloidal and smoke particle systems or structures which can be created by electric discharges in polymer materials. The latter systems belong to the class of structures that result from the aggregation processes like the idealized DLA model. It is well known [2] that the DLA model and the dielectric breakdown patterns both result from processes that are controlled by Laplace equation with appropriate boundary conditions. The point we want to make relates to the frequency at which branching occurs in such structures and we suggest that the branching probability or branching density may serve as a useful way of analyzing both the structures that results from regular or irregular tree growth as well as from random aggregation processes.

Mandelbrot observed [1] that the fractal character of tree-grown structures is associated with the branch tips that form a self-similar Cantor dust, whereas the rest of the tree is not strictly fractal. Aggregate structures on the other hand can be described by fractal mass correlation functions because each monomer unit carries the same mass. Hence a fractal dimension relating to the mass distribution can be defined, d_f. However, Coniglio and Stanley [2] have shown that perimeter sites on the DLA-type clusters also form a Cantor set which can be analyzed in multifractal terms when assigned appropriate measures or weights [3]. The regular branching trees are formed with a branching ratio of N, i.e. at each step a branch sprouts into N new branches separated by an angle Θ, and each branch being a fraction r shorter than the previous one. The fractal dimension associated with the perimeter sites of such a tree is

$$d_p = \ln(N)/\ln(1/r) . \qquad (1)$$

An example of such a computer generated tree is shown in fig. 1a.

For the corresponding perimeter sites on the DLA clusters it has been suggested [4] that their distribution can be described by the surface exponent, d_u. This exponent is related to the fractal dimension of d_f, and a simple mean field relation is [4]

$$d_u = (d_f - 1) + \tfrac{1}{2}(d - d_f) . \qquad (2)$$

Hence, a given aggregation process can be expected to be characterized by a specific relation between d_f and d_u. In other words, through relations like (1) one can define a characteristic branching ratio also for a DLA-type cluster. Below we discuss a possible way of getting to this parameter by diffraction experiments. But before doing so we would like to discuss some direct observations of naturally grown fractal tree structures like broccolis, cauliflowers and Minarets.

2. Broccolis, cauliflowers and Minarets

In his book "The Fractal Geometry of Nature" [1], Benoit Mandelbrot wonders whether "part of the difference between cauliflower and broccoli is quantified by a fractal dimension?". Such a question lends itself rather easily to experimental investigation.

Trying to map broccolis and cauliflowers onto such idealized fractal trees is not straightforward. The often haphazard arrangement of the branches and their varying lengths means that a detailed comparison would require a laborious statistical analysis. Fortunately, visual comparison of the two plants reveals three important differences which seem to capture the essential traits of broccolis and cauliflowers.

The first difference is that broccoli has a long and slender trunk, while the trunk of cauliflower is short and thick. Since the branches are scaled-down versions of the trunk, the final morphology of a fractal tree is strongly influenced by the proportions of the

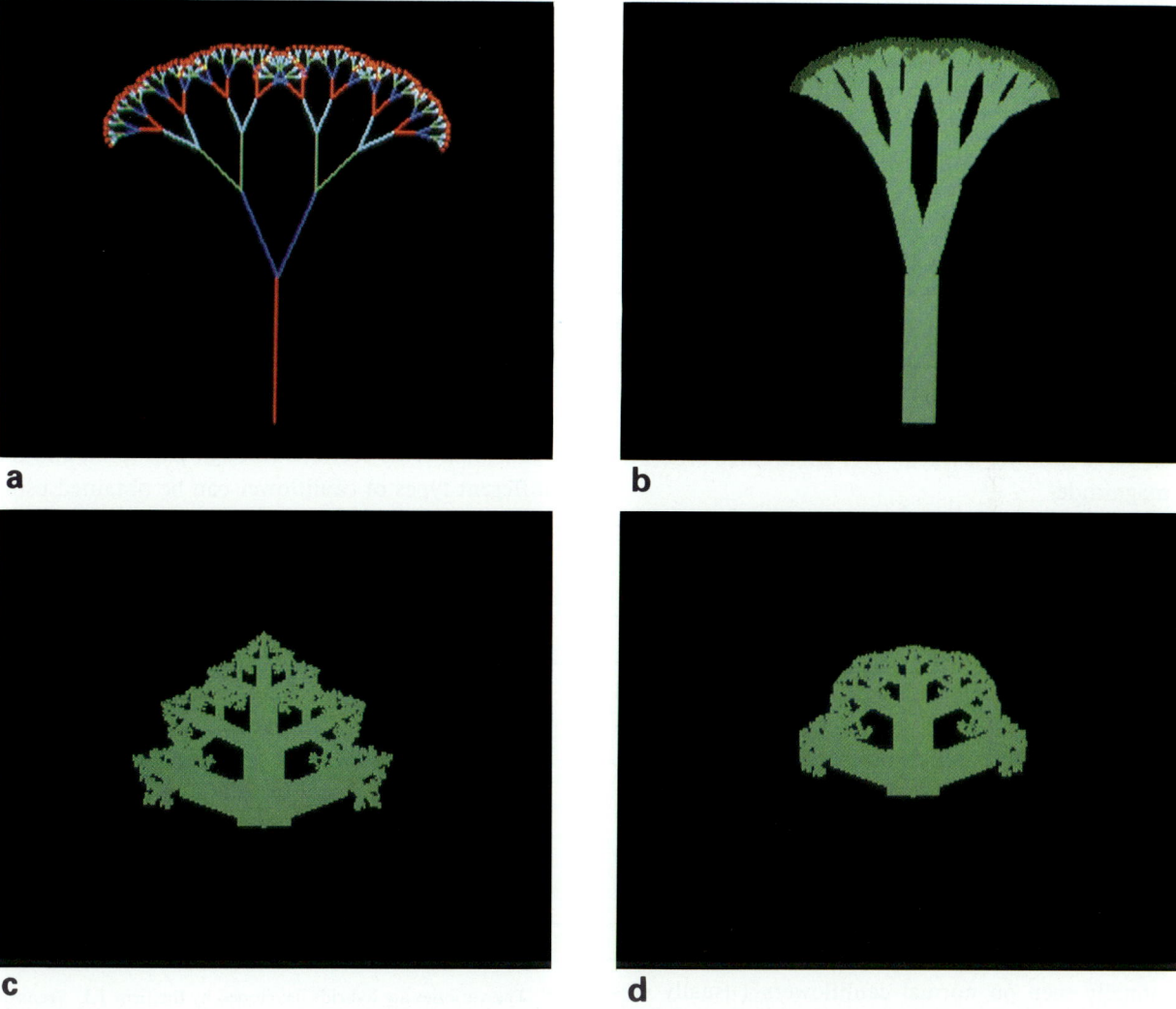

Fig. 1. (a) Example of computer model of simple tree-growth. (b) Same with trunk thickness chosen to mimic broccoli. (c) Model that resembles Minaret cauliflower. (d) Model that resembles regular cauliflower.

trunk. So even for the same N and r, one can obtain quite different-looking plants. The suspicion that a large part of the difference between broccoli and cauliflower is to do with the initial trunk is borne out by simple two-dimensional computer simulations, which show that a fractal tree can be made more broccoli-like or cauliflower-like by varying the trunk dimensions.

The second difference concerns the extent to which the plants follow a fractal-like growth. Broccoli is only weakly fractal. There are typically only two to three branching generations from the trunk to the small dark green flowers which form the head of the broccoli. These flowers sprout from the last branch in a complex pattern which is not similar to previous branchings. The branches on which the flowers grow have a typical scale of 1 cm, while the trunk is about 10 cm long. Broccoli can therefore be deemed fractal over only about one decade.

Cauliflower is, however, a full-fledged fractal tree. At least four branching generations are visible to the eye, and two further generations can be observed under the microscope. On the last branches, which have a typical size of 50 µm, the individual plant cells can be resolved. Thus this plant is fractal down to its fundamental cutoff. The typical trunk length is 5 cm, and so the cauliflower is fractal over about three orders of magnitude.

The third difference becomes apparent on dissection. The number of branches N at each node for broccoli is usually about 2–4. For cauliflower, N appears at first sight much larger (10 or more). However, closer inspection often reveals that the branches are not from a single node, but form a spiral of small pitch. Such spiral-like growth is very common in nature; consider for example pineapples, pine cones and sunflowers. In such patterns, one can distinguish two types of spiral rotating in opposite directions. It is well known that for such natural spiral patterns the ratio of the number of spirals in one direction to that in the other is equal to the ratio of two numbers in the Fibonacci series. Areas of regular spiralling are occasionally seen on normal cauliflowers (usually clearest near the top of the plant) and the observed ratio is 8:13. Spiralling can also be seen for broccoli, but is usually very irregular.

In this connection we wish to draw attention to two varieties of cauliflower which exhibit a highly regular spiral growth. The varieties sport the commercial names Alverda and Minaret [1]. Both varieties distinguish themselves from ordinary cauliflower in being light green and, interestingly enough, in having a somewhat more broccoli-like flavour. But what should definitely whet the appetite of the fractal connoisseur is the high degree of regularity in the branching. This is a case where a picture (fig. 2a) is worth a thousand words.

The Alverda looks at first sight to be simply a green cauliflower, and it is only on close inspection that the high regularity of the spiral structure is noticed. On the other hand, the Minaret of fig. 2a presents a strikingly beautiful "spiky" appearance which even your greengrocer is apt to remark upon. In fig. 2d, an electron microscope picture of a replica made of a single branch of an Alverda shows the high degree of regularity of the spiral structure of this plant, down to dimensions of 100 µm.

For detailed simulations of cauliflowers, the spiral structure should be included. But figs. 1b, 1c and 1d illustrate that quite reasonable simulations of two-dimensional cross-sections through a broccoli and the different types of cauliflower can be obtained using the simple fractal tree growth outlined at the beginning of this section. As suggested from the above discussion, the main differences between the two simulated plants are in the proportions of the starting trunk, and the number of branches at each node. A quite convincing broccoli can be obtained with $N=2$, whereas realistic cauliflowers usually require $N=3$ or more. For both trees, $r=0.6$ roughly. We are also able to vary an $N=3$ fractal cauliflower from being Alverda-like to Minaret-like, simply by increasing the length and growth rate of the central stem relative to the two side-stems, thus making the structure "spikier".

---

[1] The varieties are hybrids developed by the firm T.L. Seeds of Holland.

Fig. 2. (a)–(c) Photographs of a Minaret cauliflower showing different length scales. (d) Electron micrograph (SEM) of Alverda cauliflower on sub-millimeter scale.

3. From Minarets to aggregates

Can one make a connection between fractal tree growth and diffusion-limited aggregation (DLA)? And could this connection be of any use in understanding the latter process?

To underscore the similarity between these types of structures a picture of an electric discharge in a block of polymethylmethacrylate, PMMA, is shown in fig. 3. As mentioned above the discharge of the electron beam loaded PMMA is governed by Laplace's equation and hence belongs to the DLA family [5]. Branching at all levels is apparent and close inspection shows that most nodes contain only two branches.

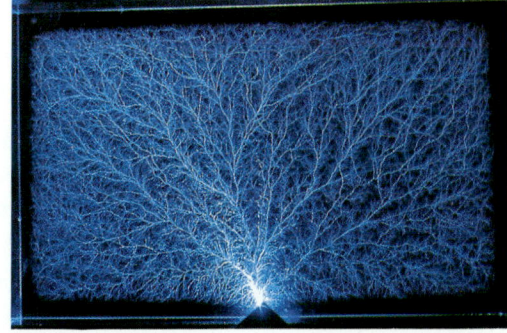

Fig. 3. Photograph of an electronic discharge in an electron beam loaded piece of PMMA (2 cm×5 cm×8 cm). The sample was prepared by A. Miller.

Hence we want to sketch an answer to the abovementioned questions. First, inspection of simulations of DLA suggest that these connected structures are reducible to tree-like structures. One approach to growing a DLA tree is as follows: let the branch length be b and let $r \to 1$, i.e. all branches are of length b. Further, if a branch tip approaches within a radius b of any other, that branch ceases to grow (this is a death law). It is then possible to place atoms of diameter b at all tree nodes, and obtain a remarkably DLA-like structure. We note that for $r \to 1$, the fractal dimension D of the tree diverges (eq. (1)). The death law is necessary to avoid this catastrophe.

Looking on DLA as a tree-like structure may well have some merits. Tree growth emphasizes the idea of nodes and of the number of branches N at each node. Along branches, the atoms have two nearest neighbours, but at the node regions the correlation number increases. One might therefore expect a relation between the average frequency of nodes, the average number of branches at each node, and the correlation function of DLA. Since it is the correlation function which is probed in scattering experiments, such a relationship could yield a simple physical picture of the underlying structure of the DLA-like object under investigation.

4. Diffraction experiments

It is well known that the scattering law for the ideal mass fractal like a DLA aggregate is

$$I(q) = F(q) S(q), \qquad S(q) = q^{-d_f}, \qquad (3)$$

where $F(q)$ is the form factor for the constituent monomer particles. This has been confirmed experimentally for a range of systems [6,7]. For values of $1/q$ near the monomer dimension, a, $S(q)$ will reflect the near-neighbour correlations. This was observed in the study by Freltoft [8], whose results for a system of colloidal silica aggregates, formed at different rates, are shown in fig. 4. A similar analysis for gold colloids has been reported by Dimon et al. [9].

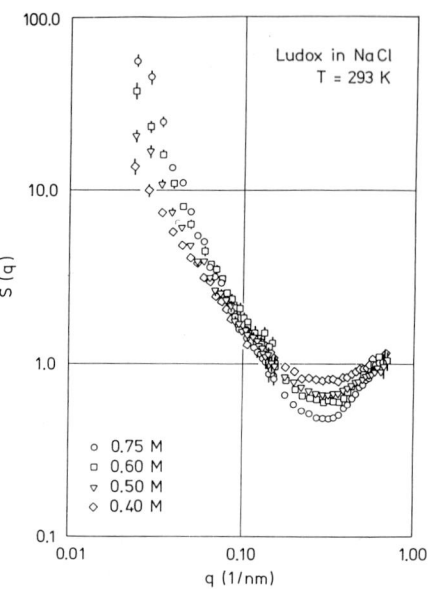

Fig. 4. Neutron scattering curves for fully aggregated samples of silica colloids in deuterated suspensions of different salinity. The data have been normalized by the spectra from the unaggregated samples giving the structure factor $S(q)$ directly.

The aggregation rates were controlled by the salinity of the suspension. The quantitative analysis of these data in terms of a structure model that explicitly included the near-neighbour shell showed, that both the fractal dimension, d_f, and the average number of nearest neighbour varied with the aggregation rate. The trend was that larger number of neighbours concurred with larger values of d_f. The values of d_f were in the range 1.9 to 2.6.

Here we suggest an alternative interpretation, namely in terms of an average branching density, assuming that more than two neighbours means a branch. We have not yet succeeded in deriving a true branching model for $S(q)$ that can be used to fit the data. The structure of such a model would be a hierarchy of factors, like form factors, reflecting the correlations on the different scales. As q gets smaller, the hierarchy would become self-similar, i.e. $S(q) \to q^{-d_f}$. Our preliminary results indicate that for the data in fig. 4, at least two levels are needed before the system becomes self-similar.

5. Conclusion

Tree-grown structures and aggregated structures have the common feature that the perimeter or growth sites form Cantor sets. The fractal dimensions of these sets depend on the branching ratio and branching density. For the aggregated structure the perimeter dimensions are also related to the Hausdorff dimension, d_f. Analysis of the neutron scattering data for colloidal silica aggregates show the expected qualitative relation between the branching density, observed as the number of nearest neighbours, and the fractal dimension, d_f. We therefore suggest that a more quantitative pursuit of this kind of analysis will be worthwhile and conject that it will establish a link between the two different classes of fractal structures.

As examples of tree-grown structures we have analyzed broccolis and cauliflowers, of which the latter display self-similarity over the largest range of length scale, up to three decades. In particular the newly developed species named Alverda and Minaret are splendid examples of regular natural fractals.

Acknowledgements

Fruitful discussions with Martine Poncet and Josianne Comon are acknowledged. Helmer Nilsson and Solveig Kjall have assisted with photography and J. Bilde Sørensen provided the electromicrograph.

References

[1] B.B. Mandelbrot, The Fractal Geometry of Nature (Freeman, San Francisco, 1982).
[2] A. Coniglio and H.E. Stanley, Phys. Rev. Lett. 52 (1984) 1068.
[3] B.B. Mandelbrot, J. Fluid Mech. 62 (1974) 331.
[4] H.E. Stanley, in: Time-Dependent Effects in Disordered Materials, R. Pynn and T. Riste, eds. (Plenum, New York, 1987).
[5] L. Pietrono, A. Erzan and C. Evertsz, Physica A 151 (1988) 207.
[6] S.K. Sinha, T. Freltoft and J.K. Kjems, in: Kinetics of Aggregation and Gelation, F. Family and D.P. Landau, eds. (North-Holland, Amsterdam, 1984).
[7] S.H. Liu, Solid State Phys. 39 (1986) 207.
[8] T. Freltoft, Thesis, Risø-M-Report 2570 (Risø, Roskilde, 1986).
[9] P. Dimon et al., Phys. Rev. Lett. 57 (1986) 595.

MULTIFRACTAL MEASURES AND STABILITY ISLANDS IN THE ANISOTROPIC KEPLER PROBLEM

Martin C. GUTZWILLER

IBM T.J. Watson Research Center, Yorktown Heights, NY 10598, USA
and Institute for Theoretical Physics, University of California, Santa Barbara, CA 93106, USA

The relation between the binary code for the trajectories in the anisotropic Kepler problem (AKP) and the coordinates in the surface of section is investigated. The binary label $0 < \eta < 1$ is found to be a strictly increasing function of the starting point $0 < X < 2$ on the heavy axis for time-reversal symmetric trajectories, excepting a single island for mass ratios between 1.5 and 1.8 which was discovered by Broucke. The function $\eta(x)$ was calculated with a step size $\Delta x = 0.0002$, and the corresponding binary label down to 2^{-48}. Relatively flat portions can be associated with trapping near the unstable Kepler-type orbit for low mass ratios, and with trapping on either side of the light axis for large mass ratios. The $f(\alpha)$ curves, fractal dimension of the set with Hölder exponent α, is unusually wide in both of these limits.

1. Introduction

The dynamical systems of classical mechanics are governed by Lagrange's or Hamilton's equations of motion, i.e., they suffer no dissipation. Hamiltonian systems with few degrees of freedom come in three varieties; their behavior can be characterized as (i) regular or integrable, (ii) soft chaos and (iii) hard chaos.

Regular systems are very exceptional, and are structurally unstable against small perturbations; but mechanics books tend to restrict themselves to examples of this kind, so that most physicists are not aware of any other types. Their motions can be described in terms of as many basic frequencies as degrees of freedom; their phase space is foliated into tori of dimension equal to the number of degrees of freedom.

At the opposite extreme is hard chaos; this kind of behavior is structurally stable against small perturbations. Individual trajectories are characterized by a sequence of symbols of infinite length which give a qualitative description of the "story" connected with the trajectory. Phase space is foliated into two transverse families of stable and unstable manifolds, each of dimension again equal to the number of degrees of freedom. Each trajectory is the intersection of a stable and an unstable manifold; the long-term behavior is extremely sensitive to the initial conditions.

The bulk of the Hamiltonian systems belongs to the category of soft chaos; their behavior is an intimate mixture of regularity and hard chaos. It is as if the phase space was divided into two subsets which interpenetrate each other in a badly fractal manner, the first regular and the second with hard chaos. The condition is generic, and represents the true jungle to be colonized by the intrepid settler.

The normal route starts from the regular region; its main guide-post is the KAM theorem. This base of operations is very narrow, however, since regular systems are not even stable against small perturbations; moreover, the KAM theorem tells us only what is left over from the regular structure, without given any description of the new territory.

The approach to soft chaos from the opposite extreme, i.e., hard chaos, has the advantage of starting from a large and secure base. In problems of geodesics on Riemannian surfaces, the two extremes are the sphere as the only representative of regularity, and the surfaces of constant negative curvature on behalf of hard chaos. There are many multiparameter families of the latter, with almost all conceivable topologies. The problem of hard chaos softening in the presence of sufficiently strong perturbations has not been treated, however, at least to the knowledge of the author.

The example of the anisotropic Kepler problem (AKP) will be used to address this issue. In particular, we will examine what happens to the foliation into stable and unstable manifolds, and to the coding of trajectories by symbol sequences, as the chaos in the AKP "softens".

The main symptom of this transition is the changing nature of the devil's staircase which connects the two invariant measures in phase space, the classical Liouville measure on one hand, and the measure defined by the symbol sequence on the other. Equivalently, the multifractal nature of the relation between these two measures will be shown to change significantly as a function of the mass ratio.

The present report is the natural outgrowth of the author's earlier collaboration with Mandelbrot [1] where multifractal analysis had been applied to Hamiltonian systems for the first time, and an entirely new class of multifractal sets was first demonstrated using examples from hyperbolic geometry. The same features, in particular a very broad $f(\alpha)$ curve, and a "slippery devil's staircase", appear again in the AKP. They can be related to the earlier work of Devaney [2–4] and the author [5–7], as well as the numerical calculations by Broucke [8]. A short history of the AKP as a prime example of a chaotic Hamiltonian system will put these developments into perspective.

2. Short history of the anisotropic Kepler problem

The AKP arises in semiconductor physics when donor impurities are investigated. An extra nuclear charge and an extra electron get embedded in the crystal lattice, and the two interact very much like a hydrogen atom. Various modifications have to be taken into account, however, such as the high dielectric constant and the symmetries of the crystal, as well as the chemical nature of the impurity.

Most difficult to understand intuitively is the anisotropy of the effective mass tensor, which also causes the classical motion to be chaotic. Without trying to explain this strange situation any further, let us write down the Hamiltonian for the electron. With the appropriately normalized Cartesian coordinates, (u, v, w) for the momentum and (x, y, z) for the position, the Hamiltonian becomes

$$H = \frac{u^2}{2\mu} + \frac{v^2 + w^2}{2\nu} - \frac{1}{(x^2 + y^2 + z^2)^{1/2}} = -\frac{1}{2}; \quad (1)$$

the masses μ and ν are chosen such that $\mu\nu = 1$, and $\mu/\nu > 1$ agrees with the experimental values, ≈ 5 for silicon and ≈ 20 for germanium. The homogeneity of the kinetic energy in (u, v, w) and of the potential energy in (x, y, z) allows to scale the value of H.

Since H has rotational symmetry around the x-axis, the angular momentum $M = yw - zv$ is a constant of motion which can be fixed from the start; the system is thereby reduced to two degrees of freedom. We shall immediately specialize to $M = 0$, and mention the more general case $M \neq 0$ only shortly at the end.

If $M = 0$, the classical motion takes place in a radial plane such as $z = 0$, so that the Hamiltonian (1) is reduced to

$$H = \frac{u^2}{2\mu} + \frac{v^2}{2\nu} - \frac{1}{(x^2 + y^2)^{1/2}} = -\frac{1}{2}. \quad (2)$$

Whereas the quantum mechanics of (1) and (2) had been studied first by Kohn and Lüttinger [9], and later on by Faulkner [10], using standard numerical approximations, the author was the first to investigate the classical mechanics of (2). The purpose was to understand the relation between the two regimes in a case where no constants of the motion besides the total energy H could be found.

Upon closer inspection, the classical trajectories seemed to be in one-to-one correspondence with the binary sequences. This strange fact could be understood by examining the trajectories in the neighborhood of the origin. The total energy H is then small compared to either its kinetic or its potential part, and the system behaves approximately as if $M=0$. The flow in phase space becomes two dimensional. This technique was worked out independently by McGehee [11] to study three-body collisions in celestial mechanics; it is now called the method of the collision manifold.

After learning about the heuristic results for the AKP, Smale put his graduate student, Robert Devaney, to work on finding mathematical proofs. Therefore, by 1977 both Devaney [2–4] and the author [7], independently and unbeknownst to each other, were able to prove the following theorem:

Given a binary sequence $a=\{...a_{-2}, a_{-1}, a_0, a_1, a_2...\}$ with $a_i=\pm 1$, there exists at least one trajectory for the Hamiltonian (2) such that its consecutive intersections with the x-axis, $\{...x_{-2}, x_{-1}, x_0, x_1, x_2...\}$, satisfy the condition $\text{sgn}(x_i)=a_i$, provided $\mu/\nu > 9/8$.

In spite of overwhelming evidence for $\mu/\nu=5$, it could not be shown that there is only one such trajectory for a given binary sequence. This paper is concerned with explaining this difficulty.

Broucke, instigated by Devaney, searched for regions in phase space where the lack of uniqueness could be demonstrated. After checking the author's results for $\mu/\nu=5$, he tried to find stable periodic orbits for much smaller mass ratios; he published the data for one such orbit for mass ratios below 2 but above 9/8. The author was made aware of Broucke's work while at the Institute of Theoretical Physics in Santa Barbara, and then set out to confirm and enlarge the scope of these investigations.

The study of a particular dynamical system may seem of limited significance. If it is a first of a large class, however, and it is also endowed with a very simple structure so that its properties can be simply stated and checked, then it can be expected to open up new insights of a general nature. Also, the basic features of the AKP are quite common in nature.

An anisotropic mass tensor together with gravitational interactions arises fairly often. Devaney [12,13] studied the isosceles three-body problem where two bodies of equal mass form an isosceles triangle in a fixed plane with a third body. The relative motion of the two equal masses has a different reduced mass from the relative motion of their center of mass with respect to the third mass. An example of this type on the atomic scale is the ammonia molecule NH_3 where laser action was first demonstrated by Townes; the three hydrogens form an equilateral triangle which is the base of a regular pyramid with the nitrogen on the top. Devaney [14] showed the close relation with binary sequences in the isosceles three-body problem, again looking at the collision manifold.

3. The Hamiltonian flow

The Hamiltonian (2) leads to the *equations of motion*

$$\frac{dx}{dt} = \frac{\partial H}{\partial u} = \frac{u}{\mu},$$

$$\frac{dy}{dt} = \frac{\partial H}{\partial v} = \frac{v}{\nu},$$

$$\frac{du}{dt} = -\frac{\partial H}{\partial x} = -\frac{x}{(x^2+y^2)^{3/2}},$$

$$\frac{dv}{dt} = -\frac{\partial H}{\partial y} = -\frac{y}{(x^2+y^2)^{3/2}}.$$

(3)

Solutions of these four ordinary differential equations are called trajectories. We consider only the ones for which the Hamiltonian (2) has the value $-\frac{1}{2}$. The others can be obtained by a simple scaling operation.

The equations of motion have three discrete *symmetries* which we will use occasionally, namely (i) time-reversal, (ii) reflection in x, (iii) reflection in y. Each of these operations generate a new solution from an old one; e.g., from the solution $[x(t), y(t), u(t), v(t)]$ the time-reversal solution $[x'(t), y'(t), u'(t), v'(t)]$ follows by setting $x'(t)=x(-t)$, $y'(t)=y(-t)$, $u'(t)=-u(-t)$, $v'(t)=-v(-t)$.

When integrating (3) to obtain a particular trajectory, it is often useful to integrate at the same time the equations for *neighboring trajectories*. These are displaced from the main trajectory by the displacements $[\delta x(t), \delta y(t), \delta u(t), \delta v(t)]$, and the latter are obtained from the linearized equations of motion

$$\frac{d\delta x}{dt} = \frac{\delta u}{\mu}, \quad \frac{d\delta y}{dt} = \frac{\delta v}{\nu}, \quad \text{etc.} \qquad (4)$$

When carrying out the numerical integrations on these equations, however, the singularity at the origin becomes more pronounced, and causes trouble.

A good way out of this difficulty consists in considering the momentum (u, v) as coordinate on an appropriate *Riemannian surface*, so that (3) become the equations for the *geodesics* after renormalizing the time. This procedure was first proposed by the author; all the numerical work in this respect uses this method.

The *numerical integration* was carried out with a standard fourth-order Runge–Kutta routine, and written in FORTRAN. The time step was chosen as 0.01, but it was crucial to reduce the time step in the neighborhood of the origin by multiplying with $(x^2+y^2)^{3/2}$. The accuracy of the calculation was checked by reversing time at the end of the trajectory so as to recover the initial condition. A less demanding, but useful check consists in calculating the value of H to see whether it stays constants.

Periodic orbits are the special solutions of (3) for which there exists a time T, called the period, such that $x(t+T)=x(t)$, $y(t+T)=y(t)$, $u(t+T)=u(t)$, $v(t+T)=v(t)$. Periodic orbits can be checked with the help of the virial theorem which says in our normalization that $\int (u\,dx+v\,dy) = T$; the integral on the left is taken over one period. This relation provides a very sensitive check on the initial conditions to see whether exact closure is actually occurring.

4. The surface of section

The (heavy) x-axis is the natural surface of section, because it gets intersected more often than the (light) y-axis. If $y=0$, then $|x|(1+u^2/\mu) \leq 2$; the momentum v is allowed to have either sign. The area in the (x, u) plane has an awkward shape, however, and will be transformed into a *rectangle* $|X| \leq 2$, $|U| \leq \sqrt{\mu}\pi/2$ with the help of the area-preserving map

$$X = x\left(1 + \frac{u^2}{\mu}\right), \quad U = \sqrt{\mu}\arctan\left(\frac{u}{\sqrt{\mu}}\right); \qquad (5)$$

we will always use these coordinates.

If the pair (X, U) is given, the initial conditions $(x, y=0, u, v)$ for the integration of a trajectory follow from (5) and (2); the sign of v is not important. The equations of motion (3) are integrated, until $y=0$ again; thus, one obtains new values (x', u') which yield (X', U') according as (5). In this manner, the Poincaré map $P:(X, U) \to (X', U')$ into itself is obtained in the usual way.

This map preserves the area, i.e., the Jacobian $\partial(X', U')/\partial(X, U) = 1$; an invariant measure is defined thereby by the element $dX\,dU$, corresponding essentially to *Liouville's theorem*. The map P is quite singular along the boundaries, $X=0$, ± 2 and $U=\pm\sqrt{\mu}\pi/2$; but it is regular inside, so that we do not have to worry about the bulk of our calculations.

In keeping with Poincaré's approach to mechanics, we will only talk about the map P, as if it contained all there is to known about the AKP in two dimensions. A trajectory is, therefore, reduced to a sequence of points in the surface of section. One of them is conveniently assigned to time $t=0$ for the start of the numerical integration, while the others are numbered by the integers in the order in which they appear along the real trajectory. Thus, we get ..., (X_{-1}, U_{-1}), (X_0, U_0), (X_1, U_1), (X_2, U_2), ...

Periodic orbits are characterized by having $(X_{2n}, U_{2n}) = (X_0, U_0)$ for some positive n. The even index $2n$ is required, because the sign of v alternates from one intersection of the trajectory with the surface of section to the next.

Since the neighborhood of (X_0, U_0) gets mapped into itself after n iterations of P, one can look at the linear displacements $(\delta X_0, \delta U_0)$ and $(\delta X_{2n}, \delta U_{2n})$. They are related by a linear map

$$\begin{pmatrix} \delta X_{2n} \\ \delta U_{2n} \end{pmatrix} = \begin{pmatrix} a & b \\ c & d \end{pmatrix} \begin{pmatrix} \delta X_0 \\ \delta U_0 \end{pmatrix}, \qquad (6)$$

whose determinant $ad-bc=1$ because of area conservation. The eigenvalues are mutually reciprocal, and lead to the three classic cases, elliptic, parabolic and hyperbolic.

The AKP leads predominantly to hyperbolic periodic orbits with real eigenvalues e^α and $e^{-\alpha}$; the exponent α is called stability exponent. The presence of elliptic periodic orbits, where the stability exponent α is imaginary, is one of the main issues of this report. Hamiltonian systems where all periodic orbits are hyperbolic are said to display *hard chaos*.

In trying to understand the Poincaré map P, one has to look at all the points in the rectangle. Such a systematic analysis seems out of the question for the time being, although some steps in this direction can be found in the author's paper [7]. Here we will restrict ourselves to the *time-symmetric trajectories* for which $X_{-j}=X_j$ and $U_{-j}=-U_j$. Their initial conditions are $|X_0|\leq 2$ with $U_0=0$.

The reflection symmetry with respect to x allows us to look only at initial values $X_0>0$ with $U_0=0$. Thus the trajectories to be examined are given by a one-sided sequence (X_1, U_1), (X_2, U_2), ...

5. Fusion and fission trajectories

The usual Kepler problem can be regularized: the exceptional trajectories which go down into the center of attraction along a straight line can be continued in time like the more general trajectories which avoid the origin; the exceptional trajectories can then be considered as limits of the ordinary ones. Such an analysis for the trajectories suffering a collision with the origin is not possible in the AKP; they cannot be continued in some sensible fashion so as to connect up with the other trajectories.

The equations of motion (3) can be examined, however, to find solutions which either leave the origin (fission) or go into it (fusion) in a finite time. These names were chosen to indicate the fact that the trajectories cannot be continued in time beyond the moment of their coalescence with the origin. Time-reversal symmetry allows us to investigate the fission

trajectories, and transfer the results to the fusion trajectories.

Either can be obtained by looking at the deviations from the *vertical trajectory*, the special solution of (3) for $x=0$, $u=0$. In terms of a parameter ζ along the trajectory, one finds the expansions

$$x=A\zeta^\alpha+\dots,$$
$$y=2\zeta+B\zeta^{2\alpha-1}+\dots,$$
$$u=\tfrac{1}{2}\tfrac{1}{2}\mu^{3/2}A\alpha\zeta^{\alpha-3/2}+\dots,$$
$$v=\tfrac{1}{2}\nu^{1/2}B(2\alpha-1)\zeta^{2\alpha-5/2}+\dots,$$
(7)

with the abbreviations

$$\alpha=\tfrac{3}{4}[1+(1-8/9\mu^2)^{1/2}],$$
$$B=\frac{3A^2}{4(2\alpha-3)(4\alpha-1)}.$$
(8)

A is the only parameter; if $A>0$, the trajectory veers to the right of the vertical, and makes its first intersection with the surface of section at $X_1>0$. The reflection symmetry with respect to the x-axis allows us to restrict ourselves to $A>0$.

The *numerical integration* of fission trajectories starts with the formulas (7), carried to the next term of the expansion in powers of ζ. By assigning some relatively small value to ζ, say 0.01, a point sufficiently far from the origin is obtained so that the ordinary Runge–Kutta routine can be used from then on. Each value of A leads then to a well-defined, one-sided sequence (X_1, U_1), (X_2, U_2), ..., which will now be discussed on the basis of the numerical computations.

Most strikingly, the value of U_j *always increases* with A, as long as X_j does not change sign. U_j starts at the lower boundary of the rectangle, $U_j=-\sqrt{\mu}\pi/2$, and moves up to the upper boundary, $U_j=+\sqrt{\mu}\pi/2$. Upon reaching the upper boundary, the sign of X_j changes, and U_j jumps back to the lower boundary. This jump occurs at a well-defined value of A, where the trajectory goes into origin at the jth intersection, i.e., it undergoes fusion, or collision with the origin.

The curves in the (X, U) rectangle which correspond to a particular intersection (X_j, U_j) of a fission trajectory can be labeled by the signs $a_i = \text{sgn}(X_i)$ for $i \leq j$. If the parameter A is allowed to vary through its full range from $-\infty$ to $+\infty$, there will be exactly 2^j such curves. Each can be assigned a parameter

$$\xi = \sum_{i=1}^{j} a_i (\tfrac{1}{2})^i, \qquad (9)$$

which is a rational number in the range $(-1, +1)$ with the denominator 2^j.

If the data are collected from all the intersections up to the nth, there will be a total of $2 + 4 + \ldots + 2^n = 2(2^n - 1)$ such curves going from the bottom to the top of the rectangle. Each half, $X \gtrless 0$, will contain $2^n - 1$ of them, labeled by $\xi \lessgtr 0$. Most remarkably, they are *ordered from left to right exactly as their label α*.

If these curves are reflected on the X-axis, one obtains their analogs for fusion rather than fission, labeled by the analogous parameter, now called η. These two sets of curves in the rectangle always *intersect transversely*, and, therefore, define a grid of mesh size 2^n by 2^n in terms of the parameters ξ and η. This new coordinate system in the rectangle can be refined arbitrarily.

6. The binary code

The trajectories in the AKP can be coded quite generally with the help of *binary sequences*, $a = \{\ldots a_{-1} a_0 a_1 a_2 \ldots\}$, where $a_i = \text{sgn}(X_i)$. In this manner the rectangle (X_0, U_0) is mapped into the space of binary sequences. According to the theorem which Devaney and the author proved, this map is surjective, i.e., each binary sequence is realized by at least one trajectory.

Before discussing the possible converse of this theorem, it is important to endow the binary sequences with a *topology*. Since we will only investigate the special class of trajectories with $U_0 = 0$, we shall limit our discussion to this case. Because $X_{-j} = X_j$, one has $a_{-j} = a_j$; moreover the x-reflection symmetry allows us to further restrict ourselves to $X_0 > 0$, and thus $a_0 = +1$. The binary code is simplified to $a = \{a_0 a_1 a_2 \ldots\}$.

If we start with a value of X_0, and we let it *increase*, the binaries a_i with low index i will remain the same for some interval of increase $\delta X_0 > 0$. If one fixes attention on the index j, there will be a critical value of δX_0 where a_j changes, whereas all binaries a_i with $i < j$ remain the same. At that point, the binary a_j goes from $a_j = -1$ below the critical δX_0, to $a_j = +1$ above. At the critical value, the trajectory which started in $X_0 + \delta X_0$ has a collision (fusion) with the origin at its jth intersection with the x-axis.

The binary sequences, immediately preceding and succeeding this fusion trajectory, are of the following kind: for δX_0 below critical, $\{a_0 a_1 \ldots a_{j-1} - + + + \ldots\}$, and for δX_0 above critical, $\{a_0 a_1 \ldots, a_{j-1} + - - - - \ldots\}$; the closer to critical the longer the string of identical binaries immediately after a_j. The limit of an infinitely long string of identical binaries a_i with $i > j$ is realized by the fusion trajectory.

The reasons for this behavior are found in a detailed discussion of the collision manifold which was mentioned in section 3. The lenght of the string of identical binaries can then be estimated in terms of the closeness of δX_0 to the critical value. This argument eventually leads to a lower limit for the Hölder exponent in the relation between an increase in X_0 and the corresponding change in the binary code; it will be presented elsewhere.

If the binary sequence $\{a_1 a_2 \ldots\}$ is assigned the number

$$\eta = \sum_{i=0}^{\infty} a_i (\tfrac{1}{2})^{i+1}, \qquad (10)$$

the fusion orbit gets the label corresponding to formula (9) for the fission orbits. Indeed, if we take $a_j = -1$ and $a_i = +1$ for $i > j$, then the summation could have been stopped at $i = j - 1$; and similarly, if we take $a_j = +1$ and $a_i = -1$ for $i > j$. Therefore, it is reasonable to give the binary code for the trajectories in the AKP the topology and measure which comes with formula (10).

Although only formula (10) will be used in the numerical work to be reported, because we limit ourselves to time-reversal invariant trajectories, it may be helpful to describe the code and its labels for the general trajectories. To the binary sequence

$a = \{\dots a_{-1} a_0 a_1 a_2 \dots\}$ are assigned the numbers
$$\xi = \sum_{i=0}^{\infty} a_{-i}(\tfrac{1}{2})^{i+1}, \qquad \eta = \sum_{i=1}^{\infty} a_i(\tfrac{1}{2})^i. \tag{11}$$

The Poincaré map P becomes simply the *shift* in the space of binary sequences $a \to a'$ where $a'_j = a_{j+1}$. The corresponding map in the square $-1 \leq \xi, \eta \leq +1$ is the *baker's map*. The curves in the (X, U) rectangle, where ξ and η take on the rational values whose denominators are powers of 2, are exactly the fission and fusion curves in the preceding section.

Periodic orbits are represented by binary sequences whose binaries have a finite cycle such that $a_{i+m} = a_i$. If the relation between the (X, U) rectangle and the (ξ, η) square can be shown to be one-to-one, then all periodic orbits can be enumerated, and a *generalized trace formula* [5, 15, 16] can be used to establish the connection between classical and statistical mechanics.

The fission and fusion curves in the (X, U) rectangle define the *foliation* of the Hamiltonian system into *stable* ($\eta = $ const.) and *unstable* ($\zeta = $ const.) manifolds. They are seen to be smooth, by construction as it were.

7. Binary sequences and initial conditions

As mentioned in section 2, Devaney never could quite accept the author's claim, based on numerical work, that the map between the surface of section and the binary sequences is one-to-one. A *counter-example* was then found by Broucke when he gave the initial conditions for a stable periodic orbit at mass ratio 1.7218, and another one (of different binary sequence) at mass ratio 1.111. How then can the existence of these stable, periodic orbits be reconciled with all the numerical, and partially analytic, experience as described in the last two sections?

As a first step, it is not difficult, but very time-consuming, to integrate the equations of motion numerically for a set of time-reversal invariant trajectories ($U_0 = 0$) by increasing the initial value $X_0 = X$ from 0 to 2 in regular steps ΔX. Each integration is stopped after a certain *number N of intersections* with the surface of section, thereby yielding the corresponding value of η according as (10) to a precision of 2^{-N-1}.

With $\Delta X = 0.0002$, i.e., a *total of $L = 10000$ trajectories*, the corresponding differences $\Delta \eta$ were never found to be negative. For mass ratios 2 and larger, every $\Delta \eta > 0$; for mass ratio 1.5, however, 83 values of $\Delta \eta$ (out of 10000) were calculated as 0 with $N = 48$. Of these, 65 came from Broucke's island, and the corresponding binary sequences showed the same periodicity with $(+ - -)$ repeating. The remaining 18 intervals with $\Delta \eta = 0$ all belong to the region of the binary sequences near the periodic orbit with the repeating unit $(+ -)$.

This particular periodic orbit which is close to a circle around the origin of the (x, y) plane is always unstable as was shown by the author [5] in his first work concerning the AKP. The stability exponent goes to zero linearly, as the mass ratio goes to 1. With mass ratio 1.5, it is $\sim 1/3$ for P^2. The displacement 0.0002 gets magnified by a factor $\sim e^{24/3} = e^8 \approx 2000$, whereas we can tolerate a factor 5000 before the 48th intersection runs into a collision with the origin. Therefore, a few degeneracies in η can be expected.

This argument assumes that the neighborhood of the periodic orbit is described by its linear approximation all the way to the boundary of the surface of section; nevertheless, it demonstrates the possibility of $\Delta \eta = 0$ in this case even after 48 intersections.

Some further isolated intervals with $\Delta \eta = 0$ come from trajectories where the binaries again alternate indefinitely, except for a single occurrence of $(+ - -)$ or $(+ + -)$ instead of $(+ -)$. All of these instances where $\Delta \eta = 0$ lead to the conclusion that Broucke's island is the only departure from a one-to-one relation between X and η for mass ratio 1.5, and there is none for mass ratios ≥ 2.

The designation "*Broucke's island*" needs a further explanation. If $U_0 = 0$, all the values of X_0 from 0.0022 to 0.0152 yield the same periodic binary sequence of units $(+ - -)$ repeating themselves 16 times out to 48 intersections. If U_0 is chosen at some fixed value different from 0, there is again an interval of X_0 with the same repeating binary sequence in the forward direction in time, but not necessarily in the backward direction since time-reversal symmetry is now lost.

All these intervals together form a domain in the (X, U) plane, called F for forward. The boundary of F is hard to obtain in this primitive manner; but the fission and fusion curves of section 5 now come to our help. If the trajectory is made to collide with the origin at its 48th intersection, rather than intersect the positive x-axis, while all the other 47 intersections maintain their $\text{sgn}(X_i)$, the initial conditions (X_0, U_0) lie on a fusion curve to the left of F. On the other hand, if there is a collision at the 48th intersection with the 47th intersection forced to be positive, while all the other 46 remain the same, the corresponding fusion curve is now to the right of F.

This construction depends, of course, on the experience of section 5. The result as shown in fig. 1 clearly shows how the two fusion curves, although extremely near in their binary label, separate to leave space for F. Broucke's island is the common domain of F and its time reversal, i.e., its mirror image with respect to the X-axis.

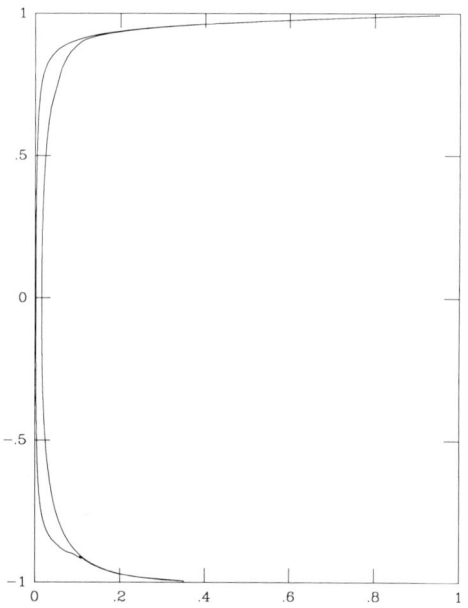

Fig. 1. "Broucke's island" in the surface of section (X, U) for mass ratio 1.5 as obtained from the fusion trajectories on either side; the coordinate U is divided by $\sqrt{\mu}\pi/2$.

8. Multifractal invariant measures

All the numerical experiments on the AKP indicate that the structure of stable and unstable manifolds as defined by the fission and fusion trajectories is maintained, although some islands of stability may be squeezed into the double foliation at low mass ratios. For mass ratios ≥ 2, and at a resolution of 0.0002 for X with $U=0$, there are no islands; there may be some for $U \neq 0$ around periodic orbits without time-reversal symmetry, but no search has been undertaken for any mass ratio.

The structure inside Broucke's island is presumably a set of nested tori, although it may be more complicated on the boundary. If η according to (10) is plotted versus X with $U=0$ at mass ratio 1.5, a *real* (*flat*) *plateau* will appear from 0.0022 to 0.0152. The remainder of the plot is not expected to show any interval without a net growth in η, although the difference $\Delta\eta$ may be so small as to vanish in fig. 2.

The dependence of η on X is continuous, but not differentiable, outside the island. The analysis of this function is only possible in terms of *multifractals*, as it was done by Mandelbrot and the author last year, for the first time in connection with a Hamiltonian system. The idea is the same, but it is now applied to the AKP instead of the motion on a surface of constant negative curvature.

The interval $0 < X < 2$ is divided into L intervals of equal length ΔX; the increase $\Delta\eta$ for each interval is obtained from the numerical integration of the equations of motion (3); the local *Hölder exponent* α from setting $\Delta\eta = (\Delta X/2)^\alpha$ is

$$\alpha = \frac{\log \Delta\eta}{\log(\Delta X/2)}; \qquad (12)$$

we use $\Delta X/2$ because X varies from 0 to 2 while η goes from 0 to 1. The local properties of the function $\eta(X)$ are best understood by examining how the values of α are distributed.

Since η comes from a binary expansion, the values of $\log \Delta\eta$ are put into *bins of width* $\frac{1}{2} \log 2$; the number of the bin is simply obtained from $[-2 \log \Delta\eta / \log 2]$, where $[z]$ is the largest integer not exceeding

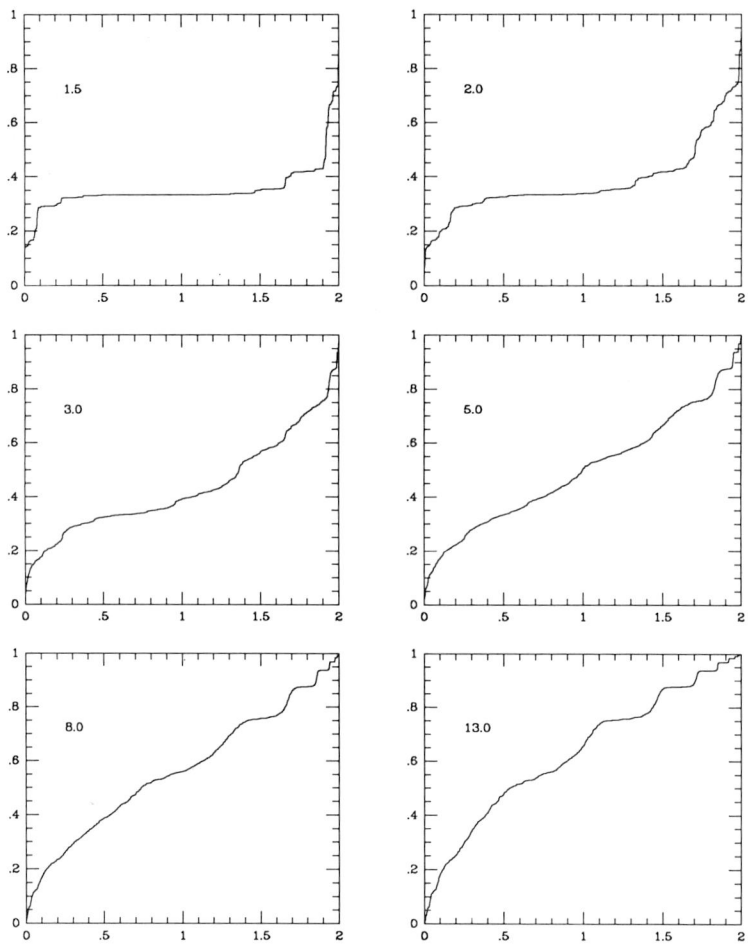

Fig. 2. Plot of the binary characteristic η versus the initial value X with $U=0$ (time-symmetric trajectories) for mass ratios 1.5, 2, 3, 5, 8, 13.

z. As explained in the preceding section, a few differences $\Delta \eta$ are smaller than 2^{-48} for mass ratio 1.5, although they can be expected to differ from 0. For mass ratios ≥ 2, all differences $\Delta \eta$ exceed 2^{-48}; nevertheless, the range of $\Delta \eta$ is enormous, considering that $\Delta X/2 = 0.0001$.

A casual look at the bin-numbers for consecutive intervals shows that they scatter wildly. The set of points in the interval $0 < X < 2$ which belong to the same bin is, therefore, assumed to be a *fractal*. Its *dimension* $f(\alpha)$ follows from the number K in the bin,

$$f(\alpha) = \frac{\log(K/\delta\alpha)}{\log L} ; \qquad (13)$$

the number K has been divided by the width of the bin $\delta\alpha = \log 2/2 \log L$, in order to obtain a number $f(\alpha)$ independent of the particular choice of bins. Of course, one has to get an infinite sample to find good values for $f(\alpha)$; the sample size $L = 10\,000$ is limited by the computing time which amounts to about 3 h of CPU time for each mass ratio.

Fig. 3 shows $f(\alpha)$ curves for several mass ratios. Two features are quite striking, and a tentative expla-

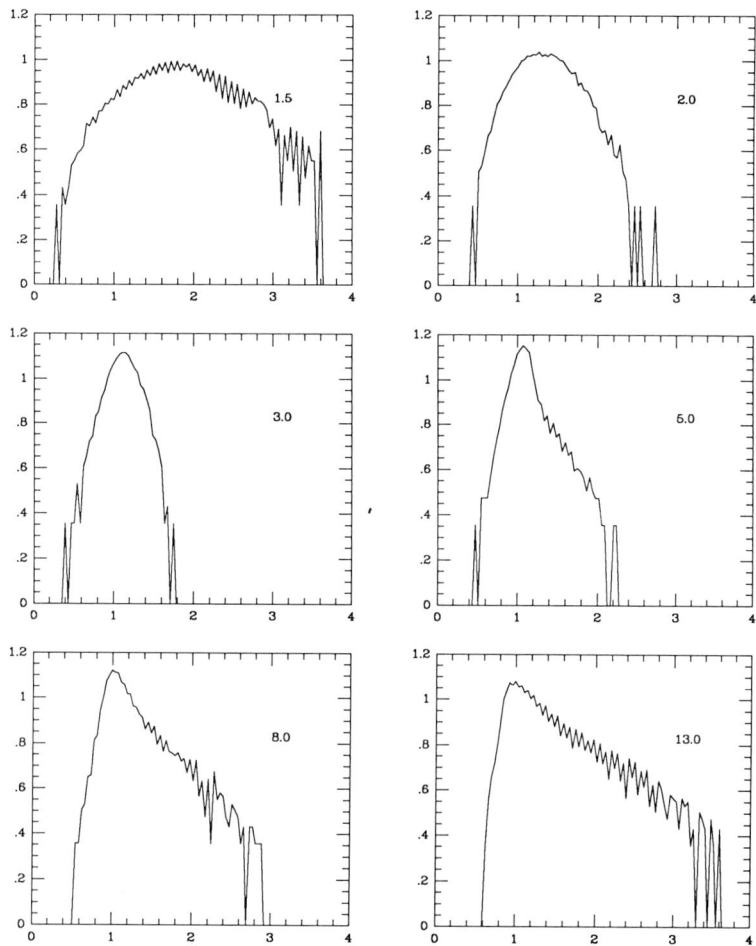

Fig. 3. Fractal dimension $f(\alpha)$ for the points with Hölder exponent α in the function $\eta(X)$ of fig. 2 for mass ratios 1.5, 2, 3, 5, 8, 13; the interval $0 < X < 2$ was divided into 10 000 intervals of equal length.

nation will be given in the next section: (i) The smallest values of α increase with the mass ratio (the statistics are necessarily poor at both ends of the distribution, and small isolated peaks correspond to no more than $K=1$ or 2 in a particular bin). (ii) The tail at the upper end of α first becomes shorter as the mass ratio increases, but then lengthens again as the mass ratio goes beyond 5.

The curves $\eta(X)$ serve to compare the description of the AKP in terms of binary sequences with the shift-mapping, and in terms of the phase space (X, U) with the Poincaré map P. Both of these mechanisms define an invariant measure, and the $f(\alpha)$ curve gives a preliminary picture of the relation between these two measures. A complete theory will have to deal with the full binary sequences, and the two-dimensional surface of section.

9. Interpretations and conclusions

A Hamiltonian system with hard chaos carries a double foliation in phase space, the stable and unstable manifolds. They can be labeled by a sequence of symbols which have a simple meaning in the "story" of a particular trajectory. This sequence defines the trajectory just as much as the point in phase space does. The map of the surface of section into itself is equivalent with the shift in the symbol sequence. The physical interpretation is found in the map which relates the surface of section with the space of all sequences.

A simplified picture of this relation is given by the function $\eta(X)$ of fig. 2 for the AKP. If this were a smooth function, it would simply define a different coordinate system in the surface of section. Instead, one finds an irregular, but still monotonically increasing dependence. Neighborhoods of slow growth indicate that the particular pattern in the symbol sequence is preferred. In extreme cases, this phenomenon leads to a "slippery devil's staircase", i.e., there are values for the coordinate in phase space X where all the derivatives in the function $\eta(X)$ vanish. The AKP does not seem present in this feature; the $f(\alpha)$ curves have a finite width. Nevertheless, α can reach values between 3 and 4 which have to be explained.

The mechanics of the AKP for low mass ratios > 1 can be expected to favor trajectories which follow roughly the classical Kepler ellipses; their symbol sequence alternates the two binaries; the value of $\eta = \frac{1}{3}$. Although this periodic orbit is unstable, its stable manifold is able to attract trajectories from a large area in the surface of section; a large variation of X yields a symbol sequence which differs not significantly from the indefinite repetition of the $(+-)$ unit. The difference may come after many periods $(+-)$, and then be arbitrary; or it may consist in the insertion of a wrong unit $(+--)$ or $(++-)$ somewhere. In the first case the trajectory seems to get lost on an unstable manifold, while in the second case the insertion of the wrong unit throws the trajectory back onto a stable manifold. The Kepler type orbit acts as a trap, but only temporarily, unlike the strange attraction in a dissipative system.

At large mass ratios, the y-axis carries so little inertia compared to the x-axis that the electron oscillates back and forth across the x-axis while moving slowly along the x-axis. Long sequences of identical binaries are now favored, leading to values of η near the rationals with a denominator 2^n. The frequent alternation of binaries is not favored any longer; the electron gets trapped for various lengths of time in one of the half-planes $x \gtrless 0$.

While these general arguments are reasonable, and confirmed by the numerical integrations, nothing seems to provide a clue for the existence of Broucke's island. Since the repetition of a binary is favored by the reduced inertia in the y-direction, why is it not enough to have a periodic orbit with a small stability exponent? Also, once the formation of an island is granted, how come there are not more of them?

The structure of the island amid the fusion and fission trajectories is fascinating; it presents a particular mechanism for going from hard chaos to regular behavior; the preferred periodic orbit, instead of acting like a temporary attractor along its stable manifold as the $(+-)$ orbit does, carves out a whole domain of stability, and pushes the former foliation out of the way, without destroying it. The corresponding binary sequence is now realized by a whole area in the surface of section; but the trajectories with binary sequences close by are not lost, and their organization by the stable and unstable manifolds seems to be maintained; this last point requires further study, though.

The binary tree of trajectories does not get mutilated; its branches get *pushed aside* to make space for the stability islands. This mechanism for going from hard to soft chaos is unexpected. One could rather think of a scenario where branches get progressively *cut*.

Such a situation may well arise when the angular momentum M in the full Hamiltonian (1) is allowed to differ from 0. If we think of y as the distance from the x-axis, the typical trajectory will try to approach the heavy x-axis as before; but it will ultimately be repelled by the centrifugal potential $M^2/2\nu y^2$.

If M^2 is very small, this close approach to the x-axis can be put into correspondence with the outright crossing for $M=0$. It is not clear, however, whether a clear-cut symbol sequence can still be defined. It appears even unlikely to have the arbitrarily long sequences of identical binaries which were typical of the collisions with the origin. Thus, some of the more delicate branches of the binary tree are eliminated. Preliminary calculations of trajectories show the typical features of soft chaos when M^2 has values ~ 1; but no study of their relation to the binary sequences has been made.

In conclusion, the trapping phenomenon in hard chaos leads to broad $f(\alpha)$ curves; but it does not necessarily lead to islands of stability. These arise in special circumstances, and are recognized by the flat portions in the $\eta(X)$ curve, similar to an ordinary devil's staircase.

Acknowledgements

Many discussions with Bruno Eckhardt, Marcos Saraceno and Dieter Wintgen have been very enlightening. Dawn Meredith was very helpful in making the figures from the computed data. This research was supported in part by the National Science Foundation under Grant No. PHY82-17853, supplemented by funds from the National Aeronautics and Space Administration, at the University of California at Santa Barbara.

References

[1] M.C. Gutzwiller and B.B. Mandelbrot, Phys. Rev. Lett. 60 (1988) 673.
[2] R.L. Devaney, J. Diff. Equ. 29 (1978) 253-268.
[3] R.L. Devaney, Invent. Math. 45 (1978) 221—251.
[4] R.L. Devaney, Lect. Notes Math. 668 (1978).
[5] M.C. Gutzwiller, J. Math. Phys. 12 (1971) 343.
[6] M.C. Gutzwiller, J. Math. Phys. 14 (1973) 139.
[7] M.C. Gutzwiller, J. Math. Phys. 18 (1977) 806.
[8] R. Broucke, in: Dynamical Astronomy, Proceedings of Second US-Hungary Workshop, V.G. Szebehely and B. Balazs, eds. (University of Texas Press, Austin, 1985), pp. 9–20.
[9] W. Kohn and J.M. Lüttinger, Phys. Rev. 96 (1954) 1488.
[10] R.A. Faulkner, Phys. Rev. 184 (1969) 713.
[11] R. McGehee, Invent. Math. 60 (1974) 249.
[12] R.L. Devaney, Invent. Math. 60 (1980) 249-267.
[13] R.L. Devaney, Cel. Mech. 28 (1982) 25-36.
[14] R.L. Devaney, Commun. Math. Phys. 80 (1981) 465-476.
[15] M.C. Gutzwiller, Phys. Rev. Lett. 45 (1980) 150.
[16] M.C. Gutzwiller, Physica D 5 (1982) 183.

FRACTALS AND PERCOLATION IN POROUS MEDIA AND FLOWS?

Etienne GUYON [1], Catalin D. MITESCU [2], Jean-Pierre HULIN and Stéphane ROUX

Ecole Supérieure de Physique et Chimie Industrielle, LHMP, CNRS UA No. 857,
10 Rue Vauquelin, 75231 Paris Cedex 05, France

We review a body of recent work, related to the research interests of our group, on the channel structure of disordered porous media, or fractured systems, where concepts of percolation and fractal geometry have proved useful in characterizing, or explaining, features of the transport properties of single and multiphase flow, or of the "dynamics" of invasion percolation.

1. Introduction

From the archetype of the labyrinth, to Hammersley's discovery of percolation, initiated by Broadbent's [1] curiosity about the clogging of gas masks, our understanding of porous media has involved notions of tortuosity, of dead-end pores, and of a multiplicity of geometric shapes and scales in the pore geometry and interfaces. All of these concepts have become highly suggestive of fractal properties for porous media. Yet, in most real cases, porous materials have no obvious geometrical fractal structure, whether we consider ground coffee in a percolator, or real sandstone! Nonetheless, we will show in this paper that fractal concepts have a broad range of application beyond mere geometrical description. Let us note, however, that in discussions of fractality in disordered porous media, a clear distinction must be made between the structure of the grains, or solid skeleton, and that of the associated pore space. We shall confine our discussion to the latter topic and briefly review below fractal porous media (section 2), media, in which an extremely broad distribution of pore sizes leads de facto to a percolation-like behavior (section 3), and multiphase flow in porous media

emphasizing particularly phenomenon of invasion percolation (section 4).

2. Fractal porous media

Though it might appear natural to look first of all to problems in filtration, we know of no direct experimental application of precise percolation concepts to filtration – whether this occurs at the surface, through the formation of filter "cake" or in depth, where pores are randomly clogged by the particles being filtered. On further thought, this is perhaps not so surprising since the process of filter clogging takes place in a very heterogeneous way, from the inlet through to the depth of the filter; moreover, there is an indirect connection between the filling and blockage of a pore and the flow in its neighbors, so that only a correlated description should apply. Percolation thus presents a first approximation to filtration. Next to experimental simulations by Payatakes and by Houi [2], some numerical and scaling work in this area has been recently carried out [3]. In the case of clogging of pores by non-Newtonian flow in heterogeneous media, it can be shown that a critical point which resembles percolation does indeed exist. However, an instability appears in the last stages of clogging which *erases out* the critical aspects of the problem [4].

---

[1] Also at: Palais de la Découverte, Avenue Franklin D. Roosevelt, 75008 Paris, France.
[2] Permanent address: Department of Physics, Pomona College, 610 N. College Avenue, Claremont, CA 91711-6348, USA.

Essays in honour of Benoit B. Mandelbrot
Fractals in Physics – A. Aharony and J. Feder (editors)

From a different viewpoint, materials of low permeability, or fractured solids, are candidates for percolation-like behavior. Experiments on low-porosity sandstones [5], as well as heavily sintered glass beads [6] show a sharp decrease of the Darcy permeability around porosities of a few percent. In this limit, there are, however, uncertainties as to the exact definition, and measurement, of the porosity, which generally serves as the control parameter for a percolation transition. It would be useful to distinguish by stereologic analysis [7] between the total porosity, and the open one, with and without "dead arms"; this has not yet been done. Such studies are of interest particularly since the ultimate stages of sintering in materials involve strong correlations between pores.

Percolation concepts have also been applied to weakly fractured rocks [8,9]. It is well known that in granitic sites the permeability can show, even in closely adjacent formations, very large fluctuations from permeable to impermeable rocks. If the fractures can be modeled by plane cracks, without much correlation in position or orientation – an assumption well established in many instances – continuum percolation models can be used. For example, with disks of radius R, randomly oriented in space, the critical percolation value for the number N_3 of disks per unit volume is given [10] by

$$\pi^2 N_3 R^3 \approx 1.8 . \tag{1}$$

This form can be understood by means of an excluded-volume argument [11]. The existence of such an invariant could be tested in principle by analysing the distribution of intercepts $n_2(l)$ by a plane in the site, where l is the length of a crack; this stereologic problem leads to an exact inversion by means of an Abel transform [12]. The analysis could be extrapolated to a system of heterogeneous disk sizes, but it is as yet unclear which third-order moment of the distribution would best characterize and fit the data – $\langle R \rangle \langle R^2 \rangle$ would be the result given by excluded-volume arguments, while an average such as $\langle R^3 \rangle$ would give, perhaps more appropriately, greater weight to the larger fractures. On the experimental side, most hydrogeology programs deal with largely interacting cracks, and there is a scarcity of data to compare with percolation models.

Other classes of fractal porous media can be envisioned. The experimental findings of Katz and Thompson [13], who deduced an internal fractal pore structure from stereologic cuts of various sandstone samples, have been the subject of some controversy. This has arisen due to the possibility of confusion between fractal pore volume (which would affect the permeability) and fractal pore surface [14]. The work of de Gennes [15], based on the possibility of the latter occurrence, discusses hypodiffusive filling through capillary films deposited on such surfaces. In a recent paper, Davis [16] analyses the possible application of this result, in terms of the capillary pressure of a thin film as a function of the film thickness h, $P_c \propto 1/h$, for cases where h is small enough so that the surface tension contribution is negligible relative to the disjoining pressure. From this, he finds that the dependence of the water saturation S_w (the fraction of pores filled by water) on P_c should be quite different for a pore surface of ordinary, and one of fractal, geometry. For the first case, it can be shown that

$$S_w = A P_c^{-2} + B P_c^{-3} . \tag{2}$$

The first term represents the contribution of wedges and drops of liquid, while the second is the effect of liquid filling concave pits. On the other hand, de Gennes has shown that for two different simple fractal pore models the saturation should follow the relation

$$S_w = A' P_c^{D-3} , \tag{3}$$

where D is the fractal dimension of the surface. Davis demonstrates that the experimental fit to eq. (2) is poor indeed (a physically unreasonable negative prefactor appears), while eq. (3) is well verified for a variety of sandstone samples and measuring methods, yielding a fractal dimension $D=2.55$.

Other theoretical papers have analyzed the permeability of deterministic fractals [17]. An experimental realization of such a structure could be obtained by an iterative Appolonian filling by spheres. In such a system, the pore size corresponds to the size of the smallest sphere in the packing. Such a problem is of practical importance in the preparation of concrete, in which a wide distribution of sizes of grains is known to lead to a desirable low-porosity solid.

In general, we do not believe that the present theoretical studies of the static permeability – an integrated property – can shed much light on the existence of a fractal structure. One can think of more specific tools, such as the following:

(a) The study of an ac permeability [18], which introduces a tunable viscous wave length $\sqrt{\nu/\omega}$, where ν is the kinematic viscosity of the fluid.

(b) The propagation of an overpressure wave – normally following a diffusion law, which allows a permeability measurement [19] – takes place with anomalous diffusion characteristics in a fractal medium, as can be shown by analog electrical simulation experiments [20].

(c) The distribution of transit times of markers in the fluid in the absence [21] or the presence of an applied flow [#1]. This last phenomenon, known as dispersion, has been much studied recently both theoretically and experimentally. There is some evidence that this "anomalous" (non-Gaussian) dispersion response might be used to analyze the fractal structure of rocks.

(d) The study of non-Newtonian fluids displaying threshold effects, discussed in more detail below, is a further possible technique. We have pointed out that the transition from a linear Darcy regime to a non-linear (Forsheimer) one is strongly influenced by pore size distribution [24].

3. Broad pore-size distributions

Let us return to systems of fractured rock, but well above a permeability threshold. It is known that the distribution of fracture widths δ can be extremely broad, ranging from centimeter to micrometer sizes. Recalling the Poiseuille dependence of the flow rate between two planes $\propto \delta^3$, we see that the flow distribution will be mostly controlled by the subset of larger fractures. The problem can be formulated theoretically by applying the classical treatment of Ambegaokar, Halperin and Langer [25] on the transport properties of disordered semiconductors. It consists of selecting a sublattice of conducting elements, selected in decreasing order of conductance (or permeance of the fractures [26]). There is then a percolation-like threshold at which the permeability becomes no longer negligible, increasing as one adds elements beyond this threshold, with the same critical exponent t which characterizes the electrical conductance in a percolation problem. The permeability saturates to the real value for the full lattice when a few additional bonds are added; the larger the distribution of bonds (or fractures), the narrower this range. We have evaluated some bounds for this problem. Recent studies by Cacas [27] in hydrogeological analyses of granite sites have confirmed the practical applicability of this concept. Independently of our work, Katz and Thompson [28] have also considered the evaluation of the permeability of sandstone, using a similar approach, based on the distribution of throat radii of pores. They reach a remarkable conclusion by connecting the permeability of a material to a single length scale Λ, which can be evaluated by various acoustic, NMR, or specific surface methods, and corresponds to the critical value introduced by the construction. More recently, Tyc and Halperin as well as Le Doussal [29] have extended this approach, showing a critical, percolation-like scaling in the case of electrical networks with an exponentially broad distribution of resistances. Thus, although a fractal lattice is not directly present in the structure, fractal concepts lead to practical results!

A different approach to estimating pore-size distributions makes use of a threshold non-Newtonian fluid flowing in the medium, as suggested above. One such fluid is "Bentonite", a clay suspension widely used in the petroleum industry. If the liquid is placed in a cylindrical tube under increasing applied pressure, it will not flow until an overpressure Δp_c, or more pre-

---

[#1] This problem is extensively discussed in the chapters by Brady and by Bacri et al. [22,23].

cisely a critical stress, it applied at the wall. Above this threshold, it flows as a viscous liquid. If a porous material is saturated with such a "Bingham fluid" and subjected to a pressure difference ΔP, it will start flowing though only above a critical value ΔP_c. Such a study is presently being carried out in our laboratory [30] [#2]. It has been simulated both experimentally and numerically by an electrical Zener diode analog, in which the pressure variable is replaced by a local voltage v_c, and the flow is represented by the electrical current [31]. The overall threshold V_c is defined by

$$V_c = \min_{\text{path } P} \sum_{i \in P} v_{c_i}, \qquad (4)$$

where the sum is evaluated over all continuous paths P connecting the electrodes over which the current flows. This solution corresponds to a minimal path estimate, which has been intensively studied, and is known to be amenable to self-affine analysis [32]. In a range above V_c, one can identify a power-law dependence of the current–voltage characteristic, which reflects the progressive enrichment of the lattice of conducting channels with increasing voltage.

4. Multiphase flow in porous media

The flow of immiscible fluids in a porous geometry – often modeled in Hele-Shaw cells with flat or rough walls – provides a vast class of applications of fractal and percolation concepts, as discussed in various papers of these proceedings. Here again, we must distinguish between two regimes. On one hand, if the velocity field is large, the behavior is controlled by the ratio of viscosities of the two fluid phases. It is possible, if the pushing fluid is of low viscosity, to obtain DLA-like structures which, as first pointed out by Paterson [33], result from the potential character of Darcy's law in the more viscous phase [#3]. We will

---

[#2] This research was carried out in collaboration with G.C. Maitland (Schlumberger Cambridge Research).

[#3] The opposite limit of a very viscous invading fluid leads to interesting crystal-like patterns [34].

here restrict our present discussion to the opposite case of very slow injection rates where capillary pressure effects completely dominate viscous ones, and where small pressure increments lead to the quasistatic progression of an invasion front [35]. For the sake of completeness, we might just mention that fascinating and subtle effects take place in the crossover region between the above two regimes, where viscous and capillary forces compete [36].

The equivalence between drainage – where a non-wetting fluid displaces under pressure a fluid wetting the pores – and percolation was recognized long ago [35] and first identified in the beautiful model experiments of Lenormand et al. [37]. It is due to the one-to-one correspondence between the capillary pressure $p(r)$ and the (randomly varying from pore to pore) radius r of a pore (expressed by the Laplace law); thus the frontier of the invading fluid is that of the percolation cluster attached to the injection face and corresponding to the given value of the applied pressure. A spectacular manifestation of the correspondence with percolation has been obtained in a vertical geometry [38] where the invading fluid was a dense liquid metal injected from below. The pressure decreases vertically because of the hydrostatic effect. This leads to a limitation of the penetration and to the existence of a stable upper front. The identification of the fractal structure so obtained can be made after letting the metal solidify once the structure has stabilized. The experiments are in excellent agreement with the numerical results of Sapoval's group on gradient percolation [39] as well as with the scaling analysis of Wilkinson [40]. The "dynamic" character of the slow drainage was stressed by Wilkinson and Willemsen [41], who coined the name "invasion percolation" for the set of rules associated with the drainage process – we use the word "dynamic" in quotes because the invasion is actually modelled at constant quasistatic flow rate, and the time variable really corresponds to the total amount of injected fluid.

We would like to dwell for the remainder of this paper on an aspect of invasion percolation, specifically the "noise" which results from the advance of the invasion front into the porous medium. The existence of "Haines jumps", or bursts of penetrating fluid has long been recognized [42], but the particular characteristics of the noise spectrum have been only recently stressed by Thompson et al. [43]. Physically, the injection of a non-wetting fluid generates a growing percolation cluster, which expands in irregular jumps associated with the filling of pockets, of larger pore radius, connected to the invading cluster by narrower necks. The spectrum of this noise should be a means to probe the pore size distribution and its spatial correlations. A recent numerical simulation of this process by Furuberg et al. [44] has revived the subject by proposing a scaling formulation of the "dynamics" of the invasion.

To explain the concept of noise, or bursts, we must analyze the evolution mechanism of invasion percolation: In this model, we assign to every bond of a lattice a random fractional probability x ($0 < x < 1$). Starting at a given origin, we allow the cluster to grow by looking, at each time step increment, for all possible neighbors of the existing cluster and choosing the one growth site with the lowest available value of x (x may, for example, be proportional to the inverse of the cross section of invasion boundary pores). In this way, we allow the cluster to grow along the path of "least resistance". The x selected by this rule as a function of time t (which can also represent the mass of the cluster) represents the local growth threshold. A plot of the resulting $x(t)$ versus t will be overall an increasing function of t, with downward fluctuations. Each valley in the graph of $x(t)$ then represents a local cluster (of smaller values of x) which is locally being filled once the threshold cutting bond has been selected. Thus fluctuations in the graph of $x(t)$ – related also physically to pressure fluctuations in the fluid – are a sensitive measure of local cluster size distribution and growth. Furuberg et al. have shown that the correlation function $N(r, T)$ of two given bonds separated by a distance r and a time interval T, can be written as a scaling function $N(r, T) = \phi(r^D/T)/r$, where D is the fractal dimension of the infinite cluster at threshold. The scaling function $\phi(u)$ has the asymptotic forms u^a (for $u \ll 1$), u^{-b} (for $u \gg 1$), and has a maximum for $u \approx 1$. This variation is qualitatively understandable if one recalls that r^D represents the number of objects within a connected region of linear size r. Thus the correlation should be maximal, for intervals T (i.e. added mass), corresponding to the dimension r of the added region, and small whenever T and r^D are not comparable. This suggests indeed that the scaling of $N(r, T)$ should be related to the static exponents of the percolation problem. Roux and Guyon [45] have more rigorously developed this idea; they have been able to express the exponents a and b in terms of percolation cluster exponents [46] and of the fractal dimension D_h of the perimeter (or hull) of the percolation cluster, which identifies the set of possible growth sites. Their derivation relates the noise spectrum of the curve $x(t)$, over a given interval T, to the distribution of percolation clusters.

5. Conclusion

In summary, we see that *percolation* and *fractal ideas* have a broad range of applicability, much beyond the more naive and seldom directly applicable situation of *real fractal geometries*. Except perhaps in some applications in the petroleum industry, much work remains to be done to establish a closer relation between the basic concepts and applications to real porous systems, particularly in the fields of chemical engineering and hydrogeology. This should involve a comparison of the property considered with the geometrical features: those obtained (i) directly from three-dimensional imaging, or study of sections (in particular those resulting in serial cuts which provide information on spatial correlation), as well as (ii) indirectly, from scattering techniques.

It is worth stressing once more that, where transport properties are involved, a single steady-state measurement is often an inadequate tool to evaluate the effects of a multiplicity of scales. Thus a variety of probes, involving the use of non-linear rheological fluids, dynamical studies, noise characteristics, or fluctuations in the measurements from site to site, must be developed, understood, and refined.

Acknowledgements

The authors wish to acknowledge support from NATO International Collaborative Research Grant RG.0126/88. We would like to thank all our colleagues from the Groupe Poreux PC: C. Baudet, E. Charlaix, E. Clément, C. Leroy, and P. Rigord, who have participated in the research reviewed here. One of us (EG) would also like to acknowledge a useful discussion with Professor J.L. Lions.

References

[1] S.R. Broadbent and J.M. Hammersley, Proc. Cambr. Phil. Soc. 53 (1957) 629.
[2] A.C. Payatakes and M. Dias, Rev. Chem. Eng. 2 (1984) 85;
D. Houi, in: Kinetics of Aggregation and Gelation, F. Family and D.P. Landau, eds. (Elsevier, Amsterdam, 1984) p. 173.
[3] A.O. Imdakm and M. Sahimi, Phys. Rev. A 36 (1987) 5304.
[4] E.L. Hinrichsen, A. Hansen and S. Roux, preprint.
[5] T. Bourbié, O. Coussy and B. Zinszner, in: Acoustique des Milieux Poreux (Technip, Paris, 1986) ch. 1.
[6] P.Z. Wong, J. Koplik and J.P. Tomanic, Phys. Rev. B 30 (1984) 6606;
E. Guyon, L. Oger and T.J. Plona, J. Phys. D 20 (1987) 1637.
[7] L. Oger, E. Guyon and T. Plona, Acta Stereol. 6 (1987) 425.
[8] S. Wilke, E. Guyon and G. de Marsily, Math. Geol. 17 (1985) 17.
[9] P.C. Robinson, J. Phys. A 16 (1986) 605.
[10] E. Charlaix, J. Phys. A 16 (1986) L533.
[11] I. Balberg, C.H. Anderson, S. Alexander and N. Wagner, Phys. Rev. B 30 (1984) 3933.
[12] E. Charlaix, E. Guyon and N. Rivier, Solid State Commun. 50 (1984) 999.
[13] A.J. Katz and A.H. Thompson, Phys. Rev. Lett. 54 (1985) 1325.
[14] J.N. Roberts, Phys. Rev. Lett. 56 (1986) 2111.
[15] P.G. de Gennes, in: Physics of Disordered Materials, D. Adler, H. Fritsche and S.R. Ovshinsky, eds. (Plenum, New York, 1985) p. 227.
[16] H.T. Davis, Europhys. Lett. 8 (1989) 629.
[17] P.M. Adler and C.G. Jacquin, Transport Porous Media 2 (1987) 553;
C.G. Jacquin and P.M. Adler, Transport Porous Media 2 (1987) 571.
[18] E. Charlaix, A.P. Kushnik and J.P. Stokes, Phys. Rev. Lett. 61 (1988) 1595.
[19] G. de Marsily, in: Hydrogéologie Quantitative (Masson, Paris, 1981) p. 85.
[20] C.D. Mitescu and J. Roussenq, Ann. Israel Phys. Soc. 5 (1983) 81;
J.P. Clerc, A.M.S. Tremblay, G. Albinet and C.D. Mitescu, J. Phys. Lett. 45 (1984) L913;
P. Rigord and J.P. Hulin, Europhys. Lett. 6 (1988) 145.
[21] W.D. Dozier, J.M. Drake and J. Klafter, Phys. Rev. Lett. 56 (1986) 197.
[22] J.F. Brady, in: Hydrodynamics of Dispersed Media, Proceedings of the 4th ESP Liquid State Conference on Hydrodynamics of Dispersed Media, Arcachon, France, May 24–27, 1988, J.P. Hulin, A.M. Cazabat, F. Carmona and E. Guyon, eds. (North-Holland, Amsterdam, 1989), in press.
[23] J.C. Bacri et al., in: Hydrodynamics of Dispersed Media, Proceedings of the 4th ESP Liquid State Conference on Hydrodynamics of Dispersed Media, Arcachon, France, May 24–27, 1988, J.P. Hulin, A.M. Cazabat, F. Carmona and E. Guyon, eds. (North-Holland, Amsterdam, 1989), in press.
[24] E. Guyon, A. Hansen and S. Roux, Trans. ASME 109 (1987) 274.
[25] V. Ambegaokar, B.I. Halperin and J.S. Langer, Phys. Rev. B 4 (1971) 2612.
[26] E. Charlaix, E. Guyon and S. Roux, Transport Porous Media 2 (1987) 31.
[27] M.-C. Cacas, Thèse, Ecole National Supérieure des Mines, Paris (1989).
[28] A.J. Katz and A.H. Thompson, Phys. Rev. B 34 (1986) 8179.
[29] S. Tyc and B.I. Halperin, Phys. Rev. B 39 (1989) 877;
P. Le Doussal, Phys. Rev. B 39 (1989) 881.
[30] M. Guillaume and A. Guillaumont, private communication.
[31] A. Gilabert, S. Roux and E. Guyon, J. Phys. (Paris) 48 (1987) 1609;
S. Roux and H.J. Herrmann, Europhys. Lett. 4 (1987) 1227.
[32] M. Kardar, G. Parisi and Y.-C. Zhang, Phys. Rev. Lett. 56 (1986) 889.
[33] L. Paterson, Phys. Rev. Lett. 52 (1984) 1621.
[34] G. Daccord and M. Wafra, private communication.
[35] P.G. de Gennes and E. Guyon, J. Mécanique 17 (1978) 403;
F.A.L. Dullien, Chem. Eng. J. 10 (1975) 1.
[36] A. Calvo, R. Chertcoff, M. Rosen and E. Guyon, Rev. Phys. Appl. 24 (1989) 553.
[37] R. Lenormand, C. Zarcone and A. Sarr, J. Fluid Mech. 135 (1983) 337 [36].
[38] E. Clément, C. Baudet, E. Guyon and J.P. Hulin, J. Phys. C 20 (1987) 608.

[39] M. Rosso, J.F. Gouyet and B. Sapoval, Phys. Rev. Lett. 57 (1986) 3195;
J.P. Hulin, E. Clément, C. Baudet, J.F. Gouyet and M. Rosso, Phys. Rev. Lett. 61 (1988) 333.
[40] D. Wilkinson, in: Physics of Finely Divided Matter (Springer, Berlin, 1985) p. 280.
[41] D. Wilkinson and J.F. Willemsen, J. Phys. A 16 (1983) 3365.
[42] F.A.L. Dullien, Porous Media: Fluid Transport and Pore Structure (Academic Press, New York, 1979).
[43] A.H. Thomson, A.J. Katz and R.A. Rashke, Phys. Rev. Lett. 58 (1987) 29.
[44] L. Furuberg, J. Freder, A. Aharony and T. Jøssang, Phys. Rev. Lett. 61 (1988) 2117.
[45] S. Roux and E. Guyon, J. Phys. A (1989), to be published.
[46] D. Stauffer, Introduction to Percolation Theory (Taylor and Francis, London, 1985).

REMARKS ON PERCOLATION AND TRANSPORT IN NETWORKS WITH A WIDE RANGE OF BOND STRENGTHS

Bertrand I. HALPERIN

Physics Department, Harvard University, Cambridge, MA 02138, USA

In conventional percolation problems, there is no simple relation between the transport exponents and the critical exponents which describe geometric properties of the percolating clusters. In several problems involving bonds with a wide range of conductance, however, it has been found that there does exist a simple relation between the transport exponent and the geometric exponent describing the divergence of the correlation length. We give a brief review of these problems.

As is well known, percolation problems are a prime example where fractal geometries play an important role in determining the macroscopic properties of a system [1]. In particular, in the limit where one approaches the percolation threshold from the conducting side, the geometry of the percolating network acquires a fractal character. The network may be characterized by various geometric exponents or fractal dimensions, which are reflected directly, or through scaling relations, in the exponents that describe the behavior of various statistical properties of the connected clusters. By contrast, the behavior of transport properties, such as the electrical conductivity of resistance network near the percolation threshold, are in most cases described by new exponents, which cannot be related directly to the more conventional geometric exponents of the percolating cluster [1,2]. There exist certain situations, however, when the transport exponents can be directly related to geometric exponents. Such situations occur in several cases where there exists a wide distribution of the strengths of the individual bonds [3–8].

For simplicity, we shall consider a simple cubic lattice in d dimensions, with lattice constant a, and with an independent probability distribution $p(g_i)$ for the conductance g_i of each bond. The standard bond percolation problem is defined by the choice

$$p(g) = p\delta(g - g_0) + (1 - p)\delta(g). \tag{1}$$

Thus there is a probability p that the bond is present, in which case the conductance has a constant value g_0. As p approaches the critical value p_c from above the fraction of bonds in the infinite percolating cluster decreases to zero as a power law [1]

$$P(p) \sim (p - p_c)^\beta. \tag{2}$$

At the same time the macroscopic conductivity decreases with a different power law,

$$\sigma(p) \sim (p - p_c)^t. \tag{3}$$

Also, as $p \to p_c$ there is a characteristic length scale ξ which diverges,

$$\xi(p) \sim |p - p_c|^{-\nu}. \tag{4}$$

For length scales R larger than ξ, the system takes on a homogeneous character, where intensive quantities approach their macroscopic values, and fluctuations are governed by the usual law of large numbers. This breaks down for length scales smaller than ξ, and one finds instead, in the region $a < R < \xi$ that the percolation network has a fractal structure. These properties are found to match smoothly at the length scale ξ, which gives rise to powerful scaling laws that relate the values of various critical exponents.

We can define a fractal dimension D by requiring that the number of bonds on the percolating network in a volume ξ^d be proportional to ξ^D. It is clear from the above arguments that

Essays in honour of Benoit B. Mandelbrot
Fractals in Physics – A. Aharony and J. Feder (editors)

$$D = d - \beta/\nu. \tag{5}$$

According to the scaling argument mentioned above, we expect that if we start from a typical point on the percolating cluster, the number of bonds on the clusters within a distance R of the given point should scale as R^D, for $a < R < \xi$.

Bonds on the infinite cluster may be immediately divided into several categories [1]. A large fraction is located on dead-end branches which carry zero current when a voltage is applied between opposite ends of the sample. Dead-end branches are characterized by the fact that they may be detached from the cluster by cutting a single bond. The remaining bonds, which carry a nonzero current are known as the percolating backbone. The fraction of bonds on the backbone is characterized by an exponent β_B which is larger than β, so the fractal dimension of the backbone, $D_\beta = d - \beta_B/\nu$, is accordingly smaller than the dimension D.

According to the "nodes, links and blobs" description [9], the percolating backbone may be described as a set of "nodes", more or less regularly spaced, with a separation ξ, connected together by tenuous "links". By definition, at least three links must be attached to each of the nodes.

A small fraction of the bonds on each link may be described as "singly connected". A singly connected bond must carry the entire current flowing through a link; also, if any singly connected bond is cut, the link can no longer carry a current. These bonds on a link which are not singly connected have at least one bond in parallel, and are said to be part of a multiply connected "blob".

Because the singly connected sites are potential bottlenecks for conduction processes, it is useful to consider that the number of singly connected bonds L_1 on a link between two nodes is a kind of metric on the percolating backbone which determines an internal "distance" between the nodes.

It has been shown by Coniglio [10] that on average,

$$L_1 \sim (p - p_c)^{-1} \sim \xi^{1/\nu}. \tag{6}$$

Thus we see that the exponent ν has a purely geometric interpretation on the percolating cluster: it relates the distance ξ in physical space to the internal metric L_1 for distance along the links.

It is clear that one obtains a lower bound to the resistance of a link by simply considering the contribution of the singly connected bonds, which are connected in series and give a resistance L_1/g_0. If we consider that the links form part of a more or less regular network with lattice constant ξ, we see that σ is bounded above by the estimate

$$\sigma_1 = g_0 \xi^{2-d}/\langle L_1 \rangle.$$

Thus we are led finally to a lower bound for the conductivity exponent t of the form $t \geq t_1$, with

$$t_1 \equiv 1 + (d-2)\nu. \tag{7}$$

For $d < 6$, it appears that the actual value of t is larger than t_1. For $d = 2$ the analysis of numerical simulations and experiments gives strong evidence that $t \approx 1.30$ while $t_1 = 1$ exactly. In $d = 3$, the values of t_1 and t are closer, but there still is a difference between the two [1]. Expansions in the variable $\epsilon \equiv 6 - d$ show that $t - t_1$ is proportional to ϵ with a small positive coefficient [2]. Thus for the standard bond percolation problem, in $d < 6$, the resistance of the "blobs" is larger than the resistance of the singly connected bonds, so it is the blob contribution which actually determines the exponent t. There is no simple relation between the conductivity exponent t and the geometric exponents such as ν or β.

The situation is different in several cases where there is a wide range of conductances for the bonds present in the sample. In these cases we find that singly connected bonds do form the bottlenecks which determine the resistivity of the network.

As one example, we consider a network where the distribution $p(g)$ has the form

$$p(g) = (1-p)\delta(g) + pf(g), \tag{8}$$

where $f(g)$ is continuous for $g > 0$, and diverges, for $g \to 0$, as a power law:

$$f(g) \sim g^{-w} \tag{9}$$

with $0 < w < 1$. Distributions of this type are found to

arise naturally in various continuum models of conduction or fluid-flow permeability in porous media [6]. It is clear that the percolation threshold p_c and the geometric properties of the percolating cluster are exactly as for the standard lattice model, eq. (1). The conductivity of the networks will be different, however. The conductivity will vanish, for $p \to p_c^+$ with an exponent t_w which may or may not be the same as the exponent t for the standard lattice problem. Early estimates of t_w were given by Kogut and Straley [11], and by Ben-Mizrahi and Bergman [12].

The lower bound to t_w that one obtains by considering only the contribution of the singly connected bonds in this case turns out to be

$$t_{1w} = t_1 + w/(1-w), \qquad (10)$$

where t_1 is given by eq. (7). It was argued by Straley [3] that t_w is precisely given by

$$t_w = \min(t, t_{1w}), \qquad (11)$$

where t is the exponent for the standard lattice problem. Thus, for w greater than the critical value

$$w_c = (t-t_1)/[1+(t-t_1)], \qquad (12)$$

we have $t=t_{1w}$, so the resistance is determined by the contribution of the singly connected bonds. For $w<w_c$, on the other hand, the resistance is determined by the blobs, and is essentially the same as for the standard percolation model.

The physical reason for this behavior is easy to state [3]. For $w>0$ there are a rather large number of bonds with arbitrarily large resistance. If one of these bonds occurs at a singly connected site, it can dominate the resistance of the entire link. If the large resistance occurs on a bond in one of the blobs, however, it has little effect since there is always at least one other parallel path in the blob which can carry the current. It is rather unlikely that there will be unusually large resistances on all of the paths in a blob. For $w>w_c$, the resistance of the weakest singly connected bond is typically larger than the sum of the resistances of the blobs, and of all the other singly connected bonds on the link, so it dominates the resistivity.

After some controversy, the validity of eq. (11) has been established recently by a renormalization group analysis correct to leading order in $\epsilon = 6-d$ [4,5].

The percolation problem can also be generalized to an elastic network, with a distribution of bond-bending and bond-stretching force constants, whose form is similar to eq. (8). Such distributions arise in continuum models of elastic percolation [6]. Again, one can establish a lower bound to the exponent for the vanishing of the network elastic constant f_w as $p \to p_c^+$, analogous to eq. (10), by considering only the singly connected bonds. The bound in this case is [6]

$$f_{1w} = dv + w/(1-w). \qquad (13)$$

Again, it seems likely that in the general case, $f_w = \min(f, f_{1w})$, where f is the exponent in the case where all occupied bonds have the same strength. It is known that $f \geq dv$ [13].

Another interesting problem arises when no bonds are actually absent in the problem, but where the range of bond resistances is wide on a logarithmic scale. Such distributions arise naturally when the transport process involves quantum mechanical tunneling and/or thermal activation over a barrier, and the barrier heights have a distribution wide compared to the temperature [14–18]. In this case the conductance of the network is largely determined by the threshold conductance g_c for percolation on the network, which is defined so that the bonds with $g_i \geq g_c$ just form a percolating network. The definition of a distribution wide on a logarithmic scale is that $gp(g)$ is approximately constant, for a wide range of g about g_c, with

$$\lambda^{-1} \equiv g_c p(g_c) \ll 1. \qquad (14)$$

The network conductivity in this case may be written, in the limit $\lambda \to \infty$, as

$$\sigma \sim C g_c a^{2-d} \lambda^{-y}, \qquad (15)$$

where C is a constant that depends on the lattice structure, and y is a critical exponent that we expect should depend only on the dimension d [8,17]. Le Doussal [7] has argued that for $d \leq 6$, the value of y is directly related to the geometric exponent v by

$$y = (d-2)v. \qquad (16)$$

(A relation equivalent to this, for the case $d=3$, was proposed in 1973 by Ambegaokar, Cochran and Kurkijärvi, in the context of a continuum model [17,18].) Le Doussal has supported his analysis by a variety of arguments, including exact solutions of an analogous problem on a hierarchical lattice, where the fractal nature of the percolating network is explicit.

In the case $d=2$, eq. (16) gives $y=0$. This is rigorously the case for a suitably defined model on a *square* lattice because of the self-duality of that problem [8,19,20]. Numerical simulations by Tyč and Halperin [8] have confirmed that for the *triangular* lattice, $|y| \lesssim 0.02$.

For $d=3$, the accepted value of ν is approximately 0.88 [1]. The estimate of y obtained by Tyč and Halperin [8] from numerical simulations was significantly smaller than this (roughly 0.6 ± 0.1), but this discrepancy may be due to anomalously large finite size corrections to the value of ν. Further work on this problem is necessary.

A heuristic derivation of the results of Le Doussal and Ambegaokar et al. can be given as follows [7]. Suppose that we simply remove from the network all bonds with conductance $g_i < g_c \, e^{-\lambda \epsilon}$, where ϵ is a positive quantity, which will be eventually chosen to be of order λ^{-1}. The conductivity of the reduced network will be a lower bound to that of the original network, but it should not be too far off if ϵ is chosen sufficiently large. It is straightforward to show that the probability that a bond is present in the reduced network is $p = p_c + \epsilon$, and the mean value of the resistance of the remaining bonds is

$$\langle g^{-1} \rangle \approx \lambda^{-1} \, e^{\lambda \epsilon} / (p_c + \epsilon) g_c. \qquad (17)$$

If ϵ is small we have a percolating network, and we may obtain a lower bound to the *resistance* of the reduced network by considering only the resistance of the singly connected bonds on the links of the conducting backbone. Although this bound would not give the correct conductivity exponent for the conventional percolation problem, it should be reliable in the present problem, because the resistance of the blobs is now much lower relative to the mean resistance of the singly connected bonds. This happens because the high-resistance bonds in the blobs have a high probability of being sorted out by a bond of lower resistance.

The mean resistance of a link is thus given approximately by $L_1 \langle g^{-1} \rangle$. If we use this value, we obtain an estimate of the conductivity of the reduced network which is

$$\sigma_{\text{red}} \approx \epsilon^{t_1} / \langle g^{-1} \rangle, \qquad (18)$$

where t_1 is given by eq. (7).

In order to estimate σ_{red} we should actually have used the "typical resistance" (roughly the median) rather than the mean resistance of a link [6]. If we choose ϵ small, so that L_1 is large, there is not much difference between these measures. If we choose ϵ too large, however, the mean will be dominated by the high-resistance tail of the distribution, which means that eq. (18) is an underestimate of σ_{red} in this case.

We have already seen that σ_{red} is a lower bound to the conductivity of the original network. Hence we should choose ϵ to maximize the right-hand side of eq. (18), which occurs for $\lambda \epsilon = t_1$. We thus find an estimate of the form (15), with $y = (d-2) \nu$.

In closing, I wish to emphasize that a considerable literature has developed in recent years on the subject of percolation and transport in systems with distribution of bond strengths, and I have not been able to give a thorough discussion of this literature here. Attention may be called to such developments as experimental studies of conductivity and elastic properties of two-dimensional continuum systems [21–23], studies of quasi-one-dimensional systems with a distribution of bond strengths [24], analyses of various problems with distributions of resistors on hierarchical lattices and other fractals [25,26], and analyses of nonlinear transport problems, such as $1/f$ noise in a continuum system near the percolation threshold [27].

The author is very grateful for the collaboration of S. Feng, P. Sen, and S. Tyč on various aspects of the work reviewed in this note, and he has benefitted greatly from conversations with P. Le Doussal, C. Lobb, and S. Chakravarty. This work was supported by the NSF through the Harvard Materials Research Laboratory and grant DMR-88-17291.

References

[1] D. Stauffer, Introduction to Percolation Theory (Taylor and Francis, London, 1985).
[2] A.B. Harris, S. Kim and T.C. Lubensky, Phys. Rev. Lett. 53 (1984) 743.
[3] J.P. Straley, J. Phys. C 15 (1982) 2333, 2343.
[4] J. Machta, Phys. Rev. B 37 (1988) 7892.
[5] T.C. Lubensky and A.-M.S. Tremblay, Phys. Rev. B 37 (1988) 7894.
[6] S. Feng, B.I. Halperin and P. Sen, Phys. Rev. B 35 (1987) 197.
[7] P. Le Doussal, Phys. Rev. B 39 (1989) 881.
[8] S. Tyč and B.I. Halperin, Phys. Rev. B 39 (1989) 877.
[9] H.E. Stanley, J. Phys. A 10 (1977) L211.
[10] A. Coniglio, Phys. Rev. Lett. 46 (1981) 250.
[11] P.M. Kogut and J.P. Straley, J. Phys. C 12 (1979) 2151.
[12] A. Ben-Mizrahi and D.J. Bergman, J. Phys. C 14 (1981) 909.
[13] Y. Kantor and I. Webman, Phys. Rev. Lett. 52 (1984) 1891.
[14] V. Ambegaokar, B.I. Halperin and J.S. Langer, Phys. Rev. B 4 (1971) 2612.
[15] B.I. Shklovskii and A.L. Efros, Zh. Eksp. Teor. Fiz. 60 (1971) 867 [Sov. Phys. JETP 33 (1971) 468].
[16] M. Pollak, J. Non-Cryst. Solids 2 (1972) 1.
[17] V. Ambegaokar, S. Cochran and J. Kurkijärvi, Phys. Rev. B 8 (1973) 3682.
[18] J. Kurkijärvi, Phys. Rev. B 9 (1974) 770.
[19] J. Marchant and R. Gabillard, C.R. Acad. Sci. (Paris) Ser. B 281 (1975) 261.
[20] J. Bernasconi, W.R. Schneider and H.J. Wiesmann, Phys. Rev. B 16 (1977) 5250.
[21] L.N. Smith and C.J. Lobb, Phys. Rev. B 20 (1979) 3653.
[22] C.J. Lobb and M.G. Forrester, Phys. Rev. B 35 (1987) 1899.
[23] L. Benguigui, Phys. Rev. B 34 (1986) 8176.
[24] S. Havlin, A. Bunde, H. Weissman and A. Aharony, Phys. Rev. B 35 (1987) 397.
[25] H. Harder, S. Havlin and A. Bunde, Phys. Rev. B 36 (1987) 3874.
[26] G.H. Weiss and S. Havlin, Phys. Rev. B 36 (1987) 807.
[27] A.M.S. Tremblay, S. Feng and P. Breton, Phys. Rev. B 33 (1986) 2077.

PROBABILITY DENSITIES OF RANDOM WALKS IN RANDOM SYSTEMS

Shlomo HAVLIN and Armin BUNDE

Department of Physics, Bar-Ilan University, Ramat Gan, Israel
and I. Institut für Theoretische Physik, Universität Hamburg, D-2000 Hamburg 36, Fed. Rep. Germany

We review recent results for the probability distribution of random walkers in random systems, where diffusion is anomalous and the mean-square displacement scales with time as $R^2(t) \sim t^{2/d_w}$, $d_w > 2$. The random systems are characterized by structural disorder and by random transition rates. In general, the mean distribution function $\langle P(r,t) \rangle$ of the random walkers is a *stretched Gaussian* and scales as $\log[P(r,t)/P(r,0)] \sim -[r/R(t)]^u$, where $u = d_w/(d_w-1)$. On random fractals, the fluctuations of the density distribution $P(r,t)$, for fixed distance r and time t, have a broad logarithmic distribution. The average moments $\langle P^q \rangle$ scale in a multifractal way as $\langle P \rangle^{\tau(q)}$, where $\tau(q) \sim q^\gamma$, $\gamma < 1$. In contrast, in chemical l-space the fluctuations of P are narrow and $\langle P^q \rangle \sim \langle P \rangle^q$.

1. Introduction

In recent years, transport properties of particles in random media have attracted much attention. Random media cover a broad range of materials, which are of both scientific and technological interest. Examples are ionic and electronic conductors, conductive polymers, semiconductor glasses and composite materials [1–2]. In general, the randomness in the systems can be characterized by two types of disorder: Structural disorder and random potentials. Models for structural disorder are percolation clusters, random walks, self-avoiding random walks, or diffusion-limited aggregation; these structures are fractals on certain length scales, and the fractal concept introduced by Benoit Mandelbrot is an important tool to characterize them [1]. Models for transport in random potentials are continuous time random walk (CTRW) [13] and regular lattices with a random distribution of transition rates [2]. The common feature of these models is that transport can be anomalous; the mean-square displacement does not obey Fick's law $R^2 \sim t$ but scales with time as $R^2(t) \sim t^{2/d_w}$, where $d_w > 2$ is the fractal dimension of the random walk.

A central role plays the probability density $P(r,t)$, which is the probability to find a random walker at time t at distance r from its starting point. In a random system, $P(r,t)$ contains information on both static disorder and the dynamical process. In homogenous systems, $P(r,t)$ is Gaussian and does not depend on the configuration considered. In random systems, $P(r,t)$ varies from configuration to configuration and depends on the starting points. To obtain a complete description of diffusion in random systems, one has to study both the configurational average of the probability density $\langle P(r,t) \rangle$ and their fluctuations.

2. The mean probability density

The form of $\langle P(r,t) \rangle$ characterizes the localization of diffusion on fractal structures, and is relevant to several other physical problems of interest such as quantum localization [14,15] or self-avoiding walks on fractals [10,16]. The variance of $\langle P(r,t) \rangle$ represents the mean-square displacement, from which the diffusion constant and the conductivity can be obtained. The Fourier transform of $\langle P(r,t) \rangle$ represents the scattering function which is also experimentally accessible.

In *structurally* disordered systems, the Euclidean distance r is not sufficient to characterize the structure. It is convenient to consider in addition the chemical distance l, which is defined as the length of the shortest path between two sites on the structure. In homogenous systems r and l are proportional to each other, while in disordered structures they are related by the scaling relation $r \sim l^{\tilde{\nu}}$; in general, $\tilde{\nu} < 1$. Since a random walker proceeds along the shortest path it is natural to consider also the density distribution in l-space $P(l, t)$, which is the probability to find the random walker at time t at chemical distance l from the origin. The mean chemical length travelled by the random walker scales with time

$$L(t) \equiv \langle l(t) \rangle \sim t^{1/d_w^l},$$

where the chemical diffusion exponent d_w^l is related to d_w by $d_w^l = \tilde{\nu} d_w$.

Since the root-mean-square displacement $R(t)$ or the mean chemical length $L(t)$ are the only characteristic length scales in the problem, both $\langle P(r, t) \rangle$ and $\langle P(l, t) \rangle$ scale as

$$\langle P(r, t) \rangle = \langle P(0, t) \rangle \, \Pi(r/R(t)) \tag{1a}$$

and

$$\langle P(l, t) \rangle = \langle P(0, t) \rangle \, \Phi(l/L(t)). \tag{1b}$$

By definition, $\Pi(0) = \Phi(0) = 1$. It is generally accepted that asymptotically both scaling functions are stretched Gaussians,

$$\Pi(x) \sim \exp(-x^u), \tag{2a}$$

$$\Phi(x) \sim \exp(-x^v), \tag{2b}$$

where the exponents u and v characterize the extension of the distribution function in r- and l-space, respectively.

The question how u and v are related to other structural and dynamical exponents has been controversially discussed in the literature [10,14–19]. In the following we will present several random systems where u and v are described by the simple relations

$$u = \frac{d_w}{d_w - 1}, \tag{3a}$$

and

$$v = \frac{d_w^l}{d_w^l - 1}. \tag{3b}$$

Note that these relations hold also for homogeneous systems where $d_w = d_w^l = 2$ and $u = v = 2$.

(a) *Self-avoiding walks*: The simplest examples of random fractals are self-avoiding walks (SAW), which serve as models for linear polymers. According to Flory, the fractal dimension of SAWs is $d_f = \frac{1}{3}(d+2)$. Since SAWs are linear fractal chains, i.e. one-dimensional in l-space, l is proportional to the mass of the SAW. Hence one has $l \sim r^{d_f}$ and $\tilde{\nu} = 1/d_f$. Along the chain $P(l, t)$ is a Gaussian,

$$P(l, t) = P(0, t) \exp\{-[cl/L(t)]^2\},$$

where $P(0, t) \sim t^{-1/2}$ and $L(t) \sim t^{1/2}$ follows the normal diffusion law. For topologically linear random fractals like the SAW, $P(l, t)$ and $L(t)$ are the same for every random configuration. Hence $\langle P(r, t) \rangle$ can be written as [10]

$$\langle P(r, t) \rangle = \sum_{l=0}^{\infty} \langle P(r|l) \rangle \, P(l, t), \tag{4}$$

where $\langle P(r|l) \rangle$ is the probability to find two sites separated by a chemical distance l and Euclidean distance r. It is known (see e.g. ref. [9]) that

$$\langle P(r|l) \rangle \sim l^{-\tilde{\nu}d} (r/l)^{-g} \exp[-(r/l^{\tilde{\nu}})^\delta], \tag{5}$$

where $\delta = (1-\tilde{\nu})^{-1}$. The parameter g characterizes the probability of having two sites separated by large l and small r. For the SAW we have $g = 1/4, 1/10$, and 0 for $d = 2, 3$, and 4, respectively. Substituting (5) into (4) and using the method of steepest descent to evaluate the integral one finds [10]

$$\log \frac{\langle P(r,t) \rangle}{\langle P(0,t) \rangle} \sim -\left(\frac{r}{\langle R(r) \rangle}\right)^{d_w/(d_w-1)}, \tag{6}$$

in agreement with (3b).

(b) *Infinite comb and regular fractal trees*: Consider a comb structure with infinite long teeth (see fig. 1a). Along the backbone of the comb, diffusion is anomalously slow, $d_w = 4$ [20], since the teeth act as temporary traps for a random walker. It was found rigorously [20] that the density distribution on the

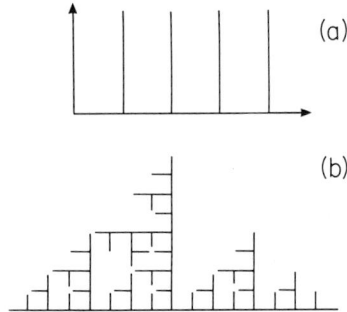

Fig. 1. (a) Two-dimensional comb with infinitely long teeth along the y-direction and the backbone in the s-direction. (b) Deterministic tree shown for the fourth generation. The fractal dimension is $\log 3/\log 2$.

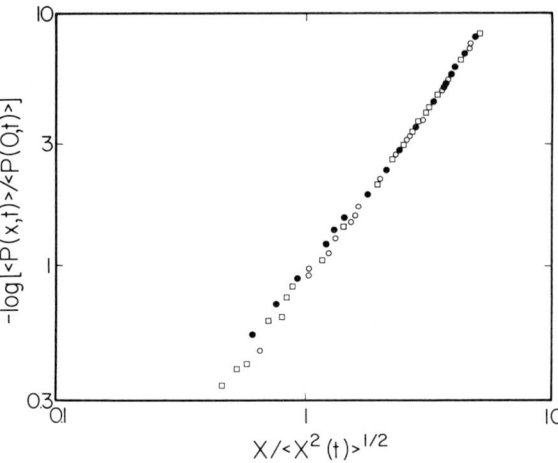

Fig. 2. Probability distributions for diffusion of hard-core particles in linear chains in the presence of a power-law distribution of transition rates, with $\alpha=1/2$. The plot shows $-\log[\langle P(x,t)\rangle/\langle P(0,t)\rangle]$ as a function of $x/\langle x^2(r)\rangle^{1/2}$ for up to 64 000 time steps and three concentrations of particles: (●) $c=0$, (□) $c=0.1$, and (○) $c=0.2$. $c=0$ represents the limit of noninteracting particles. (After ref. [21].)

backbone (x-coordinate) satisfies (2a), with u from (3a).

Diffusion on exact fractal trees (fig. 1b) has been studied by using the exact enumeration method. As has been found [10], the density distribution also follows eqs. (2) and (3).

(c) *Random transition rates*: Consider a d-dimensional lattice, where to each site a potential well is assigned. If deep potential wells are exponentially rare and jumps between neighboring sites are thermally activated, the corresponding transition rates follow a power-law distribution $P(W) \sim W^{-\alpha}$, $\alpha<1$. For $\alpha>0$ small transition rates are dominant and diffusion is anomalously slow, with d_w being a function of α [2]. The density distribution $\langle P(r,t)\rangle$ was studied numerically in one-dimensional chains. Again it was found that the form of $\langle P(r,t)\rangle$ is described by (2) and (3). Interestingly, it was found that hard-core interactions change d_w drastically but the asymptotic form of $\langle P(r,t)\rangle$ satisfies (2) and (3) with d_w for *non-interacting particles* [21] (see fig. 2).

If the waiting times are not fixed to the lattice sites, but rather chosen randomly at each time step from a power law distribution, the problem can be treated analytically [22], using the continuous time random walk technique. The results follow again eqs. (2) and (3).

(d) *Percolation clusters at criticality*: Fractal percolation clusters occur in random mixtures at the critical concentration of one component [23]. The question if (2) and (3) can be also used to describe $\langle P(r,t)\rangle$ and $\langle P(l,t)\rangle$ in percolation clusters was studied recently by extensive numerical analysis in $d=2$ [24]. Fig. 3 shows $-\log[\langle P(l,t)\rangle/\langle P(0,t)\rangle]$ and $-\log[\langle P(r,t)\rangle/\langle P(0,t)\rangle]$ as a function of $l/\langle L(t)\rangle$ and $r/\langle R(t)\rangle$, respectively, for $t=1000$. Both curves are straight lines asymptotically, with slopes $v=1.71\pm0.02$ in l-space (fig. 3a) and $u=1.54\pm0.02$ in r-space (fig. 3b). The slopes are in excellent agreement with eqs. (2) and (3). From the simulations one finds $d_w^l=2.4\pm0.05$ and $d_w=2.81\pm0.05$ for $t=1000$, and this yields $v\cong1.71$ and $u\cong1.55$.

Both probability distributions can be analytically related to each other by eq. (4), when $P(l,t)$ is substituted by its configurational average $\langle P(l,t)\rangle$. The asymptotic form of $\langle P(r|l)\rangle$ in percolation is the same as in SAWs (eq. (5)). If the method of steepest descent is used, and $\langle P(l,t)\rangle$ is assumed to have the form (2b), one obtains

$$\langle P(r,t)\rangle \sim \exp\{-[r/\langle R(t)\rangle]^u\}$$

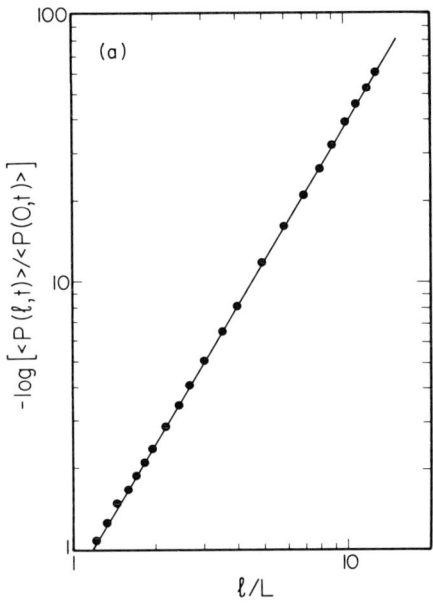

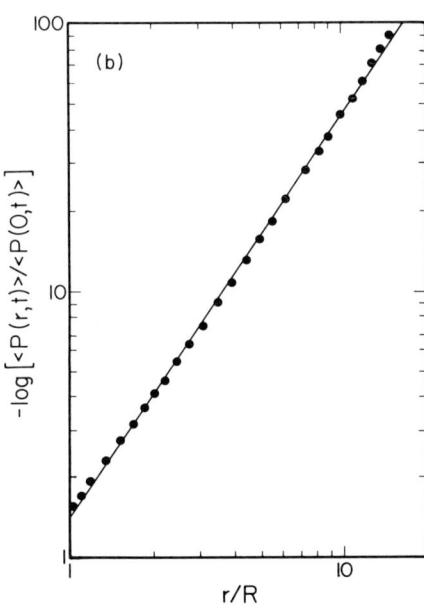

Fig. 3. Probability distributions for percolation clusters on a square lattice at criticality. (a) $-\log[\langle P(l,t)\rangle/\langle P(0,t)\rangle]$ as a function of $l/\langle L(t)\rangle$ for $t=1000$. (b) $-\log\langle [P(r,t)\rangle/\langle P(0,t)\rangle]$ as a function of $r/\langle R(t)\rangle$ for $t=1000$. Averages over 1400 configurations were performed for clusters of 300 chemical shells. (After ref. [24].)

with $u=d_w/(d_w-1)$, in agreement with fig. 3b and eq. (3b).

If in addition to the structural disorder random waiting times with power-law distribution are assigned to each site, both d_w and d_w^l depend on the exponent α. Numerical simulations have supported the validity of (2) and (3) also in this case [25].

Recently, considering random walks on percolation in $d=2$, Harris and Aharony [14] have obtained bounds for u, $d_w/(d_w-1) \leq u \leq d_w/(\tilde{\nu}d_w-1)$. According to ref. [14], u is equal to the lower bound, when one averages over *all* possible configurations, including the very rare ones, while the upper bound is obtained when averaging over typical configurations.

Let us conclude this section with three remarks.

(I) When SAWs are generated on regular lattices by a kinetic growth process, one unit mass is added at each time step and the rms displacement of SAWs is described by $d_w=1/d_f=\tilde{\nu}$. The probability density of these walks is given by eq. (5), where l by definition is identical to the mass and the number of time steps performed. Since $\delta=(1-\tilde{\nu})^{-1}=d_w/(d_w-1)$, also this distribution is characterized by eq. (3). This result holds not only for the conventional SAWs (see e.g. ref. [9]), but also for other fractal chains such as dressed self-avoiding walks [26].

(II) SAWs in percolation clusters are important model systems for polymers in random media. From the form of $\langle P(r,t)\rangle$, eqs. (1)–(3), one can derive a Flory-type theory for the end-to-end exponent ν of self-avoiding walks in percolation, giving [10,16]

$$\nu = \frac{2\tilde{\nu}d_w^{BB}-1}{d_w^{BB}(1+\tilde{\nu}d_f^{BB})-d_f^{BB}}, \qquad (7)$$

where the index BB is associated to the backbone of the percolation cluster.

(III) From $\langle P(r,t)\rangle$ one can obtain also the asymptotic behavior of the wave functions $\psi(r)$ of electrons on fractals [14]. Using scaling arguments it was found that $|\psi(r)| \sim \exp(-r^\eta)$, where $\eta=ud_w/(d_w+u)$. Substituting u from (3b) into this relation one obtains $\eta=1$, and $|\psi(r)|$ decays exponentially.

3. Fluctuations of the density distribution

To achieve a deeper understanding of the role of the configurational average $\langle \ \rangle$ on fractal structures, we discuss the fluctuations of the probability density $P(r, t)$ for fixed distance r and time t. The fluctuations can be described by the histogram $N(\log P)$ giving the number of sites with values P between $\log P$ and $\log P + d \log P$ and by the moments $\langle P^q(r, t) \rangle$, $q > 0$. Specifically, we consider random walks on percolation clusters at criticality and on linear fractals such as self-avoiding random walks (SAWs).

Very recently it was found [27] that in both fractal structures the histogram $N(\log P)$ is broad and distributed algebraically as

$$N(\log P) \sim [\log(P/P_0)]^{-\alpha}$$
$$\times \exp\left(-\frac{b}{[\log(P/P_0)]^\beta}\right). \quad (8)$$

Accordingly, the average moments cannot be described by a single exponent but show multifractal features, i.e.,

$$\langle P^q \rangle \sim \langle P \rangle^{\tau(q)}, \quad \tau(q) \sim q^\gamma, \quad \gamma < 1, \quad q > 0. \quad (9)$$

The exponents, α, β, and γ are related to standard exponents characterizing the fractal structure.

This multifractal behavior [#1] in r-space differs significantly from the normal behavior we find in chemical l-space for fluctuations of $P(l, t)$ at fixed chemical distance l and time t [#2]. In l-space, $N(\log P)$ has small constant width (which is zero for linear frac-

---

[#1] The multifractal features described by (8) and (9) are distinct from the multifractal behaviour found in turbulence, chaotic systems, or kinetic aggregation, where $\tau(q)$ approaches a straight line for $q \to \infty$. For recent reviews see e.g. refs. [28,29].

[#2] The chemical distance l between two sites on the fractal is the shortest path on the structure connecting them; the chemical dimension d_l characterizes how the mass scales with l, $M \sim l^{d_l}$. $P(l, t)$ is the probability to find the walker in chemical (topological) distance l from the origin, for a given configuration (see e.g. ref. [30]).

tals, and the average moments $\langle P^q(l, t) \rangle$ scale simply as $\langle P(l, t) \rangle^q$.

In the following we describe briefly the derivation of eqs. (8) and (9), which is rigorous for linear fractals and is strongly supported for percolation systems by numerical simulations. First we consider fractal chains.

Similar to eq. (4), the moments of the distribution functions in r- and l-space are related by

$$\langle P^q(r, t) \rangle = \int_0^\infty \langle P(r|l) \rangle P^q(l, t) \, dl. \quad (10)$$

Substituting (5) into (10) we obtain the general scaling relation

$$\langle P^q(r, t) \rangle \sim \langle P(rq^{\nu/2}, t) \rangle. \quad (11)$$

Combining (11) and (6) we obtain eq. (9), with $\gamma = \nu/(2 - \nu) = 1/(2d_f - 1)$.

In order to calculate the histogram $N(\log P)$, note that the qth moments can be also written as

$$\langle P^q \rangle = \int_0^\infty P^q N(\log P) \, d\log P. \quad (12)$$

By identifying eqs. (10) and (12) we obtain, after an appropriate change of variables, eq. (8), with $\alpha = \frac{1}{2}[(g+d)\nu + 1]$ and $\beta = \nu/2(1 - \nu)$.

The above derivation is rigorous for linear fractals where $P^q(l, t)$ is the same for every configuration. Thus the distribution of P-values for fixed l and t is a δ-function and $P^q(l, t)$ can be taken out of the average in the rhs of eq. (10). In the general case, the integrand in (10) must be substituted by $\langle P(r|l) \times P^q(l, t) \rangle$. If we assume that (a) $\langle P(r|l) P^q(l, t) \rangle$ can be decoupled into $\langle P(r|l) \rangle \langle P^q(l, t) \rangle$, (b) the moments $\langle P^q(l, t) \rangle$ scale as $\langle P(l, t) \rangle^q$, and (c) $\langle P(l, t) \rangle$ is given by eqs. (2a) and (3a), then the above considerations can be extended to general random fractals.

Next we consider in detail percolation clusters. Using assumptions (a)–(c), and following the same procedure as above for the linear fractals, one obtains

$$\langle P^q(r, t) \rangle \sim \langle P(rq^{\bar{\nu}(d_w^l - 1)/d_w^l}, t) \rangle. \quad (13)$$

From (1)–(3) and (13) we obtain eq. (9), with

$$\gamma = (d_w^l - 1)/(d_w - 1). \quad (14a)$$

Similarly, by identifying (10) with (12) we recover (8), with

$$\alpha = [(\tilde{g}+d)\tilde{\nu}(d_w^l - 1) + 1]/d_w^l \quad (14b)$$

and

$$\beta = (d_w^l - 1)/(d_w - d_w^l). \quad (14c)$$

Note that the results obtained for linear fractals are particular cases of (14), with $d_w^l = 2$.

To test the above predictions, extensive numerical studies have been carried out for percolation in $d=2$. Fig. 4a shows the results for $-\log[\langle P^q(l,t)\rangle/\langle P^q(0,t)\rangle]^{1/q}$ as a function of $l/L(t)$ for several values of q at $t=1000$ time steps. As seen in fig. 4a, *all* moments collapse asymptotically to a single straight line, supporting assumption (b). Fig. 4b shows $-\log[\langle P^q(r,t)\rangle/\langle P^q(0,t)\rangle]^{1/q}$ as a function of $r/R(t)$ for representative values of q, at time $t=1000$. In contrast to the situation in chemical space, the moments do not collapse to a single line. The asymptotic slope of all lines is 1.54 ± 0.03, in agreement with (3).

Fig. 5 shows $-\log[\langle P^q(r,t)\rangle/\langle P^q(0,t)\rangle]$ as a function of $rq^{\tilde{\nu}(d_w^l - 1)/d_w^l}/R(t)$. The data collapse supports the validity of eq. (13) and also the power-law behavior of $\tau(q)$, with the exponent γ from (14a).

Fig. 6 shows the histogram $N(\log P)$ in l- and r-space. The theoretical results, eqs. (8), (14b) and (14c) (dashed line), are in excellent agreement with the simulation and support strongly assumption (a).

It has been argued [27] that the results obtained are more general and are valid for all random fractal structures for which (5) is valid, as long as the chemical dimension $d_l \equiv \tilde{\nu} d_f$ is smaller than the fractal dimension d_f. In this case, it is always possible to find short Euclidean distances r separated by large chemical distances l, which leads to a broad distribution function $\langle P(r|l)\rangle$ and consequently to the anomalous behavior of $N(\log P)$ and the moments in r-space.

Next, consider the case when both chemical and Euclidean distance scale the same, i.e. $d_l = d_f$, which

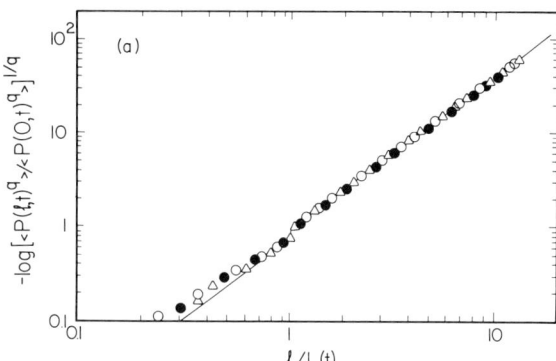

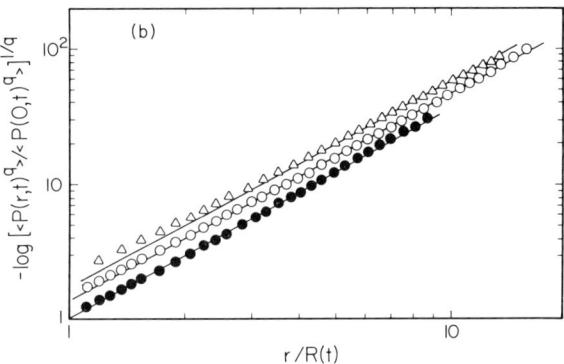

Fig. 4. Random walks on percolation clusters in $d=2$ at criticality: (a) Plot of $-\log[\langle P^q(l,t)\rangle/\langle P^q(0,t)\rangle]^{1/q}$ as a function of $l/L(t)$ for several values of q and $t=1000$. Different symbols represent different values of q: ($\triangle$) $q=0.3$, ($\bigcirc$), $q=1$, ($\bullet$) $q=4$. All data collapse to a single line asymptotically. (b) Plot of $-\log[\langle P^q(r,t)\rangle/\langle P^q(0,t)\rangle]^{1/q}$ as a function of $r/R(t)$ for $t=1000$ and representative values of q: ($\triangle$) $q=0.3$, ($\bigcirc$)$q=1$, and ($\bullet$) $q=4$. To obtain the results we considered clusters of 300 shells on a square lattice and averaged over 1400 configurations. Similar results were obtained for $t=500$, and (with less numerical accuracy) for $t=2000$ and $t=3000$. (After ref. [27].)

holds e.g. for diffusion-limited aggregates. If (5) is still valid, then $\langle P(r|l)\rangle$ becomes a δ-function at large distances and $N(\log P)$ and $\langle P^q\rangle$ scale the same in *both* l- and r-space, leading to $\gamma=1$. This conclusion follows also from (14a), anticipating that this result holds for arbitrary random fractal structures. Using the general relation $d_w^l = d_w d_l/d_f$, one can write (14a) as $\gamma - 1 = (d_f - d_l)d_w/(d_w - 1)d_f$ and the deviation of γ from the normal result $\gamma = 1$ is proportional

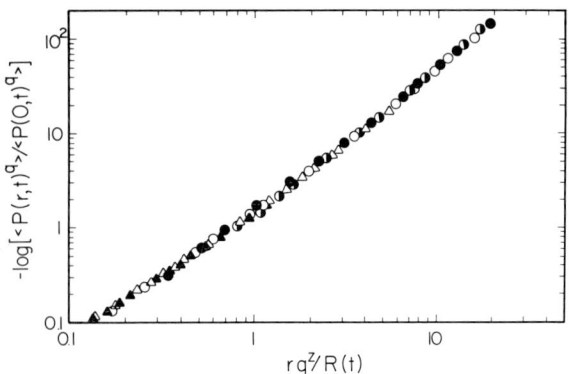

Fig. 5. Plot of $-\log[\langle P^q(r, t)\rangle/\langle P^q(0, t)\rangle]^{1/q}$ versus $rq^z/R(t)$, $z=\tilde{\nu}(d_w^l-1)/d_w^l$, for $t=1000$ and (▲) $q=0.1$, (△) $q=0.3$, (○) $q=1$, (●) $q=4$, and (◐) $q=10$. (After ref. [27].)

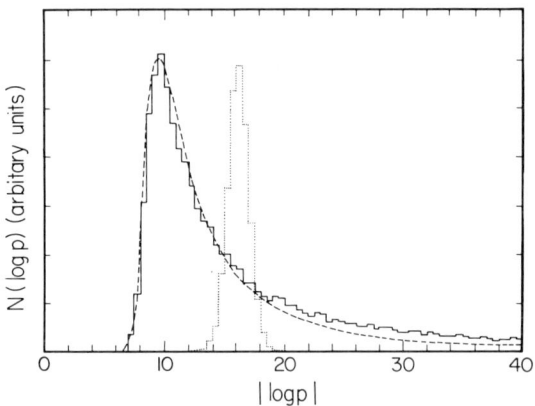

Fig. 6. Plot of the histogram $N(\log P)$ versus $|\log P|$ for fixed r and t (full line) and for fixed l and t (dotted line). The chosen representative values are $r=30$, $l=80$, and $t=1000$. Similar results were obtained for other values of r, l, and t. The dashed line represents the theoretical results, eqs. (8), (14b) and (14c), with $b=457$, and $|\log P_0|=4$. According to eqs. (14b) and (14c), we used $\alpha=2.6$ and $\beta=3.6$. P_0 is the maximum value of $P(r, t)$. (After ref. [27].)

to the difference between the chemical and the fractal dimension. Thus we can learn from the dynamics how convoluted the structure of the fractal is. The smaller γ is, the more convoluted is the structure.

In summary, we have discussed the mean probability density and its fluctuations of random walkers in a variety of random systems. In all cases, $\langle P(r, t)\rangle$ is described by a stretched Gaussian with an exponent $u=d_w/(d_w-1)$. Further characterization of the dynamics requires information on the fluctuation of P. For a more complete description of the diffusion process on random fractals we need to know the way how $N(\log P)$ and the moments of P behave in r-space. This is described by the new exponents α, β, and γ, which in the cases considered here are related to standard structural and transport exponents.

Acknowledgements

We like to thank A. Aharony, S. Alexander, A. Brooks Harris, and D. Stauffer for valuable discussions. This work was supported by MINERVA.

References

[1] B.B. Mandelbrot, Fractals: Forms, Chance, and Dimension (Freeman, San Francisco, 1977); The Fractal Geometry of Nature (Freeman, San Francisco, 1983).
[2] S. Alexander and R. Orbach, J. Phys. (Paris) 43 (1983) L652;
S. Alexander, J. Bernasconi, W.R. Schneider and R. Orbach, Rev. Mod. Phys. 53 (1981) 175.
[3] A. Aharony, in: Directions of Condensed Matter Physics, G. Grinstein and G. Mazenko, eds. (World Scientific, Singapore, 1986).
[4] A. Blumen, J. Klafter and G. Zumofen, in: Optical Spectroscopy of Glasses, I. Zschokke, ed. (Reidel, Dordrecht, 1986), p. 199.
[5] B. Sapoval, M. Rosso, and J.F. Gouyet, in: Superionic Conductors and Solid Electrolytes, A. Laskar and S. Chandra, eds. (Academic Press, New York, 1988).
[6] H.E. Stanley and N. Ostrowsky, eds., Random Fluctuations and Pattern Growth (Kluwer, Dordrecht, 1988).
[7] A. Bunde, Adv. Solid State Phys. 26 (1986) 113.
[8] R. Engelman and Z. Jaeger, eds., Fragmentation Form and Flow in Fractured Media (IPS, Bristol, 1986).
[9] P.G. de Gennes, Scaling Concepts in Polymer Physics (Cornell Univ. Press, Ithaca, 1979).
[10] S. Havlin and D. Ben-Avraham, Adv. Phys. 36 (1987) 695.
[11] S. Feder, Fractals (Plenum, New York, 1988).
[12] D. Avnir, ed., The Fractal Approach to the Chemistry of Disordered Systems, Polymers, Colloids and Surfaces (Wiley, New York, 1989).

[13] E.W. Montroll and G.H. Weiss, J. Math. Phys. 6 (1970) 167.
[14] A.B. Harris and A. Aharony, Europhys. Lett. 4 (1987) 1355.
[15] Y.E. Levy and B. Souillard, Europhys. Lett. 4 (1987) 233.
[16] A. Aharony and A.B. Harris, J. Stat. Phys. 54 (1989) 1091.
[17] B. O'Shaughnessy and I. Procaccia, Phys. Rev. A 32 (1985) 3073.
[18] S. Havlin, D. Movshovitz, B.L. Trus and G.H. Weiss, J. Phys. A 18 (1985) L719;
H. Harder, S. Havlin and A. Bunde, Phys. Rev. B 36 (1987) 3874.
[19] R.A. Guyer, Phys. Rev. A 32 (1984) 2324.
[20] R. Ball, S. Havlin and G.H. Weiss, J. Phys. A 20 (1987) 4055.
[21] E. Koscielny-Bunde, A. Bunde, S. Havlin and H.E. Stanley, Phys. Rev. A 37 (1988) 1821; preprint.
[22] H. Weissman, G.H. Weiss and S. Havlin, J. Stat. Phys., in press.
[23] D. Stauffer, Introduction to Percolation Theory (Taylor and Francis, London, 1985).
[24] H.E. Roman, A. Bunde and S. Havlin, Ber. Bunsenges. Phys. Chem., in press.
[25] H. Harder, S. Havlin and A. Bunde, Phys. Rev. B 36 (1987) 3874.
[26] J.F. Gouyet, H. Harder and A. Bunde, J. Phys. A 20 (1987) 1795;
J.F. Gouyet and S. Havlin, J. Phys. A 21 (1988) 1921.
[27] A. Bunde, S. Havlin and H.E. Roman, preprint.
[28] B.B. Mandelbrot, in: Superionic Conductors and Solid Electrolytes, A. Laskar and S. Chandra, eds. (Academic Press, New York, 1988).
[29] H.E. Stanley and P. Meakin, Nature 335 (1988) 405.
[30] S. Havlin and R. Nossal, J. Phys. A 17 (1984) L427.

FRACTAL DETERMINISTIC CRACKS

H.J. HERRMANN

SPhT, CEN Saclay, 91191 Gif sur Yvette, France

Cracks are grown in an elastic medium submitted to external shear by solving the *full* equations of motion on a two-dimensional lattice. One finds that deterministic fracture patterns are in general branched and can be fractal. This effect is due to the competition between the direction of global stress and the local growth direction imposed by the lattice anisotropy. Viscoelastic effects are also discussed.

1. Introduction

Since the pioneering work of Mandelbrot et al. [1] it is known that cracks in nature can be fractal. The big open question remains, how this fractality comes about. The mechanisms leading to fracture are highly material dependent and have been studied quite extensively [2]. Despite the diversity of experimental situations one can hope to find generic features due to the underlying instabilities and their interplay with noise, anisotropy or memory effects. Recently two new approaches in this direction have been proposed [3], one inspired from random resistor networks [4] and another using DLA [5] as a guideline [6,7].

Some progress has been made by modelling the growth of a single, connected crack. It was found in numerical simulations of central-force media with a breaking probability proportional to the elongation of the springs that the cracks formed are fractal [6,7]. The fractal dimension of these cracks seems to depend strongly on the type of external force that is applied (uniaxial tension, shear, uniform dilatation) but since only very small cracks can be grown, precise statements are difficult to make. Here, we will investigate some of the origins of this fractal behaviour and obtain much better accuracy by considering deterministic models. In addition, in our studies we solve the full elastic equations which contain automatically angular forces, two elastic moduli and the possibility of local rotations, as compared to the central-force model that has been studied before and which can show pathologies on some lattices.

2. Modelling of fractures

The elastic medium is usually described by the field of displacement vectors $\boldsymbol{u}$ which obeys the equation of motion also called the Lamé equation [8],

$$(\lambda+\mu)\nabla(\nabla\cdot\boldsymbol{u})+\mu\nabla^2\boldsymbol{u}=0, \tag{1}$$

where λ and μ are the Lamé coefficients. On the external boundary an imposed displacement is fixed. Suppose that in this medium one has already a crack and one wants to study how this crack grows. Then one will have on the surface of the crack the boundary condition that the stress normal to the surface of the crack is zero and the crack will grow in the direction perpendicular to the surface at the point where the strain parallel to the surface is largest. The detailed growth law depends on the microscopic mechanism like how the elastic energy is transported away from the growing tip. We will use in this work a normal growth velocity v_n of the form

$$v_n \propto [\,(\partial_\| u_\|)^2 + q\, \partial_\|^2 u_\perp\,]^\eta, \tag{2}$$

where q and η are material-dependent parameters. For $\eta = 1$ this growth law is inspired by the von Mises yielding criterion [1], but we cannot derive it from first principles. Physically q is the affinity of the breaking process to the bending mode (second term on rhs of eq. (2)) as compared to cleavage (first term on rhs of eq. (2)). Time derivatives in eq. (1), like inertia terms, are neglected. We do neither consider plasticity nor non-linear elasticity.

In the above approach we will usually assume that the relaxation of the local strain due to the growth of the crack is much faster than the velocity of the crack, which means that we do not consider viscoelastic effects. One can, however, easily introduce a finite relaxation time of the elastic system with respect to the typical time to grow the crack by a given length if one uses numerical relaxation techniques as will be discussed later and in this way viscoelasticity can be effectively implemented.

We discretized the elastic equations on a square lattice by using the beam model [9]. In fact the beam model is even richer than eq. (1) because it allows for local rotations in the medium, i.e. discretizes the equations of motion of asymmetric elasticity [10]. The discretization introduces anisotropy and a cutoff at small length scales, two physical effects that are very often present in real materials.

A detailed description of the beam model is given in ref. [2]. We implement eq. (1) as follows: for each beam that is eligible for being broken one calculates the quantity p defined by

$$p = [f^2 + q \max(|m_1|, |m_2|)]^\eta, \qquad (3)$$

where f is the traction (and/or compression) force applied on the beam and m_1 and m_2 are the moments that are acting at the two ends of the beam; this p determines according to eq. (2) whether the beam will be broken. Each time a beam is broken the shape of the crack and consequently the boundary condition of the equation of motion has changed and one has to solve the discretized equation again if one wants to know which beam to break next.

We consider a finite square lattice of linear size L, with periodic boundary conditions in the horizontal direction. On top and on bottom we impose an external shear. We remove one beam in the center of the lattice which represents the initial microcrack. Next we consider the six nearest-neighbor beams of this broken beam. These include the two beams that are parallel to the broken beam and the four perpendicular beams that touch a common site with the broken beam. This choice of nearest neighbors comes from the fact that the actual crack consists of the bonds that are dual to the set of broken beams [2]. Other connectivity conditions have also been used [7]. The Lamé equation is solved by a conjugate gradient method [11] to very high precision (10^{-20}) and the p's of eq. (3) are calculated for each of the nearest-neighbor beams. We set $p=0$ for a beam that is not a nearest neighbor to the crack. Now various criteria for breaking are possible: I. One breaks the beam with the largest value of p. II. One breaks the beam for which $q_0 = p + f_0 p_{-1}$ is largest, where p_{-1} is the value of p that this beam had before the previous beam was broken; f_0 is a memory factor. III. On each beam of the lattice we put a counter c which is set to zero in the very beginning. Each time one has obtained the p's, one calculates $\alpha = (1-c)/p$ and breaks the beam which has the smallest α, namely $\alpha_{\min}$. After the beam has been broken each counter c is set to $c = \alpha_{\min} p + f c_{-1}$, where c_{-1} is the value the counter had before the breaking and f is another memory factor.

All three breaking criteria described above are deterministic. Criterion III corresponds for $f=1$ to the limit of infinite noise reduction [12]. Noise reduction was invented to reduce statistical noise in DLA simulations and has also been applied recently to central-force breaking [13]. What noise reduction certainly does is to introduce a memory effect with long-range time correlations. Physically the three breaking criteria defined above correspond to three different situations. Criterion I describes ideally brittle and fast rupture. Criterion II contains a short-time memory one would expect in cracks that propagate slower and produce strong local deformations at the tip of the crack as happens in many realistic situations. One could imagine applying this criterion to situations with very fast stress corrosion effects. Criterion III could be applied to situations of relatively slow stress corrosion, crazing of polymers or static fatigue. The memory factors f_0 and f measure the strength of these time correlations. In criterion III the limit $f\to 0$ gives criterion II with $f_0=1$, and in criterion II the limit $f_0\to 0$ gives criterion I.

3. Results

If one breaks according to criterion I, cleavage tends to have the crack grow in the diagonal direction while the bending mode favours a horizontal rupture. The competition between these two effects can lead to complex branched structures. The exact shape of these cracks depends strongly on q and the system size. For any finite q the horizontal rupture will eventually win if the system is large enough while for $q=0$ one obtains diagonal cracks with eventual kinks. For this reason the cracks will not be fractal. However, we point out that in the analogous scalar model (i.e. DLA) only straight lines will be formed in criterion I; the different behaviour here is due to the fact that competing directions are possible in a vectorial model.

Let us now consider cracks grown using criterion II. We see in figs. 1 and 2 cracks with $q=0$ and $\eta=1.0$ obtained in a system of size $L=118$. Over 4 h on a Cray XMP processor were needed to generate each of these structures. The crack in fig. 1 is obtained by breaking only through tension while in fig. 2 a beam can also break under compression. In fig. 2 we show the whole crack which has a four-fold symmetry because it is grown deterministically, i.e. no random numbers are used and the result is independent of the roundoff errors of the computer. In fig. 1 we only show the upper part of the crack, the lower part being reflexion symmetric. The very slight curvature of the crack is a finite-size effect of the lattice which is, however, such a weak effect that its influence cannot be noticed quantitatively, for instance, in the value of the fractal dimension that we will discuss next.

If we count the number of broken beams inside a box of length l around the first broken beam and plot it as a function of l in a log–log plot ("sand box method") we find lines with slopes larger than unity, which means that the cracks are fractals. In system

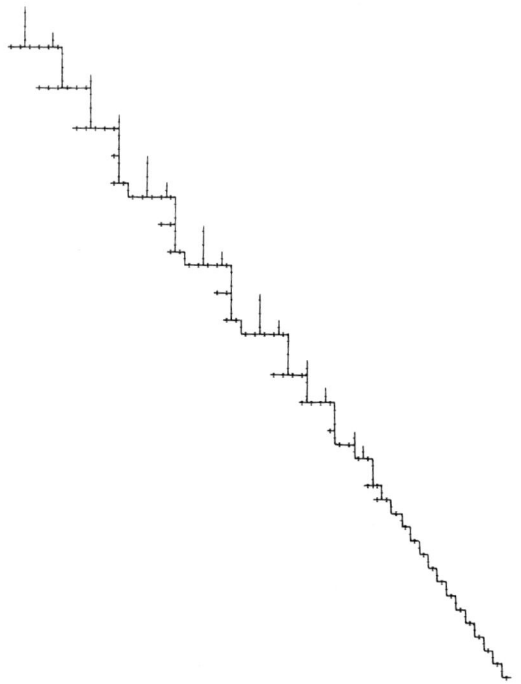

Fig. 1. Upper half of a crack grown in a 118×118 system under external shear using criterion II with $f_0=1$, $\eta=1.0$, $q=0$ (breaking through traction only) and vertical first broken beam.

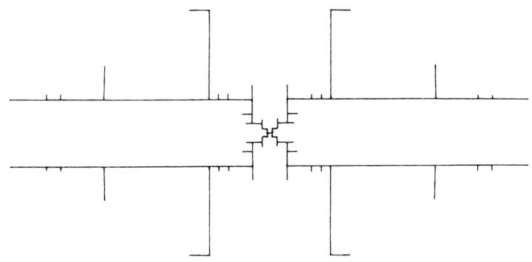

Fig. 2. Crack grown in a 118×118 system under external shear using criterion II with $f_0=1$, $\eta=1.0$, $q=0$ and vertical first broken beam. Here breaking occurs through traction and compression.

sizes of $L=118$ we find for the fractal dimensions d_f values that depend on η: $d_f=1.3$ for $\eta=1.0$, $d_f=1.25$ for $\eta=0.7$, $d_f=1.15$ for $\eta=0.5$ and $d_f=1.1$ for $\eta=0.2$. The structures are self-similar around the origin and probably directed fractals [14]. Changing the elastic constants (i.e. the Lamé coefficients) just changes the opening angle of the crack. If in eq. (3) one uses an exponential instead of a power law, the structures seem to be dense [15].

The effect that using criterion II gives fractal structures is novel and very distinct from what is seen in the scalar case of DLA. It shows that neither noise nor long-range time correlations are necessary to obtain fractal breakdown. The origin of fractality is the competition between a global stress perpendicular to the diagonal and a local stress that tends to continue a given straight crack due to tip instability. Again we see the important role of the interplay of different directions which is only possible in a truly vectorial model. The relevance of a short memory in criterion II indicates that there might be a relation between this case and the models that have been put forward for snow flakes [16].

In figs. 3 and 4 we show cracks grown using criterion III for $q=0$, $f=1$, $\eta=1$ and $L=118$. The physical situation is similar to that seen in criterion II, only the fractal dimensions are higher. This case can be directly compared to results obtained for DLA in the limit of infinite noise reduction [13,17] where needles, not fractals are predicted.

Let us next investigate viscoelastic effects. For this

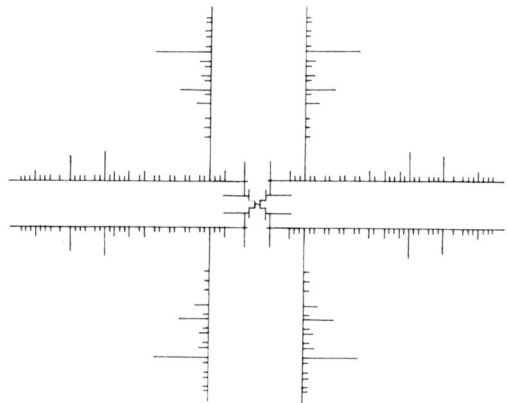

Fig. 4. Crack grown in a 118×118 system using criterion III with $f=1$, $\eta=1.0$, $q=0$ and vertical first broken beam. Here breaking occurs through traction and compression.

purpose we solve the elastic equations with the standard relaxation technique instead of using the more efficient conjugate gradient method. For a sufficiently large number of relaxation steps n one finds, of course, the same result as with the conjugate gradient method; to do so one needs about $n=10000$ steps for $L=60$. If instead one takes a much smaller number of steps, the displacement field is not yet in equilibrium when the next beam is broken, which can be interpreted as a finite relaxation time of the elastic medium as compared to the speed of the crack. In fig. 5 we see what happens to the case of fig. 3 if one uses in a 60×60 system $n=10$, 20 and 100. Clearly the shapes are still quite different from the one for complete relaxation of fig. 3 and the difference gets larger for smaller n. We see that the crack initially has a horizontal part which increases with decreasing n and which for fixed n actually increases with the system size L faster than L. Therefore if n is kept fixed the cracks cannot be asymptotically fractal. The very different behaviour that is found for fixed n as compared to the case of completely equilibrating the displacement field must be taken into account in cases where only a small n has been used as in ref. [6].

In fig. 6 we compare a deterministic crack with an experimental example of stress corrosion cracking in an alloy [18]. Due to the heuristic nature and simplicity of our model it makes no sense to compare

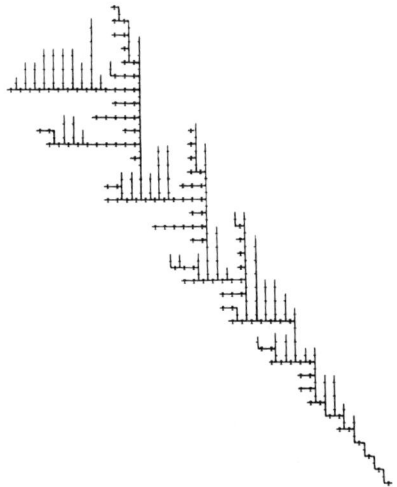

Fig. 3. Upper half of a crack grown in a 118×118 system using criterion III with $f=1$, $\eta=1$, $q=0$ and vertical first broken beam when the beams break under traction only.

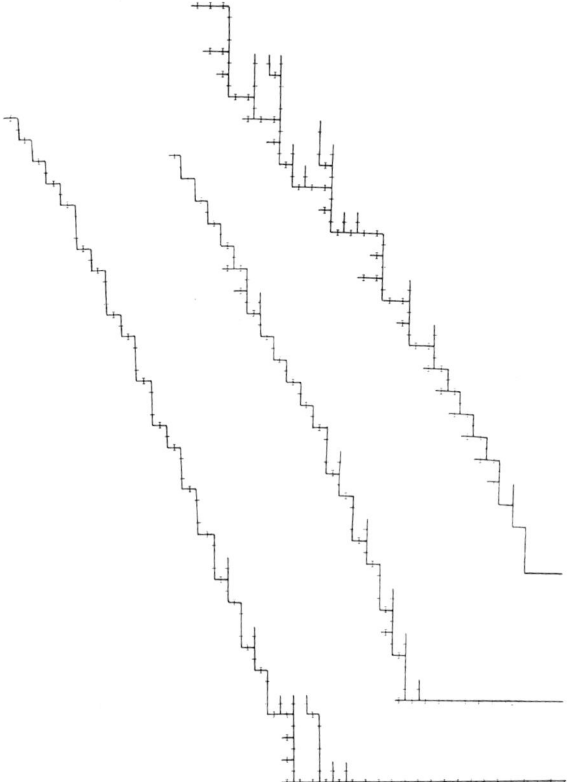

Fig. 5. Upper halves of cracks grown in a 60×60 system using criterion III with $f=1$, $\eta=1$, $q=0$ and vertical first broken beam. The beams break only under traction and we use standard relaxation of n steps to solve the elastic equations with relaxation steps $n=100$ for the upper crack, $n=20$ for the middle crack and $n=10$ for the lower crack. The origin of the cracks has been vertically shifted.

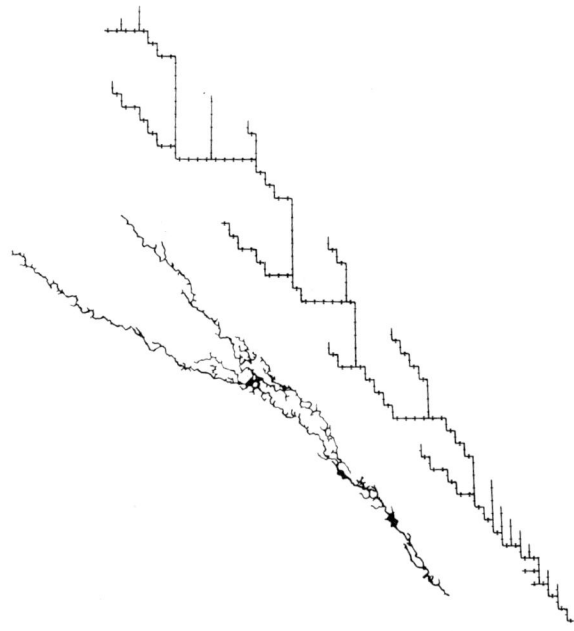

Fig. 6. Numerical and experimental cracks. The upper shape is the upper half of a crack grown in a 118×118 system under external shear with $q=0$ and a vertical initially broken beam using criterion II with $f_0=1$ and $\eta=0.2$. The lower picture shows the morphology of cracking in Ti–11.5 Mo–6 Zr–4.5 Sn aged 100 h at 750 K and tested in 0.6 M LiCl in methanol at −500 mV under increasing intensity (taken from ref. [18]).

numerical values of fractal dimensions. It seems also clear that the inhomogeneities of the medium in the experimental crack are important. The vague similarity that one can see between the patterns in fig. 6 seems to indicate that the effects found in our model may explain to a certain degree the branching behaviour of experimental cracks.

4. Conclusion

In conclusion, we have treated the problem of deterministic single-crack propagation for finite square lattice samples. The patterns of cracks can become very complex and in particular they can be fractal with a fractal dimension that depends on the breaking criterion, i.e. the effect of memory. This phenomenon is due to the competition between the direction of global stress and the direction of local growth imposed by the lattice anisotropy. Therefore the *vectorial* nature of the elastic medium leads to crucially different results from what is known to occur in the *scalar* case of DLA. These differences must be taken into account also when randomness is introduced either quenched in the form of random elastic constants [19] or annealed in the form of breaking probabilities [6,7]. The dependence of shapes and fractal dimensions on the various parameters as well as the influence of noise, i.e. of a probabilistic growth rule, will be described elsewhere [15]. In the viscoelastic case when the relaxation of the elastic medium has a finite characteristic time which is not much shorter than a typical growth time of the crack the structures are asymptotically not fractal.

Acknowledgement

I thank my collaborators Lucilla de Arcangelis and Janos Kertész.

References

[1] B.B. Mandelbrot, D.E. Passoja and A.J. Paulley, Nature 308 (1984) 721;
B.B. Mandelbrot, The Fractal Geometry of Nature (Freeman, San Fransisco, 1983) p. 459.
[2] H. Liebowitz, ed., Fracture, Vols. I–VII (Academic Press, New York, 1984).
[3] H.J. Herrmann, in: Random Fluctuations and Pattern Growth, H.E. Stanley and N. Ostrowsky, eds. (Kluwer, Dordrecht, 1988) p. 149.
[4] L. de Arcangelis, S. Redner and H.J. Herrmann, J. Phys. (Paris) 46 (1985) L585;
P.M. Duxbury, P.D. Beale and P.L. Leath, Phys. Rev. Lett. 57 (1986) 1052;
P.M. Duxbury and P.L. Leath, J. Phys. A 20 (1987) L411;
P.D. Beale and D.J. Srolovitz, Phys. Rev. B 37 (1988) 5500, and references therein;
B. Kahng, G.G. Batrouni, S. Redner, L. de Arcangelis and H.J. Herrmann, Phys. Rev. B 37 (1988) 7625;
H.J. Herrmann, A. Hansen and S. Roux, Phys. Rev. B 39 (1989) 637, and references therein.
[5] T.A. Witten and L.M. Sander, Phys. Rev. Lett. 47 (1981) 1400;
L. Niemeyer, L. Pietronero and H.J. Wiesmann, Phys. Rev. Lett. 52 (1984) 1033.
[6] E. Louis, F. Guinea and F. Flores, in: Fractals in Physics, L. Pietronero and E. Tosatti, eds. (North-Holland, Amsterdam, 1986) p. 117;
E. Louis and F. Guinea, Europhys. Lett. 3 (1987) 871;
P. Meakin, G. Li, L.M. Sander, E. Louis and F. Guinea, J. Phys. A, to be published.
[7] E.L. Hinrichsen, A. Hansen and S. Roux, Europhys. Lett. 8 (1989) 1.
[8] L.D. Landau and E.M. Lifshitz, Theory of Elasticity (Pergamon, Oxford, 1986).
[9] S. Roux and E. Guyon, J. Phys. (Paris) 46 (1985) L999.
[10] W. Nowacki, Theory of Asymmetric Elasticity (Pergamon, Oxford, 1986).
[11] G.G. Batrouni and A. Hansen, J. Stat. Phys. 52 (1988) 747.
[12] C. Tang, Phys. Rev. A 31 (1985) 1977;
J. Szép, J. Cserti and J. Kertész, J. Phys. A 18 (1985) L413;
J. Nittmann and H.E. Stanley, Nature 321 (1986) 661;
J. Kertész and T. Vicsek, J. Phys. A 19 (1986) L257.
[13] J. Fernandez, F. Guinea and E. Louis, J. Phys. A 21 (1988) L301.
[14] B.B. Mandelbrot and T. Vicsek, J. Phys. A, to be published.
[15] H.J. Herrmann, J. Kertész and L. de Arcangelis, in preparation.
[16] F. Family, D.E. Platt and T. Vicsek, J. Phys. A 20 (1987) L1177.
[17] J.P. Eckmann, P. Meakin, I. Procaccia and R. Zeitak, Phys. Rev. A, in press.
[18] M.J. Blackburn, W.H. Smyrl and J.A. Feeney, in: Stress Corrosion in High Strength Steels and in Titanium and Aluminium Alloys, B.F. Brown, ed. (Naval Res. Lab., Washington, 1972) p. 344.
[19] H. Takayasu, in: Fractals in Physics, eds. L. Pietronero and E. Tossatti (North-Holland, Amsterdam, 1986) p. 181;
H. Takayasu, Phys. Rev. Lett. 54 (1985) 1099.

FRACTALS AND SELF-ORGANIZED CRITICALITY IN DISSIPATIVE DYNAMICS

Terence HWA and Mehran KARDAR

Department of Physics, Massachusetts Institute of Technology, Cambridge, MA 02139, USA

Motivated by recent models of Bak, Tang and Wiesenfeld, we study fractal structures arising from dissipative transport in open systems. A simple continuum equation is constructed to describe fluctuations around a steady-state in a flowing "sandpile". Formation of fractals is understood in terms of a conservation law in dynamics. A dynamic renormalization-group calculation allows us to determine fractal dimensions and various critical exponents *exactly* in all dimensions.

1. Introduction

A wide variety of systems in nature exhibit self-similarities over extended ranges of spatial and temporal scales. The concept of "fractal" is first introduced by Mandelbrot [1] to describe self-similar spatial organizations such as mountain ranges, river basins, and coastlines that are ubiquitous in nature, while $1/f$-like power spectra have been used to characterize self-similar temporal sequences as found in star flicker, earthquakes and traffic flows [2]. Bak, Tang and Wiesenfeld [3] (BTW) have suggested that there may be an intimate connection between the scale invariances in the spatial and temporal domains. Indeed at the critical point of a second-order phase transition, the phenomena of critical opalescence (power-law decay of spatial correlation functions) and critical slowing down (lack of a characteristic time scale) are fundamentally related [4]. Since in nature (and in the cellular automata studied by BTW) there is no tuning parameter that can be adjusted to achieve fractal structures, BTW have dubbed such phenomena "self-organized criticality".

BTW note that a common feature of most situations exhibiting $1/f$-noise is an underlying *dissipative transport in an open system*. They in fact propose a cellular automaton model describing the flow of "sand" in an open box. Under the action of simple dynamic rules, their model naturally evolves into a stationary state that lacks characteristic time or length scales (i.e. is critical). Here we point out an important principle which is necessary (and sufficient) to ensure that the steady state configurations are fractal. This important condition is that the dynamics should satisfy a *conservation law*. By identifying appropriate symmetries and conserved quantities it should then be possible to write continuum field equations describing such processes (in the hydrodynamic limit) and to extract their critical behavior [4].

In this paper, we first apply the above program to a sandpile model, and construct the simplest continuum equation for height transport. We explicitly demonstrate how self-organized criticality emerges naturally from height conservation. Next, dynamic renormalization-group methods are used to obtain flow trajectories and fixed points in an $\epsilon = 4 - d$ expansion. Below the upper critical dimension of 4, spatial anisotropy is dynamically generated and is described by an exponent ζ. This exponent, along with the usual dynamic scaling exponent z, and the anomalous spatial correlation exponent χ can in fact be calculated *exactly* for all dimensions. Subject to certain assumptions we can then obtain the exponents ϕ for the decay of the power spectrum ($S(f) \sim f^{-\phi}$), τ for cluster-size distributions ($D(s) \sim s^{1-\tau}$), and fractal dimension D_f of clusters ($s \sim L^{D_f}$), by relating them to z, χ, and ζ. We conclude by briefly describing other processes (and universality classes) exhibiting "self-organized criticality".

2. Model and analysis

Following BTW we consider a discrete sandpile model as follows: the state of sand is completely specified by an integer height function (equal to the number of "sand" grains above each point) defined on a *finite d*-dimensional lattice. To simulate the effect of gravity, the dynamics is simply to transport a grain of sand from a lattice position x to its nearest neighbor $x+a$ if the height difference $\Delta f = f(x) - f(x+a)$ exceeds some threshold value. This is a *dissipative* relaxation mechanism as it is accompanied by loss of gravitational potential energy. The system is *open* as sand is randomly added (resulting in sandpiles and avalanches) and is allowed to escape from the boundaries. A particular *transport* direction is selected by the choice of boundary conditions – open at one edge and closed at the opposite edge. Thus our model is inherently anisotropic and somewhat different from the model actually simulated by BTW; the difference will be made clear later on.

We make the plausible hypothesis that the system achieves a steady-state which is *on average* a flat surface. What is the nature of fluctuations around this steady state? To study the long time and large distance behavior (hydrodynamic limit), we may coarse-grain the lattice and study the "terrain" of sandpiles as a d-dimensional surface embedded in $d+1$ dimensions. The surface fluctuations are now described by a function $h(x, t)$, measuring deviations from the average steady-state surface as shown in fig. 1. The component of gravity parallel to the surface picks out a direction of transport $\hat{T}$. Let $x_\parallel \equiv (\hat{T} \cdot x) \hat{T}$ and $x_\perp \equiv x - x_\parallel$; then the system clearly has rotational invariance in $x_\perp$ and translation invariance in $x_\perp$ and $x_\parallel$; but it lacks reflection symmetry in $x_\parallel$ or h because of the presence of a preferred direction $\hat{T}$. However, there is a joint inversion symmetry $h \to -h$ and $x_\parallel \to -x_\parallel$ as can be seen from fig. 2.

The most common method for constructing dynamical equations is to consider variations of an underlying Hamiltonian [4]. Since we are dealing with an open system it is not clear what should be included in such a Hamiltonian, and we shall instead

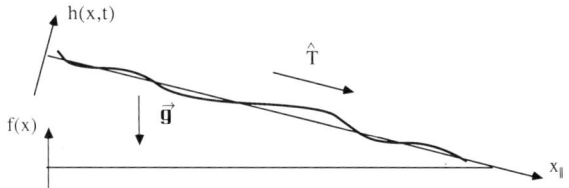

Fig. 1. The height function is defined as deviation from the flat steady-state sand profile. Gravity drives sand along the transport direction $\tilde{T}$.

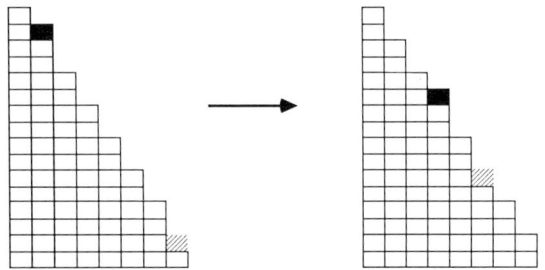

Fig. 2. The $x_\parallel \to -x_\parallel$, $h \to -h$ symmetry: $+h$ (black block) moves to the right, while $-h$ (shaded block) moves to the left.

construct the evolution equation based only on considerations of symmetry and conservation laws [#1]. Symmetries obeyed by the sandpile were outlined previously. The important conservation law is that (*excluding* the sand added from *outside the system*) the relaxation dynamics during avalanches does not change the number of sand particles. Given the above constraints, the simplest equation of motion including leading non-linearity is

$$\frac{\partial h(x, t)}{\partial t} = v_\parallel \partial_\parallel^2 h + v_\perp \nabla_\perp^2 h - \tfrac{1}{2} \lambda \partial_\parallel h^2 + \eta(x, t), \quad (1a)$$

where the first terms describe relaxation of the height through surface tension (the Laplacian term is split into parts parallel and perpendicular to the transport direction $\hat{T}$). The third term is present due to the absence of $x_\parallel \to -x_\parallel$ symmetry and hence related to the presence of transport, and η represents a stochastic noise.

Eq. (1a) is a "driven-diffusion equation" [6]. In

[#1] For similar calculations in a related problem see ref. [5].

the absence of noise ($\eta=0$), there is a local height-conservation law and a transport current $\boldsymbol{j}(\boldsymbol{x}, t)$ satisfying $\partial_t h + \nabla \cdot \boldsymbol{j} = 0$, where $\boldsymbol{j} = -v_\perp \nabla_\perp h - v_\parallel \partial_\parallel h \hat{\boldsymbol{T}} + \frac{1}{2}\lambda h^2 \hat{\boldsymbol{T}}$. (In the original discrete model, there is a threshold for initiating sand flow. The non-linearity here represents an analytic term trying to mimick the discrete step function.) Addition of sand particles from outside destroys the local height conservation rule. Although in steady state the balance of drainage from the boundaries and flux added particles implies $\langle \eta(\boldsymbol{x}, t) \rangle = 0$, the leading moment of noise in the hydrodynamic limit is given by

$$\langle \eta(\boldsymbol{x}, t) \eta(\boldsymbol{x}', t') \rangle = 2D\delta^d(\boldsymbol{x}-\boldsymbol{x}')\delta(t-t'). \quad (1b)$$

The conservative nature of the *deterministic* part of the dynamics rules out terms such as $-h/\tau$ in the equation of motion. (Such a term would introduce a characteristic time τ and characteristic length scales in the problem that would remove scale invariance.) *Thus self-similar fluctuations and the resulting fractal structures are direct consequences of the conservation law.* We shall see that the stochastic equation preserves this condition under renormalization. Also, the joint inversion symmetry illustrated in fig. 2 removes the usual convective term $\partial_\parallel h$.

To study quantitatively the scaling behavior of fluctuations, we do a dynamic renormalization-group calculation [4] (DRG). We solve eq. (1) by perturbation series in powers of λ^2. For $d<4$, terms in perturbation series diverge due to the infrared limit of momentum integrations. The divergences are circumvented by integrating out only momenta in an outer shell $e^{-l}\Lambda_\parallel < q_\parallel < \Lambda_\parallel$ and $e^{-\zeta l}\Lambda_\perp < |\boldsymbol{q}_\perp| < \Lambda_\perp$, where $l \ll 1$ is the infinitesimal shell thickness and Λ's are the short-distance cutoffs in the two directions. The remaining modes are inflated back to the size of the original Brillouin zone by the homogeneous scaling transformation

$$h(x_\parallel, \boldsymbol{x}_\perp, t) = e^{\chi l} h(x_\parallel e^{-l}, \boldsymbol{x}_\perp e^{-\zeta l}, t e^{-zl}), \quad (2)$$

where z, χ and ζ are respectively the dynamic exponent, roughening exponent, and spatial-anisotropy exponent. (Alternatively we can describe scaling in terms of the transverse distance $\boldsymbol{x}_\perp$, in which case $z_\perp = z/\zeta$ (and $\chi_\perp = \chi/\zeta$). The rescaled modes obey eqs. (1) with renormalized parameters satisfying recursion relations of the form [5]

$$\frac{dv_\parallel}{dl} = v_\parallel [(z-2) + u I_{v_\parallel}^{(1)} + u^2 I_{v_\parallel}^{(2)} + ...], \quad (3a)$$

$$\frac{dv_\perp}{dl} = v_\perp [(z-2\zeta) + u I_{v_\perp}^{(1)} + u^2 I_{v_\perp}^{(2)} + ...], \quad (3b)$$

$$\frac{d\lambda}{dl} = \lambda [(z+\chi-1) + u I_\lambda^{(1)} + u^2 I_\lambda^{(2)} + ...], \quad (3c)$$

$$\frac{dD}{dl} = D[(z-2\chi+(1-d)\zeta-1) + u I_D^{(1)} + u^2 I_D^{(2)} + ...]. \quad (3d)$$

The quantities in parentheses are the bare dimensions of the parameters,

$$u = \frac{2 S_{d-1}}{(2\pi)^d} \frac{\lambda^2 D}{v_\parallel^{3/2} v_\perp^{3/2}}$$

is the perturbation parameter or the effective coupling constant (S_d being the surface area of a unit d-dimensional sphere), and the I's represent numerical values of various diagrams involved in the perturbation calculation.

In general, evaluations of exponents and fixed points of flow described by eqs. (3) involve tedious calculation of integrals I. However, in this particular case, as the non-linearity is proportional to the *external* momentum $k_\parallel$, the parameters $v_\perp$, and D are not renormalized by the loop diagrams in the hydrodynamic ($k_\parallel \to 0$) limit, i.e. $I_{v_\perp}^{(n)}$, $I_D^{(n)}$ are all zeroes. Furthermore, eq. (2) is invariant under a "Galilean" transformation [5]: $x_\parallel \to x_\parallel - \epsilon \lambda t$, $t \to t$ if $h \to h + \epsilon$; this insures that the parameter λ is scale invariant and $I_\lambda^{(n)} = 0$. Therefore the fixed points and flow can be found once $I_{v_\parallel}^{(n)}$'s are calculated. To the one-loop order we have

$$\frac{du}{dl} = u\left((4-d) - \frac{7-d}{2}\frac{3\pi}{32}u\right).$$

For $d>4$ the stable fixed point is at $u^*=0$ and ordinary diffusion behavior is expected. For $d<4$, the stable fixed point moves to $u^* = (64/9\pi)(4-d)$. Even though u^* is known only perturbatively, the exponents are determined *exactly* for all dimensions using eqs. (3b)-(3d),

$$z = \frac{6}{7-d}, \quad \chi = \frac{1-d}{7-d}, \quad \zeta = \frac{3}{7-d}. \quad (4)$$

The exactness of these exponents is a result of non-renormalization conditions on $v_\perp$, λ and D which removes anomalous dimensions; it is *not* affected by higher-order corrections to the position of the fixed point. For $d<4$, $\zeta<1$ and anisotropic scaling is generated by DRG.

The exponents z, χ, and ζ are the fundamental scaling dimensions of this system; other useful quantities can in principle be calculated from them. In connection with $1/f$ noise, an interesting property is the frequency dependence of some macroscopic response functions. A natural choice is the total output current, $J_s(t) = \int d^{d-1} x_\perp j(L_\parallel, x_\perp, t)$, with integration at the system edge at $x_\parallel = L_\parallel$. The temporal correlations $\langle J_s(t) J_s(0) \rangle$ are computed by using $\langle j(x_\perp, t) j(0, 0) \rangle_c \sim \langle h^2(x_\perp, t) h^2(0, 0) \rangle_c \sim |x_\perp|^{4\chi/\zeta} f(x_\perp^{z/\zeta}/t)$. (Note that in $d<4$ scaling is dominated by the h^2 part of the current, and the ∂h component is less relevant.) Dimensional counting eventually leads to $\langle J(t) J(0) \rangle \sim L_\perp^{d-1} t^{4\chi+(d-1)\zeta/z}$, which upon Fourier transformation yields a power spectrum $S_s(f) \sim 1/f^{\phi_s}$ with $\phi_s = 1/z$ (using results in eq. (4)). Recent experiments on real sand [7], however, indicate that there is a typical time scale between sand flow events, i.e. $S_s(f)$ peaks at some characteristic frequency. This is thought to result from the streaming motion due to the small kinetic friction of sand [7]. This non-dissipative kinetic motion of sand can be included in eq. (1a) by adding a term proportional to $\partial_t^2 h$. The behavior of the system in the long-time limit is still dominated by the $\partial_t h$ term, but the length scale at which the crossover from kinetic to dissipative behavior occurs may well exceed possible experimental setups.

The exponent ϕ_s for decay of the transport current interpolates between $\frac{1}{2}$ and 1 as d changes from 4 to 1. This differs from the result of BTW, who find [3] an exponent that increases from 1 as dimension is lowered from 4. Actually, the response function studied by BTW is the total energy $E_b(t)$ dissipated throughout the system at time t. Assuming that the local energy dissipated is proportional to the amount of sand transported [3] (i.e. $E_b(x, t) \sim j(x, t) \cdot \hat{T} \sim h^2(x, t)$), we have $E_b(t) \sim \int d^d x\, h^2(x, t)$. Using the same procedure as before we find $\langle E_b(t) E_b(0) \rangle \sim L^d t^{4\chi+(d-1)\zeta+1/z}$, and an energy dissipation power spectrum $S_b(f) \sim 1/f^{\phi_b}$ with $\phi_b = 2/z$. The mean-field behavior of this spectrum is indeed $1/f$ noise, with ϕ_b increasing to 2 as spatial dimension is lowered to 1.

For sand, we are most interested in the fractal structure of clusters and avalanches, especially since it is easily defined and measured numerically and experimentally. Consider a system in steady state: A perturbation at $(0, 0)$ will produce a response (avalanche) at (x, t) which should scale as $h_R(x, t) \sim x_\parallel^\chi f(x_\perp^\zeta/x_\parallel^\zeta, t/x_\parallel^z)$. The avalanche will spread over a spatial extent $L_\parallel$ in the transport direction and last a period of time $T \sim L_\parallel^z$. Following BTW, we define the size of the avalanche (cluster size s) as the total number of slidings due to the perturbation. Again using $\langle h^2 \rangle_c$ as a local measure of dissipation, we find $s \sim L_\parallel^{2z}$. If we choose this scaling to define a fractal dimension D_f for the anisotropic clusters, we conclude $D_f = 2z$. Knowing D_f, we can relate the distribution of cluster $D(s) \sim s^{1-\tau}$ to the bulk-dissipation power spectrum via the exponent identity $\phi_b = D_f(3-\tau)/z$ quoted by BTW [3], yielding $\tau = \frac{5}{2} - \frac{1}{6}(4-d)$. For ease of reference, numerical values calculated for all these exponents for spatial dimensions 1 through 4 are listed in table 1. At this point we should warn readers that while the calculations for χ, z, and ζ (and to some extent ϕ_s) are exact, the ones leading to ϕ_b, τ, and D_f are somewhat heuristic. We have attempted to follow the procedure of BTW as much as possible.

We now return to the cellular automaton model studied by BTW. It is interesting that various exponents obtained numerically are all within 15% of those in table 1. The agreement may be fortuitous for two reasons: Firstly, the clusters we find inherently satisfy *anisotropic* scaling, while at least in their numerical analysis BTW do not consider such anisotropy. Secondly, in their simulations BTW appear to use as their basic dynamical variable an averaged slope z_i. Now in dimensions greater than 1 the knowledge of

Table 1
Numerical values for exponents defined in the text

	$d \geq 4$	$d=3$	$d=2$	$d=1$
z	2	1.50	1.20	1
χ	−1	−0.50	−0.20	0
ζ	1	0.75	0.60	−
ϕ_s	0.5	0.67	0.83	1
ϕ_b	1	1.33	1.67	2
τ	2.5	2.33	2.17	2
D_f	4	3	2.4	2

a slope $z=\nabla h$ does not uniquely determine the height profile. In any case the evolution rules employed by BTW do conserve the "slope" z_i locally in the absence of external perturbations. Thus a continuum version of this quantity must satisfy an equation of the form $\partial_t z = \nabla \cdot \boldsymbol{j} + \eta(\boldsymbol{x}, t)$. Again the absence of a $-z/\tau$ term is responsible for lack of a time scale and self-similarity of the resulting patterns. Since any non-linearity in $\boldsymbol{j}$ appears in the equations proportional to a gradient, renormalization can only modify the form of $\boldsymbol{j}$, without changing the conservative nature of the deterministic part.

Different universality classes for "self-organized criticality" thus depend on the possible choices of the current $\boldsymbol{j}$, and on the spectrum of the applied noise η. For local dynamic rules both these quantities are determined by symmetries and conservation laws. For example if the noise in eq. (1a) is itself conservative (i.e. $\langle \eta(\boldsymbol{x}) \eta(\boldsymbol{x}') \rangle \sim \nabla^2 \delta^d(\boldsymbol{x}-\boldsymbol{x}')$) then we regain the driven-diffusion equation originally studied in the context of forced particle diffusion [6]. This problem has an upper critical dimension of 2 and exponents given by eq. (4) with d replaced by $d-2$. If there is no net transport current and the system is isotropic, the simplest non-linear current is $\boldsymbol{j}(h) = -v\nabla h + \lambda \nabla h^2$. With non-conservative noise the corresponding stochastic equation has a critical dimension of 2; but a structure that as in the case of the Burger's equation allows a non-trivial strong coupling behavior in all dimensions. Recently, Obukhov [8] has also studied a similar problem with an upper critical dimension of 4; but we are not clear to which universality class his model belongs.

Finally it is interesting to note that the natural roughness of the surface in $d=1$ goes away once a transport current is switched on (i.e. χ changes from $\frac{1}{2}$ to 0 once λ is non-zero). This type of behavior has been observed numerically [9] for the interface of an Ising model subject to a transverse "electric field". This approach may thus also provide a clue to some apparently unrelated problems.

Acknowledgement

We would like to thank P. Bak for initiating our interest in this problem. This research was supported by the National Science Foundation through the MIT Center for Material Science Grant No. DMR-84-18718. MK acknowledges support from the A.P. Sloan Foundation.

References

[1] B. Mandelbrot, The Fractal Geometry of Nature (Freeman, San Francisco, 1982).
[2] W.H. Press, Comm. Mod. Phys. C 7 (1978) 103;
P. Dutta and P.M. Horn, Rev. Mod. Phys. 53 (1981) 497.
[3] P. Bak, C. Tang and K. Wiesenfeld, Phys. Rev. Lett. 59 (1987) 381;
C. Tang and P. Bak, Phys. Rev. Lett. 60 (1988) 2347, and references therein.
[4] S.-K. Ma, Modern Theory of Critical Phenomena (Benjamin/Cummings, Menlo Park, 1976);
P.C. Hohenberg and B.I. Halperin, Rev. Mod. Phys. 49 (1977) 435, and references therein.
[5] E. Medina, T. Hwa, M. Kardar and Y.-C. Zhang, Phys. Rev. A, press.
[6] H.K. Janssen and B. Schmittmann, Z. Phys. B 63 (1986) 517.
[7] H.M. Jaeger, C. Liu and S.R. Nagel, Phys. Rev. Lett. 62 (1988) 40.
[8] S. Obukhov, in: Random Fluctuations and Pattern Growth, eds. H.E. Stanley and N. Ostrowsky (Kluwer, Dordrecht, 1988).
[9] K. Leung, K.K. Mon, J.L. Valles and R.K.P. Zia, Phys. Rev. Lett. 61 (1988) 1744.

BOUNDARY LAYER INSTABILITY IN A COUPLED-MAP MODEL

Mogens H. JENSEN

Nordita, Blegdamsvej 17, DK-2100 Copenhagen, Denmark

We describe a simple coupled-map lattice model of a convective system with a boundary layer. At a critical point, the convective effects exceed the diffusive effects and the boundary layer becomes unstable and emits thermal plumes. This instability is studied within a mean field approximation. Also frequency spectra of time signals and temperature fluctuations are considered.

When a fluid or a gas in a container is heated from below a convective state will commence as the temperature gradient exceeds a critical value. In the convective state, the hot fluid (or gas) moves upwards and the cold fluid downwards but close to the bottom and top plates there can be a small region, a thermal boundary layer, where heat is not transported by convection by only by diffusion [1]. As the temperature gradient is increased further up to high values of the Rayleigh number ($\approx 10^8$), this boundary layer becomes unstable and begins to emit patches ("hot plumes") into the laminar convective regime [2]. The motion of these plumes causes the large temperature fluctuations which are associated with a "hard" turbulent state. These phenomena have recently been studied experimentally and theoretically by the Chicago group [2].

Here we discuss a very simple phenomenological model which shows features qualitatively similar to the experiment. The model is a coupled-map lattice system [3] introduced in ref. [4]. We make use of maps that exhibit both turbulent and laminar behavior and are on the form [5]

$$f_r(x) = rx, \quad x \leq 0.5,$$
$$= r(1-x), \quad 0.5 < x \leq 1, \quad (1)$$
$$= x, \quad x > 1.$$

The motion is chaotic when $x \leq 1$, which plays the role of a turbulent (or "hot") state, and the motion is laminar for $x > 1$. When coupled diffusively in 1-D and 2-D, Chaté and Manneville found that as the coupling strength ϵ exceeds a critical value ϵ_c a turbulent site will percolate through the system much like in directed percolation [5]. When ϵ is below ϵ_c any initial state will turn into a laminar state.

Our model also uses a diffusive ϵ-coupling but in addition we introduce a convective term of strength v that displaces "hot" fluid (i.e. sites with $x \leq 1$) upwards:

$$x_{n+1}^{(i,j)} = f_r(x_n^{(i,j)}) + \tfrac{1}{4}\epsilon[f_r(x_n^{(i-1,j)}) + f_r(x_n^{(i+1,j)})$$
$$+ f_r(x_n^{(i,j-1)}) + f_r(x_n^{(i,j+1)}) - 4f_r(x_n^{(i,j)})]$$
$$+ v[f_r(x_n^{(i,j-1)}) - f_r(x_n^{(i,j)})]. \quad (2)$$

(i, j) is a point on a $N \times N$ lattice and n is the time step.

Next a boundary condition at the bottom is introduced in order to play the role of a constant "hot" temperature at the bottom plate. Since the sites with $x \leq 1$ are the "hot" sites we enforce this condition by the constraint

$$x^{(i,1)} = x_B, \quad x_B < 1. \quad (3)$$

This means the x-value in the first row is kept fixed at x_B (in the following we set $x_B = 0$). The model is simulated on a computer. It is initiated in a laminar state $x_0^{(i,j)} = 1.1 + \eta$ (where η is a small-amplitude noise term). The time evolution can be visualized by marking the hot sites, i.e. sites where $x^{(i,j)} \leq 1$. In that way it is easy to observe the hot plumes that travel through the laminar regime. For small value of the convection term (i.e. where v is less than a critical value v_c), the simulation shows that the "hot" boundary condition at the bottom will introduce a "hot" boundary layer of depth of a few lattice lengths. Above the boundary layer the system is in its laminar state. This thermal boundary layer is identified by calculating the average number of hot sites in each layer above the bottom. This number shows a sharp gradient over the width of the boundary layer and then goes to zero in the laminar state, just like the sharp temperature gradient in experimentally observed thermal boundary layers.

As the strength of the gradient term exceeds a critical value v_c, the boundary layer becomes unstable and starts to emit patches (hot plumes) into the laminar regime. The shapes and sizes of the plumes can vary a lot. Fig. 1 shows a snapshot of the simulation. For $\epsilon = 0.12$ used in fig. 1 we find that $v_c \approx 0.018$ by monitoring the value of v for which the boundary layer begins to emit plumes. These observations are qualitatively similar to the experimental findings. In the visualization of the experiment one observes that the boundary layer becomes unstable against small amplitude traveling waves and as the Rayleigh number is increased, the waves may detach as convective plumes [2].

In order to understand the boundary layer instability in more detail we consider the variable $x_n^{(i,j)}$ within a simple mean-field approximation. A mean-field value m_j of $x_n^{(i,j)}$ is defined at each layer in the lattice and is calculated in the following way (for specific values of r, ϵ and v):

$$m_j = \frac{1}{T}\frac{1}{N} \sum_{i=1}^{N} \sum_{n=1}^{T} x_n^{(i,j)} . \qquad (4)$$

In other words, we average $x_n^{(i,j)}$ both over space (N sites) and time (T time steps) to obtain a mean-field value m_j for layer j. By definition, $0 \leq m_j \leq \frac{1}{2}r$, and of course $m_1 = x_B$. Within this mean field we compare the diffusive effects against the convective effects. At a specific layer in the lattice, $\bar{j}$, chosen in the top of the boundary layer, we calculate the following two terms

$$\mathrm{Di}(r, \epsilon, v, \bar{j}) = \tfrac{1}{4}\epsilon(m_{\bar{j}-1} + m_{\bar{j}+1} - 2m_{\bar{j}}) , \qquad (5a)$$

$$\mathrm{Co}(r, \epsilon, v, \bar{j}) = v(m_{\bar{j}-1} - m_{\bar{j}}) . \qquad (5b)$$

These two terms are the diffusive and convective

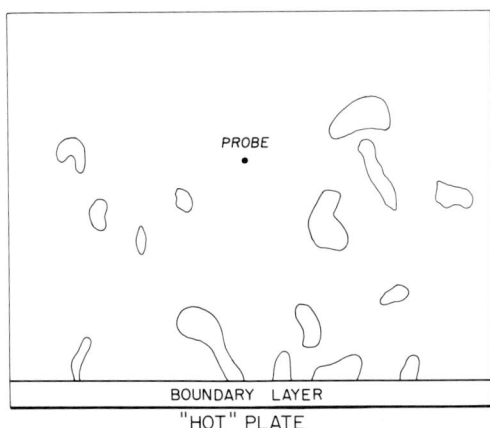

Fig. 1. A snapshot of the simulation at $\epsilon = 0.12$, $v = 0.04 > v_c \approx 0.018$, $r = 3.0$, and $x_B = 0$ on a 50×50 lattice. The patches are the "hot plumes" for which $x^{(i,j)} < 1$. The plumes are released from the boundary layer and drift upwards by convection.

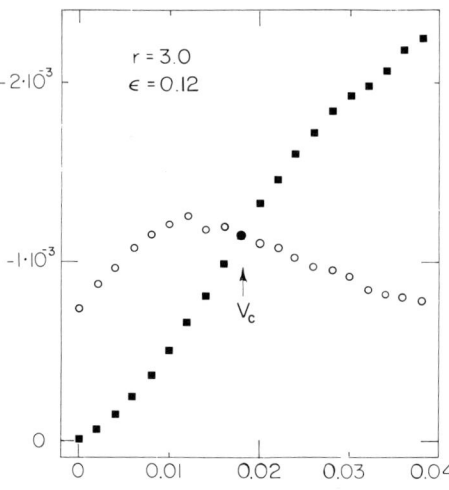

Fig. 2. A plot of the diffusive term $\mathrm{Di}(3, 0.12, v, 3)$ (eq. (5a)), shown by circles, and the convective term $\mathrm{Co}(3, 0.12, v, 3)$ (eq. (5b)), shown by squares, versus the gradient strength v. The curves cross at the point, v_c, where the boundary layer becomes unstable. The size is $N = 100$.

terms in the local mean field. For $\epsilon = 0.12$, the top of the boundary layer is around $\bar{j} = 3$ which we use in eq. (5). Fig. 2 shows $\text{Di}(\bar{j}, 0.12, v, 3)$ and $\text{Co}(3, 0.12, v, 3)$ plotted versus v. The two curves cross at $v \approx 0.018$. This is the point where, for increasing value of v, the convective term becomes larger the diffusive term. Above this point the boundary layer will not only be a diffusive layer but will be unstable against convection such that patches from the top of the boundary layer will be released into the laminar regime. Therefore this cross point is the critical point for the boundary layer instability, v_c. By magnifying around the crossing point and averaging over many time steps (≈ 20000) we estimate $v_c = 0.018025 \pm 0.000015$.

In ref. [4] the total number of "hot" sites at the first layer of the system was calculated numerically in order to mimic the total heat flux into the system. As a function of the gradient v a power scaling law for the heat flux was found, but this scaling extended only over one decade. Here, we shall study the heat flux due to the plumes and seek therefore a "critical" scaling law around the value v_c. That is, we measure the "heat" released on top of the boundary layer due to the plume motion [#1]. This heat is estimated as the excess fraction of "hot" sites on the top of the boundary layer ($j = 3$). Let $H(\epsilon, v, j)$ denote the number of hot sites in layer j. Averaged over many time steps (≈ 25000) we then investigate a scaling law

$$\frac{\langle H(0.12, v, 3) \rangle - \langle H(0.12, v_c, 3) \rangle}{N} \sim (v - v_c)^\gamma. \quad (6)$$

Fig. 3 shows a plot of eq. (6) on a log–log scale. The error bars are due to the large fluctuations when v is close to v_c. There is reasonably good scaling over around four decades. For large v-values outside the figure, the plumes start to percolate and we begin to see a deviation from the scaling. Surprisingly, the numerics indicate that $\gamma \approx 1$ meaning that the plume flux grows linearly with the distance from the instability point. Actually, we also observe from fig. 2 that the convective term is roughly proportional to v over

[#1] I thank K. Kaneko for his suggestion.

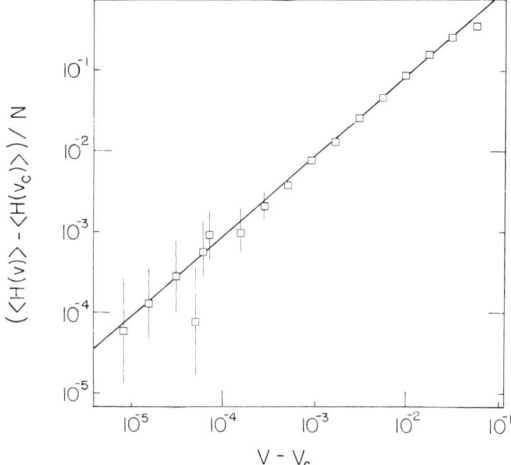

Fig. 3. A plot of $(\langle H(0.12, v, 3) \rangle - \langle H(0.12, v_c, 3) \rangle)/N$ versus $v - v_c$ on a log–log scale. The error bars are due to fluctuations close to v_c. The straight line has slope 1.0. $N = 100$.

some range and this term drives the plume emission.

As in the experiment [2], we shall characterize the turbulent state quantitatively by the temperature fluctuations in the laminar regime. Those are measured by placing a probe at a specific point in the center of the cell (here at $(i, j) = (25, 18)$ with $N = 50$, see fig. 1). Next the number of time steps, t_p, for each plume to pass the probe, is measured. This time plays the role of a temperature fluctuation, i.e. a long time (large plume) will likely give rise to a large fluctuation in the temperature. As the system evolves many plumes sweep intermittently across the probe. The corresponding distribution of times, $D(t_p)$, is plotted in fig. 4. The straight line indicates an exponential distribution,

$$D(t_p) \sim \exp(-at_p). \quad (7)$$

To check whether this law is robust to changes in the parameters, the distribution is calculated for four different values of v and ϵ. When normalized as in fig. 4 there does not seem to be any significant dependence in the constant a on the parameters v and ϵ. Similar behavior has been observed experimentally in the "hard" turbulent regime [2]. As argued in ref. [2], one expects an exponential distribution if the behav-

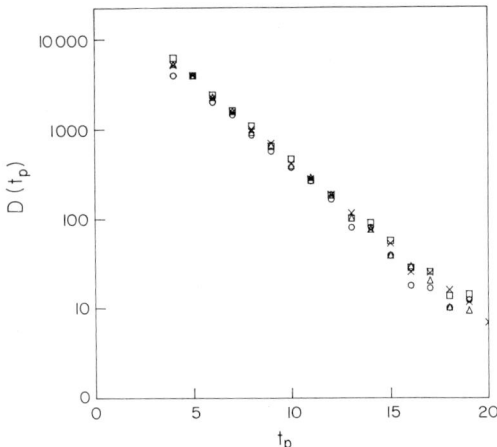

Fig. 4. The distribution of passage times, $\log D(t_p)$, versus t_p. Circles: $\epsilon=0.12$, $v=0.035$; triangles: $\epsilon=0.12$, $v=0.05$; squares: $\epsilon=0.12$, $v=0.08$; crosses: $\epsilon=0.14$, $v=0.05$. The different curves are normalized to the same value of $D(t_p)$ at $t_p=5$. Measurements for $t_p \leq 3$ are disregarded. Each calculation is performed over $\approx 10^6$ time steps.

ior around the probe can be considered as a series of randomly distributed events. This is very much in accordance with our visualization of the motion around the probe; plumes pass in an intermittent fashion and one event is little influenced by the previous events.

Finally, the frequency spectra of a time signal recorded at various probes in the system are considered. At a point (i, j) on the lattice we monitor a signal $y_n^{(i,j)}$ defined as

$$y_n^{(i,j)} = 0 \quad \text{if } x_n^{(i,j)} \leq 1 \qquad (8)$$
$$= 1 \quad \text{if } x_n^{(i,j)} > 1 .$$

The signal is digitized to avoid "noise" arising from the chaotic jumping $x_n^{(i,j)}$ around the interval [0,1]. Time sequences (≈ 5000 steps) have been numeri-

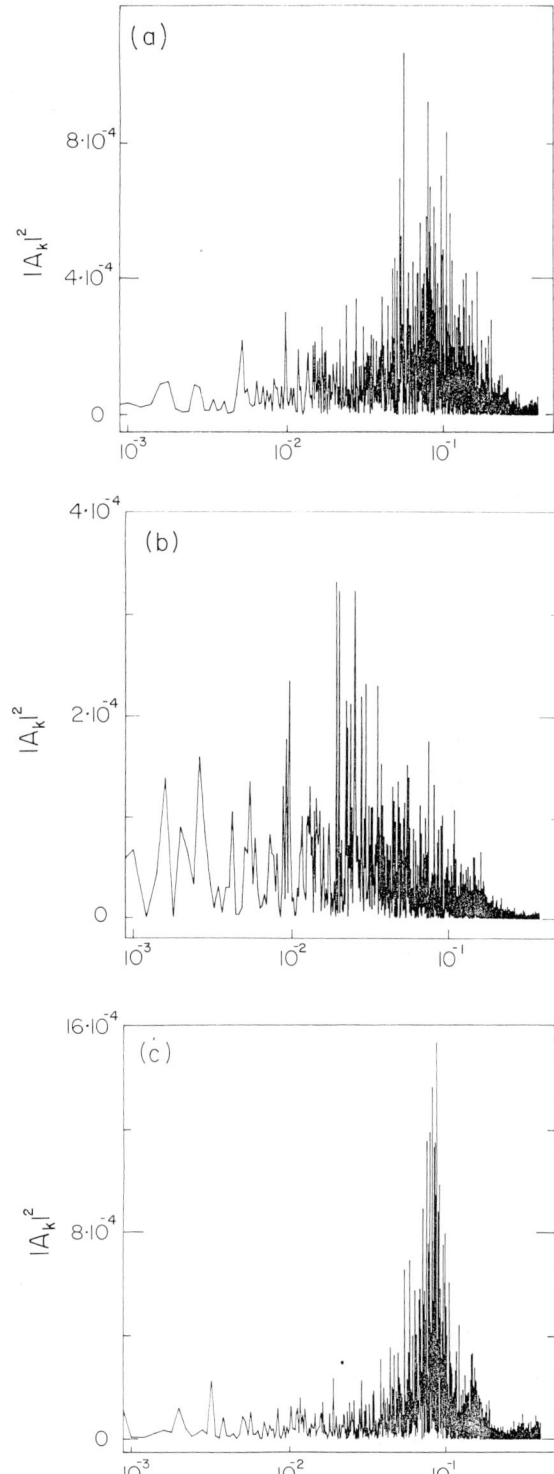

Fig. 5. (a,b) The power spectra of time signals, eq. (8). Signal (a) is recorded in the boundary layer at $(i,j)=(25, 3)$ (with $N=50$) and signal (b) is recorded in the center of the cell $(i,j)=(25, 18)$. Note the characteristic frequencies of the spectra. (c) The power spectrum for the "mean field" time signal, eq. (10), recorded in the boundary layer. The characteristic frequency is reasonably close to the one in (a).

cally extracted at two probes, one close to the bottom plate, at $(i,j) = (25, 3)$, and one at the central probe $(i, j) = (25, 18)$. The corresponding Fourier spectra are shown in fig. 5. They all show a clear maximum, indicating a characteristic frequency of the plume motion. In agreement with the experimental observations [2], the characteristic frequency at the center, ω_c (fig. 5b), is lower than the characteristic frequency at the bottom, ω_b (fig. 5a). We can obtain an approximation to this time signal from iterating a single map in its local mean field. In a specific layer j we define a "mean field map" of the variable $\bar{x}_n^{(j)}$ as

$$\bar{x}_{n+1}^{(j)} = f_r(\bar{x}_n^{(j)}) + \tfrac{1}{4}\epsilon(m_{j-1} + 2m_j + m_{j+1} - 4\bar{x}_n^{(j)})$$
$$+ v(m_{j-1} - \bar{x}_n^{(j)}) . \qquad (9)$$

The time signal $\bar{y}_n^{(j)}$ defined as

$$\bar{y}_n^{(j)} = 0 \quad \text{if } \bar{x}_n^{(j)} \leq 1$$
$$= 1 \quad \text{if } \bar{x}_n^{(j)} > 1 , \qquad (10)$$

now plays the role of a mean field time recording in a layer. To compare with fig. 5a we consider the boundary layer ($j=3$) and estimate numerically the fields; $m_2 = 0.7465$, $m_3 = 0.9142$, $m_4 = 0.9720$. Fig. 5c shows the corresponding Fourier spectrum and we observe a characteristic frequency very close to ω_b found from the simulation. Thus a simple mean field model appears to provide a good approximation for the time evolution.

In conclusion, we have discussed a simple coupled-map lattice model for boundary layer induced turbulence. The model helps to develop some intuition about the temporal aspects of such turbulence and shows qualitative similarities with experiments. Quantitative agreement is not to be expected because the model has many limitations: for instance, there are no conserved quantities, there is no back-flow from the top, and there is only one field (and not two fields, namely temperature and velocity). We are in the process of investigating improvements of the model along such directions.

References

[1] L.D. Landau and E.M. Lifshitz, Fluid Dynamics (Pergamon, Oxford, 1985).
[2] B. Castaing, G.H. Gunaratne, F. Heslot, L. Kadanoff, A. Libchaber, S. Thomae, X.-Z. Wu, S. Zaleski and G. Zanetti, J. Fluid Mech., in press;
F. Heslot, B. Castaing and A. Libchaber, Phys. Rev. A 36 (1987) 5870.
[3] K. Kaneko, Prog. Theor. Phys. 74 (1985) 1033;
J.D. Keeler and J.D. Farmer, Physica D 23 (1988) 415;
J. Crutchfield and K. Kaneko, in: Directions in Chaos, Vol. 1, ed. H. Bai-lin (World Scientific, Singapore, 1988).
[4] M.H. Jensen, Phys. Rev. Lett. 62 (1989) 1361.
[5] H. Chaté and P. Manneville, Phys. Rev. A 38 (1988) 4351; Physica D 32 (1989) 409.

GEOMETRICAL OPTICS IN FRACTALS

Remi JULLIEN and Robert BOTET

Physique des Solides, Bâtiment 510, Université Paris-Sud, Centre d'Orsay, 91405 Orsay, France

A numerical Monte Carlo method is presented to study the scattering of polarized light by fractal aggregates built with identical tangent spheres in the geometrical optics limit where the radius of the spheres is larger than the wavelength of the incident beam. The method is applied to both deterministic and random fractal aggregates containing up to several thousands of spheres. Simple scaling laws are obtained and the difference between transparent ($D<2$) and opaque ($D>2$) aggregates is stressed.

1. Introduction

Since the discovery by Forrest and Witten of the fractal structure of aerosols [1], several theoretical models have been built to describe mechanism of cluster aggregation [2] and successful quantitative comparisons with experiments are now available [3]. Most of the experimental determinations of the fractal dimension, D, of aggregates are based on small angle scattering experiments with X-rays, neutrons or light. Generally, D is extracted from the q^{-D} law [4] (where q is related to the scattered angle θ, and the wavelength λ, through the formula $q=(4\pi/\lambda)\times\sin(\frac{1}{2}\theta)$). However this law has a restricted range of validity, since it is based on a simple Fourier transform which neglects multiple scattering as well as proximity or shadowing effects. In particular, for "opaque" fractal aggregates ($D>2$) such effects cannot be avoided [5]. An approximate treatment, able to include multiple scattering, has been introduced [6] which is restricted to small values of the product ka (where $k=2\pi/\lambda$ and a is the radius of the spheres).

Here we present a method entirely built using geometrical optics. An inconvenience compared with all other methods is that, in this limit, only addition of incoherent waves (i.e. addition of their intensities and not of their amplitudes) has a physical meaning. Thus coherence effects cannot be taken into account. In particular, we cannot recover the q^{-D} law. Another inconvenience is that only transmitted and reflected rays are considered, i.e. diffracted light is neglected [7]. In practice, this method can be applied to aggregates made of spherical particles whose radius is larger than λ (i.e. for large ka values), under the condition that the angular dependence of the diffracted light can be neglected. Some of the results could eventually be extended to small ka values, if there exist some physical reasons to kill coherence effects. On the other hand, the present approach has the considerable advantage to fully take into account multiple scattering and shadowing effects. In the following, we present the main principles of the method and some of the results of its application to fractal aggregates. More details can be found elsewhere [8].

2. Main principles of the method

The aggregate is characterized by the refraction index (relative to the external medium), n, and the radius, a, of the spheres, and by a list of coordinates for their centers. One generates a set of incident rays parallel to a given direction $\boldsymbol{u}_0$, with a given polarization and issued from points A_0 (outside the aggregate) regularly spaced on a square lattice of parameter δ, in a plane perpendicular to $\boldsymbol{u}_0$ (i.e. there is one ray per area δ^2). These rays are followed all along their trajectory inside the aggregate and the intensity (i.e. the square of the electric field intensity), $I(\theta)$, is collected at the end as a function of θ, θ being the angle between the scattered direction, $\boldsymbol{u}_s$, and $\boldsymbol{u}_0$ (more precisely $I(\theta)\sin\theta\,d\theta$ is the intensity scattered between θ and $\theta+d\theta$).

After each reflexion/refraction, the impact point A_i, the new direction $\boldsymbol{u}_i$ and the new electrical field are exactly determined, using Snell–Descartes laws and Fresnel formulae. However, on diopter i, instead of collecting both rays (reflected *and* refracted) issued from a given incident one, a reflexion *or* a refraction is selected using a random number (between 0 and 1): if this number is lower than a given number p_i, a reflexion is chosen and, if not, a refraction is chosen. One then collects q_i which is equal to p_i, if a reflexion is chosen, or equal to $1-p_i$, if not. At the end, $I(\theta)$ is divided by the cumulative product of the q_i's. This procedure gives the exact result in the limit of an infinite number of incident rays. We have checked that the results do not depend on the choice of the p_i's. (In practice, we have chosen p_i equal to the mean reflexion coefficient for the intensity.)

To take care of polarization effects, the interesting intensities to be collected are I_{VV}, I_{HH}, I_{HV}, where the first index refers to the incident ray and the second one to the scattered ray, V denotes a polarization perpendicular and H, a polarization parallel to the scattered plane. The scattered plane, which is parallel to both $\boldsymbol{u}_0$ and $\boldsymbol{u}_s$, can only be known at the end, when the ray is collected in the "θ-detector". Then the projections of both $\boldsymbol{E}_0$ and $\boldsymbol{E}_s$, perpendicular and parallel to the scattered plane, are determined. One knows that in general these projections are related by [7]

$E_{s,per} = a_{VV} E_{0,per} + a_{HV} E_{0,par}$,

$E_{s,par} = a_{VH} E_{0,per} + a_{HH} E_{0,par}$,

with $a_{HV} = a_{VH}$, I_{VV}, $I_{HV} = I_{VH}$ and I_{HH} are the squares of the matrix elements a_{VV}, a_{HV} and a_{HH}. To calculate these quantities, each ray has been sent twice, with the same A_0 and $\boldsymbol{u}_0$, but with two different polarizations, to obtain four linear equations which are then solved. In all the results presented hereafter the optical index has been chosen equal to $n=1,33$, rays have been sent four times more to improve the statistics,

intensities have been averaged over 256 directions regularly distributed in space and δ has been chosen to be several hundred times smaller than the linear size of the aggregates.

3. Application to fractal aggregates

The method has been applied to several deterministic and random fractals. Typical examples are shown in fig. 1. The deterministic ones were built using an algorithm similar to the one introduced by Vicsek [9] starting from spheres stuck in n directions $\boldsymbol{v}_i$ around a given one, this process being repeated iteratively. The fractal dimension is $D=\log(n+1)/\log 3$. We have considered $n=6$ ($\boldsymbol{v}_i=(\pm 1, 0, 0)$ and circular permutations), $n=8$ ($\boldsymbol{v}_i=(\pm 1, \pm 1, \pm 1)$) and $n=12$ ($\boldsymbol{v}_i=(0, \pm 1, \pm 1)$ and c.p.) leading to $D=1.77...$, 2 and $2.33...$, respectively. In each case we have considered four generations, i.e. aggregates containing up to 2401, 6561 and 28561 spheres, respectively. To complete this series with compact aggregates ($D=3$), we have also considered cubes of increasing edges, containing tangent spheres (up to $24\times 24\times 24$) regularly disposed on a simple cubic lattice.

The "random" fractals have been built using diffusion-limited aggregation algorithms in three dimensions and variants [2]. We have used the three-dimensional particle–cluster and cluster–cluster tip-to-tip models [10] leading to fractal dimensions $D=1$ and $D=1.4$, respectively, as well as the off-lattice ballistic cluster–cluster model [11], whose fractal dimension is $D=1.9$. We have also considered three-dimensional on lattice Witten–Sander aggregates whose fractal dimension is known to be $D=2.5$ [2]. In each case we have considered clusters containing $N=16, 32, 64, ..., 2048$ particles. We have, before all, calculated the apparent fractal dimension of our clusters from the log–log plot of their radius of gyration versus their number of particles. We have found $D=1.1, 1.4, 1.9, 2.4$, with an error bar of order 0.08. To complete the set with a compact random aggregate, we have used a simple procedure [12] in which spheres are added at random to the aggregate under

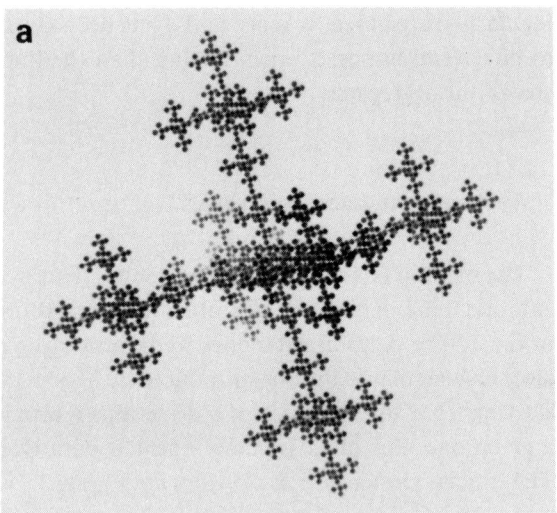

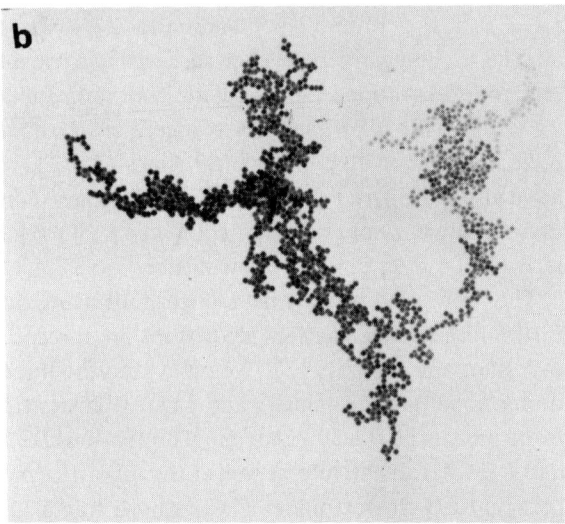

Fig. 1. Typical examples of fractal aggregates used in the calculation. (a) Deterministic fractal aggregate with $N=2401$, $D=1.77$ (Vicsek's algorithm [9]); (b) random fractal aggregate with $N=2048$, $D=1.9$ (ballistic cluster–cluster aggregation [11]).

the conditions that they must be tangent to three other spheres and that their distance to the first sphere must be minimum. For all these aggregates the full $I(\theta)$ curves have been determined (using 64 "detectors" regularly spaced in $\cos\theta$ from -1 to $+1$) and typical examples can be seen in ref. [8]. Here, we only report on the scattering cross-section as well as on depolarization effects.

The cross-section, normalized in order to be equal to one for a single sphere, is calculated by $\sigma=N_d\delta^2/\pi a^2$, where N_d is the total number of rays effectively detected (i.e. having hit at least one sphere). It is proportional to the total scattered intensity integrated over all the scattered directions (this is due to the fact that we have not considered any absorption here). We give the results as a log–log plot of σ as a function of N, the number of particles in the aggregate, in fig. 2. Our results are consistent with σ proportional to N for $D<2$ and varying as $N^{2/D}$ for $D>2$. This is expected from a very simple argument: when the aggregate is transparent ($D<2$), shadowing effects are negligible and almost all the particles feel the incident light, while when it is opaque ($D>2$), only a number of particles proportional to the square of the radius (number of particles seen on a projected area) feel the light. It can be noticed that there might be some logarithmic corrections near $D=2$, where a slope slightly smaller than one is recovered. The same general conclusions were already reached by Meakin et al. [13].

The depolarization of the scattered light can be quantitatively estimated by the relative importance of the cross intensities I_{VH} and I_{HV} compared to I_{VV} and I_{HH}. We have calculated I_{VV}/I_{VH} for forward ($\theta=0$) as well as for backward ($\theta=180°$) scattering (for these angles $I_{VV}=I_{HH}$). For the backward scattering, I_{VV}/I_{VH} is almost constant (extremely slightly decreasing) with increasing sizes and no significant change of behavior is observed around $D=2$. For the forward scattering the results are reported in fig. 3. One again clearly sees, on these curves, a change of behavior for $D=2$. While I_{VV}/I_{VH} does not vary too much with size for $D<2$, it decreases appreciably when increasing N for $D>2$. If one assumes a linear behavior for $D>2$, the estimated slope seems to be very close to $-(D-2)/D$. (Again there is some ambiguity for fractals with fractal dimension close to two.) This result is consistent with the fact that, as long as I_{VH} remains small compared to I_{VV}, the ratio I_{VH}/I_{VV} would be proportional to the mean number

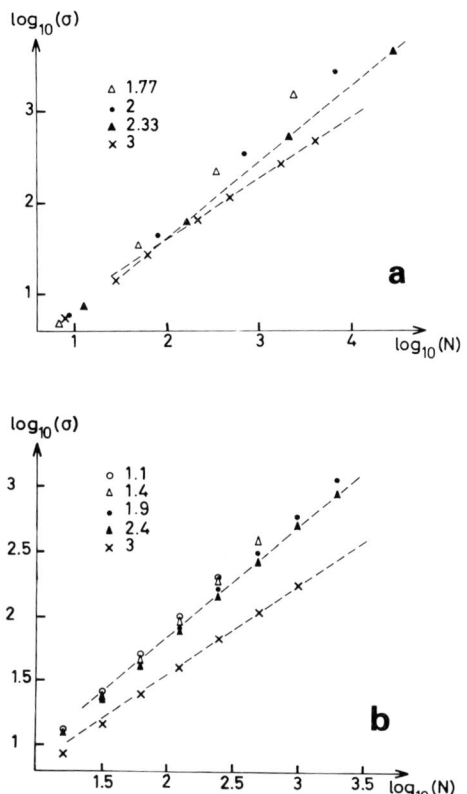

Fig. 2. Plot of the scattering cross-section σ as a function of N for deterministic (a) and for random (b) fractal aggregates (log–log plot). In the case of deterministic fractals the symbols $\triangle$, $\bullet$, $\blacktriangle$, $\times$, correspond to fractal dimensions $D=1.77$, 2, 2.33 and 3, respectively. In the case of random fractals the symbols $\circ$, $\triangle$, $\bullet$, $\blacktriangle$, $\times$, correspond to $D=1.1$, 1.4, 1.9, 2.4 and 3, respectively. For $D>2$, a straight line of slope $2/D$ has been drawn through the data (this is not a least-squares fit).

of spheres crossed by a ray passing through the aggregate. We have found the same kind of scaling for the ratio between forward and backward scattering intensities.

4. Conclusion

We have presented a simple method able to calculate the intensity of polarized light scattered by a fractal aggregate made of identical spheres. This method is restricted to the geometrical optics limit

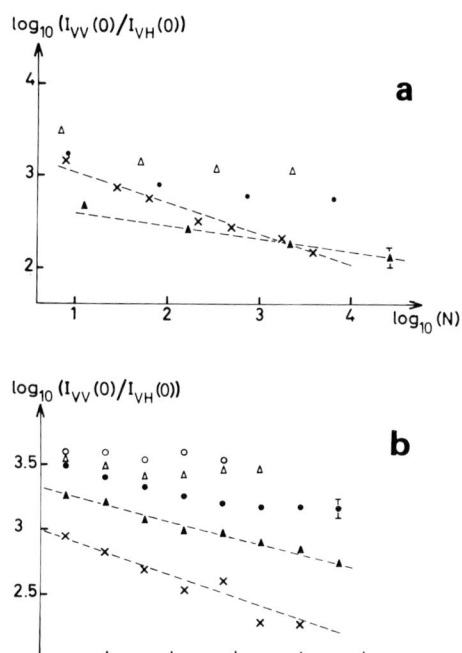

Fig. 3. Plot of $I_{VV}(0)/I_{VH}(0)$ as a function of N for deterministic (a) and for random (b) fractal aggregates (log–log plot). For $D>2$, a straight line of slope $-(D-2)/D$ has been drawn through the data (this is not a least-squares fit). The symbols are the same as those used in fig. 2.

where the radius of the spheres is much larger than the wavelength of the light and where coherence effects as well as diffraction phenomena are not considered. The method has been applied to fractal aggregates containing up to several thousands of particles and some general scaling properties have been found.

Acknowledgements

We acknowledge very fruitful discussions with P. Adam and J.M. Flesselles. Numerical calculations were done at CIRCE (Centre Inter-Régional de Calcul Electronique), Orsay, and supported by contract DRET No. 87/1354.

References

[1] S. Forrest and T.A. Witten, J. Phys. A 12 (1979) L109.
[2] R. Jullien and R. Botet, Aggregation and Fractal Aggregates (World Scientific, Singapore, 1987).
[3] D. Weitz, M. Lin, and C. Sandroff, Surf. Sci. 158 (1985) 147;
M. Axelos, D. Tchoubar and R. Jullien, J. Phys. (Paris) 47 (1986) 1843;
R. Botet, R. Jullien and A. Skjeltorp, Recherche 18 (1987) 1246;
G. Helgesen, A. Skjeltorp, P. Mors, R. Botet and R. Jullien, Phys. Rev. Lett. 61 (1988) 1736.
[4] D. Schaeffer, J. Martin, P. Wiltzius and D. Cannell, Phys. Rev. Lett. 52 (1984) 2371;
S.H. Chen and J. Teixeira, Phys. Rev. Lett. 57 (1986) 2583;
M. Axelos, D. Tchoubar, J. Bottero and F. Fiessinger, J. Phys. (Paris) 46 (1985) 1587.
[5] M. Berry and I. Percival, Opt. Acta 33 (1986) 577.
[6] J. Frey, J.J. Pinvidic, R. Botet and R. Jullien, J. Phys. (Paris) 49 (1988) 1969.
[7] C.F. Bohren, Absorption and Scattering of Light by Small Particles (Wiley, New York, 1983).
[8] R. Jullien and R. Botet, J. Phys. (Paris), in press.
[9] T. Vicsek, J. Phys. A 16 (1983) L647.
[10] R. Jullien, Phys. Rev. Lett. 55 (1985) 1697; J. Phys. A 19 (1986) 2129.
[11] R. Ball and R. Jullien, J. Phys. (Paris) 45 (1984) L1031.
[12] C.H. Bennet, J. Appl. Phys. 43 (1972) 2727.
[13] P. Meakin, B. Donn and G.W. Mulholland, Langmuir, in press.

FRACTALS AND MULTIFRACTALS IN AVALANCHE MODELS

Leo P. KADANOFF

The Research Institutes, The University of Chicago, 5640 S. Ellis Avenue, Chicago, IL 60637, USA

A contribution to the Birthday Celebration for Benoit Mandelbrot

This note reports work by Kadanoff, Nagel, Wu, and Zhou which has been submitted to Physical Review A. In this paper, we examine extensions of the sandslide (or avalanche) model originally proposed by Bak, Tang, and Wiesenfeld to see the nature of the probability distributions produced by the dynamical processes. In particular, we wish to see (a) whether the distributions have a well-defined scaling limit; (b) whether the distributions have a simple (Widom) scaling behavior or are instead multifractal and (c) whether they fail into universality classes. We answer these questions by doing simulational studies of cases in which the models describe one- or two-dimensional arrays of stacks. For the one-dimensional case the answers are: "yes, hope that an analogous construction might perhaps provide a renormalization group argument for the determination of the probability distribution in the avalanche examples".

1. The sand model

As noted by Bak, Tang, and Wiesenfeld [1], in steady state many dynamical systems will organize themselves to be in a critical state. In particular, we look at a model in which sand is placed on a one-dimensional shelf of width L. The sand is placed in L stacks, with the jth site having a stack containing h_j grains of sand. By convention, h_j is zero for $j > L$. One version of this model is defined via an algorithm in which the processes are as follows:

(A) Add a grain of sand at a randomly selected site. (Now the avalanche begins.)

(B) If the slope (i.e. the height difference between adjacent stacks) is greater than a constant value NF, then NF grains from the stack fall over onto adjacent stacks. (The model is non-trivial if NF > 2.)

Grains that reach beyond the right-hand end of the pile fall off and are seen no more.

We continue process (B) until no more stacks are unstable. (Now the avalanche ends.)

(C) Return to (A) so that the process can continue.

It is important to notice that when L is large, avalanches of all sizes can and do occur. To describe the event, we use two basic variables:

(i) the drop number, D, which is the number of grains which drop off the end of the piles between two additions of new grains of sand, and

(ii) the flip number, F, which is the total number of grains of sand which move between two addition events.

In terms of these quantities we then define probability distributions, $\rho(D, L)$ and $\rho(F, L)$, which are the probabilities that D grains will drop or F grains will move in between two addition events.

In this model, step by step motions over short distances produce some very large-scale events. In critical phenomena, also, short-range interactions produce large-scale correlations. In critical phenomena, we are interested in probabilities (derived from partition function) and correlation functions. Call the quantities we wish to calculate $X = D$ or F. Then the usual structure of critical phenomena gives Widom [2] scaling (in the finite-sized scaling version):

$$\rho(X, L) = L^{-\beta} \rho^*(X/L^\nu) \quad \text{for } L, X \gg 1. \tag{1}$$

The two different X's of course give different values of β and ν. We call this result *simple scaling*. The other possibility is a *multifractal* [3] behavior in which

$$(\ln \rho(X, L))/\ln L = f(\alpha), \qquad (2)$$

$$\alpha = (\ln X)/(\ln L).$$

One question of considerable interest is whether the probabilities obey law (1) or law (2). Numerical analysis of the one-dimensional examples indicate that they obey the multifractal law and, in addition, that the results obey some kind of *universality*. In particular, we see that the function $f(\alpha)$ is independent of NF.

2. The binomial distribution: a simple multifractal example

Since the work of Billingsley [4], it has been known that the binomial distribution provides an example of a probability distribution which has a rich asymptotic structure. In modern terms, it does not obey simple scaling but is instead multifractal. Consider Q objects are distributed randomly between 2 bins. The total number of ways that this can be done is $L = 2^Q$. The probability that P objects will show up in the first bin is, of course,

$$\rho(P, Q) = Q!/(P!(Q-P)! 2^Q).$$

Let $P, Q \gg 1$. From the Stirling approximation for factorials we then find that $\rho(P, Q)$ has a multifractal form (2) with

$$f(\alpha) = -\ln 2 - \alpha \ln \alpha - (1-\alpha) \ln(1-\alpha). \qquad (3)$$

Near the peak (at $\alpha = 1/2$) the distribution is Gaussian, but the result (3) also works far into the wings of the distribution.

To understand the sand models one might try to construct a fundamental theory which would, perhaps by renormalization methods, give answers for the important quantities. One might aim for example to calculate $f(\alpha)$. I do not know how to do this for the sand-pile. But I can do it for the binomial distribution. Perhaps this calculation will give some insight into the methods which could work for the sand pile.

Recall Pascal's triangle for binomial coefficients, which is the statement

$$\rho(P, Q) = \tfrac{1}{2}[\rho(P-1, Q-1) + \rho(P, Q-1)]. \qquad (4)$$

If one substitutes the multifractal form (2) into this equation and then expands in $1/P$ and $1/Q$ one then finds the simple renormalization fixed point equation

$$2 = \exp[-f(\alpha) - (1-\alpha)f'(\alpha)]$$
$$+ \exp[-f(\alpha) + \alpha f'(\alpha)]. \qquad (5)$$

Of course, the $f(\alpha)$ of eq. (3) is the correct solution to this equation.

Can one attack the sandslide problem in this fashion? I cannot yet see how to achieve this but ...

References

[1] P. Bak, Ch. Tang and K. Weisenfeld, Phys. Rev. Lett. 59 (1987) 381.
[2] B. Widom, J. Chem. Phys. 43 (1965) 3892, 3898.
[3] B. Mandelbrot, J. Fluid Mech. 62 (1974) 331;
U. Frisch and G. Parlai, in: Turbulence and Predictability in Geophysical Fluid Dynamics and Climate Dynamics, M. Ghil, R. Benzl and G. Parisi, eds. (North-Holland, Amsterdam, 1985), p. 84.
[4] P. Billingsley, Ergodic Theory and Information (Wiley, New York, 1965), p. 139.

ENERGETIC AND ENTROPIC ELASTICITY OF THE SIERPIŃSKI GASKET

Yacov KANTOR

School of Physics and Astronomy, Raymond and Beverly Sackler Faculty of Exact Sciences, Tel Aviv University, Tel Aviv 69978, Israel

Two complementary (energetic and entropic) models of the problem of elasticity on the Sierpiński gasket are compared and discussed. Both models are known to exhibit a power-law scaling of the elastic constants. These results are compared with the properties of random systems, and it is shown that the Sierpiński gasket is significantly more stable than a "typical" random system: in the energetic elasticity model the elastic constants decay very slowly with increasing length scale, while in the entropic elasticity model the excluded-volume effects are able to stabilize the shape of the gasket, i.e. its relative shape fluctuations decay with increasing system size.

1. Introduction

In recent years there has been a renewed interest in the elastic properties of inhomogeneous materials. In many such systems one can identify a correlation ξ, above which the material can be treated as homogeneous. The linear elastic response of *homogeneous* solids is characterized by the elastic stiffness tensor C_{ijkl}. This tensor is simply related to the volume and shape dependence of the free energy F of the solid [1]: $C_{ijkl} = (1/V) \partial^2 F / \partial \epsilon_{ij} \partial \epsilon_{kl}$, where V is the volume of the system, while ϵ_{ij} is the applied strain. At finite temperature T, both the energy U and the entropy S contribute to the shape dependence of $F = U - TS$. For the purpose of the theoretical research one can *roughly* divide the inhomogeneous systems into two groups: (a) Materials whose elementary grains are very large will belong to the *energetic elasticity* group. The weakly compacted sandstone is a typical member of such a group. Under applied external distortion the entropy of these materials remains unchanged, and the change in the free energy F can be attributed to the increase in the energy U. (b) Materials consisting of very small "building blocks", e.g. polymeric systems such as rubbers and gels [2], belong to the *entropic elasticity* group. The distortion of a polymeric network reduces the available phase space thus decreasing the entropy of the system and increasing its free energy [3].

The "borderline" between the two groups is not sharp, and depends on such properties as grain size, geometry and temperature. One may quantify the difference between the types of the systems by comparing their relative fluctuations [4].

Frequently, at length scales $L < \xi$ the geometrical properties of the systems can be characterized by a fractal dimension d_f [5] which relates the mass N of the object to its linear size L: $N \sim L^{d_f}$. In the fractal regime one may expect a power-law dependence on length of the various physical properties such as resistance or force constants. These power laws can be used to infer the properties on length scales $L > \xi$ by assuming a smooth crossover from the fractal regime to a homogeneous regime. In some cases, such as colloidal aggregates [6,7] there is no crossover to the homogeneous regime, and the fractal behavior is limited by the finite size of the aggregate. In these systems one directly measures the length-scale dependence of the elastic properties [7]. The theoretical analysis of the vibrations in the fractal regime led to the discovery [8] of modes denoted "fractons", with a peculiar power-law scaling of the density of states. More recently, these states have been observed experimentally [9].

Essays in honour of Benoit B. Mandelbrot
Fractals in Physics – A. Aharony and J. Feder (editors)

In this paper, I compare several approaches [10–12] to the elasticity of a particular regular fractal – the Sierpiński gasket [5]. I show that certain properties of that object differ from what one could call "a typical random fractal". Nevertheless, the system permits one to quantify several important concepts of the elastic behavior of inhomogeneous systems.

2. The Sierpiński gasket – basic properties

The Sierpiński gasket is a deterministic fractal defined by the following iterative procedure [5]. A triangle is subdivided into four triangles by lines connecting the mid-points of its edges, followed by removal of the central triangle and inflation of the entire system by a factor of 2. The procedure is repeatedly applied to each of the remaining triangles. The fractal dimension of the gasket is $d_f = \ln 3/\ln 2 \approx 1.585$.

A considerable number of investigations [13,14] dealt with the properties of various models defined on the Sierpiński gasket, hoping that it will provide a qualitative insight into the general behavior of random fractals, as well as supply some quantitative estimates for critical exponents. Gefen et al. [13] investigated the scaling of the conductivity of a resistor network having the topology of the Sierpiński gasket: equal resistors have been placed along the edges of the present triangles of the gasket, and it has been shown that the resistance of a two-dimensional gasket increases by a factor 5/3 after each iteration, leading to the power-law dependence of the two-point conductance $\Sigma \sim L^{-\tilde{\zeta}_r}$, with $\tilde{\zeta}_r = \ln(5/3)/\ln 2 \approx 0.737$. (For a d-dimensional gasket $\tilde{\zeta}_r = \ln[(d+3)/(d+1)]/\ln 2$.) Since the fractal dimension of the Sierpiński gasket is close to the fractal dimension of the backbone of the percolating cluster, it has been suggested [13], that the value of $\tilde{\zeta}_r$ obtained for the gasket could approximate the conductivity exponent of a random system. Later, an appreciable discrepancy in the numerical values has been found. However, the gasket remained one of the simplest, but, nevertheless, nontrivial models for testing the properties of the random systems.

3. Energetic elasticity models

The behavior of elasticity and conduction in a continuum are governed by similar differential equations. Thus, it was frequently assumed that the critical behavior of both properties is identical. Bergman has suggested [15] to use the Sierpiński gasket to look for possible differences between the problems. In its simplest form, one replaces the resistors described in the previous section by springs [10], i.e. uses central forces between the neighboring nodes of the network, and measures the changes in the elastic response of the system under the rescaling. Different elastic moduli can be measured by applying a variety of displacements to the external corners of the gasket. It has been found [10] that all possible force constants k (which are the analogues of the conductance Σ in the electric case) halve after each iteration, leading to the scaling relation $k \sim L^{-\tilde{\zeta}_e}$, with $\tilde{\zeta}_e = 1$. (This result remains unchanged if one considers a d-dimensional gasket.)

The behavior of the central force model is governed by a single force constant, which describes the properties of the elementary spring. A more general case was considered by introducing a three-terminal element at each node of the gasket [10], as well as by considering three-body bending forces on the gasket [11]. Both models have the same effect of adding an energetic cost to the change of the angle between the bonds. However, these modifications did not change the behavior of the effective elastic moduli of the gasket, and $\tilde{\zeta}_e$ remained 1.

The immediate conclusion from the above results is that the critical behavior of elasticity differs from the behavior of conductivity. The subsequent investigations of the energetic elasticity of percolating systems revealed significant differences between the predictions of the Sierpiński gasket model and the behavior of random systems: in the presence of central forces the rigidity threshold was found to be significantly larger than the percolation threshold [16], i.e. the geometric continuity was insufficient to support stress. In the presence of bending forces $\tilde{\zeta}_e$ is *significantly* larger than $\tilde{\zeta}_r$, and its value is primarily determined by the bending forces [17]. The differences in the behavior can be attributed to the fact that the Sierpiński gasket consists of the "most stable structures" – triangles, while the elastic response of a percolating system is primarily determined by the long tortuous paths [17].

So far we considered a purely energetic model. What happens to the system at finite temperature T? The fluctuations δ of the gasket size can be related to the force constant k via [18]

$$k\delta^2 \approx k_B T. \tag{1}$$

Therefore the relative fluctuations of the Sierpiński gasket decay with increasing length scale:

$$\delta/L \approx (\sqrt{k_B T/k})/L \sim L^{\tilde{\zeta}_e/2-1} = L^{-1/2}.$$

By contrast, for percolation clusters, where [17] $\tilde{\zeta}_e \approx 3$, the relative fluctuations increase with increasing length scale. Thus, for most tenuous structures large fluctuations will (on sufficiently large length-scales) eventually break the validity of the energetic approach which essentially assumed small fluctuations [4].

4. Elasticity of polymeric networks

Linear and branched polymers are examples of entropy-dominated tenuous structures. They do not have a "ground-state shape", and can be defined only by the *connectivity*, while their relative shape fluctuations are of order unity. The radius of gyration (rms size) R_g of such a system can be usually related to its *internal* linear size L (in the case of a linear polymer L is the number of monomers, while in the case of a polymeric surface L is the linear size of a stretched surface) by a power law: $R_g \sim L^\nu$. In such a situation one cannot use the regular elastic stiffness tensor. However, the scaling $k \sim L^{-\tilde{\zeta}_e}$ of a typical force constant k can be easily determined: since the typical fluctuation δ satisfies eq. (1), and δ is of order R_g, the force constant [19] $k \approx k_B T/R^2 \sim L^{-2\nu}$. Thus, the elasticity exponent $\tilde{\zeta}_e = 2\nu$ for polymeric structures.

The success of the scaling theory in polymer physics rests on detailed investigation of linear polymers [2]. The treatment of more complicated objects usually relies on approximate theories. It is believed that on sufficiently long length-scale a tenuous network without self-avoidance (i.e. without the excluded-volume, or steric, interactions) can be correctly described by a network of harmonic springs which have a vanishing equilibrium length ("Gaussian network"). It can be shown [20,21] that the squared radius of gyration R_{g0}^2 of such a polymeric network is proportional to the mean resistance of a resistor network, which has the same topology as the polymeric network. Therefore, in the absence of the self-avoidance in polymer having the connectivity of a Sierpiński gasket we should expect $R_{g0} \sim L^{\nu_0}$, with $\nu_0 = \tilde{\zeta}_r/2 \approx 0.368$. (For the gasket we define L as the length of a stretched fractal, i.e. it is proportional to 2^n, where n is the number of iterations of the gasket.)

Excluded-volume interactions cause an expansion of the system (compared with the case without self-avoidance) and an increase in ν. From a dimensional analysis one finds that the self-avoidance becomes irrelevant when the structure is embedded in space dimension [20] $d > d_c \equiv 4d_f/\tilde{\zeta}_r$. For a structure with the connectivity of a two-dimensional Sierpiński gasket the upper critical dimension $d_c \approx 8.6$. For $d < d_c$ we can approximate the free energy F by a Flory-type expression [22]:

$$\frac{F}{k_B T} = \left(\frac{R_g}{R_{g0}}\right)^2 + v\left(\frac{N}{R_g^d}\right)^2 R_g^d, \tag{2}$$

where we omit the dimensionless prefactors of order unity. The first term on the right of (2) is the elastic (entropic) free energy of a network (R_{g0} is the radius of gyration of the same network without self-avoidance). The second term is an estimate of the repulsive interaction energy (the squared density of the monomers $(N/R_g^d)^2$ is a mean-field-type estimate of the number of pairs of monomers coming into close contact with each other in a unit volume). By minimizing (2), we find $R_g \sim L^{\nu_F}$, with $\nu_F = (\tilde{\zeta}_r + 2d_f)/(d+2)$.

For a polymer with the connectivity of the Sierpiński gasket in $d=2$, $\nu_F = 0.977$. Notice, that for $d < d_k \equiv \tilde{\zeta}_r + 2d_f - 2$, we have $\nu_F > 1$. In general, the definition of the internal size L is somewhat arbitrary (the value of ν depends on that definition) and ν can exceed unity. However, with the particular definition of L for the Sierpiński gasket, such result would mean that the system is overstretched, and the Flory approximation fails. For the gasket the approximation breaks down at space dimension $d_k \approx 1.907$. We should keep in mind that the expression for d_k has been obtained within an approximate treatment, and it is quite possible that even at $d=2$ the system is already completely stretched.

5. Monte Carlo simulation of the entropic model

Duering and Kantor [12] considered a two-dimensional model system which is described by the Hamiltonian

$$\frac{H}{k_B T} = \sum_{\langle i,j \rangle_{nn}} V_{att}(|\mathbf{r}_i - \mathbf{r}_j|) + \sum_{\{i,j\}} V_{rep}(|\mathbf{r}_i - \mathbf{r}_j|), \quad (3)$$

where $\mathbf{r}_i$ is the position of the ith atom. The attractive potential V_{att} acts only between pairs $\langle i,j \rangle$ of atoms, which are nearest neighbors on a regular Sierpiński gasket, thus ensuring the *connectivity* (but *not the shape*) of the gasket. $V_{att}(r) = 0$, for $r < b$, and ∞ otherwise. (This type of "tethering potential" has been previously used to investigate self-avoiding surfaces [23].) The excluded-volume interaction was implemented by a hard-core repulsive potential V_{rep}, which acts between *any pair* $\{i, j\}$ of atoms. The physical properties of the system are determined by the entropy, and the force constants are strictly proportional to T.

The configuration space of the structure has been sampled using the Monte Carlo (MC) method [24] which consisted of randomly picking an atom and attempting to displace it in a randomly chosen direction. To verify the standard assumption that without the excluded volume the system should approach the behavior of a Gaussian network, an equilibration was performed for a modified version of the model in which V_{rep} was restricted to act only between the nearest neighbors of the network, and an excellent agreement with the theoretical prediction has been found.

In the MC simulation R_g was measured for a sequence of gasket sizes L. For small L the effective value of ν was somewhat smaller than unity, but it increased with increasing L and tended towards $\nu = 1.002 \pm 0.005$, leaving the Flory estimate outside the error bars. Since $\nu \leq 1$, the result suggests that $\nu = 1$ *exactly*. This conclusion also follows naturally from a simple inspection of fig. 1: we notice that on short length scales the structure is quite featureless. However, on larger length scales the "regular" shape of the gasket becomes apparent. This behavior, resembling homogeneous structure, indicates that $\nu = 1$.

The "shape stability" permits introduction of a regular elastic compliance tensor S_{ijkl}, and its MC measurement from the strain fluctuations [25]:

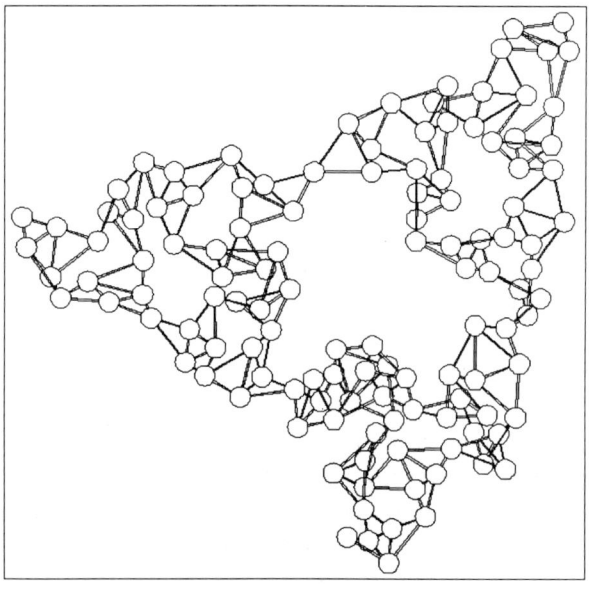

Fig. 1. A typical equilibrium configuration of a polymer which has the connectivity of the Sierpiński gasket.

$S_{ijkl} = (A/k_B T) \langle \epsilon_{ij} \epsilon_{kl} \rangle$, where A is the area of the system, while ϵ_{ij} is the thermally induced strain. All non-vanishing elements of S_{ijkl} had the same L-dependence, and were consistent with the (expected) isotropic symmetry, thus reducing the results to two independent constants with the same scaling properties. From the L-dependence of the elastic moduli one obtains an estimate of the entropic elasticity exponent $\tilde{\zeta}_e = 0.90 \pm 0.15$, which is close to the value 1 expected in the energetic elasticity models of the gasket [10,11] with central forces (with or without bending forces), and somewhat exceeds the value 0.737 which would follow from a scalar elasticity model [11]. The result is, obviously, inconsistent with $\tilde{\zeta}_e = 2\nu = 2$, which would be expected in a polymeric system.

The most surprising part of the results described in this section is the fact, that the excluded-volume effects are able to stabilize the shape of the fractal. Once we accept such stability, the scaling of the elastic constants becomes quite natural: One could assume that beyond a certain length scale, the fluctuations are small and the entire system can be approximately described by a simply *energetic* Hamiltonian. Since we have shown that the relative fluctuations of the energetic elasticity model on a Sierpiński gasket do not grow, such an effective description will remain valid also on even larger length-scales.

6. Discussion

In this work several elasticity models on Sierpiński gasket have been compared. It has been shown, that the gasket is significantly more stable than "a typical random fractal". The indications of the exceptional behavior came from the fact that the relative fluctuations of energetic models do not grow, and that the excluded volume in the entropic model stabilizes the structure. Both behaviors are related to an exceptionally high connectivity and to the fact that the fractal consists of stable structures – triangles. Only further investigation of additional models will show in a more quantitative way, which properties of fractals cause such an exceptional behavior.

From the theoretical point of view the results indicate that some additional parameter (besides the fractal and spectral dimensions) might enter the theory of fractal polymers. The results also imply that one should not use the Sierpiński gasket as a prototype for explaining the properties of gels [26], unless one has experimental evidence for the presence of such connectivity in a gel. The exceptional geometry of the Sierpiński gasket, however, does not mean that such "highly triangulated" structures are rare, since in certain random aggregation processes such stable structures may be preferably created.

Acknowledgement

This research was supported by the US–Israel Binational Science Foundation through Grant No. 87-00010, and by the Bat Sheva de Rothschild Foundation.

References

[1] J.H. Weiner, Statistical Mechanics of Elasticity (Wiley, New York, 1983).
[2] P.G. de Gennes, Scaling Concepts in Polymer Physics (Cornell Univ. Press, Ithaca, 1979).
[3] S.F. Edwards, J. Phys. A 1 (1968) 15.
[4] Y. Kantor and T.A. Witten, J. Phys. (Paris) 45 (1984) L 675.
[5] B. B. Mandelbrot, Fractals: Form, Chance and Dimension (Freeman, San Francisco, 1987); The Fractal Geometry of Nature (Freeman, San Francisco, 1982).
[6] A.I. Medalia and F.A. Hechman, J. Colloid Interface Sci. 36 (1971) 173;
J. Eisenlauer and E. Killman, J. Colloid Interface Sci. 74 (1980) 108;
K. Kusaka, N. Wada and A. Tasaki, Japan. J. Appl. Phys. 8 (1969) 599;
S.R. Forest and T.A. Witten, J. Phys. A 12 (1979) L 309.
[7] L. Ye, D.A. Weitz, P. Sheng, S. Bhattacharya, J.S. Huang and M.J. Higgins, Exxon preprint (1989).
[8] S. Alexander and R. Orbach, J. Phys. (Paris) 43 (1982) L 625;
R. Rammal and G. Toulouse, J. Phys. (Paris) 44 (1983) L 13.

[9] J. Fricke, Sci. Am. 256 (1988) 92;
E. Courtens, J. Pelous, J. Phalippou, R. Vacher and T. Woignier, Phys. Rev. Lett. 58 (1987) 128;
E. Courtens, R. Vacher, J. Pelous and T. Woignier, Europhys. Lett. 6 (1988) 245;
T. Freltoft, J. Kjems and D. Richter, Phys. Rev. Lett. 59 (1987) 1212.
[10] D.J. Bergman and Y. Kantor, Phys. Rev. Lett. 53 (1984) 511.
[11] S. Alexander, J. Phys. (Paris) 45 (1984) 1939.
[12] E. Duering and Y. Kantor, Tel Aviv University preprint (1989).
[13] Y. Gefen, A. Aharony, B.B. Mandelbrot and S. Kirkpatrick, Phys. Rev. Lett. 47 (1981) 1771.
[14] Y. Gefen, B.B. Mandelbrot and A. Aharony, Phys. Rev. Lett. 45 (1980) 855;
M. Stephen, Phys. Lett. A 32 (1981) 67;
R. Rammal and G. Toulouse, Phys. Rev. Lett. 49 (1982) 1194;
S. Alexander, Phys. Rev. B 27 (1983) 1541;
S. Alexander and E. Halevi, J. Phys. (Paris) 44 (1983) 805;
E. Domany, S. Alexander, D. Bensimon and L.P. Kadanoff, Phys. Rev. B 28 (1983) 3110;
R. Rammal, Phys. Rev. B 28 (1983) 4871;
S. Alexander, Phys. Rev. B 29 (1984) 5504.
[15] D.J. Bergman, Santa Barbara Workshop on Disordered Systems (1983).
[16] S. Feng and P.N. Sen, Phys. Rev. Lett. 52 (1984) 216.
[17] Y. Kantor and I. Webman, Phys. Rev. Lett. 52 (1984) 1891.
[18] L. Landau and I. Lifshitz, Statistical Physics (Pergamon Press, London, 1958) ch. 12.
[19] P.G. de Gennes, Polymers 9 (1976) 587.
[20] M.E. Cates, Phys. Rev. Lett. 53 (1984) 926; J. Phys. (Paris) 46 (1985) 1059.
[21] Y. Kantor, in: Proceedings of the 5th Jerusalem Winter School on Statistical Mechanics of Membranes and Surfaces, D.R. Nelson, T. Piran and S. Weinberg, eds. (World Scientific, Singapore), to be published.
[22] P. Flory, Principles of Polymer Chemistry (Cornell Univ. Press, Ithaca, 1971).
[23] Y. Kantor, M. Kardar and D.R. Nelson, Phys. Rev. Lett. 57 (1986) 791.
[24] A. Baumgärtner, in: Application of the Monte Carlo Method in Statistical Physics, K. Binder, ed. (Springer, Berlin, 1984), p. 145.
[25] M. Parinello and A. Rahman, Phys. Rev. Lett. 45 (1980) 1196.
[26] J. Bastide and F. Boue, in: Statphys 16, ed. H.E. Stanley (North-Holland, Amsterdam, 1986), p. 251.

GROWTH OF SELF-AFFINE SURFACES

János KERTÉSZ [a,1] and Dietrich E. WOLF [b]

[a] *Institute for Theoretical Physics, University of Cologne, Zülpicher Strasse 77, D-5000 Cologne 41, Fed. Rep. Germany*
[b] *IFF of KFA, P.O. Box 1913, D-5170 Jülich, Fed. Rep. Germany*

Growing rough surfaces are a natural illustration of self-affinity, a term coined some years ago by Benoit Mandelbrot. The Eden model, ballistic deposition and polynucleation growth are well-known examples where such structures emerge. Using the concept of the intrinsic surface width associated with the physical lower cutoff we explain important corrections to scaling and show how anomalous roughening at morphological transitions develops.

1. Introduction

When discussing self-affinity, Benoit Mandelbrot says in his Scripta paper [1] that the investigation of surface structures led him to the study of this field. In fact, besides "records of function", surfaces often provide beautiful examples of objects which are statistically invariant under an affine transformation (instead of a simple dilation as in the case of self-similarity) [2,3].

Here we concentrate on *growing* surfaces [4] which under appropriate circumstances become rough, i.e. the fluctuations characterized by the surface width increase with the size of the aggregate. Ideally the only characteristic length is the linear extension L of the $(d-1)$-dimensional surface, and the surface width scales as $w \sim L^\zeta$, i.e. the affinity transformation which leaves the infinite surface statistically invariant is a diagonal matrix with $d-1$ equal elements (λ) and one different (λ^ζ). In growth models one is also interested in the time evolution of the surface width: starting from a flat substrate the width increases with time t as a power law $w \sim t^{\zeta/z}$ for $t \ll L^z$. The general scaling behavior can be summarized in the form

$$w = L^\zeta f(t/L^z), \tag{1}$$

where $f(c) \sim c^{\zeta/z}$ for $c \ll 1$ and $f(c) \sim$ const. for $c \gg 1$. Eq. (1) was first introduced in this context by Family and Vicsek [5].

In reality there is often a second length scale ξ interfering with the simple scaling picture above and representing a lower cutoff for self-affinity. In lattice models this can be the lattice constant leading to finite-size corrections. Associated with this is an *intrinsic width* w_i which describes the fluctuations of the surface within the interval of length ξ. For example in most growth processes deviations from single-step solid-on-solid (SOS) configurations occur, leading to a height distribution within one lattice constant. In section 2 we summarize how the intrinsic width influences the scaling behavior of the surface and discuss noise reduction as a method suitable for suppressing it [6–8].

More interesting are examples where the second length ξ diverges itself giving rise to a divergence of the intrinsic width $w_i \sim \xi^{\zeta'}$. This happens at transitions between two surface morphologies characterized by different roughness exponents ζ_1, ζ_2 in which case ζ' characterizes the *anomalous roughening* at the critical point [9]. This problem will be reviewed in section 3.

The paper ends with a short summary in section 4.

---

[1] On leave from Institute for Technical Physics, H-1325 Budapest, Hungary.

2. Intrinsic width and scaling

When studying the Eden model numerically, it turned out that strong corrections to scaling hindered the verification of (1) especially in higher dimensions. A major source of the corrections is the intrinsic surface width [6,10]. In ref. [6] we introduced the method of noise reduction to handle this problem: in contrast to the original Eden model where a randomly picked perimeter is immediately added to the cluster, we put counters on the perimeters and occupy a site only after m trials. Fig 1 presents a couple of snapshots of noise-reduced Eden clusters.

Noise reduction improves scaling since it suppresses the formation of holes, overhangs and high steps and thus reduces the corrections to scaling due to the intrinsic width. These corrections can be excellently represented by the convolution approximation [7,8]:

$$w^2 = w_{sc}^2 + w_i^2, \qquad (2)$$

where w is the total width, w_{sc} its scaling part for which (1) should be valid and w_i the intrinsic width independent of L.

Noise reduction not only suppresses w_i, it also causes a time delay [7,11] which can be understood if substrate effects are considered. For large values of m the evolution of the surface starts with layerwise growth. In fact, for $d=2$ we showed numerically that the number of layers of this early stage is proportional to m, and in ref. [11] we argue that it should be so for higher dimensions as well. The large scale fluctuations are induced by noise and therefore develop when the layerwise growth is over.

Combining all information into a general scaling form we get [7]:

$$w^2(t, L, m) = [a(m)L^\zeta f(t/mL^z)]^2 + w_i^2(m), \qquad (3)$$

where $a(m)$ represents the m-dependence of the amplitude of the scaling term. Fig. 2 shows our two-dimensional results for the $t \to \infty$ case which is in full

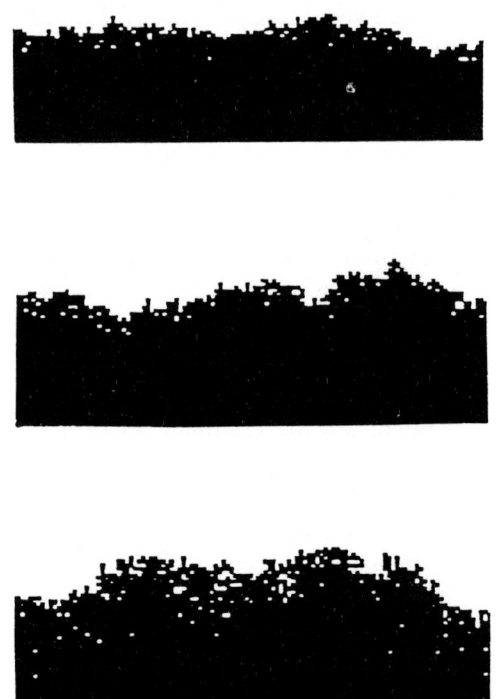

Fig. 1. Snapshots of Eden surfaces grown with different noise reduction parameter m (top: $m=4$, middle: $m=2$, bottom: $m=1$).

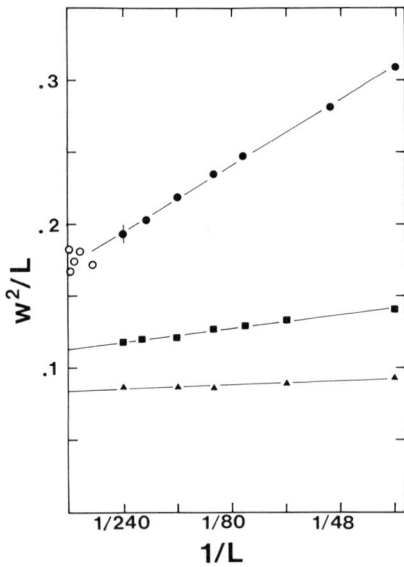

Fig. 2. Test of (3) for the two-dimensional Eden model ($\zeta = 1/2$): w is the $t \to \infty$ limit of the surface width, the slope of the curves is w_i^2 and is decreased by noise reduction (●: $m=1$, ■: $m=2$, ▲: $m=4$).

agreement with assumption (3). (Here we would like to mention that the deviations from (3) found in a modified Eden model [12] are probably due to the positioning of the counters.) Using the method of noise reduction we could reproduce the exactly known two-dimensional exponents with high accuracy and relatively small computing effort [6].

The dimensional dependence of the exponents is interesting because there were contradicting theoretical suggestions: Kardar and Zhang [13] predicted dimension-independent superuniversal exponents and Meakin et al. [14] got for a seemingly related model $\zeta = 0(\log)$ in three dimensions. Using noise reduction we obtained for the Eden model [8] $\zeta = 0.33 \pm 0.01$ for $d = 3$ and $\zeta = 0.24 \pm 0.02$ for $d = 4$ (which looks like $\zeta = 1/d$). The three-dimensional value $\zeta = 1/3$ occurs also in subsequent theoretical studies [15]. Recently the numerical investigation of a related model with zero intrinsic width led Kim and Kosterlitz [16] to conjecture that $\zeta = 2/(d+2)$. The numerical discrepancies should be either due to unrevealed corrections to scaling, or the models belong to different universality classes.

3. Anomalous roughening in surface growth

The second characteristic length ξ may diverge for example at a morphological transition. This is the case in a class of stochastic growth models to be discussed below, for which there exists a smooth as well as a rough phase. Transitions from a smooth surface (with finite width, $\zeta = 0$) to a rough one (with diverging width, $\zeta > 0$) are well known from thermal systems. Equilibrium shapes of three-dimensional crystals provide examples for these roughening transitions. By analogy we call morphological transitions between a smoothly growing surface and a rough one kinetic roughening transition [9].

Suppose there is a transition at a critical value p_c of a parameter p between two morphologies $i = 1, 2$ characterized by the exponents ζ_i, z_i and that close to the transition point a new diverging length $\xi \sim |\epsilon|^{-\nu}$ influences the roughness ($\epsilon = p - p_c$). The intrinsic width for the considered system is in this case not only due to the local structure of the surface but it is physically built up over a longer range ξ and is L-independent only as long as $L \gg \xi$. We propose the following scaling form of the surface width:

$$w(\epsilon, L, t) \sim \xi^{\zeta'} f_i(L/\xi, t/\xi^{z'}) , \qquad (4)$$

with the scaling function f_1 (f_2) above (below) the transition. Implicitly we assumed that ζ' as well as z' are the same for $\epsilon > 0$ and $\epsilon < 0$.

Eq. (4) describes the crossover between the two phases. Since the width remains finite for any fixed L and t, the divergent factor $\xi^{\zeta'}$ in (6) must be compensated by a suitable power law behavior of f_i in the limit $\xi \to \infty$. Therefore, at the critical point the surface shows roughening with anomalous exponents ζ', z':

$$w(0, L, t) \sim L^{\zeta'} g(t/L^{z'}) . \qquad (5)$$

In a class of stochastic growth models the physics underlying the morphological transition is sufficiently well understood that the anomalous exponents can be predicted. These are models with a maximal velocity $v_{\max}$ by which the uppermost point of the surface can propagate. Furthermore, the mass increase (growth rate) must be tunable independently. For small growth rate the surface propagates with a velocity smaller than the maximal one, so that it is expected to become rough. If the growth rate is increased until the surface propagates with maximal velocity, it always feels the global constraint and cannot get rough anymore. Therefore one has a morphological transition between phases with $\zeta_1 > 0$ and $\zeta_2 = 0$. The transition is triggered by directed percolation [17]: the stochastic growth process defines an effective local transition probability of reaching the level corresponding to the maximum velocity at every time step, i.e. an "occupation probability" for percolation through directed paths. Directed percolation produces self-affine fractal clusters at the critical point. In the $(d-1)$-dimensional space-like directions the correlation length ξ_r scales with the exponent ν_r while in the direction of the anisotropy $\xi_t \sim |\epsilon|^{-\nu_t}$ [17]. It is natural to identify ξ_r and ξ_t with ξ and $\xi^{z'}$, respectively, so that

$$\nu = \nu_r, \quad z' = \nu_t/\nu_r. \tag{6}$$

The order parameter V in the rough phase is:

$$V = v_{\max} - v, \quad \text{with } v = \partial_t \bar{h}(t), \tag{7}$$

where $\bar{h}(t)$ is the average height. We make the following scaling assumption:

$$V(\epsilon, L, t) = \xi^{-\beta'/\nu} \psi(L/\xi, t/\xi^{z'}). \tag{8}$$

(We introduce the exponent β' reserving β for the usual percolation order parameter.)

Now we present an argument why β' should be equal to ν_t. This is the essential simplification which allows us to predict the anomalous roughening exponent ζ'. In the stationary state for $p > p_c$ the surface moves with maximal velocity: $\bar{h}(t_0 + t) = \bar{h}(t_0) + v_{\max} t$. However, for $p < p_c$ there exists a characteristic time τ after which the surface stays behind by one lattice constant compared to the position it would have reached with the maximal velocity: $\bar{h}(t_0 + \tau) = \bar{h}(t_0) + v_{\max} \tau - 1$. This is just the time for which directed percolation correlations survive, i.e. $\tau \sim \xi_t$. Hence, $V = 1/\tau \sim \xi^{-\nu_t/\nu}$. Comparison with (8) shows that $\beta' = \nu_t$. As a consequence one gets at p_c

$$V \sim 1/t. \tag{9}$$

Obviously, the surface has to fit into the interval between the average height and $y = v_{\max} t$, so that

$$w < \text{const.} \int V \, dt \sim \log t. \tag{10}$$

Thus ζ'/z' and therefore ζ' have to be zero.

Detailed calculations for the anomalous roughening were carried out for a two-dimensional model confirming these predictions [9,18]. The surface width squared was found to increase logarithmically in agreement with (10) and has logarithmic size dependence (cf. fig. 3):

$$w^2 \sim \log t, \quad w^2 \sim \log L. \tag{11}$$

For more details see refs. [9,18].

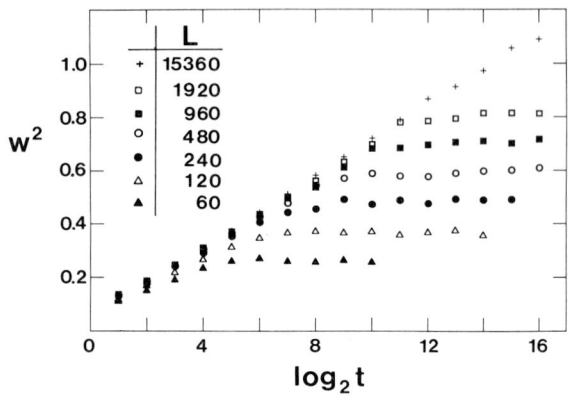

Fig. 3. Anomalous roughening [9] in polynucleation growth [19] on a square lattice at the critical point.

4. Summary

We have discussed the influence of a second characteristic length ξ (in addition to the extent L of the surface) in self-affine surface growth processes. It is intimately related to the concept of an intrinsic width. In the Eden model the intrinsic width is responsible for important corrections to the leading scaling behavior, but it can be damped by noise reduction. If the second length diverges, the intrinsic width may show anomalous roughness dominating the surface morphology if $\xi \gg L$. This was shown to happen at a kinetic roughening transition triggered by directed percolation, where the anomalous roughening is logarithmic.

Acknowledgement

Part of this work was performed within the SFB 341 of the DFG. JK thanks the Humboldt Foundation for support.

References

[1] B.B. Mandelbrot, Physica Scripta 32 (1985) 257.
[2] B.B. Mandelbrot, in: Fractals in Physics, eds. L. Pietronero and E. Tosatti (North-Holland, Amsterdam, 1986) pp. 3–16, 17–20, 21–28.

[3] B.B. Mandelbrot, The Fractal Geometry of Nature (Freeman, San Francisco, 1983) p. 182.
[4] T. Vicsek, Fractal Growth Phenomena (World Scientific, Singapore, 1989) p. 182.
[5] F. Family and T. Vicsek, J. Phys. A 18 (1985) L75.
[6] D.E. Wolf and J. Kertész, J. Phys. A 20 (1987) L257.
[7] J. Kertész and D.E. Wolf, J. Phys. A 21 (1988) 747.
[8] D.E. Wolf and J. Kertész, Europhys. Lett. 4 (1987) 651.
[9] J. Kertész and D.E. Wolf, Phys. Rev. Lett., in press.
[10] J.G. Zabolitzky and D. Stauffer, Phys. Rev. A 34 (1986) 1523.
[11] D.E. Wolf and J. Kertész, preprint.
[12] P. Devillard and H.E. Stanley, Phys. Rev. A 38 (1988) 6451.
[13] M. Kardar and Y.-C. Zhang, Phys. Rev. Lett. 58 (1987) 2087.
[14] P. Meakin, R. Ramanlal, L.M. Sander and R.C. Ball, Phys. Rev. A 34 (1986) 5091.
[15] T. Halpin-Healy, Phys. Rev. Lett. 62 (1989) 442;
T. Natterman, preprint.
[16] J.M. Kim and J.M. Kosterlitz, Phys. Rev. Lett. 62 (1989) 2289.
[17] W. Kinzel, in: Percolation Structures and Processes, G. Deutscher, R. Zallen and J. Adler, eds., Ann. Israel. Phys. Soc., Vol. 5 (Hilger, Bristol, 1983) p. 425.
[18] D.E. Wolf and J. Kertész, in preparation.
[19] F.C. Frank, J. Crystal Growth 22 (1974) 233.

JOHNSON–NYQUIST NOISE DERIVED FROM QUANTUM MECHANICAL TRANSMISSION

Rolf LANDAUER

IBM T.J. Watson Research Center, Yorktown Heights, NY 10598, USA

Expressions for the conductance of a sample with elastic scattering, in terms of the transmissive behavior of the sample, have come into widespread use. We show that thermal equilibrium noise currents, calculated within the same framework, obey the Johnson–Nyquist formula.

1. Introduction

The past decade has seen continually increasing attention to expressions for conductance in terms of the transmissive behavior of the sample [1]. Do these models allow a direct calculation of thermal agitation noise, and does it obey the well known Johnson–Nyquist [2,3] noise formula? If the conductance expressions are correct, it would be most remarkable if we could not rederive Nyquist noise from that model. (Johnson's experimental work came first. But theoreticians generate papers more readily and common usage, over the decades, has increasingly emphasized Nyquist.) The derivation, however, will be a good test of the model, and our understanding of it. We will concentrate on the simplest possible case where the sample acts as a purely elastic and quantum mechanically coherent source of scattering, and the irreversible events are limited to the reservoirs to which the sample is connected.

To calculate conductance we evaluate the current flow when a purely elastic scatterer is tied through perfect leads to reservoirs with differing electrochemical potentials.

We will evaluate the current fluctuations for the circuits of fig. 1. In that case we assume that the ends of the *resistor* are tied into far apart portions of the same reservoir, or else into two different reservoirs. In the latter case, we assume that the capacitance between the two reservoirs is large enough to be an ef-

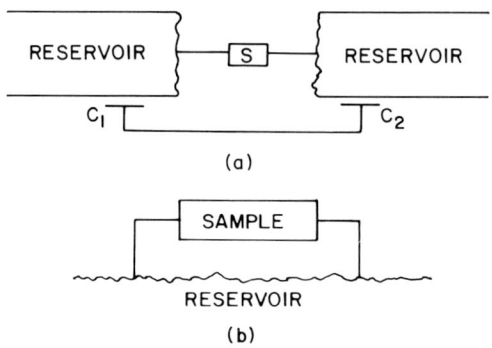

Fig. 1. (a) Sample S, with purely elastic scattering tied via perfectly conducting leads to two reservoirs. These, in turn, are connected via large capacitances, C_1 and C_2, constituting an effective short at frequencies of interest. (b) Two ends of the sample connected via perfectly conducting leads into far apart portions of the same reservoir.

fective short at the frequencies of concern. We will also assume that we are dealing with frequencies low enough to allow us to neglect all other reactances. That includes the inductance defined by the fact that any current flow path is associated with a magnetic field. The neglected reactances also include the capacitive effects which can carry alternating current from one part of a resistor to another part, or to the reservoir.

We confine our analysis to a one-dimensional model. This is, after all, a conceptual exercise and warrants only minimal complexity. The question discussed here was originally posed by M. Büttiker, and this author at first assumed that the path to the answer was trivial. This turned out not to be the case. To minimize minus signs, we will assume that our carriers have a positive charge e.

2. The Maxwell-Boltzmann case

The net current flow from left to right, if the sample is attached to two reservoirs which are not in equilibrium, is given by

$$j = e \int v(E) \frac{dn_R}{dE} [f_L(E) - f_R(E)] T(E) \, dE. \quad (1)$$

The integration is over energy and dn_R/dE is the density of states of electrons moving in one direction (e.g. to the right), but allowing for both spins. $T(E)$, with an explicit argument is a transmission probability, whereas T without argument will denote the temperature.

$f_L(E)$ and $f_R(E)$ are the occupation probabilities which apply way inside the two reservoirs to which the sample is attached. $T(E)$ is the energy-dependent transmission coefficient, and v the electron velocity in the ideal conductor leads connecting the sample to the reservoirs. A detailed discussion of the physics of reservoirs and the range of applicability of eq. (1) is given in ref. [1]. Much of the existing literature gives inadequate attention to such questions. In the usual fashion, $v(E) \, (dn_R/dE) = 1/\pi \hbar$, allowing for a spin degeneracy of two. We assume that spin is not a significant variable in this problem. Spins are preserved in the transmission processes and have no effect on the transmission probability. For Maxwell-Boltzmann statistics and a potential difference δV

$$f_L(E) - f_R(E) = \frac{e\delta V}{kT} \exp[-(E-E_F)/kT]. \quad (2)$$

This yields

$$G = \frac{e^2}{kT} \frac{1}{\pi \hbar} \int T(E) \exp[-(E-E_F)/kT] \, dE. \quad (3)$$

Now, let us turn to the noise. In the case of Maxwell-Boltzmann statistics the electrons are uncorrelated. The electron stream entering each reservoir exhibits shot noise. The two terms together constitute the thermal equilibrium noise. The electrons arriving at the right give a current

$$I_R = e \int v(E) \frac{dn_R}{dE} \exp[-(E-E_F)/kT] T(E) \, dE, \quad (4)$$

and similarly for I_L. The total current magnitude producing shot noise is

$$I_S = |I_R| + |I_L|$$
$$= 2e \int v(E) \frac{dn_R}{dE} T(E) \exp[-(E-E_F)/kT] \, dE. \quad (5)$$

Using $v(dn_R/dE) = 1/\pi\hbar$ this becomes

$$I_S = \frac{2e}{\pi\hbar} \int T(E) \exp[-(E-E_F)/kT] \, dE. \quad (6)$$

Using eq. (3) this becomes

$$I_S = \frac{2kT}{e} G. \quad (7)$$

Shot noise gives us a fluctuating noise current, j, in the frequency range Δf, of rms magnitude $2eI_S \Delta f$. Thus

$$\langle j^2 \rangle_{\Delta f} = 2e\left(\frac{2kT}{e} G\right) \Delta f = 4GkT \, \Delta f. \quad (8)$$

But that is the Nyquist [3] formula for the noise current through a short-circuited conductance of magnitude G.

3. The Fermi-Dirac case

In eq. (4) we integrated over all occupied levels. They contribute independently in proportion to their occupation. The Fermi-Dirac case will be different. Fully occupied levels, well below the Fermi-level, will not contribute to the noise. In the Fermi-Dirac case eq. (1) still applies. Now, however, instead of eq. (2) we find

$$f_L(E) - f_R(E) = -e\delta V(\partial f/\partial E) = \frac{e\delta V}{kT} f(1-f), \quad (9)$$

and eq. (1) yields

$$G = \frac{1}{\pi\hbar} \frac{e^2}{kT} \int f(1-f) T(E) \, \mathrm{d}E \, . \qquad (10)$$

Now for the noise. If f is not small compared to 1, we no longer have shot noise. Indeed, for $f=1$ the electron stream is regular, there is no noise. Thus, noise cannot increase monotonically with f. Consider a very narrow energy range ΔE. Make it small enough so that $T(E)$ is effectively constant within ΔE. Consider successive orthonormal wave packets following each other, in space or time, all limited to ΔE. These states do not have sharp ends and may resemble Wannier functions. A detailed discussion of such states has been provided by Stevens [4]. The exact nature of our states does not really matter, as long as successive states develop the same way in time, as they propagate.

The fact that ΔE is made small enough so that $T(E)$ is essentially constant over this range implies that our wave packets are long in duration compared to any kinetic time associated with traversal of the sample [5]. Now the Pauli principle can couple the transmission of packets incident from the left to the reflection of packets incident from the right; after all, they are filling the same departing wave packet states. To simplify this problem we will assume that the wave packets incident from the left and those incident from the right are correlated in their timing; a wave packet transmitted from the left need be related to the occupation of only one wave packet reflected from the right.

The probability that both arriving wave packets are occupied is f^2, and in that case no net charge transport through the sample occurs. The probability that both arriving wave packets are empty is $(1-f)^2$, and again no net charge transport occurs. The probability that only the wave packet arriving from the left is present is $f(1-f)$. If the transmission probability is T, then the probability that an electron has been transferred from left to right is $Tf(1-f)$. Similarly the probability that an electron has been transferred from right to left is $Tf(1-f)$. This situation is illustrated, somewhat symbolically, in fig. 2, showing a

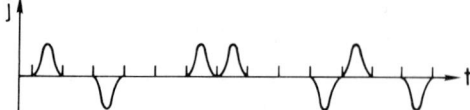

Fig. 2. Symbolic sequence of current pulses, due to electrons transferred through the sample.

sequence of current pulses. For each pulse the current flow, $j(t)$, obeys $\int |j| \, \mathrm{d}t = e$. The rate at which wave packets arrive, say from the left, is $\Delta E/\pi\hbar$. Part of the *symbolism* in fig. 2 consists in the depiction of non-overlapping current pulses. If our decomposition into energy ranges really utilizes energy ranges with exact and sharp boundaries, then the current pulses will overlap. Overlap will not, however, affect our subsequent considerations.

What is the noise contribution from the train of pulses shown in fig. 2? The mean squared current in the frequency range Δf is given by [6]

$$\langle j^2 \rangle_{\Delta f} = \lim_{\Theta \to \infty} \frac{2|S(f)|^2}{\Theta} \Delta f, \qquad (11)$$

where Θ is a time interval, and

$$S(f) = \int_0^{\Theta} j(t) \exp(-\mathrm{i}\omega t) \, \mathrm{d}t \, . \qquad (12)$$

Let each of the pulses in fig. 2 have the form $j_0(t-n\tau)p_n$, where $p_n = \pm 1$ for the time slots with a current pulse, and zero otherwise. As already stated $\tau = \pi\hbar/\Delta E$. $j_0(u)$ will be taken to be a positive pulse. Then

$$S(f) = \int_0^{\Theta} \exp(-\mathrm{i}\omega t) \sum_n j_0(t-n\tau) p_n$$

$$= \sum_n \exp(-\mathrm{i}\omega n\tau) p_n \int_0^{\Theta} \exp(-\mathrm{i}\omega u) j_0(u) \, \mathrm{d}u \, . \qquad (13)$$

The summation over n extends over Θ/τ time slots. Consider the integral shown in the final right-hand-side term of eq. (13). We will limit our concern to frequencies such that a period is long compared to a current pulse. That, in turn, will require this period to be long compared to any kinetic time associated with the transit through the sample. (At higher frequencies we can no longer expect white noise, nor can we expect the sample to be purely resistive.) Then $\exp(-i\omega u)$ is essentially unity during the pulse and the final right-hand side integral in eq. (13) reduces to the electron charge e. Therefore

$$S(f) = \sum_n \exp(-i\omega n\tau) p_n e, \qquad (14)$$

and

$$|S(f)|^2 = \sum_m \exp(-i\omega n\tau) \exp(i\omega n'\tau) p_n p_{n'} e^2. \qquad (15)$$

Now $\langle p_n \rangle = 0$. Furthermore p_n and $p_{n'}$ are uncorrelated, unless $n = n'$. Thus, eq. (15), after averaging over the ensemble containing all the allowable sequences for p_n, yields

$$|S(f)|^2 = \sum_n \langle p_n^2 \rangle e^2. \qquad (16)$$

p_n is ± 1 for a fraction $2f(1-f)T(E)$ of the time slots, and vanishes otherwise. Therefore

$$|S(f)|^2 = 2Ne^2 f(1-f) T(E), \qquad (17)$$

where N is the number of time slots, equal to Θ/τ, or $\Theta \Delta E / \pi \hbar$. Thus, invoking eq. (11)

$$\frac{2|S(f)|^2}{\Theta} = \frac{4}{\pi \hbar} e^2 f(1-f) T(E) \Delta E. \qquad (18)$$

This is the mean squared current contribution, in the frequency range Δf, from the energy range ΔE. The total noise current is obtained by integrating over energy, yielding

$$\langle j^2 \rangle_{\Delta f} = \frac{4\Delta f}{\pi \hbar} e^2 \int f(1-f) T(E) \, dE. \qquad (19)$$

Comparing this to eq. (10) we once again obtain

$$\langle j^2 \rangle_{\Delta f} = 4kTG \, \Delta f; \qquad (20)$$

our desired result.

4. Conclusion

Our verification of the usual thermal equilibrium noise equations is neither unexpected, nor particularly significant. There still are, however, some who are uncomfortable with a viewpoint which separates the elastic scattering responsible for the size of the resistance from the irreversible events which are necessary for the occurrence of a dissipative process. These reservations continue despite the very successful extensions and refinements of the viewpoint, in recent years [7,8]. It can be hoped that this additional demonstration may help the remaining skeptics.

My debt to M. Büttiker for posing this question has already been stated. Interaction with Y. Gefen led to the present formulation of section 2, replacing a more awkward earlier approach.

In a collection dedicated to Benoit Mandelbrot, a word about the connection to his work seems in order. Fluctuations, noise and Brownian motion are, of course, an integral part of his domain. Beyond that, however, the connection ceases. This note represents the kind of concern with the detailed kinetics which Mandelbrot long ago realized was not essential for the classification and description of noise. Too many of us, including this author, foolishly underestimated the significance of Mandelbrot's work, because he understood that one could get away from the grubby details.

References

[1] R. Landauer, Conductance determined by transmission: probes and quantized constriction resistance, J. Phys. C, to be published; in: Nanostructure Physics and Fabrication, W.P. Kirk and M. Reed, eds. (Academic Press, New York), to be published.
[2] J.B. Johnson, Phys. Rev. 29 (1927) 367.
[3] H. Nyquist, Phys. Rev. 32 (1928) 110.
[4] K.W.H. Stevens, J. Phys. C 20 (1987) 5791.
[5] M. Büttiker and R. Landauer, IBM J. Res. Dev. 30 (1986) 451.
[6] S.O. Rice, Bell Syst. Techn. J 23 (1944) 282; see also reprint, in: Noise and Stochastic Processes, N. Wax, ed. (Dover, New York, 1954), p. 133.
[7] Y. Imry, in: Directions in Condensed Matter Physics, G. Grinstein and G. Mazenko, eds. (World Scientific, Singapore, 1986), p. 101.
[8] M. Büttiker, IBM J. Res. Dev. 32 (1988) 317.

APPLICATIONS OF FRACTAL CONCEPTS IN PETROLEUM ENGINEERING

Roland LENORMAND

Institut Français du Pétrole, BP 311, 92506 Rueil Malmaison Cedex, France

The different models of fluid transport in porous media are summarized and their applications in oil reservoir simulation are analysed. It is shown that the problems for application of these models and of fractal concepts are of three types: no validity of the models due to viscous forces always present in both fluids at large scale, difficulty to use fractal geometry as a model of reservoir and finally the need for analytical or differential equations.

1. Introduction

Ask a scientific data bank for a list of papers with keywords "fractals" and "porous media", you obtain several hundred answers, and a large fraction have been published by research laboratories of oil companies over the last decade. Now, ask a reservoir engineer in any of these companies how many programs use the concept of fractal, generally he would answer that he does not know what you mean. Having worked for several years at the frontier between basic and applied research in oil industry, I am always puzzled by the huge gap between theory and applications in this domain.

The purpose of this paper is to try to analyse this problem at the scale of an oil reservoir. However, fractal geometry has been used at a microscopic scale. Even if the fractal nature of the internal surface of the pores has been experimentally demonstrated [1,2], the relation between permeability and fractal dimension has never been clearly established [3,4]. This fractal geometry should be very useful for describing the flow of a wetting fluid by capillary effects along the roughness of the grains when the bulk of the pores is filled by a nonwetting fluid (see de Gennes [5]). Such a displacement can occur during injection of immiscible gas and oil recovery by gravity forces.

I will now consider fluid transport properties in a porous medium above the pore size (from mm to km). The different statistical approaches will be summarized and their applicability in reservoir engineering will be discussed. Finally, I will try to present the needs in petroleum engineering concerning transport properties in porous media.

2. Modeling flow through porous media

The complexity of displacements of one fluid by another is due to the superposition of phenomena at pore scale (capillary effects, mixing of flow lines, molecular diffusion in dead zones...) and the effects of heterogeneities at large scale.

There are mainly two types of approaches. The first is a "physical" model: the medium is replaced by a more simple geometry like networks of capillaries or lattices of blocks of various permeabilities. Displacement is calculated step by step using the physical laws related to capillarity, gravity and viscosity. All the randomness and the correlations of the real rock are contained in the geometry of the medium.

The "statistical" approach uses the analogy between the fluid patterns (non-miscible) or profiles (miscibles) obtained during injection and the results of statistical models such as diffusion-limited aggregation (DLA), invasion percolation (IP), random walks (RW), ... In this approach, the randomness is introduced in the growth mechanism and different "rules" are developed to reproduce the observed mechanisms.

The equivalence between the two approaches is still a problem, but we will discuss the applications of these models rather than the theoretical background.

2.1. Physical models

The geometrical support is two-dimensional with axes X and Y. It represents, for instance, a square network of capillaries with various radii or a matrix of porous elements. In both cases, the local permeability K is a function of X and Y, and the spatial distribution can take different forms:

2.1.1. Isotropic medium ($X = Y$)

Constant K: the permeability of the medium is uniform.

Random K (random "noise"): the permeability follows a given probability law (Gaussian distribution, for example) and is distributed at random locations (no correlation).

"Fractional noise": this case is a generalization of the random noise. The main difference is the presence of correlation at all length scales, depending on a parameter H ($H = 1/2$ corresponds to random noise). Application of this type of correlation for the description of petroleum reservoir has been proposed by Hewett [6]. A more complete study of fractional noise has been published by Mandelbrot [7] and a summary can be found in Feder's book on fractals [8].

The construction of the permeability field uses the porosity logs recorded in two wells and the technique of successive random addition for interpolation (Voss [9]). The resemblance with geological structures is striking. However, in Hewett's paper the way the horizontal correlation is obtained is not clear to me.

Fractal K: permeability takes only two values, 0 and 1, and the points with value 1 are located on a fractal network. It can be a percolation cluster or a regular geometrical fractal (Sierpiński...). Redner et al. [10] have proposed a self-similar hierarchical model without dead ends suitable for analytical calculation of convective flow.

Multifractal K: An application of multifractal to represent the structure of a porous medium has been proposed by Meakin [11]. The geometrical construction of the multifractal lattice is the following. It starts with a square divided in four with four values K_1, K_2, ... of the permeability (or probability). Then, each square is divided in four and a new permeability is calculated for each subsquare. In the first large square, permeabilities are $K_1 K_i$, where i is chosen at random between 1 and 4. Then each of the new squares is divided and so on. The result is a random field of permeabilities with correlation at all scales, the value at a given point keeping the "memory" of all the previous squares of different sizes.

2.1.2. Anisotropic medium ($X \neq Y$)

Anisotropy of a geological structure can be represented by using a different model for X and Y. For instance "constant K" in X and "random K" in Y lead to a random layered structure.

2.1.3. Modeling fluid displacement

The principle is to solve "standard" flow equations in the heterogeneous permeability field, taking into account capillary (if the fluids are immiscible), viscosity and gravity effects. When the geometrical model represents the pores of the medium, the simulator is called a network simulator [12]. At large scale, this is a reservoir simulator [13]. In both cases, the pressure field and the flow rates are calculated at each node. The principle is analogous to the calculation of electrical potentials in a network of resistors. However, the "hydraulic resistances" are functions of the nature of the fluids and consequently, the simulation has to be performed step by step, the new values of resistances being calculated at each step.

2.2. Statistical models

Instead of modeling the medium, the statistical approach is concerned with the displacement itself. As a consequence, a different statistical model corresponds to each type of displacement. Generally, the medium is assumed to be homogeneous but, some heterogeneities can be introduced via a space-dependent probability (this is the case for the multifractal DLA simulation studied by Meaking [11]).

2.2.1. Diffusion-limited aggregation (DLA)

This model is used for the study of viscous fingering when a zero-viscosity fluid pushes a high-viscosity fluid without any capillary effects (Paterson [14]). It consists in releasing particles which move at random on a grid and stick to a seed (at the beginning) or on the aggregate. A reverse model (anti-DLA) can also describe the injection of a high-viscosity fluid.

DLA has also been used to describe the flow of a reactive fluid, such as the dissolution of a carbonated rock by HCl solution (Daccord and Lenormand [15]).

2.2.2. Invasion percolation (IP)

This approach is related to capillary mechanisms which take place at the microscopic (pore) scale. IP is suitable for the description of immiscible displacements performed at very low flow rate (Wilkinson [16]): capillary forces prevent the non-wetting fluid from spontaneously entering a pore, and when a pressure is applied, the injected fluid invades the subnetwork formed by the largest throats, which is, in fact the percolation cluster.

2.2.3. Random walks

The domain of application is the transport of tracers in porous media: only one fluid is flowing and at a given time, a pulse or a step of tracer is injected in the fluid without modification of flow conditions. In a homogeneous medium, a pulse of tracer is dispersed and the concentration follows a Gaussian distribution with a variance proportional to the time. The spreading or "dispersion" of tracer is described by the standard diffusion equation, which gives the probability in a random walk.

The random walk approach is often used to study the effect of correlations, for instance when two adjacent steps are correlated (Scheidegger [17]). Weatcraft and Tyler [18] have used a "fractal random walk" model in order to explain time-dependent dispersivity in heterogeneous media. A review of statistical models of dispersion is given in this paper and an analytical calculation of the dispersion coefficient is proposed by using fractal streamtubes (obtained by adding randomness in a Koch curve). In this domain, continuous time random walk and Lévy walk [19] should lead to more rigourous results.

3. Discussion

All these different models are in good agreement with small-scale experiments but generally cannot be used at large scale (reservoir simulation).

Lenormand et al. [12] have shown that both physical models used at pore scale (network simulators) and statistical models (DLA and percolation) agree with laboratory experiments. For instance, in the case of viscous fingering, fig. 1 shows the comparison between experiments in micromodels, network simulations and DLA. Experimental work using 2D glass-bead packs (Måløy et al. [20]) or micromodels (Chen et al. [21]) have shown the same good agreement, even when the displacement is performed in a fractal geometry (Oxall et al. [22]).

However, these models are difficult to use at large scale, for the following reason:

(1) *Cross-over regimes*

In the case of immiscible fluids, it has been shown that statistical models are valid when one type of forces is dominant, for instance capillary forces for invasion percolation, viscous forces in the displaced fluid for DLA. However, for real fluids with finite viscosities, viscous forces are proportional to the length of the injected pattern, and they increase with the size of the sample. We have proposed (Lenormand [23]) a phase diagram which gives the limit of validity of these models as functions of fluids properties and flow rate. For instance, for DLA, with a viscosity ratio of 10^{-5}, the limit size of the network

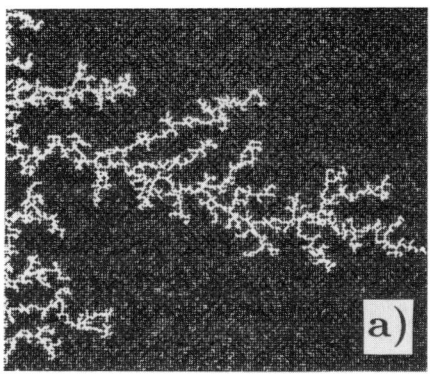
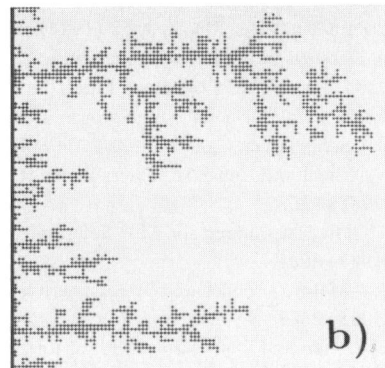
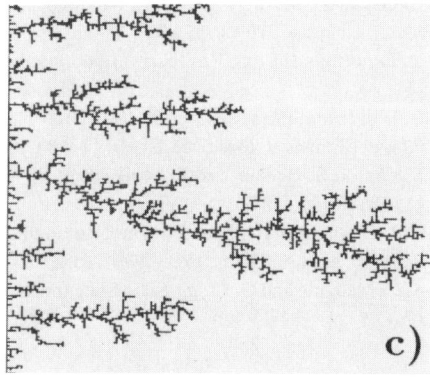

Fig. 1. Viscous fingering: (a) experiment in micromodels, (b) network simulation and (c) statistical model (DLA).

is of the order of 1000×1000 pores. A similar analysis has been done by Wilkinson [24] for the effect of gravity on invasion percolation. This result explains why statistical models cannot be used at the scale of an oil reservoir.

(2) *Fractal geometry*

For the transport of tracers, anomalous dispersion which is observed at large scale (variance of the effluent concentration proportional to t^γ) can be modeled by a fractal medium, such as a percolation cluster (de Gennes [25], Koplik et al. [26]). A reservoir engineer would be very reluctant to use a percolation cluster as a geometry for simulations, even, if he understands that the presence of heterogeneities at all scales in a reservoir is related to a fractal geometry.

(3) *Need for "microscopic equations"*

Reservoir simulators are based on convection-diffusion equations. Even in the case of multiphase flow, the formalism of relative permeabilities and capillary pressure curves lead to this type of equation. Consequently, even if a fractal approach is valid, it has to be written in a suitable form, for instance as a concentration of one fluid function of the space coordinates and time and taking into account the physical parameters of the problem.

So far, only one case of fractal displacement has been described by analytical equations. It is the case of the flow of a reactive fluid which dissolves the porous matrix [15]. Due to the very high contrast between the permeability of the dissolved channels and the matrix, a DLA-type approach is valid, even at large scale (a few meters). It has been possible to calculate the evolution of microscopic variables (pressure, invasion of acid, ...) by introducing the fractal dimension of DLA. This model has now been tested in the field and is being used for the design of acidizing processes.

4. Conclusion

So far, porous media have been a rich source for theoretical studies but we have seen that the existing results cannot be used directly for applications in petroleum engineering (and also in hydrology or chemical engineering). I think that fractal geometry is the best way for taking into account heterogeneities at all scales but further studies should be oriented in the two following directions:

(1) Improvement of geometrical models, in order to account for anisotropy and correlation at all scales. With a few tunable parameters, this model should be

able to reproduce simple geometries (random, layered, ...) as limiting cases.

(2) Improvement of microscopic equations either by introducing time-dependent parameters or developing new types of equations. For instance, if we consider the random walk approach, it has been shown that the correlation change the order of the derivative, second order in time with short-range correlation (Scheidegger [17]), or m order in time with CTRW (Klafter et al. [19]). In addition, definitions of fractional random walk and "fractal" or non-integer derivatives [27] are very similar. This kind of equations have been proposed by Le Méhauté and Crépy [28] a few years ago in the case of transport through a fractal interface, and should be investigated for the transport of a fluid through heterogeneous media.

References

[1] D. Farin and D. Avnir, in: Characterization of Porous Solids, K.K. Unger, D. Behrens and H. Kral, eds. (Elsevier, Amsterdam, 1988), and references therein.
[2] A.J. Katz and A.H. Thompson, Phys. Rev. Lett. 54 (1985) 1325.
[3] J.P. Hansen and A.T. Skjeltorp, Phys. Rev. B 38 (1988) 2635.
[4] C.G. Jacquin and P.M. Adler, Transport Porous Media 2 (1987) 571.
[5] P.G. de Gennes, in: Physics of Disordered Materials, D. Adler, H. Fritzsche and S.R. Ovshinsky, eds. (Plenum, New York, 1985).
[6] T.A. Hewett, paper SPE 15386 presented at the 61st Conference of the Society of Petroleum Engineers, New Orleans (1986).
[7] B.B. Mandelbrot and J.R. Wallis, Water Resource Res. 5 (1969) part 1–3, p. 228.
[8] J. Feder, Fractals (Plenum, New York, 1988).
[9] R.F. Voss, in: Fundamental Algorithms for Computer Graphics, R.A. Earnshaw, ed. (Springer, Berlin, 1985) p. 805.
[10] S. Redner, J. Koplik and D. Wilkinson, J. Phys. A 20 (1987) 1543.
[11] P. Meakin, Phys. Rev. A 36 (1987) 2833.
[12] R. Lenormand, E. Touboul and C. Zarcone, J. Fluid Mech. 189 (1988) 165.
[13] M.A. Christie, paper SPE 16005 presented at 1987 Symposium Reservoir Simulation of the Society of Petroleum Engineers.
[14] L. Paterson, Phys. Rev. Lett. 52 (1984) 1621.
[15] G. Daccord and R. Lenormand, Nature 325 (1987) 41.
[16] D. Wilkinson, in: Physics of Finely Divided Matter, N. Boccara and M. Daoud, eds. (Springer, Berlin, 1985) p. 280.
[17] A.E. Scheidegger, Can. J. Fluids 36 (1958) 649.
[18] W. Wheatcraft and S.W. Tyler, Water Resources Res. 24 (1988) 566.
[19] J. Klafter, A. Blumen and M.F. Shlesinger, Phys. Rev. A 35 (1987) 3081.
[20] K.J. Måløy, J. Feder and T. Jøssang, Phys. Rev. Lett. 55 (1985) 2688.
[21] J.D. Chen and D. Wilkinson, Phys. Rev. Lett. 55 (1985) 1892.
[22] U. Oxaal, M. Murat, F. Boger, A. Aharony, J. Feder and T. Jøssang, Nature 329 (1987) 32.
[23] R. Lenormand, Proc. Roy. Soc. (London) A 423 (1989) 159.
[24] D. Wilkinson, Phys. Rev. A 30 (1984) 520.
[25] P.G. de Gennes, J. Fluid Mech. 136 (1983) 189.
[26] J. Koplik, S. Redner and D. Wilkinson, Phys. Rev. A 37 (1988) 2619.
[27] K.B. Oldham and J. Spanier, The Fractional Calculus, R. Bellman, ed. (Academic Press, New York, 1970).
[28] A. Le Méhauté and G. Crepy, Solid State Ionics 9/10 (1983) 17.

FRACTURE AS A GROWTH PROCESS

E. LOUIS
Departamento de Fisica Aplicada, Universidad de Alicante, Apartado 99, E-03080 Alicante, Spain

F. GUINEA
Instituto de Ciencia de Materiales (CSIC), Universidad Autónoma, Cantoblanco, E-28049 Madrid, Spain

We discuss some aspects of models developed to investigate the eventual fractal geometry of growing cracks in solids. The emphasis is placed upon two questions: (i) relationship with other growth processes, in particular those occurring in scalar fields, and, (ii) practical implications. As regards the first point, the similarities and differences between the different growing patterns are stressed; for instance, it is shown that an analysis of field singularities at the tips leads to fractal dimensions which are always lower for mechanical breakdown than for Laplacian fractals. Recent efforts carried out to connect models with actual cracking of materials, which exhibit a rich phenomenology, are commented on. Finally, several implications of the fractal nature of fracture surfaces, having practical significance, are stressed.

1. Introduction

The concept of fractal [1] is being greatly useful in identifying a hidden symmetry in a wide variety of objects and phenomena in nature. In particular, many growth processes far from equilibrium show well-defined fractal structures [2,3]. The most outstanding examples of growing structures are diffusion-limited aggregation (DLA) [4] and dielectric breakdown (DB) [5]. More recently, crack propagation has also been considered from this point of view [6,7]. Fracture is one of the most intriguing phenomena in materials science [8], and has a great technological importance. Many factors play a relevant role: grain boundaries, microcracks, particles and impurities, temperature, ... [9], leading to a rich phenomenology ranging from cleavage to ductile fracture. The first suggestion regarding the eventual fractal character of fracture surfaces in materials is due to Mandelbrot et al. [10]. Since then, the efforts made to understand fracture as a growth process [6,7,11–17] and identify the practical implications of its fractal nature [18–21], have been continuously increasing.

The aim of this paper is to discuss several aspects of the models developed to simulate crack propagation in solids and investigate its fractal character. Many features of the simulations seem to play a key role, among which we mention, the description of the elastic media, the boundary conditions and the growth law. Models introduced in this field do in fact show important differences in those three aspects. We shall attempt to stress both basic questions of interest to the general field of pattern growth and fractal geometry, and points of practical significance in the field of fracture, a major branch of materials science.

2. Models and numerical simulations

We will consider a single crack, growing in an otherwise defect-free medium. This amounts to assuming that the kinetics of growth overcomes nucleation kinetics; as this might not always be the case in real materials, it should be considered as a first step. We shall further assume that crack propagation is slow enough as to allow the stresses to relax, in such a way that the material is always at equilibrium; again this assumption is not of general applicability. We shall

Essays in honour of Benoit B. Mandelbrot
Fractals in Physics – A. Aharony and J. Feder (editors)

consider isotropic two-dimensional systems characterized by two Lamé coefficients λ and μ; some comments on anisotropy and plasticity will also be made. Thus, the equilibrium equations which relate the stress and strain fields are the Lamé equations [22]

$$(\lambda+\mu)\delta_i\left(\sum_j \delta_j u_j\right)+\mu\left(\sum_j \delta_j^2\right)u_i=0, \quad (1)$$

where u_i is the ith component of the displacement field. Here we already note an important difference with respect to DLA or DB, namely, the vectorial nature of the displacement field as compared to the scalar field of Laplacian fractals. Furthermore, (1) cannot be reduced to the Laplace equation but in the unphysical limit of a null bulk modulus. However, a general characteristic of the solutions of the Lamé equations, is that they can be written in terms of functions which satisfy the biharmonic equations [22], suggesting a formal connection with DLA or DB.

Solving the Lamé equations requires in most cases to discretize them following a given prescription, adequate for the actual physical problem. The specific way in which this has been done in most crack growth studies carried out up to now, is either by taking a particular lattice and defining an interaction between the nodes [6,7,11–16] or by using the so-called beam model [17]. Although we shall be mainly concerned with the first approach, we will also make some comments on the latter. In choosing a particular lattice, it should be noted that some 2D lattices (square and honeycomb) show a pathological feature, which consists in having a null shear modulus when the interaction between nodes is described by means of the central-force Hamiltonian. This has driven most authors to use the triangular lattice. On the other hand, by taking this lattice as a basis, a wide variety of elastic media can be described by using different Hamiltonians. A Hamiltonian that accounts for both central and angular forces, and which in the triangular lattice describes an isotropic medium, is

$$H=\tfrac{1}{2}\sum_{\langle ij\rangle}\alpha_{ij}\,[\hat{r}_{ij}(\boldsymbol{u}_i-\boldsymbol{u}_j)]^2$$
$$+\tfrac{1}{2}\sum_{\langle ijk\rangle}\beta_{ijk}[\hat{r}_{ik}(\boldsymbol{u}_i-\boldsymbol{u}_j)+\hat{r}_{ij}(\boldsymbol{u}_i-\boldsymbol{u}_k)]^2, \quad (2)$$

where i, j and k are nearest-neighbor nodes, $\hat{r}_{ij}$ is the unit vector in the direction i–j, α_{ij} is the bond-stretching force constant and β_{ijk} the bond-bending force constant of the angle $\angle ijk$. If α_{ij} and β_{ijk} are taken equal, for all bonds and angles, to α and β respectively, the continuum limit of this Hamiltonian gives $\lambda/\mu=(\alpha-4\beta)/(\alpha+8\beta)$ [29]. An alternative way of varying λ/μ is based on the central-force part of (2), and consists in choosing a unit cell with two different force constants; a reconstruction $\sqrt{3}\times\sqrt{3}$, has been found to be particularly convenient [7]. Similarly, anisotropic media can be described by combining the central-force Hamiltonian with a triangular lattice in which the spring constants of one of the directions is taken different from those of the other two [12].

In order to allow a crack to grow, we need to generate a finite distribution of stresses within the medium. This introduces a new variable in the study of growing cracks, namely, the boundary conditions. Several can be considered and their role in determining the shape of the cracks might be crucial. Among the most common we mention shear deformation and uniform dilation [7,15], and uniaxial deformation [16]. On the other hand, these boundary conditions can be implemented by fixing, at the sample surface, either the displacements or the forces. In this context it is also interesting to comment on the effect of the size of the samples used to simulate the growth of cracks. We note that the logarithmic nature of the Green's function (Green's tensor in the case of elasticity) suggests that great care has to be devoted to this question. To illustrate this point we have calculated the transversal stresses at the surface of a circular crack propagating in a circular sample (circular ring), under uniform deformation. This stress depends on the type of boundary condition; when the displacements at the surface are fixed, the result is [24]

$$\sigma_{\theta\theta}(r_0)=\frac{4\mu(\lambda+\mu)c}{r_1}\,[\mu+(\lambda+\mu)(r_0^2/r_1^2)]^{-1}, \quad (3)$$

where r_0 and r_1 are the radii of the inner and outer circles respectively, and $c=u_r(r_1)$. Thus, the tangential tension monotonically decreases with the size of the crack. Instead, when the radial stress is fixed at the surface, $\sigma_{rr}(r_1)=p$, the result is

$$\sigma_{\theta\theta}(r_0)=2p(1-r_0^2/r_1^2)^{-1}. \quad (4)$$

In this case $\sigma_{\theta\theta}$ increases with r_0. This strong dependence of the transversal stresses at the crack surface on the type of boundary condition could be relevant in determining the shape of cracks, particularly when a mixed growth law such as that of eq. (5) below is used. Nonetheless, for sharp cracks we expect a continuous increase of the stresses at the tips, at least when they are far enough from the sample surface. In fig. 1 we have plotted the maximum stress at the surface of the crack for the case of fixed displacements. The stress first increases, as expected for infinitely large samples, and thereafter, as the boundary effects predominate, it decreases as predicted by (3). To reduce the effects of the boundaries, one may take a boundary moving at the pace of the crack [7], as is also illustrated in fig. 1. It should be stressed that the role of these size effects on the growth of the cracks is not yet clear.

We turn now to discuss the growth law. The most simple law which can be chosen is the one equivalent to that already utilized in DB simulations [5]. In the present case this amounts to assume that the growth speed v_n is proportional to a power of the tangential tensions (τ_i) at the surface; note that no stresses are propagated transversal to the crack surface. When this power is equal to unity, the DLA law is rescued [4]. However, several important cracking processes in nature seem to obey more complicated rules. To cover such cases it has been recently suggested [24] to use a mixed law as follows

$$v_n \propto \tau_t + q_1 \tau_t^{\eta_1}. \quad (5)$$

The effects of varying q_1 and η_1 [24] will be discussed below. The implementation of the above growth law within the central-force approximation is rather straightforward, as one identifies the tangential tension with the absolute value of the stresses in each bond. Instead, it is not so evident if angular forces are also included. In this case the elastic energy due to the angular term should be in some way distributed among the bonds forming the angle; it may also be assumed that only central forces enter into the growth law [23].

More recently, and within the framework of the beam model [17], it has been suggested to use a growth law based on the von Mises yielding criterion [8]; in this case the growth speed is of the form

$$v_n \propto [(\delta_t u_t)^2 + q_2 \delta_t^2 u_n]^{\eta_2}. \quad (6)$$

It has been claimed [17] that different rupture criteria based on (6), may describe failure processes ranging from brittle fracture to slow stress corrosion.

The simplest growth law mentioned above, that is, the growth speed proportional to the tangential tension, allows an analytical study of the characteristics of the patterns in terms of field singularities, as already done by Turkevich and Scher [25] for DB. In the case of a sharp angular notch of included angle ϕ, the components of the stress tensor which give the tangential forces ($\sigma_{\theta\theta}$ and $\sigma_{\theta r}$) vary, as a function of

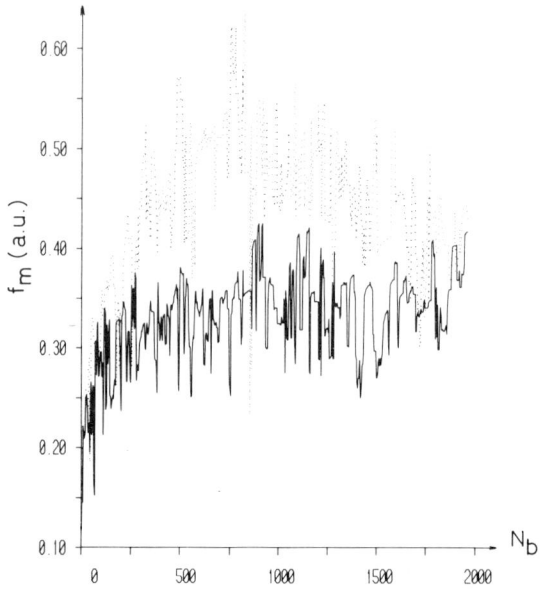

Fig. 1. Maximum force (f_m) at the crack surface versus crack size (number of broken bonds, N_b) for uniform dilation and fixed displacements at the boundary. Fixed boundary (dotted line) and moving boundary (continuous line).

the distance to the tip r, as [26] $r^{\epsilon-1}$, where ϵ is the solution of the equation

$$\sin[\epsilon(2\pi-\phi)] = -\epsilon\sin(2\pi-\phi). \quad (7)$$

The fractal dimension of the growing pattern is then predicted to be $D=1+\epsilon$. In table 1 we report the values of D obtained through this analysis for selected values of ϕ; fractal dimensions in the case of dielectric breakdown [25] are also given. It is noted that the predicted fractal dimensions for crack growth are always lower than those for DB, in agreement with simulations [15]. Moreover, D is far more stringently bounded, $1.5 < D < 1.62$, than in the case of DB.

All these ingredients can be used to generate a growth process in a similar manner to that followed in DLA or DB. The crack is initiated by making the force constants of a given bond, of the lattice, $i=1$ and $j=m$, equal to zero, that is, α_{lm} and $\beta_{lmk} = 0$, for all nearest-neighbors k. Then, the nodes of the lattice are displaced to achieve equilibrium; the Lamé equations have been solved using standard relaxation methods [7,15–17]. The chosen growth law is used to decide how the crack propagates. This can be done either in a deterministic way [17], or stochastically; in both cases the breaking criterium is defined in terms of the growth speed (see above). Here we comment on a technical point that has recently been of some concern; we refer to the definition of the surface of the crack, more specifically, to the bonds adjacent to the broken ones, to be considered as candidates for breaking [15,16]. It has been found that the fractal dimension increases with the width of the surface [15]. This is partly due to the boundary conditions required to produce the finite distribution of stresses required for crack propagation; as a consequence, the stresses on the bonds do not vanish far from the surface of the crack, but rather they tend to a constant (whose absolute value depends on the boundary condition).

3. Characteristics of the growing cracks

A typical crack grown by means of the models described previously is shown in fig. 2. The pattern clearly shows the tip slitting characteristic of fractal growth, and the screening effects, also present in DLA and DB [2], which are a consequence of the accumulation of stresses at the tips of the crack. The most extensive simulations carried out up to now [15] give a fractal dimension (D) for patterns such as that of fig. 2, of 1.55, whereas in the case of shear it raises up to 1.60; both values are lower than those obtained in similar DLA [4] or DB [5] simulations. These results correspond to a model in which the ten next-nearest neighbors of the already broken bonds were considered as candidates for breaking [15].

Patterns generated within the central-force approximation show a remarkable independence of the elastic constants [7]. Instead, it has recently been found [23] that angular forces sharply increase D; however, this is more likely a consequence of the topology of the Hamiltonian than of the change in the elastic constants produced by the angular force term. For anisotropic media it has been found that the patterns become elongated and loose their self-similarity [12]; in this case, their characterization may require the use of the concept of self-affinity [1]. Similar results have been obtained in DLA simulations which assumed an anisotropic sticking coefficient [28]. Anisotropic patterns can also be generated by averaging out fluctuations; reduction of noise allowed producing cracks showing a regular anisotropic structure [13], as already found in DLA [2].

The effects of plasticity on the growing cracks have also been investigated [14]. The way in which this has been done consists in replacing the broken bonds by a constant force, which is a fixed fraction (γ) of the stress they stored at the moment of failure; this is

Table 1
Values of the fractal dimension for different characteristic angles ϕ taken as representative of different lattices (see text). MB stands for mechanical breakdown

	$\phi=0$	$\phi=\pi/3$	$\phi=\pi/2$	$\phi=2\pi/3$
DB	1.5	1.6	1.67	1.75
MB	1.5	1.51	1.55	1.62

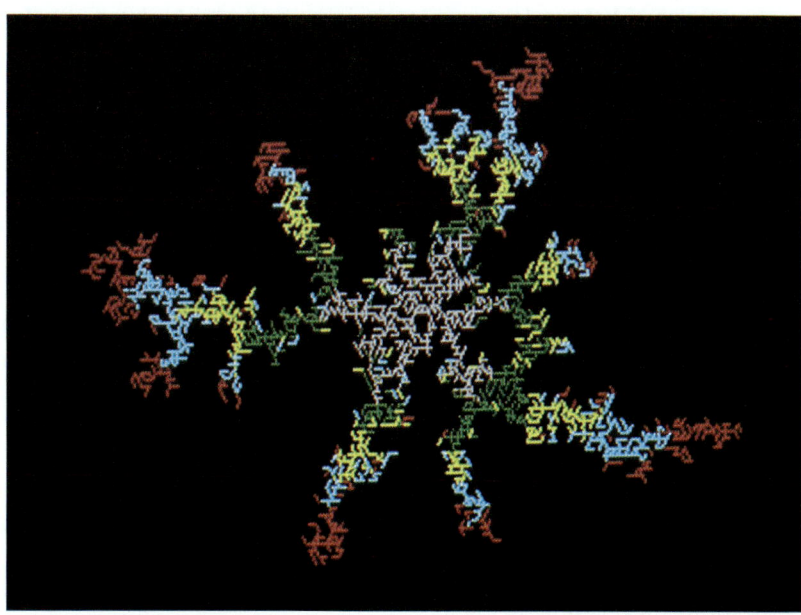

Fig. 2. Typical crack (broken bonds) generated by means of the models described in the text and under the following specific conditions: uniform dilation (fixed displacements), central-force Hamiltonian in the unreconstructed triangular lattice ($\lambda=\mu$), and $q_1=0$ and $\eta_1=0$ in (5).

a convenient description of an elasto-plastic solid. For $\gamma=0$, the previous case was reduced [7], whereas for $\gamma=1$ the patterns turned out to be Eden clusters; in between, the patterns had dimensions in the range 1.6–2.0. An interpretation of these results can be found in the continuum limit. In the present case the failed region does not completely loose its ability to sustain stresses, and it exerts forces on the unfailed elastic part. This physical situation can be described by assuming that the rate of change of forces across the surface is a constant fraction of the force itself; in terms of the stress tensor this can be expressed as

$$\frac{\delta \ln(\sigma_{ni})}{\delta n} = \frac{1-\gamma}{\gamma}, \quad i=\text{n, t}. \tag{8}$$

When $\gamma=0$, (8) can only be satisfied by $\sigma_{ni}=0$, i.e. the boundary condition at the crack surface in an elastic medium. In the opposite limit, $\gamma=1$, (8) is similar to that for a volume element in an elastic medium [22]; in particular the initial stress distribution is always a solution, giving a uniform growth probability which leads to the Eden model.

Recently, some efforts have been addressed to developing models which may account for the rich phenomenology of failure processes. Cracks in real materials are in many cases far less compact than those of fig. 2. To describe such cases it has been suggested to use the growth law of (5), with a finite q_1 [24]. If η_1 is chosen greater than one, a crossover from a DLA-like pattern to a more diluted structure could be expected. The size of the crack at which the crossover takes place, and its abruptness, depends on q_1, η_1 and the absolute value of the displacement at the boundary; note that using (5) makes the results dependent on the latter parameter. The resulting patterns [24] can describe processes in which the crack frist nucleates in a dense inner region, and then rapidly propagates in the form of thin branches. Lately, the beam model, in combination with the growth law of (6), has been used to investigate crack propagation of a single deterministic crack [17]. In order to describe a phenomenology ranging from brittle fracture to slow stress corrosion cracking, several fracture criteria, which included memory effects at different levels, were considered. The patterns resulted to be very complex, having in some cases a fractal character. Comparison with experimental results for stress corrosion cracking in a metallic alloy was very promising [17].

4. Practical implications of the fractal geometry of fractured surfaces

Although the use of the fractal concept in the study of failure in real materials is rather new [10,18–21], some significant achievements have already been attained; we shall briefly comment on some of them. As remarked in section 1, Mandelbrot et al. [10] were first in pointing out that fractured surfaces could have a fractal geometry. Later analyses of failure in materials [18] not only confirmed the earlier results but indicated that the surface energy (E), as the fracture roughness (R), also shows a fractal behavior. However, they also illustrated the difficulty in obtaining reliable results, as these may crucially depend on the way the actual morphology of the fractured surfaces is analyzed. Several consequences of the fractal geometry of fractured surfaces have been found. For instance it has recently been shown that the fatigue threshold in steels is a linear function of the fractal dimension [19]. An upper bound for the fracture toughness has even been predicted, in terms of the fractal dimension characteristic of the fracture surface in a given material [20].

The linear relationship between $\log(R)$, or $\log(E)$, and the log of the scale of observation (L) inherent to the fractal behavior mentioned above, has recently been questioned [21]. In trying to find universalities in a wide range of materials, several authors have concluded that a plot of $\log(R)$ against $\log(L)$ exhibits a sigmoidal behavior. These results have been interpreted in terms of the different mechanisms governing the failure process in the different materials, and have been analyzed by means of multifractals.

In summary, we have discussed the application of fractals to the study of fracture, one of the most complicated and important phenomena in materials research. Significant progress has already been produced in two directions. One attempts to characterize the fractured surfaces by means of the fractal geometry, and find out practical implications. The second one looks at fracture as a growth process. Although there remains a long way to go, one may glimpse important advances in the field of fracture being derived from these studies.

Acknowledgement

The financial support of the "Comisión Interministerial para la Ciencia y la Tecnología", Spain (through contract No. PB85-0437-C02-01) is gratefully acknowledged.

References

[1] B.B. Mandelbrot, The Fractal Geometry of Nature (Freeman, San Francisco, 1983).
[2] P. Meakin, in: Phase Transition and Critical Phenomena, A.C. Domb and J.L. Lebowitz, eds. (Academic Press, New York, 1988), p. 336.
[3] H.E. Stanley and N. Ostrowsky, eds., Random Fluctuations and Pattern Growth (Kluwer, Dordrecht, 1988).
[4] T.A. Witten and L.M. Sander, Phys. Rev. B 27 (1988) 5686.
[5] L. Niemeyer, L. Pietronero and H.J. Wiesmann, Phys. Rev. Lett. 52 (1984) 1033.
[6] E. Louis, F. Guinea and F. Flores, in: Fractals in Physics, L. Pietronero and E. Tosatti, eds. (North-Holland, Amsterdam, 1986), p. 177.
[7] E. Louis and F. Guinea, Europhys. Lett. 3 (1987) 871.
[8] H. Liebowitz, ed., Fracture, Vols. I–II (Academic Press, New York, 1984).
[9] R.M. Latanision and J.R. Pickens, eds., Atomistic of Fracture (Plenum, New York, 1983).
[10] B.B. Mandelbrot, D.E. Passoja and A.J. Paullay, Nature 308 (1984) 721.
[11] Y. Termonia and P. Meakin, Nature 320 (1986) 6061.

[12] F. Guinea, O. Plá and E. Louis, Fragmentation, form and flow in fractured media, Ann. Israel Phys. Soc. 78 (1986) 587.
[13] L. Navas, F. Guinea and E. Louis, J. Phys. 21 (1988) L301.
[14] M.P. López-Sancho, F. Guinea and E. Louis, J. Phys. A 21 (1988) L1079.
[15] P. Meakin, G. Li, L.M. Sander, E. Louis and F. Guinea, J. Phys. A. (1988), in press.
[16] E.L. Hinrichsen, A. Hansen and S. Roux, Europhys. Lett. 8 (1989) 1.
[17] H.J. Herrmann, J. Kertész and L. de Arcangelis, Phys. Rev. B 39 (1989) 637; preprint.
[18] C.S. Pande, L.E. Richards, N. Louat, B.D. Dempsey and A.J. Schwoeble, Acta Metall. 35 (1987) 1633; A.R. Rosenfield, Scripta Metall. 21 (1987) 1359.
[19] Z.G. Wang, D.L. Chen, X.X. Jiang, S.H. Ai and C.H. Shih, Scripta Metall. 22 (1988) 827.
[20] J.C.M. Li, Scripta Metall. 23 (1988) 837.
[21] E.E. Underwood and K. Banerji, Mater. Sci. Eng. 80 (1986) 1; R.E. Williford, Scripta Metall. 22 (1988) 197, 1749.
[22] L.D. Landau and E.M. Lifshitz, Theory of Elasticity (Pergamon, New York, 1959).
[23] E. Louis, R. García-Molina, F. Guinea, P. Meakin and L.M. Sander, presented at the EPS 9th General Conference of the Condensed Matter Division (1989); to be published.
[24] O. Plá, F. Guinea, E. Louis, G. Li, H. Yan, L.M. Sander and P. Meakin, presented at the APS March Meeting (1989); to be published.
[25] L.A. Turkevich and H. Scher, Phys. Rev. Lett. 55 (1985) 1026.
[26] C. Atkinson, J.M. Bastero and J.M. Martínez-Esnaola, Eng. Fracture Mech. 31 (1988) 637, and references therein.
[27] R.C. Ball, R.M. Brady, G. Rossi and B.R. Thompson, Phys. Rev. Lett. 55 (1985) 1406.

FRACTAL DIMENSION OF THE FRACTURED SURFACE OF MATERIALS

C.W. LUNG [a,b,c] and S.Z. ZHANG [b,c]

[a] *International Centre for Theoretical Physics, Trieste, Italy*
[b] *International Centre for Materials Physics, Academia Sinica, 110015 Shenyang, People's Republic of China*
[c] *Fundamental Physics Centre, University of Science and Technology of China, Hefei, Anhui, People's Republic of China*

The fractal dimension of the fractured surface of materials is discussed in order to show that the origin of the negative correlation between D_0 and the toughness lies in the method of fractal dimension measurement with the perimeter–area relation and also in the physical mechanism of crack propagation.

1. Introduction

Mandelbrot et al. [1,2] firstly showed that fractured surfaces are fractals in nature and that the fractal dimensions of the surfaces correlate well with the toughness of the materials. The values of measured fractal dimension D_m decrease smoothly with an increase of the toughness of materials. One of the present authors [3] suggested that the effective critical crack extension force calculated from a fractal surface would be larger than that calculated from a flat fracture surface due to the areas of the fractal surfaces being actually larger than flat ones. The fractured surface can be approximately considered as a fractal surface. He analyzed this problem in connection with fracture mechanics. Mu and Lung [4] investigated the relationship between the fractal dimension of the fractured surface and the fracture toughness of materials. The values of fractal dimension decrease linearly with an increase of the logarithm values of fracture toughness K_{1C}. At the same time, many authors have pointed out [2,4–6] that the correlation between D_m and toughness (dynamic tear energy, fracture toughness, etc.) is a negative one (i.e. higher toughness for smaller D_m). This, however, is difficult to explain.

The aim of the present paper is to show that the origin of the negative correlation lies in the method of fractal dimension measurement and also in the physical mechanism of crack propagation.

2. The perimeter–area relation of the Koch island with a finite number of generations

The mathematical fractal model might have an infinite number of generations without upper and lower limits; otherwise coarse graining of the initiator or fine graining of the lowest limit of generation would not give the same shape as the original one. But the realistic fractals are always of a finite number of generations. As an example, we may consider the triadic Koch island in fig. 1. The initiator is a triangle with side length $\epsilon_0 = 1$. The coastline dimension is $D_0 = \log 4 / \log 3 = 1.2618$.

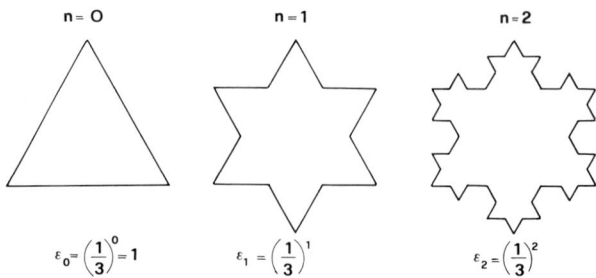

Fig. 1. The triadic Koch island with a triangle initiator.

Considering the perimeter–area (P–A) relation of the triadic Koch island, the nth generation perimeter P_n and the area A_n are

$$P_n/P_0 = (4/3)^n,$$

$$A_n/A_0 = 1 + 3*(1/3)^2 + 3*4*(1/3)^4$$
$$+ ... 3*4^{n-1}*(1/3)^{2n}.$$

Therefore

$$D(n) = D(\epsilon)$$
$$= \frac{(1-D_0)\log\epsilon + \log\alpha}{\log\alpha + [(1-D_0)/D_0]\log\epsilon + 0.5\log A_m(\epsilon)}, \quad (1)$$

where P_0 and A_0 are the perimeter and area of the initiator, respectively; $\epsilon = (1/3)^n$. From eq. (1) [6], we see that $D(n)$ drops to a minimum value as $n=3$ at first and then goes up to a limit value 1.2618 as n increases to infinity (fig. 2). The fractal dimension defined by the P–A relation depends on the generation. This means that the apparent value of fractal dimension measured by the P–A relation depends on the length of the yardstick. This example also illustrates that the P–A relation may be expected to hold only in the limit of small yardsticks. This is consistent with the arguments given by Feder [5] and Lung et al. [6] recently. Unfortunately, the fractal structure of a fractured surface may only be of a few generations. We should pay attention to the characteristic of the first several generations.

Note that if we were to terminate the construction at some level n_{max}, such that there could be no further structure to be revealed by refining our ruler, we would find in units of P_0, for $n > n_{max}$ (fig. 3),

$$P_n/P_0(n > n_{max}) = P_n/P_0(n_{max}), \quad (2)$$

where P_0 corresponds to the perimeter of the initiator. Then, P_n would hold a constant value even under further refining of our ruler and $D(n)$ would have the value

$$D(n > n_{max}) = D(n_{max}). \quad (3)$$

In all physical situations, such a lower cutoff $P_n/P_0(n_{max})$ will be present, and an upper cutoff will also be present.

According to these relations,

$$P_n^{1/D(n)} = \alpha_n A_n^{1/2},$$
$$2\log P_n = D(n)(2\log\alpha_n + \log A_n). \quad (4)$$

For the triadic island, we can calculate α_n, by eq. (4) with P_n, A_n and $D(n)$ from eqs. (1)–(3).

Fig. 4a gives us the P–A relation as a function of n. It shows that the $\log P$–$\log A$ relation is not linear especially in the first several generations. Why do P_n and A_n data, measured by many authors, with the slit–island method (SIM), look linear? We think this is a false impression. In the slit–island method, we usually measure many different sizes of initiators (islands). Then

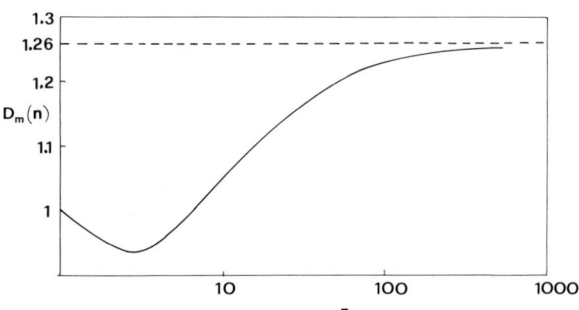

Fig. 2. The relationship of measured fractal dimension $D_m(n)$ versus the generation number n.

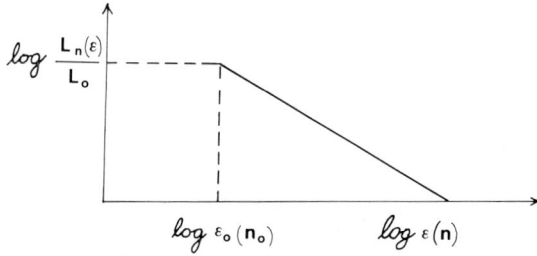

Fig. 3. The relationship of $\log(P_n/P_0)$ versus $\log\epsilon(n)$ in finite construction of fractal. n_0, the number of generation after which $\log(P_n/P_0)$ remains constant as $n > n_0$.

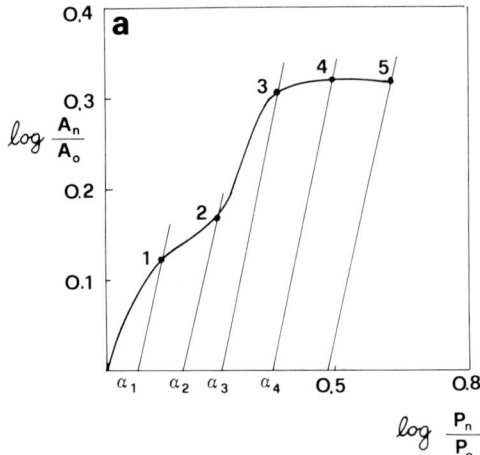

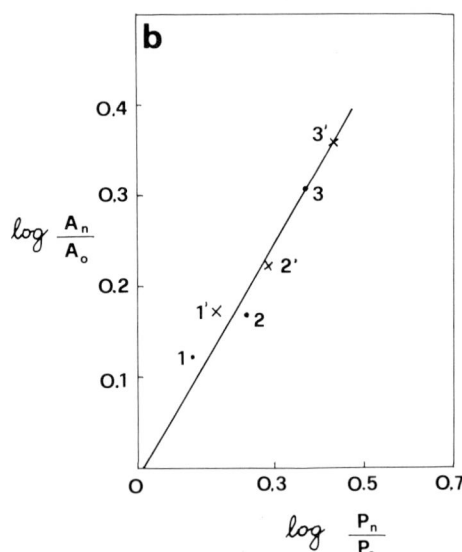

Fig. 4. The perimeter–area relation as a function of n.

$$\log(P_{ni}/P_0) = \log(P_{ni}/P_{0i}) + \log(P_{0i}/P_0)$$
$$= \log[P_{ni}(n)] + \log r_i, \qquad (5)$$
$$\log(A_{ni}/A_0) = \log(A_{ni}/A_{0i}) + \log(A_{0i}/A_0)$$
$$= \log[A_{ni}(n)] + 2\log r_i,$$

where P_{0i} and A_{0i} are the values of the perimeter and area of the initiator of the ith island, respectively. We notice that $\log(P_{ni}/P_{0i})$ and $\log(A_{ni}/A_{0i})$ are functions of n and that $\log(A_{0i}/A_0) = 2\log(P_{0i}/P_0)$ $= 2\log r_i$ ($r_i = P_{0i}/P_i$), because the initiators are normal triangles. We may easily choose the values of P_{0i}/P_0 such that, in fig. 4b, data 1' ($n=1$), 2' ($n=2$), and 3' ($n=3$) can be obtained. A linear relationship seems to be obtained by combining them with the original points 1 ($n=1$), 2 ($n=2$), and 3 ($n=3$). It seems questionable to determine the fractal dimension with the slope of this line because the reasonable lines would be α_1 1, α_2 2 and α_3 3 in fig. 4a.

We think that the fractal dimension measured with the P–A relation is not the real fractal dimension D_0 of the fractured surfaces and that it is one of the origins of the negative correlation between measured D_m and the toughness of materials. One of the present authors and his coworker measured D_m with different lengths of yardsticks. For sufficiently small yardstick length, a positive correlation between D_m and K_{1C} was observed [6].

The best way to measure the fractal dimension of fractured surfaces may be the relation [8]

$$L(\epsilon) = L_0 \epsilon^{1-D_0}. \qquad (6)$$

We may measure the total length of crack propagation with different lengths of yardsticks. Then D_0 can be obtained by

$$D_0 = 1 - \log[L(\epsilon)/L_0]/\log \epsilon. \qquad (7)$$

3. The micromechanism of fracture

Are there any other reasons that the correlation between D_0 and K_{1C} is negative, other than the SIM of fractal dimension measurement? One of the present authors analyzed the critical crack extension force with the fractal model for the intergranular fracture case [3]. A positive correlation between D_0 and K_{1C} was found. In this paper we analyze the distance between two large inclusions and the number of grains over the distance. We show that in this case the correlation between D_0 and fracture toughness could be a negative one.

Let us suppose that new segments of microcracks of grain size were superimposed on the preceding

Table 1
The variation of fractal dimension of D_0 with the change of segment number N and the angle φ

φ (deg)	N							
	2	3	4	5	6	7	8	9
90	2.0000	1.3652	1.3333	1.2549	1.2398	1.2091	1.2000	1.1833
100	1.6247	1.2631	1.2380	1.1862	1.1747	1.1538	1.1470	1.1355
120	1.2619	1.1292	1.1158	1.0932	1.0873	1.0778	1.0743	1.0690
140	1.0986	1.0526	1.0470	1.0384	1.0360	1.0323	1.0308	1.0278

larger segment due to large inclusions. The fractal dimension may be calculated as [7]

$$D_{0a} = \log(N)/\log[N\sin(\varphi/2)],$$
$$2\sin^{-1}(1/N^{1/2}) < \varphi < \pi, \quad N = \text{even},$$
$$D_{0b} = \log(N)/\log[1 + (N^2-1)\sin^2(\varphi/2)]^{1/2},$$
$$2\sin^{-1}[1/(N+1)^{1/2}] < \varphi < \pi, \quad N = \text{odd}.$$
(8)

Here, we have assumed the angles between two adjacent segments to be equal. We also know that the value of the angle φ depends on the grain configurations. In this case, the fractal dimension D_0 may decrease with an increase of the segment number N (table 1).

Decreasing either the test temperature or the tempering temperature would produce materials which have a high yield strength. High strength materials have a smaller critical crack length for propagation due to their low K_{1C}. It may induce more smaller cracks (inclusions) to propagate. This makes the crack propagation between two small inclusions easier. Then, the segment number N decreases, and the fractal dimension of the fractured surface increases. We therefore come to the conclusion that the fractal dimension of fractured surfaces increases with the decrease of fracture toughness. The correlation between D_0 and K_{1C} is therefore negative.

Real fracture processes are quite complicated. A wide variety of mechanisms play relevant roles: grain boundaries, inclusions, second phase particles, etc. Experimental results are the total effects of many elementary processes. We should take care to select the main factor. Even then, it seems hopeful to use fractals to characterize fractured surfaces with which material parameters can be correlated.

Acknowledgements

One of the authors (C.W.L.) would like to thank Professor Abdus Salam, the International Atomic Energy Agency and UNESCO for hospitality at the International Centre for Theoretical Physics, Trieste, and Professor S. Lundqvist for his advice and encouragement on our work on fractals at Trieste.

References

[1] B.B. Mandelbrot, The Fractal Geometry of Nature (Freeman, San Francisco, 1983) pp. 25, 29, 459.
[2] B.B. Mandelbrot, D.E. Passoja and A.J. Paullay, Nature 308 (1984) 721.
[3] C.W. Lung, in: Fractals in Physics, L. Pietronero and E. Tosatti, eds. (North-Holland, Amsterdam, 1986), p. 189.
[4] Z.Q. Mu and C.W. Lung, J. Phys. D 21 (1988) 848.
[5] J. Feder, Fractals (Plenum, New York, 1987), pp. 200–202.
[6] C.W. Lung and Z.Q. Mu, Phys. Rev. B 38 (1988) 11781.
[7] S.Z. Zhang and C.W. Lung, J. Phys. D, in press.
[8] Q.Y. Long, Z.Q. Mu and C.W. Lung, private communications.

ON THE SELF-AFFINITY OF VARIOUS CURVES

Mitsugu MATSUSHITA [a] and Shunji OUCHI [b]

[a] *Department of Physics, Chuo University, Kasuga, Bunkyo-ku, Tokyo 112, Japan*
[b] *Institute of Geosciences, Chuo University, Kasuga, Bunkyo-ku, Tokyo 112, Japan*

A numerical method is discussed which enables us to scale-independently analyze the self-affinity of various curves such as noise and topographical curves. Curve length, N, and standard deviations for two appropriately chosen coordinates, X and Z, of a curve in two dimensions are measured between many arbitrary pairs of points on the curve by using the smallest fixed scale (yardstick). The self-affinity is confirmed by checking the scaling of the form $X \sim N^{\nu_x}$ and $Z \sim N^{\nu_z}$ with $\nu_x \neq \nu_z$. The method is applied to transect profiles of real mountain topography, which is found to be self-affine with $\nu_x = 1$ for the horizontal and $\nu_z < 1$ for the vertical coordinate variations. The consistency of a scaling relation ($D = 2 - H$) between a self-similar contour line with the fractal dimension D and a self-affine transect profile with $H = \nu_z$ in the same area is also discussed.

1. Introduction

Fractal geometry [1] is now widely known to provide a nice description of seemingly complex forms seen in nature. Shapes such as coastlines and clouds often possess a statistical but remarkably simple invariance under changes of magnification (scale invariance) [1–3]. Suppose we approach a beautiful planet, the Earth, from afar and look at it through a spaceship window. We may easily notice some coastline with huge peninsulas and gulfs, familiar from the world atlas. As we approach more closely, we may start noticing smaller peninsulas and gulfs which were invisible before, on the coasts of the huge ones. Approaching closer and closer, smaller and smaller peninsulas and bays may become visible. If we look at coastlines through a small window, with a rather limited visible angle, it may happen that they look more or less the same and we cannot judge how high we are above the earth surface. Coastlines seem to have no characteristic length scales, at least within some length ranges. This self-similarity can be quantitatively characterized by a fractal dimension. For instance, many Japanese rias such as Sanriku Coast are known to have the fractal dimension of about 1.3.

It is, however, rather misleading to say that a landscape is a fractal with a fractal dimension of, say, 2.3. However rough the earth surface is, it may look almost flat when viewed from 100 km above. However jagged mountains are when viewed from their foot, they look like a slightly wavy, horizontal line when viewed from far away over a huge plain. Landscapes do not seem to be self-similar. Interesting questions thus immediately arise as to what kind of fractals they are and if so, how they can be characterized.

It should be noted that in the case of a landscape (or the silhouette of a chain of mountains which is very similar to a transect profile of the landscape) the vertical direction is a special one due to the existence of gravity. The vertical (or altitude) variation of a landscape must be scaled differently from any horizontal one. By contrast, in the case of a coastline (or, more generally, a contour line of a landscape) there are no preferred horizontal directions. The contours of a landscape may be isotropic and self-similar, whereas the vertical ones may be anisotropic and not self-similar. When given patterns are scaled differently in different directions (or scaled anisotropically), they are called self-affine fractals [1–5].

It is now known that many growth processes give rise to self-affine fractals. For instance, the surface of a cluster or a pattern produced by the growth process of Eden [6–9] and ballistic deposition [6] models is self-affine. An entire pattern grown from an appropriate substrate such as a line (or a fiber) and a plane can often be subdivided into individual clusters (or trees) grown from the substrate. The individual clusters in the case of Scheidegger's river network [10,11], Eden and ballistic deposition [11] models have self-affine structures: The root-mean-square (rms) height from the substrate h_s of a cluster of size s (the number of particles or constituent units forming the cluster) and the rms width w_s both scale as

$$h_s \sim s^{\nu_h} \quad \text{and} \quad w_s \sim s^{\nu_w}, \tag{1}$$

but the exponents ν_h and ν_w are different. Self-similar fractals such as individual clusters of diffusion-limited deposition models, on the other hand, can be characterized by only one exponent $\nu_h = \nu_w = \nu = D^{-1}$, where D is the fractal dimension.

In this paper we propose, based on the same idea as in eq. (1), a useful way to quantitatively characterize various curves such as noise $V(t)$ and topographical altitude variations $z(x)$. In particular, we apply the method to a real landscape and confirm the self-affinity. We also discuss the topographical implication of the scaling relation derived.

2. Method

Suppose that we have a curve in two-dimensional space, $y = f(x)$, as in fig. 1. The extension to higher dimensions is straightforward. Let us first define the smallest length scale or unit length scale a_0 ($=1$) and measure by this scale the curve length Na_0 ($=N$) between two arbitrary points A and B on the curve. (This is equivalent to regarding the curve as consisting of particles of diameter a_0 and counting the number of particles between A and B.) We then calculate x- and y-variances X^2 and Y^2 of all measured points on the curve between the two points A and B;

$$X^2 = \frac{1}{N} \sum_{i=1}^{N} (x_i - x_c)^2,$$
$$Y^2 = \frac{1}{N} \sum_{i=1}^{N} (y_i - y_c)^2 \tag{2a}$$

with

$$x_c = \frac{1}{N} \sum_{i=1}^{N} x_i, \quad y_c = \frac{1}{N} \sum_{i=1}^{N} y_i, \tag{2b}$$

where (x_i, y_i) is the coordinate of the ith measured point P_i on the curve. The standard deviations X and Y indicate the approximate size of the part of the curve. Repeat the measurement procedures described above for many pairs of points on the curve and check by log–log plots of X and Y versus N whether they scale as

$$X \sim N^{\nu_x}, \quad Y \sim N^{\nu_y}, \tag{3}$$

where the exponents ν_x and ν_y are in general different. If so, they are then related to each other as

$$Y \sim X^H, \tag{4}$$

where the exponent H is given by

$$H = \nu_y / \nu_x. \tag{5}$$

The exponents ν_x and ν_y in eq. (3) are analogous to those of anisotropic correlation lengths in critical phenomena such as in liquid crystals. It should be noted that the method described above can directly apply to any self-similar fractal curves as well. In this case $\nu_x = \nu_y = 1/D$, where D is a fractal dimension of the curve. We confirmed this by applying the method to a Koch curve and obtained $\nu_x \approx \nu_y \approx 0.78$ ($D \approx 1.28$). Also note the consequence of $H = 1$ from eq. (5) for any self-similar fractals, which means that the exponent H is useless when discussing both self-affine and self-similar fractals in the same context.

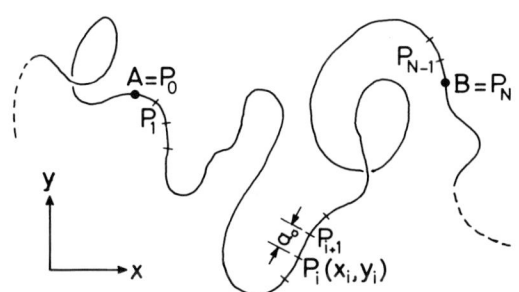

Fig. 1. Measuring the curve length Na_0 between a pair of points A and B on a given curve by the smallest fixed length scale a_0.

3. Simple application

The simplest example to be applied is a fractional Brownian motion (fBm) trace $x_H(t)$ [1–5]. If we choose a one-step length in the x_H–t plane as the unit length, the curve length N between any two points on the curve is always proportional to the corresponding time interval T, i.e. $T \sim N^{\nu_t}$ and $\nu_t = 1$. Further, the vertical standard deviation X_H is known to scale as in eq. (4), i.e. $X_H \sim T^H$, which yields $\nu_x = H$ from (5).

Fig. 2 shows a plot of the displacement of the usual one-dimensional random walk ($H = 1/2$) as a function of time. In fig. 3 we show the plot of results obtained by applying the method described in section 2, i.e. the horizontal and vertical standard deviations T and X_H as a function of curve length N for the curve shown in fig. 2. It gives $\nu_t = 1.00$ and $\nu_x \approx 0.50$, which were obtained by employing a least-squares fitting.

Fig. 2. A typical example of fractional Brownian motion (fBm) traces $x_H(t)$ with $H = 1/2$ (usual one-dimensional random walk).

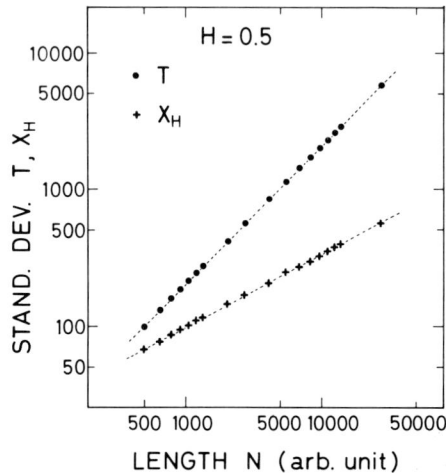

Fig. 3. Dependence of the standard deviations of horizontal and vertical coordinates T and X_H on the curve length N between various pairs of points on the fBm trace shown in fig. 2. The slopes yield the self-affine exponents $\nu_T = 1.00$ and $\nu_x \approx 0.50$.

We also confirmed that ν_t and ν_x are very insensitive to a change of vertical and horizontal length scales and to the unit length scale used to measure the curve length. It should be noted that the direct application of the box-counting method to this $x_H(t)$ yields values of fractal dimension between 1 and H, depending on the length scales [12]. (Typical values are $D_G = 1$, $D_L = 2 - H$ and $D_\ell = 1/H$, which are called global, local and latent dimensions, respectively.)

4. Topographical curves

Let us next apply the method to a real mountain topography. Fig. 4 shows a transect profile of the well-dissected Mount Yamizo (1022 m high) area in Fukushima Prefecture, Japan. The data were taken by drawing a line on a 1/25 000 scale map of the area (see also fig. 6 below) and noting the intersection points with contour lines. Regarding this curve as being given, we calculated the curve length N, standard deviations of horizontal and vertical coordinates X and Z, respectively, between two arbitrary points on the curve. We repeated the procedure for many pairs of points. The results are shown in fig. 5, where X and Z are plotted against N. As is expected, the standard deviation of the horizontal coordinates is proportional to the curve length, i.e. $X \sim N$ and $\nu_x = 1$. On the other hand, that of the vertical coordinates shows an approximate dependence of $Z \sim N^{\nu_z}$ with $\nu_z \approx 0.55$. Averaging the values of ν_z obtained from other vertical contours in this area yields $\nu_z \approx 0.59$.

The fact that transect profiles such as shown in fig. 4 are approximately self-affine with $\nu_x = 1$ and $\nu_z < 1$ means that they can be represented by fBm with

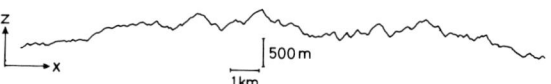

Fig. 4. Transect profile near Mt. Yamizo in Fukushima Prefecture. The altitude variation is exaggerated by doubling vertical scales.

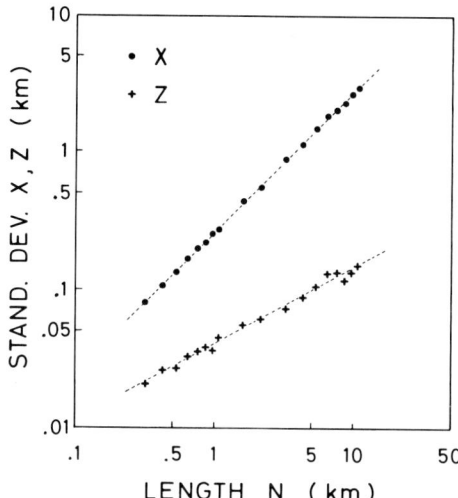

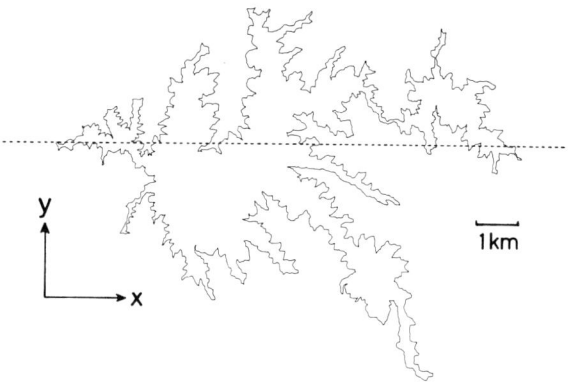

Fig. 6. Contour line at 700 m taken from the 1/25 000 scale map of the Mt. Yamizo area. A dotted line indicates that the transect profile shown in fig. 4 was taken from it.

Fig. 5. Dependence of the standard deviations of horizontal and vertical coordinates X and Z on the curve length N between many pairs of points on the curve shown in fig. 4. The slopes yield the self-affine exponents $\nu_x = 1.00$ and $\nu_z \approx 0.55$.

$H = \nu_z$. This is why artificial surfaces based on the fBm [1–3] look so natural to us. The continuum limit of the probability of having the displacement between x and $x + dx$ at time t for fBm is given by

$$w(x,t)\,dx = \frac{1}{(4\pi D t^{2H})^{1/2}} \exp\left(\frac{x^2}{4Dt^{2H}}\right) dx, \quad (6)$$

where D is the diffusion constant. The probability of returning to the origin is then given by $p_r(t) \sim t^{-H}$ and the number of returns within the time interval $(0, T)$ is

$$N_r \sim \int_0^T p_r(t)\,dt \sim T^{1-H}.$$

This means that for the fBm, such as that shown in fig. 2, the set of points which satisfy $x_H(t) = 0$ is self-similar with the fractal dimension $D_0 = 1 - H$.

Now suppose that we have a topographical surface, whose transect profiles, such as those shown in fig. 4, are self-affine with $\nu_x = 1$ and $\nu_z = H \ (<1)$. Then the set of level-crossing points with the same altitude on the transect profile is self-similar with the fractal dimension on $D_0 = 1 - H$, and the contour line of the same altitude is also self-similar with the fractal dimension

$$D = D_0 + 1 = 2 - H, \quad (7)$$

because there are no preferred horizontal directions. Note that this fractal dimension is not the local dimension D_L of self-affine fBm curves.

Fig. 6 shows the contour line of 700 m taken from the 1/25 000 scale map of the Mt. Yamizo area. (A dotted line on the figure indicates where the transect profile shown in fig. 4 was taken.) We applied the method described above to the curve and confirmed that the contour line is self-similar with $\nu_x \approx \nu_z \approx 0.73$ ($D \approx 1.37$), as shown in fig. 7. Taking $H = \nu_z \approx 0.59$ in this area into account, eq. (7) is found to hold well.

One more example is shown in fig. 8, which was taken from the 1/25 000 scale map of the Mt. Shirouma (2932 m high) area in the Japanese Alps. We again calculated the dependence of the standard deviations of horizontal and vertical coordinates of this curve, X and Z, on the curve length N, as in figs. 3 and 5. The results are shown in fig. 9. One interesting point, compared with fig. 5, is that the standard deviation of the vertical variation Z exhibits a clear crossover around $N_c = 2$ km. Let us call it the cross-

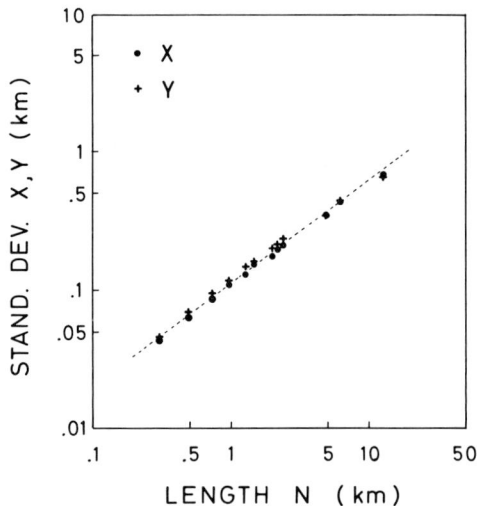

Fig. 7. Dependence of the standard deviations of two horizontal coordinates X and Y on the curve length N between many pairs of points on the curve shown in fig. 6. The slopes indicate that the curve in fig. 6 is self-similar with $\nu_x \approx \nu_y \approx 0.73$ ($D \approx 1.37$).

Fig. 8. Transect profile near Mt. Shirouma in the Japanese Alps. The altitude variation is doubly exaggerated.

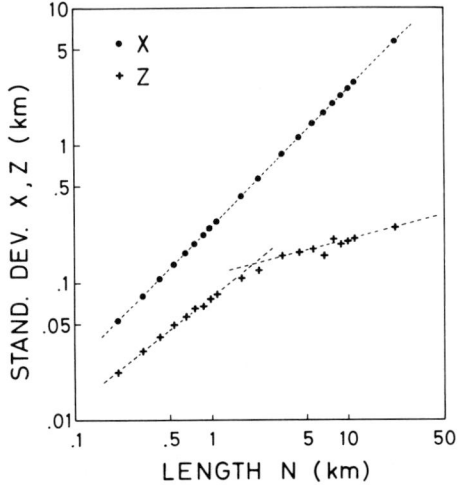

Fig. 9. Dependence of the standard deviations of horizontal and vertical coordinates X and Z on the curve length N between many pairs of points on the curve shown in fig. 8.

over from local to global altitude variation. Furthermore, fig. 9 ensures that within some length range of both $N < N_c$ and $N > N_c$ the transect profile shown in fig. 8 is self-affine with $\nu_{zL} \approx 0.73$ locally for $N < N_c$ and $\nu_{zG} \approx 0.26$ globally for $N > N_c$.

5. Summary and discussion

We have proposed a simple but useful method to analyze the self-affinity of various curves. We have applied it to fractional Brownian motion traces and real mountain topography, and confirmed their self-affinity. We have also checked the scaling relation between the exponents characterizing the self-affine transect profiles and self-similar contour lines in the same mountain area.

Fig. 9 clearly suggests that a real topographical surface is characterized by, at least, two regimes: local and global. Local structures are caused by rather small-scale erosion due to, e.g., floods and earthquakes, and local geological properties may determine the value of the exponent ν_{zL}. On the other hand, global structures may be brought about by large-scale folds due to the plate tectonics, and global geological properties may determine ν_{zG}. In fact, a simple extrapolation of $Z \sim N^{\nu_{zG}}$ to the range $N \approx 2000$ km in fig. 9 gives $Z \approx 2000$ m, which seems reasonable for the Mainland of Japan.

Eq. (3) together with (7) may be very useful to investigate the self-affinity of various surfaces such as fractured surfaces of rocks, minerals and metals. It would also be very interesting to reinvestigate river networks, especially Hack's empirical law, from the present viewpoint.

References

[1] B.B. Mandelbrot, The Fractal Geometry of Nature (Freeman, San Francisco, 1982).
[2] J. Feder, Fractals (Plenum, New York, 1988).
[3] H.-O. Peitgen and D. Saupe, eds., The Science of Fractal Images (Springer, Berlin, 1988).
[4] T. Vicsek, Fractal Growth Phenomena (World Scientific, Singapore, 1988).

[5] B.B. Mandelbrot, in: Fractals in Physics, L. Pietronero and E. Tosatti, eds. (North-Holland, Amsterdam, 1986), pp. 3, 17, 21.
[6] F. Family and T. Vicsek, J. Phys. A 18 (1985) L75.
[7] R. Jullien and R. Botet, Phys. Rev. Lett. 54 (1985) 2055; J. Phys. A 18 (1985) 2279.
[8] P. Freche, D. Stauffer and H.E. Stanley, J. Phys. A 18 (1985) L1163.
[9] P. Meakin, R. Jullien and R. Botet, Europhys. Lett. 1 (1986) 609.
[10] H. Kondoh, M. Matsushita and Y. Fukuda, J. Phys. Soc. Japan 56 (1987) 1913.
[11] P. Meakin, J. Phys. A 20 (1987) L1113.
[12] R.F. Voss, in: The Science of Fractal Images, H.-O. Peitgen and D. Saupe, eds. (Springer, Berlin, 1988), ch. 1.

THE GROWTH OF SELF-AFFINE FRACTAL SURFACES

Paul MEAKIN

Central Research and Development Department, E.I. du Pont de Nemours and Company, Wilmington, DE 19880-0356, USA

Rough surfaces generated by a wide range of different processes can be described quite well in terms of self-affine fractal geometry. Here ballistic deposition models are used to illustrate some of the fractal characteristics that are common to many rough surfaces. The models also show how computer simulations and theoretical approaches have been used in a complementary manner to obtain a better understanding of the structure and growth of random rough surfaces. Some results obtained for ballistic deposition at oblique incidence and for spatially correlated ballistic deposition represent some of the recent advances made in these directions.

1. Introduction

In the surge of interest in applications of the concepts of fractal geometry to problems in the physical sciences following the publication of The Fractal Geometry of Nature [1] in 1982 and earlier books by Mandelbrot [2,3], attention was focused primarily on simple self-similar fractals that can be characterized by a single "all purpose" [4] fractal dimensionality (D). For such structures almost any reasonable procedure for measuring the fractal dimensionality will lead to essentially the same results if the fractal scaling regime extends over a sufficiently wide range of length scales. In practice, of course, some approaches give more reliable estimates than others for the asymptotic fractal dimensionality.

Many fractal structures found in nature exhibit a self-affine [1] fractal geometry. Considerable difficulties and confusion were encountered when attempts were made to measure the fractal dimensions of self-affine structures using the approaches that worked well for self-similar fractals. Fortunately, confusion of self-affine and self-similar fractals is now much less common as a result of the much broader dissemination of the principles of fractal geometry [1-3] and the efforts that have been made by Mandelbrot [4-6] and others [7-9] to clarify the basic nature of self-affine fractals.

This paper is concerned with the formation of rough surfaces that appear to be smooth on very long length scales but are very rough on shorter length scales. The Brownian process $B(t)$ that describes the distance moved by a Brownian particle during the time t provides a valuable paradigm for such self-affine fractal surfaces. The Brownian process can be rescaled onto itself (in a statistical sense) by the transformation

$$B(t) \to b^{-1/2} B(bt) . \qquad (1)$$

The more general fractal Brownian processes $B_H(t)$ can be rescaled by the transformation

$$B_H(t) \sim b^{-H} B_H(bt) , \qquad (2)$$

where the exponent H is the Hurst [10] exponent [1].

In many practical processes self-affine fractal surfaces are formed by the addition of material to or the removal of material from an initially smooth surface at a more or less constant rate. The geometry of these surfaces can be described by the function $h(x)$ that specifies the vertical distance (height) from the rough surface to the original smooth surface at the position x on that surface. In these cases where $h(x)$ is not a single value function of (x), it is often physically appropriate to consider only the largest value of the surface height at each position (x). After a time t the rough surface exhibits fractal scaling on length scales up to a length of $\xi_\perp$ in a direction perpendicular to the surface (perpendicular to the original smooth surface) and up to a length $\xi_\parallel$ in a direction parallel to the surface. The correlation length $\xi_\perp$ corresponds to the "thickness" of the rough surface and $\xi_\parallel$ is a persistence length for the surface roughness. In general, for self-affine surfaces $\xi_\perp$ and $\xi_\parallel$ grow differently with increasing time,

$$\xi_\parallel \sim t^{1/z}, \tag{3a}$$

$$\xi_\perp \sim t^\beta, \tag{3b}$$

$$\xi_\perp \sim \xi_\parallel^\alpha \sim \xi_\parallel^{\beta z}. \tag{3c}$$

A self-affine fractal surface grown in this fashion can be described by the height different correlation function $C_h(r)$ defined as

$$C_h(r) = \langle h(\mathbf{x}) - h(\mathbf{x}+\mathbf{r}) \rangle_{|\mathbf{r}|=r}. \tag{4}$$

This correlation function is expected to have the power law form

$$C_h(r) \sim r^{-H} \tag{5}$$

for distances r larger than an inner cut-off length r_1 and smaller than $\xi_\parallel$ (fig. 1).

In computer simulations of the processes leading to the formation of fractal surfaces, it is often convenient to use strips ($d_s=1$, where d_s is the substrate Euclidean dimensionality) of width L or columns ($d_s=2$) of area $L \times L$ with periodic boundary conditions. In this event $\xi_\parallel$ and $\xi_\perp$ grow according to eqs. (3a) and (3b) until $\xi_\parallel$ reaches a value of L. At this stage both $\xi_\parallel$ and $\xi_\perp$ stop growing and $\xi_\perp$ saturates at a value given by eq. (3c),

$$\xi \simeq \xi_\perp \sim L^\alpha. \tag{6}$$

Here ξ is a measure of the surface thickness. It is often convenient to measure the dependence of the surface thickness (ξ) on both t and L and in many cases the variance of the surface height is used to provide a quantitative characterization of the surface thickness. Family and Vicsek [11] showed that for self-affine fractal surfaces generated by two-dimensional ballistic deposition models the dependence of ξ on L and t could be represented by the scaling form

$$\xi(L, t) \sim L^\alpha f(t/L^{\alpha/\beta}). \tag{7}$$

In the limit $x \gg 1$, where x is the argument of the scaling function $f(x)$, $f(x) \to$ const., so that $\xi \sim L^\alpha$ (compare with (3c)). In the limit $x \ll 1$, $f(x)$ increases as a power of x and the condition that ξ should depend only on t indicates that $f(x) \sim x^\beta$ or $\xi \sim t^\beta$ (compare with eq. (3b)). These results are generally valid for the growth of self-affine rough surfaces from smooth flat surfaces. The exponent α in eqs. (3), (6) and (7) corresponds to the Hurst exponent H in a fractal Brownian process.

2. The formation of self-affine surfaces

A variety of different processes including corrosion, erosion, wear, fracture, growth, deposition and dissolution lead to the formation of rough surfaces that appear to be self-affine. For some processes the self-affine geometry has been established as a result of experimental studies and/or computer simulations. Here attention is focussed on the ballistic deposition process in which particles are added, one at a time, to a growing surface via linear or ballistic trajectories. Processes of this type have been simulated for more than 30 years beginning with the pioneering work of Vold [12]. Another process leading to the formation of self-affine surfaces that has been studied extensively by computer simulations is the Eden [13] growth process in which all positions on the

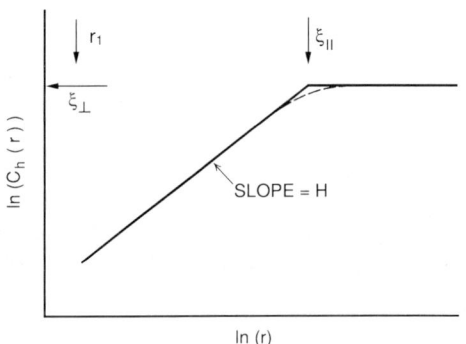

Fig. 1. Schematic representation of the height difference correlation function ($C_h(r)$) for a self-affine fractal surface.

surface can grow with probabilities that are either constant or distributed over a small range of probabilities that depend on the local environment. The Eden growth process and the ballistic deposition process are believed to belong to the same universality class in the sense that all of the exponents that have been used to describe their structures have the same values [14,15].

3. Ballistic deposition

The most simple (two-dimensional) model for ballistic deposition onto a flat substrate is illustrated in fig. 2. A column of a square lattice at position i along a linear substrate is selected at random and the site in that column with a height h_i (y coordinate) given by

$$h'_i = \max(h_{i-1}, h_i + 1, h_{i+1}) \tag{8}$$

is filled. Here h_i is the height of the highest occupied site in the ith column and h'_i is the height of the active zone site in that column. In this model the active zone consists of all of those sites that might be filled by the next growth event. Since this is a simple algorithm that requires relatively little information storage (only the heights h'_i of the active zone or the maximum heights h_i are needed to follow the dynamics of the surface growth process) large structures (up to 10^{10} or more sites [14]) can be grown in order to explore the near-asymptotic fractal scaling associated with the deposit surface. Fig. 3 shows part of a deposit grown in this fashion.

The height difference correlation functions obtained from a square lattice model using a strip width (L) of 2^{18} (262 144) lattice sites is shown in fig. 4a at several stages during the deposition process. The correlation functions appear to have the form shown in fig. 1 and the slope (H or α) for distances r in the range $r_1 < r < \xi_\parallel$ has a value of about 0.45. Fig. 4b shows how these correlation functions can be scaled using the scaling form

$$C_h(r) = M^\beta g(r/M^{\beta/\alpha}) \tag{9a}$$

or

$$C_h(r) = t^\beta g'(r/t^{\beta/\alpha}), \tag{9b}$$

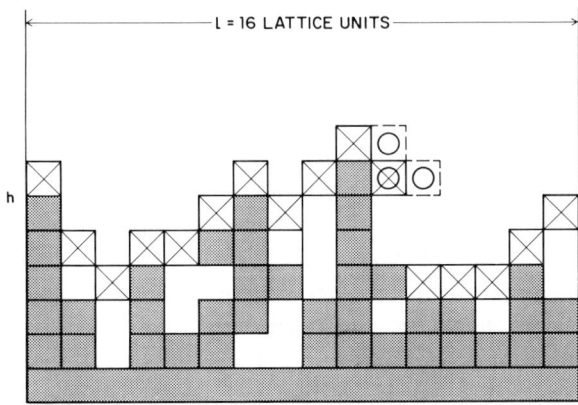

Fig. 2. This figure shows a square lattice model for ballistic deposition onto a line. Filled sites are shaded and active zone sites are indicated by "×". If the active zone site indicated by "⊗" is filled, two new active zone sites indicated by "○" will be created and the old active zone sites below them will no longer be part of the active zone.

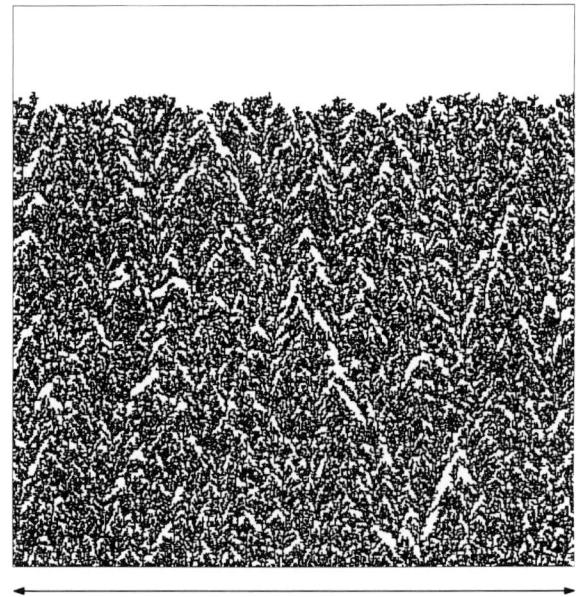

Fig. 3. Part of a deposit grown using the model illustrated in fig. 2.

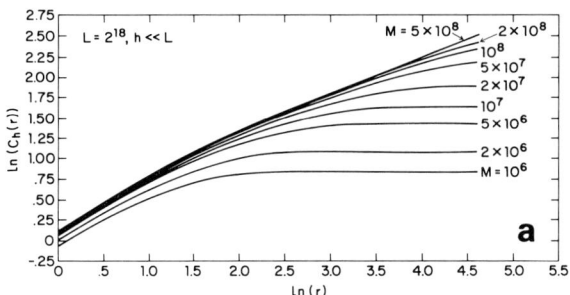

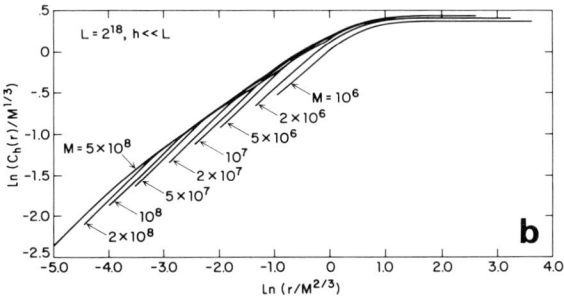

Fig. 4. Height difference correlation functions obtained from square lattice model simulations onto a line of length 2^{18} (262 144) lattice sites. (a) shows the correlation function at several stages during the simulation (M is the total number of filled sites in the deposit). (b) shows how these correlation functions can be scaled using the scaling form given in eq. (9).

where M is the total mass (number of sites) deposited. Here the theoretical values [15] ($\alpha=1/2$, $\beta=1/3$) were used to obtain the data collapse which seems to be quite satisfactory for $M \geq 5 \times 10^6$ ($t \geq 20$) and distances greater than $r_1 \simeq 8$ lattice units. Most simulations of this type have been analyzed using the dependence of ξ or ξ' on h and L where ξ^2 is the variance in the maximum heights and ξ'^2 is the variance of the active zone heights. The exponents α and β can then be measured from the dependence of ξ (or ξ') on L in the limit $t \gg L^{\alpha/\beta}$, from the dependence of ξ (or ξ') on t in the limit $t \ll L^{\alpha/\beta}$ or by using the Family–Vicsek scaling form [11] (7) in the crossover region. In this way values of about 0.47 and 0.33 have been obtained for α and β respectively [13]. A value for α closer to the theoretical [15] value of $1/2$ ($\alpha \simeq 0.500 \pm 0.002$) was obtained from a related

model in which the step heights in the surface were all restricted to be ± 1 lattice unit [14].

Using a cubic lattice model for deposition onto a planar substrate, values of about 0.33 and 0.24 were obtained for α and β respectively. These values satisfy, approximately, the theoretical scaling relationship [14,16]

$$\alpha + \alpha/\beta = 2 . \quad (10)$$

Despite considerable recent efforts [14,16–19] the exponents that characterize the self-affine surfaces of three dimensional ballistic deposits are still not precisely known.

4. Ballistic deposition at non-normal incidence

In most simulations that have been carried out to explore the fractal properties of rough surfaces the particles or lattice sites have been added at an angle of incidence of 0° (i.e., perpendicular to the initially smooth surface). Non-normal incidence can be conveniently investigated using both lattice and off-lattice models by deposition in a strip or channel of width L onto an inclined substrate [20,21]. As the angle of incidence increases, both α and β increase; until as $\theta \to \pi/2$, α approaches a value of about 1.0 and β approaches a value of about $1/2$. The increase in α and β with increasing θ appears to be continuous but is quite small until θ reaches an angle of about 60°. Fig. 5 shows a portion of a large scale simulation of off-lattice ballistic deposition on a line with an angle of incidence of 87.5°.

It is evident from fig. 5 that, for large angles of incidence, the deposit breaks up into a pattern of individual clusters that are compact and do not branch. These clusters "compete" for the incoming flux of particles and, if one cluster becomes completely shaded by another, it ceases to grow forever. Consequently, the evolution of the patterns can be understood in terms of the motion of the cluster tips [22].

The columnar structure obtained at large angles of incidence can be represented by an array of rods (fig. 6) growing at a fixed angle of ϕ with respect to the

Fig. 5. A portion of a deposit generated by a large scale off-lattice simulation of ballistic deposition onto a line with an angle of incidence (θ) of 87.5°. One of the large steps on the surface with a height δ is indicated by a vertical double headed arrow.

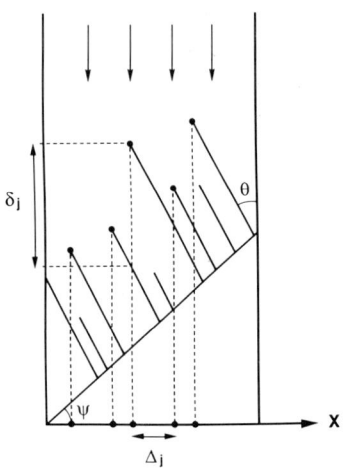

Fig. 6. Schematic representation of a deposit of independently growing rods which interact only through screening. The mapping of the tip positions onto a system of coalescing particles on a line is indicated.

surface normal. The tip of each rod that is still active moves with a constant velocity in the direction of growth (along the axis of the rod). Fluctuations in the particle flux cause an additional random-walk-like displacement of the rod tips. Consequently, the growth of the rod tips projected onto the x axis (fig. 6) can be represented as a drift in the $-x$ direction with a superimposed diffusion. In a frame moving with the drift velocity the projected cluster tips follow Brownian trajectories and the shading of one cluster by another can be represented by the coalescence of two Brownian particles. A number of exact results [23–26] are known for this system and they can be used to calculate all of the exponents characterizing the surface structure and internal structure [15] of the deposit [22]. The mapping of the problem of deposition onto a line at oblique incidence onto the coalescence of Brownian particles leads to the theoretical values of 1 for α and 1/2 for β. In addition, values of 1/3, 2/3 and 4/3 are obtained for the exponents $\nu_\perp$, $\nu_\parallel$ and τ [15] that describe the size dependence of the width, the length and the size distribution of the clusters. In addition, the distribution of the large step on the surface (see fig. 5) is given by

$$N(\delta, t)/L = \delta^{-2} h(\delta/t^\beta) , \qquad (11)$$

where $N(\delta, t)$ is the number of steps in the surface height of size δ at time t. The scaling function $h(x)$ is predicted to have the form

$$h(x) = ax^3 \exp(-bx^2) . \qquad (12)$$

All of the predictions of this theoretical model have been confirmed by large scale simulations [22].

A similar approach can be used to develop an understanding of the structure formed by ballistic deposition onto an inclined plane (fig. 7). At large angles of incidence the pattern consists of overlapping but not contacting clusters that resemble somewhat the skin of a fish. Each of the clusters or scales have a characteristic length (l), width (w) and thickness ($\bar{t}$) These lengths are related to the cluster size (s) by

$$l \sim s_\parallel^\nu , \qquad (13a)$$

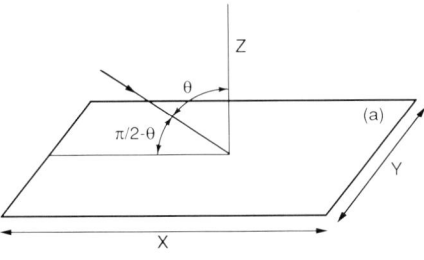

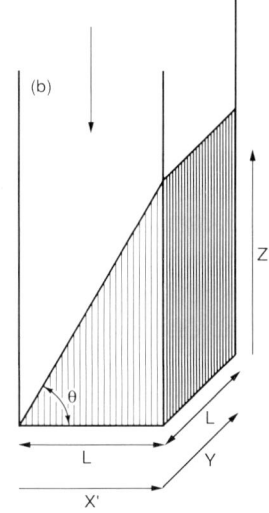

Fig. 7. The coordinate system used in simulations of ballistic deposition onto a surface at oblique incidence.

$$w \sim s^{\nu_y}, \qquad (13b)$$

$$\bar{t} \sim s^{\nu_x}, \qquad (13c)$$

where $\nu_\parallel \simeq 1/2$, $\nu_x \simeq 1/6$ and $\nu_y \simeq 1/3$ [27]. A cross section through the deposit in the (yz) plane resembles quite strongly the two-dimensional pattern shown in fig. 5. The exponent τ describing the cluster size distribution

$$N_s \sim s^{-\tau} \qquad (14)$$

has a value of about 3/2. Here N_s is the number of scales or clusters of size s. The deposit surface and that part of the substructure grown at time t can be described by the correlation lengths ξ_x and ξ_y whose time dependence is given by

$$\xi_x \sim t^{\sigma_x}, \qquad (14a)$$

$$\xi_y \sim t^{\sigma_y}, \qquad (14b)$$

and the surface thickness $\xi_\perp$ grows according to (3b). The exponent σ_y has a value of about 2/3 and $\sigma_x = \beta \simeq 1/3$ for angles of incidence that approach $\pi/2$.

Fig. 8 shows the location of the large steps in the z' direction (projected onto the $(x'y)$ plane (fig. 7b). These large steps are associated with the advancing edges of the substructure clusters. The distribution of these large step sizes can be described by the scaling form given in eq. (11). The scaling function $h(x)$ in this case has the form

$$h(x) = a'x^2 \exp(-b'x^m), \qquad (15)$$

where m is approximately 2–3.

The growth of the deposition pattern can be understood in terms of the dynamics of the advancing cluster edges. Projected onto the $(x'y)$ plane the leading edges of the clusters form a network of domain

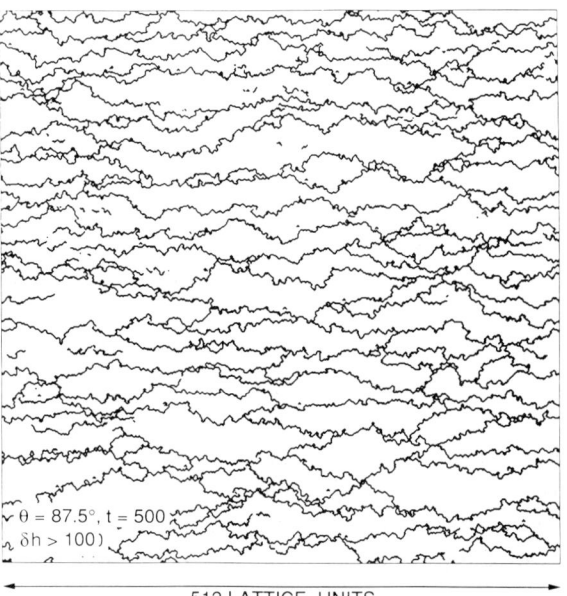

Fig. 8. This figure shows the location of the large steps (those with a height > 100 lattice units) projected on the $(x'y)$ plane obtained from simulations carried out in a column of size $L \times L = 512 \times 512$ lattice units. The pattern formed by the large steps (leading edges of the surviving clusters) at the time $t = 500$ is shown here.

boundaries whose motion is an Eden [13] growth process. The overshadowing of one cluster by another corresponds to the coalescence of their corresponding domain boundaries. Consequently, the pattern formation process can be described in terms of the coalescence of domain boundaries whose kinetics is an Eden growth process in much the same way that the two-dimensional pattern formation process can be understood in terms of the coalescence of Brownian particles. It follows from the analytical solution for the growth of the self-affine Eden model interfaces ($\xi_\perp \sim t^{1/3}$, $\xi_\parallel \sim t^{2/3}$) that $\sigma_x = 1/3$ and $\sigma_y = 2/3$. These values can then be used to calculate the other exponents describing the evolution of the pattern using scaling relationships.

The main simplifying feature at near grazing incidence appears to be the breakup of the deposit into non-contacting clusters, which allows a description of the d-dimensional surface in terms of independently moving ($d-1$)-dimensional objects.

5. Spatially correlated ballistic deposition

Most simulations of ballistic deposition are concerned with systems with a random, but otherwise uniform, flux of particles onto the growing surface. However, the effects of spatial correlations in the deposition density on the deposit structure is an interesting subject that has been investigated theoretically [28] and by computer simulations [29,30]. In the computer simulations [29,30] the particles are imagined to be dropped at the ends of the steps in a Levy flight [1] taking place in a horizontal line ($d_s = 1$) or plane ($d_s = 2$) above the growing deposit. The position at which the nth particle (or site) is deposited is given by

$$X_n = X_{n-1} + \delta X, \qquad (16)$$

where δX is a randomly selected vector with a length given by the power law distribution

$$P(|\delta X| > X_0) = X_0^{-f}. \qquad (17)$$

Here $P(|\delta X| > X_0)$ is the probability that the step length is greater than X_0 and f is an exponent corresponding to the fractal dimensionality of the Levy flight. Periodic boundary conditions are used in the lateral direction (s).

Fig. 9 shows a structure grown using a lattice model (here X_n is truncated to an integer to select the column in which growth will occur) for deposition onto a line. A value of 1.0 was used for f in this simulation which is on the same scale as that shown in fig. 3 for ordinary ballistic deposition ($f \to 0$). The deposit in fig. 9 has a lower density and a "rougher" surface. Large scale simulation results suggest that, for deposition onto a line, α (or H) = $1/2$ for $f < 1/2$ and $\alpha = 1/2 + (f - 1/2)/3$ for $f > 1/2$. The scaling reaction $\alpha + \alpha/\beta = 2$ appears to be obeyed. This is in accord with the theoretical results of Medina et al. [28]. If restructuring (movement of the deposited particle to the nearest local minimum) is included, the simulation results indicate that $\alpha = (1 + f/2)/2$ and $\beta = \alpha/2$.

The results obtained from three-dimensional sim-

Fig. 9. Part of a deposit generated using a square lattice model for spatially correlated ballistic deposition with a correlation exponent (f) of 1.0. This figure should be compared with that shown in fig. 3 for uncorrelated random deposition.

ulations are qualitatively similar [25]. For $f=0.5$ and 1.0, the exponent α (or H) has an effective value of about 0.36 (essentially the same value as that obtained from random uncorrelated ballistic deposition models). For $f=1.5$ and 2.0, α has a value of about 0.44 and 0.50, respectively. With restructuring, $\alpha=0$ for uncorrelated random deposition (ξ depends logarithmically on L). Values of about 0.28, 0.51, 0.72 and 0.85 were measured for $f=0.5$, 1.0, 1.5 and 2.0 respectively.

6. Summary

Self-affine fractal surfaces are characteristic of many systems of both scientific and practical importance. Here some recent advances concerned with the structure of rough surfaces generated by ballistic deposition are described. Both computer simulations and theoretical analysis have contributed to the recent progress in this area. In particular, the work on ballistic deposition at oblique angles illustrates how these two approaches complement each other. The computer simulation results motivated the theoretical models but cannot be used to obtain exact results. Exact results (that can be tested by comparison with simulations) are obtained from the theoretical analysis which also leads to a deeper understanding of the pattern formation process. The ballistic deposition models described here are certainly over-simplified. However, they do lead to structures that resemble those formed by vapor depositions onto a cold substrate. A careful quantitative analysis of the structure of deposits formed in this way using the concepts of fractal geometry would be a valuable contribution that is, so far, lacking.

Acknowledgement

The work described here was carried out in collaboration with R.C. Ball, R. Jullien, J. Krug, P. Ramanlal and L.M. Sander.

References

[1] B.B. Mandelbrot, The Fractal Geometry of Nature (Freeman, New York, 1982).
[2] B.B. Mandelbrot, Les Objets Fractals: Forme, Hasard et Dimension (Flammarion, Paris, 1975).
[3] B.B. Mandelbrot, Fractals: Form, Chance and Dimension (Freeman, San Francisco, 1977).
[4] B.B. Mandelbrot, in: Fractals in Physics, L. Pietronero and E. Tosatti, eds. (North-Holland, Amsterdam, 1986), pp. 3, 17, 21.
[5] B.B. Mandelbrot, Physica Scripta 32 (1985) 257.
[6] B.B. Mandelbrot, Encyclopedia of Physical Science and Technology, Vol. 5 (Academic Press, New York, 1987), p. 579.
[7] R.F. Voss, in: Scaling Phenomena in Disordered Systems, NATO ASI Ser. B 133, R. Pynn and A.T. Skjeltorp, eds. (Plenum, New York, 1986).
[8] S. Alexander, in: Transport and Relaxation in Random Materials, J. Klafter, R.J. Rubin and M.S. Shlesinger, eds. (World Scientific, Singapore, 1986).
[9] J. Feder, Fractals (Plenum, New York, 1988).
[10] H.E. Hurst, R.P. Black and Y.M. Simaika, Long Term Storage: An Experimental Study (Constable, London, 1965).
[11] F. Family and T. Vicsek, J. Phys. A 18 (1985) L75.
[12] M.J. Vold, J. Colloid Sci. 14 (1959) 168.
[13] M. Eden, Proc. Fourth Berkeley Symposium on Mathematics, Statistics and Probability, Vol. 4, F. Neyman, ed. (University of California Press, Berkeley, 1961) p. 223.
[14] P. Meakin, P. Ramanlal, L.M. Sander and R.C. Ball, Phys. Rev. A 34 (1986) 5091.
[15] P. Meakin, J. Phys. A 20 (1987) L1113.
[16] J. Krug, Phys. Rev. A 36 (1987) 4565.
[17] M. Kardar, G. Parisi and Y.C. Zhang, Phys. Rev. Lett. 56 (1986) 442.
[18] T. Halpin-Healy, Phys. Rev. Lett. 62 (1986) 442.
[19] J.M. Kim and M. Kosterlitz, preprint.
[20] P. Meakin and R. Julien, J. Phys. (Paris) 48 (1987) 1651.
[21] P. Meakin, Phys. Rev. A 38 (1988) 994.
[22] J. Krug and P. Meakin, Phys. Rev. A (1989), to be published.
[23] M. Bramson and D. Griffeath, Ann. Prob. 8 (1980) 183.
[24] C. Doering and D. ben-Avraham, Phys. Rev. A 38 (1988) 3035.
[25] J.L. Spouge, Phys. Rev. Lett. 60 (1988) 871.
[26] M. Bramson and D. Griffeath, Z. Wahrscheinlichkeitstheorie verw. Gebiete 53 (1980) 183.
[27] P. Meakin and J. Krug, preprint.
[28] E. Medina, T. Hwa, M. Kardar and Y.-C. Zhang, Phys. Rev. A 39 (1989) 3053.
[29] P. Meakin and R. Julien, Europhys. Lett. (1989), to be published.
[30] P. Meakin and R. Julien, preprint.

FRACTALS AND THE AC CONDUCTIVITY OF DISORDERED MATERIALS

G.A. NIKLASSON

Physics Department, Chalmers University of Technology, S-41296 Göteborg, Sweden

The frequency dependence of the ac conductivity of disordered insulators and semi-conductors often displays a power-law behaviour at high frequencies. This power law crosses over to a dc conductivity, or to another power law, at low frequencies. The power-law regions can be interpreted in terms of conduction on fractal structures or may be due to fractal time processes. We review the various theories that have been put forward for the dielectric response based on the concept of fractals. In experimental data both temperature-dependent and temperature-independent power law exponents have been observed. The physical mechanisms behind the ac response may be a power-law distribution of transition rates, conduction on fractal and percolation clusters or Coulomb interactions, but a detailed theory allowing us to discriminate between these is still lacking.

1. Introduction

The very slow relaxations that occur upon charging and discharging of dielectric materials have been studied for more than a century [1]. From experiments conducted over long periods, von Schweidler [2] found that the discharge current decays approximately as a power law in time. Much later Scher and Lax [3] interpreted the conduction in disordered solids in terms of a distribution of waiting times between the transitions of the charge carriers. It soon became evident that this distribution in many cases must have a power law form, in order to fit experimental data [4]. For effects of this type, Mandelbrot [5] introduced the concept of fractal time [6]. During recent years fractal time processes and fractal structures have been widely used to interpret the dielectric response of disordered materials [7].

In this paper we review recent progress in the understanding of the relationship of fractals to the frequency dependence of the ac conductivity. Some examples of experimental data that can be interpreted in terms of fractals are also given. We mainly consider materials where the ac conductivity is constant at low frequencies and shows a power law in frequency at high frequencies. The anomalous low-frequency dispersion (ALFD) [8], which exhibits two power laws in different frequency ranges, was thoroughly treated in a recent review [7], and will only be briefly mentioned here.

2. The "universal response"

The ubiquitous occurrence of power laws in the dielectric response of almost all materials has been called the "universal response" by Jonscher [9]. In this paper we are concerned with disordered materials where the dielectric response arises from charge carriers. For materials which display a dc conductivity at low frequencies the following empirical relations are well established:

$$\epsilon_1(\omega) = \epsilon_1(0), \quad \epsilon_2(\omega) = \sigma_{dc}/\omega\epsilon_0, \quad \omega < \omega_c, \quad (1a)$$

$$\epsilon_1(\omega) - \epsilon_\infty \sim \epsilon_2(\omega) \sim \omega^{n-1}, \quad \omega > \omega_c. \quad (1b)$$

Here $\epsilon(\omega) = \epsilon_1(\omega) + i\epsilon_2(\omega)$ is the complex frequency-dependent dielectric permittivity, σ_{dc} is the dc conductivity, ϵ_0 is the permittivity of vacuum, ϵ_∞ is the contribution to the permittivity from other processes at higher frequencies, and ω_c is a crossover frequency. The exponent n can take values between zero and unity. In the case of an ALFD eq. (1a) has to replaced by [8]

$$\epsilon_1(\omega) - \epsilon_\infty \sim \epsilon_2(\omega) \sim \omega^{-p}, \quad \omega < \omega_c, \qquad (2)$$

where the exponent p also can display values between zero and unity.

In experimental situations it is of vital importance to distinguish a true power law from something that looks like a power law over a limited range of frequencies. The occurrence of a power law is most accurately established by taking advantage of the information contained in both the real and the imaginary part of the permittivity. A convenient quantity to study is the loss tangent, or its inverse $\tan \phi$, which is given by

$$\tan \phi = [\epsilon_1(\omega) - \epsilon_\infty]/\epsilon_2(\omega) . \qquad (3)$$

In a frequency range where power-law behaviour prevails, the inverse loss tangent has a constant value. In the case of eq. (1b) the value becomes [9] equal to $\tan(n\pi/2)$. This relation makes possible an accurate determination of the exponent. However, it must be emphasized that in order to use eq. (3) for this purpose, the high-frequency permittivity ϵ_∞ must also be known. A possible way to determine this quantity may be inferred from recent work by Long [10]. If eq. (1b) holds, we can take the derivative of ϵ_1 with respect to $\ln \omega$ and obtain ϵ_∞ from the relation

$$\epsilon_\infty = \epsilon_1(\omega) - \frac{1}{n-1} \frac{d\epsilon_1(\omega)}{d \ln \omega}, \qquad (4)$$

when evaluated at an arbitrary frequency within the power-law region. We believe that this procedure will yield reasonable estimates of ϵ_∞ also when the frequency response is an approximate power law.

Unfortunately procedures such as these were not often followed in earlier work, in analyses of experimental data on the ac conductivity and dielectric permittivity. In our discussion of experimental data below, we will only consider data where the power-law character has been established by the method outlined above, or by the alternative method of Kramers–Kronig analysis.

3. Theory

In this section we briefly review the relationship between fractal structures, fractal time processes and the dielectric permittivity. First we treat conduction on fractal structures and percolation clusters. Subsequently we consider some physical mechanisms that can give rise to a fractal time response.

The dielectric permittivity is proportional to [7] the complex frequency-dependent diffusion coefficient divided by the frequency. Furthermore the diffusion coefficient has been related to the mean-square displacement of a random walk of the charge carriers by Scher and Lax [3]. Hence the dielectric response of a fractal structure depends not only on the fractal dimension, D_f, but also on the dimensionality of a random walk on the structure, D_w. In fig. 1a we depict the iterative construction rule for a fractal cluster built of small particles. In real materials the building blocks may be conducting particles or alternatively localized states partaking in a hopping conduction process. In many cases the fractal structure persists to a certain correlation length, while the structure be-

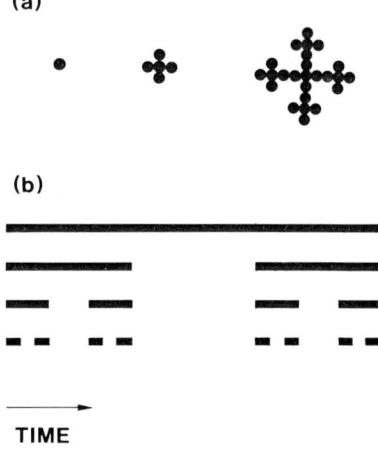

Fig. 1. (a) An iterative construction rule that builds up a fractal cluster of particles is illustrated. (b) The Cantor bar construction for a fractal time process. The set of points obtained after applying the construction rule infinitely many times represents the times at which a charge carrier in a disordered material makes a transition.

comes regular at longer length scales. The dielectric response of such a structure follows eq. (1) where [11] $n=1-D_f/D_w$ and ω_c corresponds to the correlation length.

On the other hand, if there exists a distribution of fractal clusters in a material, the situation may become more complicated. Percolation theory [12], which has been the subject of intense interest recently, is an example where the size distribution of clusters is a power law. Percolation theory should be valid if the conducting particles or localized states are randomly distributed in the material. There exist two different theories for the ac properties of percolation materials, which arrive at eq. (1). In the theory of anomalous diffusion [13] the power-law exponent for $\omega > \omega_c$ is obtained from $n=(1+D_w-D_f)/D_w$. However, this theory neglects capacitances between the different clusters. If these are important, the exponent n is instead given [14,15] by $n=t/(s+t)$, where t is the power-law exponent for the dc conductivity above the percolation threshold and s is the exponent of the dielectric constant near the threshold. The diffusion of charge carriers on fractals and percolation clusters have recently been thoroughly reviewed [16].

In fig. 1b we illustrate an iterative procedure for obtaining the distribution of events pertinent to a fractal time process [6]. The ac response of a material exhibiting such a sequence of transitions of charge carriers is a power law with exponent $n=1-D_t$ [7]. Here D_t denotes the fractal dimension of the process. The simplest example of a fractal time process arises when hopping takes place in an exponential distribution of activation energies, or equivalently from multiple trapping in an exponential band tail. In these cases $D_t=T/T_0$, where T is the temperature and T_0 is a parameter that characterizes the steepness of the distribution.

During the last decade evidence has been presented suggesting that Coulomb interactions between the charge carriers can give rise to fractal time processes. An interesting argument is given by the screened hopping model of Jonscher [17]. A transition of a charge carrier is accompanied by subsequent transitions of the screening charges around it. If the screening is incomplete so that the screened charge is a fixed fraction of the unscreened one, it can be shown [17] that the ratio of energy stored to energy lost per cycle is a constant. Furthermore, this is equivalent to stating that the loss tangent is constant [9], which is a signature of a power-law response, as mentioned above. Another indication that interactions can give rise to a fractal time process can be obtained from recent work by Funke [18]. He developed a rate equation formalism of the hopping process allowing for Coulomb interactions and obtained a power-law dielectric response over a range of frequencies, depending on the parameters of the calculation.

A more general situation occurs when a fractal time process is operating on a fractal structure. In this case the random walk exponent D_w is modified by the presence of another scaling process. Harder et al. [19] have studied this subject in detail and arrived at the results

$$D_w = D_{w0} + D_f(1-D_t)/D_t \quad \text{when } D_f < D_w,$$

and

$$D_w = D_{w0}/D_t \quad \text{when } D_f > D_w.$$

Here D_{w0} denotes the random walk dimension in the absence of the fractal time. In the case of percolation theory it is established [7] that the conductivity exponent t depends on the value of D_t. It should be noted that the crossover frequency of the fractal time needs not be equal to that pertaining to conduction on the fractal structure. This allows for the existence of two power laws with different values of the exponents in different frequency ranges. A dielectric response of this kind, the ALFD, is often seen in experiments.

4. Experiments

In this section we present some experimental results that can be interpreted in terms of the theories mentioned above. In order to establish a connection between structure and ac properties, many studies have considered materials with accurately known structural characteristics. Song et al. [20] studied composites of carbon and Teflon close to the percolation threshold. The composition dependence of the dc properties was in excellent agreement with percolation theory, but the ac exponent n was found to be 0.86, which is larger than the percolation prediction $n=t/(s+t)=0.7$. A fractal time process, possibly Coulomb interactions, appears to be operating on the percolation clusters thus leading to a higher value of n.

Disordered insulators and semiconductors are an interesting class of materials that exhibit a dc or "quasi-dc" conductivity at low frequencies crossing over to a power-law response at high frequencies. The exponent n is in certain cases temperature dependent, but in many cases independent of temperature. In fig. 2 we depict n as a function of temperature for thin films of WO_3 [21], ScO_x [22], and amorphous P [23]. For these data the existence of a power law has been established by studies of the inverse loss tangent, or by Kramers–Kronig analysis. A temperature-dependent n has also been claimed in other materials, but it is often not clear whether an exact power law was observed or whether different parts of a slow crossover were detected as the temperature was changed. The temperature dependence of n in fig. 2 is reminiscent of the behaviour associated with hopping in an exponential density of states or with an exponential distribution of activation energies. However, n approaches unity at a finite temperature, which is not in agreement with the theoretical expression $n=1-T/T_0$. The reason for this discrepancy must be left as an open question at present.

It is also quite common to find cases where the exponent n does not vary noticeably with temperature. As an example we mention some sputtered SiO_2 and silicon oxynitride films recently studied by us [24]. In fig. 3 we exhibit the inverse loss tangent of SiO_2 in a wide frequency range. A power-law response with $n=0.95$ and a slow crossover towards lower frequencies is clearly seen. The data in fig. 3 cover a sufficiently wide frequency range so that comparison with analytical theories of the dielectric response are worthwhile. The value of n was used as input in these calculations. It is seen that an average over a Pareto (i.e. power-law) distribution of transition rates cannot explain our measurements. On the other hand a recent theory of Dissado and Hill [25] is in good agreement with our data, although the crossover to dc behaviour at low frequencies is broader in the experiment. This theory is built on the concept of an infrared divergent response and can be interpreted in terms of fractal dynamics [26]. We propose that the power law in fig. 3 is due to a fractal time process operating on a percolation structure. We feel that the most likely cause of the fractal time in this case is interaction effects, since n was found to be independent of temperature.

Finally, the ALFD, which consists of two power laws has been amply documented before [7,8,25] and will not be treated here.

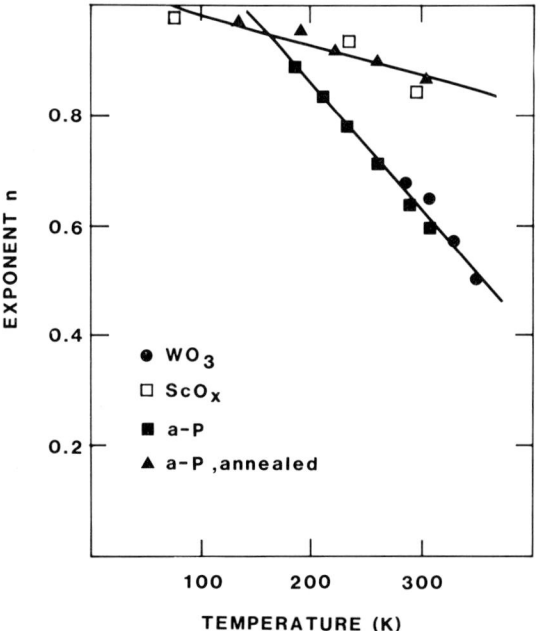

Fig. 2. The power-law exponent of the ac conductivity at high frequencies, n, is depicted as a function of temperature for thin films of tungsten trioxide, scandium oxide and amorphous phosphorous, as shown in the inset. The lines are drawn as a guide for the eye.

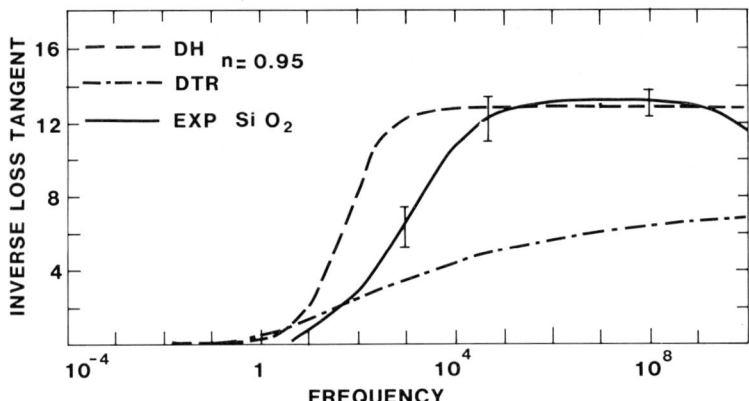

Fig. 3. Inverse loss tangent as a function of frequency for sputtered silicon dioxide thin films. The frequency scale was normalized by putting the onset of dc conductivity to unity. Full line with error bars shows experimental results, which display a power law with $n=0.95$ at high frequencies. Dashed line shows calculations by the Dissado–Hill (DH) theory, while dash-dotted line shows a calculation using a Pareto distribution of transition rates (DTR). The calculations were carried out with $n=0.95$.

5. Conclusion

In this paper we have applied the concept of fractals in order to interpret the dielectric properties of disordered solids. The occurrence of power laws in the dielectric response suggests such an interpretation. We have reviewed various theories of the ac properties of materials exhibiting fractal dynamics and applied them to experiments on some disordered thin film systems. In general it appears difficult to make a prediction of the measured power-law exponents from an assumption of a fractal or percolation structure only. It seems that in addition another physical mechanism is operating in a variety of disordered insulators and semiconductors. Hence this additional mechanism should be of a general nature and should exhibit fractal time behaviour. We suggest that Coulomb interactions between the charge carriers is a good candidate. However, it is uncertain whether interaction effects can explain both temperature-independent and temperature-dependent power-law exponents.

Acknowledgements

This research was financially supported by the National Swedish Board for Technical Development and by ABB Corporate Research, Västerås, Sweden. I wish to thank K. Brantervik, T.S. Eriksson and C.G. Granqvist for discussions and assistance.

References

[1] R. Kohlrausch, Pogg. Ann. Phys. Chem. 91 (1854) 56, 179.
[2] E.R. von Schweidler, Ann. Phys. 24 (1907) 711.
[3] H. Scher and M. Lax, Phys. Rev. B 7 (1973) 4491, 4502.
[4] H. Scher and E.W. Montroll, Phys. Rev. B 12 (1975) 2455.
[5] B.B. Mandelbrot, Fractals: Form, Chance and Dimension (Freeman, San Francisco, 1977).
[6] M.F. Shlesinger, Ann. Rev. Phys. Chem. 39 (1988) 269.
[7] G.A. Niklasson, J. Appl. Phys. 62 (1987) R1.
[8] A.K. Jonscher, Phil. Mag. B 38 (1978) 587.
[9] A.K. Jonscher, Dielectric Relaxation in Solids (Chelsea Dielectrics, London, 1983).
[10] A.R. Long, Adv. Phys. 31 (1982) 553.
[11] J.P. Clarc, A.-M.S. Tremblay, G. Albinet and C.D. Mitescu, J. Phys. Lett. 45 (1984) L913.
[12] D. Stauffer, Introduction to Percolation Theory (Taylor and Francis, London, 1985).
[13] Y. Gefen, A. Aharony and S. Alexander, Phys. Rev. Lett. 50 (1983) 77.
[14] A.L. Efros and B.I. Shklovskii, Phys. Stat. Sol. (b) 76 (1976) 475.
[15] D. Stroud and D.J. Bergman, Phys. Rev. B 25 (1982) 2061.
[16] S. Havlin and D. Ben-Avraham, Adv. Phys. 36 (1987) 695.

[17] A.K. Jonscher, Phys. Stat. Sol. (b) 84 (1977) 159.
[18] K. Funke and I. Riess, Z. Phys. Chem. 140 (1984) 217;
K. Funke, Z. Phys. Chem. 154 (1987) 251.
[19] H. Harder, S. Havlin and A. Bunde, Phys. Rev. B 36 (1987) 3874.
[20] Y. Song, T.W. Noh, S.-I. Lee and J.R. Gaines, Phys. Rev. B 33 (1986) 904.
[21] A.M. Solodukha and O.K. Zhukov, Fiz. Tverd. Tela 28 (1986) 579 [Sov. Phys. Solid State 28 (1986) 323].
[22] G.E. Pike, Phys. Rev. B 6 (1972) 1572.
[23] P. Extance, S.R. Elliot and E.A. Davis, Phys. Rev. B 32 (1985) 8148.
[24] G.A. Niklasson, T.S. Eriksson and K. Brantervik, Appl. Phys. Lett. 54 (1989) 965.
[25] L.A. Dissado and R.M. Hill, J. Chem. Soc. Faraday Trans. II 80 (1984) 291.
[26] L.A. Dissado and R.M. Hill, Chem. Phys. 111 (1987) 193.

FRACTON DYNAMICS

R. ORBACH

Department of Physics, University of California, Los Angeles, CA 90024, USA

The dynamics of fractal networks can be expressed in terms of elementary excitations termed "fractons". Examples are short length scale vibrational excitations in silica aerogels and magnetic excitations in site-diluted antiferromagnets. Fracton properties have been developed in terms of scaling theories and direct numerical simulations. Scattering structure factors are available within an effective medium approximation. Use of scaling forms for the fracton "velocity" and "width" allow for direct comparison with experiment. Fracton wave functions can be exhibited explicitly for percolating networks at critical concentration. Specific fracton realizations fall off sharply at the edges, while ensemble averages follow a super-localization form, with length scales satisfying scaling, and an exponent $d_\phi \simeq 2.3$.

1. Introduction

The concept of fractal structure, developed by Mandelbrot [1], has proven to be of great utility because many structures that appear purely random can be described within a geometric mathematical framework. Fractal concepts describe not only the static geometrical properties of such structures but also their dynamical properties and interactions with external measurement probes. It is the purpose of this article to explore the latter with regard to recent experimental studies of vibrational and magnetic structures which have been found to exhibit fractal structure (self-similar geometry [1]) at short length scales. In section 2, we explore the nature of fracton excitations, and discuss the properties of the scattering structure factor. In addition, very recent computational modelling [2] of the vibrational dynamics of a percolating network has generated explicit realizations of the excitations, to which we have appended the name "fractons" [3]. We examine the nature of the fracton wavefunction in section 3.

2. Fracton excitations and the scattering structure factor, $S(q, \omega)$

The two physical systems which we shall discuss, the silica aerogels and the site diluted antiferromagnet $Mn_xZn_{1-x}F_2$, exhibit self-similar geometry at length scales shorter than a characteristic length which we shall denote by ξ. The mass density is length scale dependent in this regime, varying as

$$\rho(r) = M(r)/V(r) = Br^D/Cr^d \propto r^{D-d}. \quad (1)$$

Here, D is the fractal dimension [1], and d is the embedding (or Euclidean) dimension. For length scales longer than ξ, the geometry crosses over to Euclidean, with $\rho(r)$ constant. For a percolating network, ξ depends upon the concentration p for site (or bond) occupancy: $\xi \simeq a/(p-p_c)^\nu$ where $\nu = 4/3$ ($d=2$) and $\nu \simeq 0.88$ ($d=3$) [4], and p_c is the critical percolation concentration. This allows us to vary ξ at will by altering p. Such is the case, for example, for site diluted magnets where the magnetic ions can be substituted by nonmagnetic cations (e.g. $Mn_xZn_{1-x}F_2$ [5]). For the silica aerogels, which are definitely not a percolating network, ξ can be changed over a wide range by varying the microscopic sample density.

Dynamics can be introduced by making an assumption concerning the length dependence of the force constant [6] or exchange coupling of the same general form as for the density (1). For the diluted antiferromagnet,

$$J \simeq r^{-\theta/2}, \quad r < \xi, \qquad (2)$$
$$= \text{constant}, \quad r > \xi.$$

This follows from mapping the antiferromagnetic excitation problem onto the diffusion equation [7], and making use of the introduction of the quantity θ by Gefen et al. [8]. Follow a scaling analysis of the same form as Rammal and Toulouse [9], the excitation frequency ω varies with wave vector q as

$$\omega(q) = J(1/q) q \propto q^{1+\theta/2} \equiv q^{D/\bar{d}}, \quad r < \xi, \qquad (3)$$

where we have introduced the fracton dimensionality [3] $\bar{d} = 2D/(2+\theta)$. For $r < \xi$, $\bar{d} \leq D \leq d$, with equalities for $r > \xi$ ($\theta = 0$ for $r > \xi$). The value of $\bar{d}$ was conjectured to equal 4/3 for $d \geq 2$ [3]. Most numerical estimates and an ϵ expansion [#1] show that this is a (good) approximation, but not exact. A series expansion in $d=2$ by Essam and Bhatti [11] finds it close to exact.

The availability of a dispersion law (3) enables us to define a crossover frequency ω_c [3],

$$\omega_c \propto \xi^{-D/\bar{d}}, \qquad (4)$$

such that for $\omega > \omega_c$ the excitations are fractons, but for $\omega < \omega_c$ the excitations are Euclidean (e.g. phonons and magnons).

The issue of fracton lifetime was addressed recently by Aharony et al. [12]. They suggested, under conditions of a single length scale, that the fracton linewidth $\Gamma \equiv 1/\tau(\omega)$ behaves as

$$\Gamma \equiv 1/\tau(\omega) \propto \omega^{d+1}/\omega_c^d, \quad \omega \leq \omega_c, \qquad (5)$$
$$\propto \omega, \quad \omega \geq \omega_c.$$

The behavior in the $\omega \leq \omega_c$ regime is termed Rayleigh; in the $\omega \geq \omega_c$ regime it is termed Joffe–Regel scattering. It must be emphasized that fractons, in the harmonic approximation, are well defined quantum states with infinite inelastic lifetimes. The widths which appear in (5) result from elastic scattering, and in the absence of anharmonicity are those measured with plane wave probes (e.g. neutron or light scattering). Only recently has anharmonicity been introduced to calculate fracton inelastic lifetimes [13].

It is necessary to connect the dispersion law (3) and linewidth (5) expressions through the crossover regime between the high (fracton) and low frequency (phonon, magnon) limits in order to predict the form for the scattering structure factor $S(q, \omega)$. There is as yet no satisfactory theoretical method for so doing (the effective medium approximation gives much too sharp crossover behavior). Using an effective velocity of sound, $c(\omega)$, Courtens et al. [14] have introduced the smooth phenomenological forms for vibrational excitations,

$$c(\omega) = c_0 [1 + (\omega/\omega_c)^m]^{z/m}, \qquad (6a)$$

$$\Gamma(\omega) = \frac{\omega^4}{\omega_c^3} \frac{1}{[1+(\omega/\omega_c)^m]^{3/m}}. \qquad (6b)$$

Here, the exponent m characterizes the sharpness of the crossover, and z is defined by $c(\omega) = \omega^z$ at $\omega > \omega_c$, whence $z = 1 - \bar{d}/D$. The beauty of (6) is that it can be used to obtain the scattering structure factor $S(q, \omega)$ over the full frequency range right through the (phonon, magnon)–fracton crossover region.

An integral form for $S(q, \omega)$ for fractons has been obtained by Aharony, Entin-Wohlman and Orbach [15]. Defining $\omega_x = \max\{\omega_c, q^{D/\bar{d}}\}$, the asymptotic values for $S_{\text{fr}}(q, \omega)$ are

$$S_{\text{fr}}(q, \omega) \propto \omega, \quad \omega \ll \omega_x, \qquad (7)$$
$$\propto 1/\omega^3, \quad \omega \gg \omega_x.$$

The only method known to obtain a closed form expression for $S(q, \omega)$ is through the effective medium approximation. This has been accomplished for vibrational excitations by Polatsek and Entin-Wohlman [16] and for magnetic excitations in diluted antiferromagnets by the authors of ref. [7]. It is interesting to compare the two. They differ in form even though the dispersion laws are similar because of the canonical transformation necessary for the latter.

[#1] See ref. [10] for a full discussion.

$$S_{\text{vib}}(q,\omega) = \frac{c(\omega)q}{\omega^2}$$
$$\times \left(\frac{\tau^{-1}(\omega)}{[\omega-c(\omega)q]^2+\tau^{-2}(\omega)} \right.$$
$$\left. - \frac{\tau^{-1}(\omega)}{[\omega+c(\omega)q]^2+\tau^{-2}(\omega)} \right), \quad (8a)$$

$$S_{\text{mag}}(q,\omega) = \frac{c(\omega)q^2}{\omega}$$
$$\times \left(\frac{\tau^{-1}(\omega)}{[\omega-c(\omega)q]^2+\tau^{-2}(\omega)} \right.$$
$$\left. + \frac{\tau^{-1}(\omega)}{[\omega+c(\omega)q]^2+\tau^{-2}(\omega)} \right). \quad (8b)$$

Courtens et al. [14] used (6) in (8a) to fit their Brillouin scattering curves for silica aerogels over the full phonon and well into the fracton regimes. The fits were excellent, and have enabled them to obtain independent values for D and $\bar{d}$ for a number of aerogels prepared for different densities and under different preparation regimes. Whereas they obtained $D \simeq 2.4$ and $\bar{d} \simeq 1.3$ for a series of neutrally reacted silica aerogels which had been aged for a long time in the alcogel state before hypercritical drying, they found [17] $D \simeq 2.33$ and $\bar{d} \simeq 1.8$ for hypercritical drying soon after gelation. They reached the conclusion that "$\bar{d}$ is not a universal dimension, but rather an additional one which is sensitive to the microstructure of the fractal".

As yet, use has not been made of (6) in (8b) to attempt to fit the measured $S(q,\omega)$ for $Mn_{0.5}Zn_{0.5}F_2$ by Uemura and Birgeneau [5]. It would be interesting because these authors found, near $q \sim 1/\xi$, a double peak in $S(q,\omega)$ which they attributed to separate well-defined magnon and fracton excitations. Such structure has not been seen for vibrational excitations in the silica aerogels, though use of (6) in (8a) fits the scattering data well.

3. Fracton wave function

The specific form for the fracton wave function is required for dynamical calculations. Use has been made of a superlocalized form [18] to calculate localized moment relaxation in fractal lattices [19] and for specific evaluation of the effects of lattice anharmonicity [13]. The availability of supercomputers has now allowed us to check the validity of this form.

Use of algorithm of Williams and Maris [20] has enabled Nakayama and Yakubo [2] to solve for the eigenfunctions of a site-percolating network consisting of N atoms with unit mass and linear springs connecting two nearest-neighbor atoms for systems with $N > 10^5$. The equations of motion of the system are

$$\ddot{u}_i(t) + \sum_j K_{ij} u_j(t) = 0, \quad (9)$$

where u_i is the scalar displacement of the atom on the ith site. The force constant is taken to be $K_{ij}=0$ ($i \neq j$) if either sites i or j are unoccupied, and $K_{ij}=1$ otherwise. Diagonal elements satisfy the relation $K_{ii} = -z_i$ where z_i is the coordination number of the site i. Should we have put $K_{ii}=0$, we would have created the quantum percolation problem. For the purposes of this manuscript, u_i has only a single component.

The realization of an eigenmode at a specific eigenfrequency will in general have a shape different from another eigenmode at the same frequency in a random structure. It is necessary to take ensemble averages to obtain a smooth wave function whose shape can be parameterized by a minimum set of parameters. We have ensemble averaged over 100 eigenfunctions each at five different frequencies for percolating vibrational networks at $p_c=0.593$ in $d=2$ on a 700×700 square lattice, and at $p_c=0.312$ in $d=3$ on a $90 \times 90 \times 90$ simple cubic lattice.

We have analyzed the $d=2$ results using the suggested [18] form for the fracton wave function

$$\phi_{\text{fr}}(r) \sim \exp\{-[r/\Lambda(\omega)]^{d_\phi}\}, \quad (10)$$

where $\Lambda(\omega) \sim q^{-1}(\omega)$ from the "dispersion law" (3), and r is a radial distance from the center. In this sense, we are taking explicit note of the superlocalized form [18] for the fracton wave function, identifying $\Lambda(\omega)$ as the localization length. The use of (3) provides a strong prediction on the spatial extent of the wave function as a function of fracton frequency.

The absolute value of the wave function at $\omega=0.01$ ($q=1/a$, the lattice constant, at $\omega=1$) was averaged over 129 realizations. The resultant shape is exhibited in fig. 1 as a function of the distance r from the center (open circles), and the log-log plot of r and $-\ln(\langle|\phi_{\rm fr}|\rangle)$ (filled circles). The straight line through the filled circles is a least square fit, while the solid line through the open circles represents the same line on a linear scale. The gradient of the straight line generates d_ϕ for the averaged fracton wave function. We find $d_\phi=2.3\pm0.1$, while the localization length $\Lambda(\omega=0.01)=17.2$. The two quantities were also extracted at four other frequencies, $\omega=0.005$, 0.006, 0.007, and 0.008, excited on five $d=2$ percolating networks at $p_{\rm c}$. The results are given in table 1. The values of $\Lambda(\omega)$ are plotted as a function of frequency ω on a log-log scale in fig. 2. The straight line through the filled circles is a least square fit to $\Lambda(\omega)\sim\omega^{-\lambda}$, with $\lambda=0.71$. This value of the exponent is in close agreement with the scaling prediction for the fracton dispersion law for a $d=2$ percolating network ($\lambda=\bar{d}/D=0.705$). This agreement between scaling

Table I
The value of the localization length $\Lambda(\omega)$ in units of the atomic spacing ($a=1$), and the geometrical exponent d_ϕ for various frequencies ω

ω	$\Lambda(\omega)$	d_ϕ
0.005	28.13	2.25
0.006	24.18	2.24
0.007	23.49	2.34
0.008	19.41	2.26
0.010	17.20	2.31

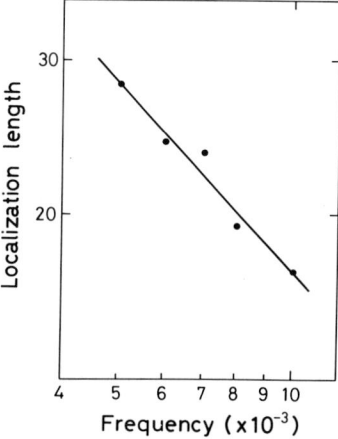

Fig. 2. The values of the localization length $\Lambda(\omega)$ plotted as a function of frequency on a log-log scale. The straight line $[\Lambda(\omega)\propto\omega^{-0.71}]$ is a least squares fit, showing that fractons in our ensemble follow the correct fracton dispersion law $[\Lambda(\omega)\propto\omega^{-\bar{d}/D}$, with $\bar{d}/D=0.705]$.

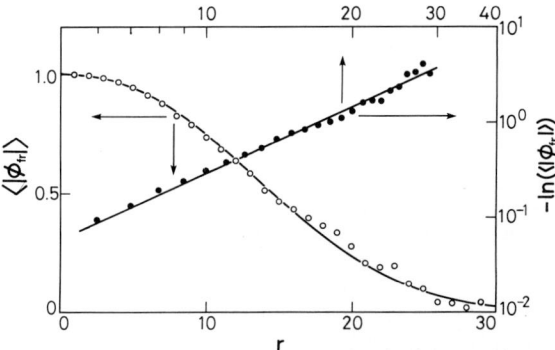

Fig. 1. Filled and open circles are numerical results of the ensemble averaged shape of the fracton wavefunctions $|\phi_{\rm fr}|$ for 129 specific fracton realizations with $\omega=0.01$. The filled circles are plotted on a log-log scale, in which the upper abscissa ($\ln r$) gives the distance r from the center of the fracton in units of the atomic spacing $a=1$. The right ordinate $[-\ln(\langle|\phi_{\rm fr}|\rangle)]$ measures the fracton amplitude in a logarithmic scale. The straight line is a least squares fit for the filled circles. This line gives the value for the exponent $d_\phi\simeq2.3$ as well as $\Lambda(\omega=0.01)=17.2$. The solid curve is drawn on a linear scale with $d_\phi=2.3$ and $\Lambda(\omega=0.01)=17.2$.

theory and the numerical simulations suggests that the form of $\langle\phi_{\rm fr}\rangle$ in (10) involving the ratio $r/\Lambda(\omega)$ is correct. The exponent d_ϕ is essentially independent of ω as seen from table 1. It is also much larger than the upper bound found for localized electronic states [21], suggesting that the two problems are different in detail. In addition to the ensemble average, it is possible to view a single realization. Figs. 3 and 4 are fractons with $\omega=0.01$ excited on $d=2$ and $d=3$ percolation networks at $p_{\rm c}$, respectively. The individual fractons possess very sharp edges as can be seen in fig. 5, a cross section through the fracton exhibited in fig. 3 in two different directions. Thus, it is only the

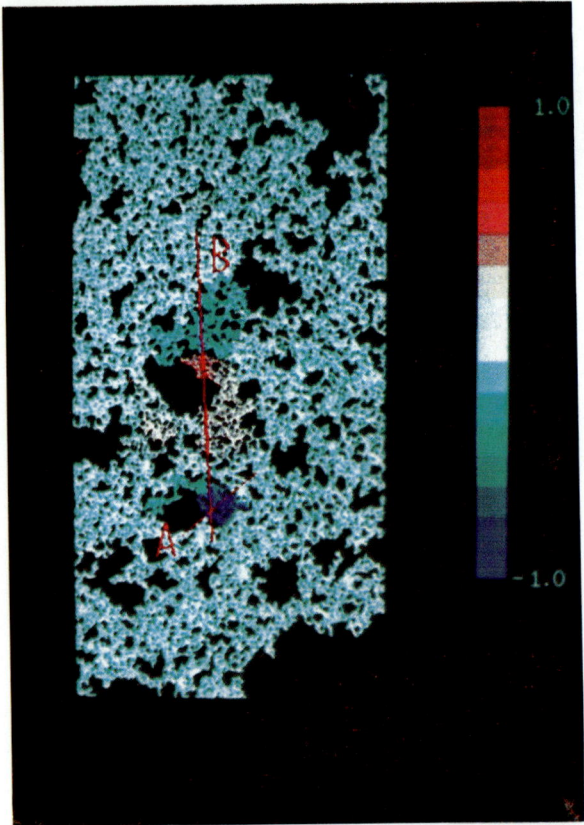

Fig. 3. A specific realization of a fracton with frequency $\omega=0.01$ excited on a $d=2$ percolating network at p_c with 169 576 sites formed on a 700×700 square lattice. The colors indicate the amplitude of vibration normalized with the maximum equal to unity. Note the two lines A and B drawn on the figure. The amplitude of the fracton along these lines is plotted in fig. 5.

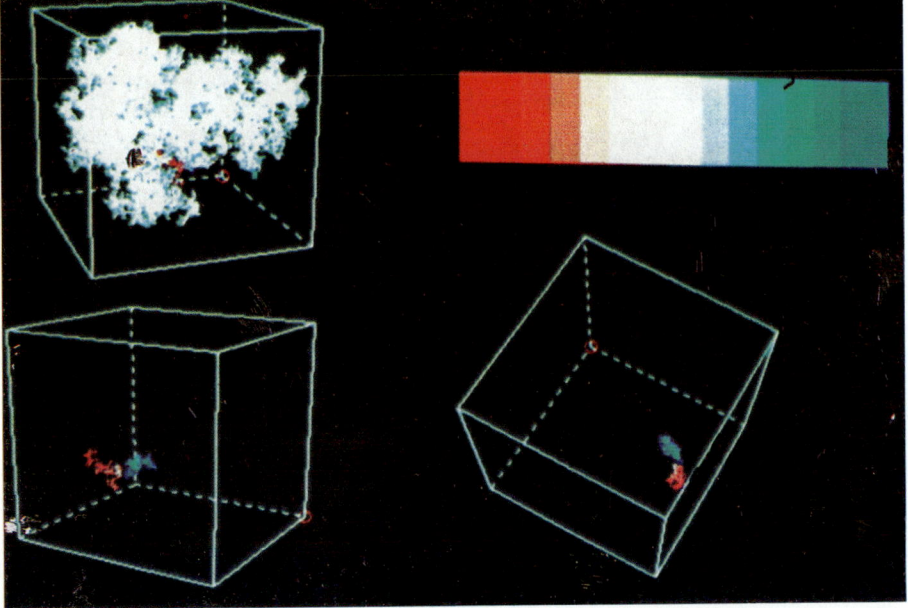

Fig. 4. A specific realization of a fracton with frequency $\omega=0.01$ excited on a $d=3$ percolating network at p_c with 18 970 sites on a $90\times 90\times 90$ simple cubic lattice. The left upper box shows the whole system containing the excited fracton. The lower two boxes are figures in which only the fracton's amplitude is exhibited. A small red circle indicates the origin of the system. The colors indicate the amplitude of vibration normalized with the maximum equal to unity.

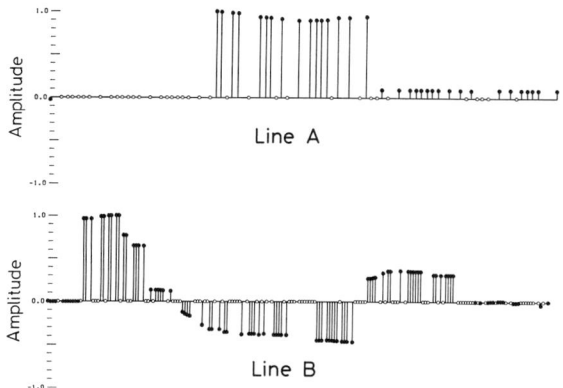

Fig. 5. Cross sections of the fracton in $d=2$ exhibited in fig. 3. The upper figure corresponds to the line A in fig. 3. The lower figure corresponds to the line B in fig. 3. The sign of the amplitudes is exchanged for convenience.

ensemble average of the fracton which can be described by superlocalization. Individual realizations have edges much sharper than that described by (10). This will have profound implications for calculations of fracton dynamics, and represents a challenge for explicit calculations of physical quantities. We are currently exploring anharmonic couplings of fractons through these simulations [22].

4. Summary

We have described the nature of fracton excitations from a scaling perspective, generating the dispersion law and elastic lifetime. We have shown how these can be incorporated into an expression for the scattering structure factor $S(q, \omega)$. Next we have used numerical simulations in order to extract the specific fracton wave function. We have found that it falls off sharply at the edges of the vibrational excitation, and that only the ensemble studies give promise of actual calculation of fracton dynamics, including lattice vibrational relaxation of localized centers and fracton–phonon anharmonic interactions important for thermal transport [13].

Acknowledgements

This work was supported by a grant from the U.S. National Science Foundation, DMR 88-05443. The author gratefully acknowledges very helpful conversations with Professors A. Aharony, S. Alexander, O. Entin-Wohlman, T. Nakayama, Dr. E. Courtens and Dr. R. Vacher, and Mr. G. Polatsek and Mr. K. Yakubo.

References

[1] B.B. Mandelbrot, The Fractal Geometry of Nature (Freeman, San Francisco, 1983); Ann. Israel Phys. Soc. 5 (1983) 59; J. Stat. Phys. 34 (1984) 895.
[2] K. Yakubo and T. Nakayama, J. Phys. Soc. Japan 58 (1989) 1504; Phys. Rev. B (1989), in press;
T. Nakayama, K. Yakubo and R. Orbach, J. Phys. Soc. Japan 58 (1989) 1891.
[3] S. Alexander and R. Orbach, J. Phys. (Paris) 43 (1982) L625.
[4] P. Grassberger, Math. Biosci. 62 (1986) 157; J. Phys. A 19 (1986) 1681.
[5] Y.J. Uemura and R.J. Birgeneau, Phys. Rev. Lett. 57 (1986) 1947; Phys. Rev. B 24 (1987) 7024.
[6] Kin-Wah Yu, P. Chaikin and R. Orbach, Phys. Rev. B 28 (1983) 4831.
[7] R. Orbach and Kin-Wah Yu, J. Appl. Phys. 61 (1987) 3689; G. Polatsek, O. Entin-Wohlman and R. Orbach, J. Phys. (Paris) 49 (1988) C8-1191; Phys. Rev. B (1989), in press.
[8] Y. Gefen, A. Aharony and S. Alexander, Phys. Rev. Lett. 50 (1983) 77.
[9] R. Rammal and G. Toulouse, J. Phys. (Paris) 44 (1983) L13.
[10] S. Havlin and D. Ben-Avraham, Adv. Phys. 36 (1987) 695.
[11] J.W. Essam and F.M. Bhatti, J. Phys. A 18 (1985) 3577.
[12] A. Aharony, S. Alexander, O. Entin-Wohlman and R. Orbach, Phys. Rev. Lett. 58 (1987) 132.
[13] S. Alexander, O. Entin-Wohlman and R. Orbach, Phys. Rev. B 34 (1986) 2726;
A. Jagannathan, R.Orbach and O. Entin-Wohlman, in: Cooperative Dynamics in Complex Physical Systems, H. Takayama, ed. (Springer, Berlin, 1988), p. 183; Phys. Rev. B, in press.
[14] E. Courtens, R. Vacher, J. Pelous and T. Woignier, Europhys. Lett. 6 (1988) 245.
[15] A. Aharony, O. Entin-Wohlman and R. Orbach, in: Time-Dependent Effects in Disordered Materials, R. Pynn and T. Riste, eds. (Plenum, New York, 1987), p. 233.

[16] G. Polatsek and O. Entin-Wohlman, Phys. Rev. B 37 (1988) 7726.
[17] R. Vacher, E. Courtens, G. Coddens, J. Pelous and T. Woignier, Phys. Rev. B (1989), in press.
[18] Y.-E. Levy and B. Souillard, Europhys. Lett. 4 (1987) 233;
O. Entin-Wohlman, S. Alexander and R. Orbach, Phys. Rev. B 32 (1985) 8007;
S. Alexander, O. Entin-Wohlman and R. Orbach, Phys. Rev. B 34 (1986) 2726.
[19] S. Alexander, O. Entin-Wohlman and R. Orbach, J. Phys. (Paris) 46 (1985) L549, L555; Phys. Rev. B 32 (1985) 6447, 8007; B 33 (1986) 3935; B 35 (1987) 1166.
[20] M.L. Williams and H.J. Maris, Phys. Rev. B 31 (1985) 4508.
[21] A.B. Harris and A. Aharony, Europhys. Lett. 4 (1987) 1355.
[22] T. Nakayama, K. Yakubo and R. Orbach, unpublished.

THE FRACTAL GALAXY DISTRIBUTION

P.J.E. PEEBLES

Joseph Henry Laboratories, Princeton University, Princeton, NJ 08544, USA

The space distribution of galaxies on scales less than about 50 million light years is remarkably well approximated as a fractal with dimension $D = 1.23 \pm 0.04$. The history of the discovery of this effect is reviewed, and then some issues that arise in attempts to explore the details, are mentioned.

1. Historical remarks

It is a pleasure to take part in the celebration of the work of a truly creative individual. The story of how he arrived at the idea that the space distribution of galaxies might be described as a fractal, and predicted [1] the character of the hierarchy of galaxy N-point correlation functions, is best left to be told by Benoit Mandelbrot himself. This personal account of how the same ideas evolved in parallel in astronomy should not be confused with a proper historical study of what all the main actors were doing.

The old astronomical phrase for a fractal is a "clustering hierarchy". Charlier [2], in the first decades of this century, applied it to the distribution shown in fig. 1. Many of the dots represent nearby galaxies of stars, comparable in size to our Milky Way galaxy. (Others turned out to be bright objects in our galaxy.) The band across the center is nearly empty because dust in the plane of our galaxy blocks the light. There are more galaxies in the top half of the map than the bottom; we are on the edge of the concentration de Vaucouleurs [3] later called the Local Supercluster. Within this concentration there are many condensations and subcondensations. The result certainly resembles a fractal.

Einstein [4] granted that an unbounded clustering hierarchy is a logical way to construct a universe, but rejected it on the grounds that it violates his interpretation of Mach's principle, that inertial frames are definable in a consistent way across all space because there is about as much mass everywhere to define inertial reaction by gravity. This led him to prefer a homogeneous universe. In the 1930s that became the standard picture, perhaps in part because of Einstein, certainly also because Hubble's [5] deeper galaxy surveys seemed to show that clustering of the sort observed in the Local Supercluster fades into a homogeneous distribution when averaged over larger length scales. As noted below, I think the evidence for this is now close to compelling, though, it is fair to say, not definitive. But in any case, that does not dispose of the evidence for a clustering hierarchy on smaller scales. A few people, such as de Vaucouleurs [6], kept the idea alive.

I decided to study the statistics of the galaxy distribution (starting in a serious way in 1972) because it seemed to me that if we could measure the relatively small fluctuations from homogeneity on large scales it should not be too hard to trace their evolution back in time and so learn something about the primeval conditions responsible for the formation of galaxies and clusters of galaxies. The program now is popular, though my early estimates of the ease of application were optimistic.

The natural first choice of measure of the galaxy distribution is the autocorrelation function, or its transform, the power spectrum. Early examples of the

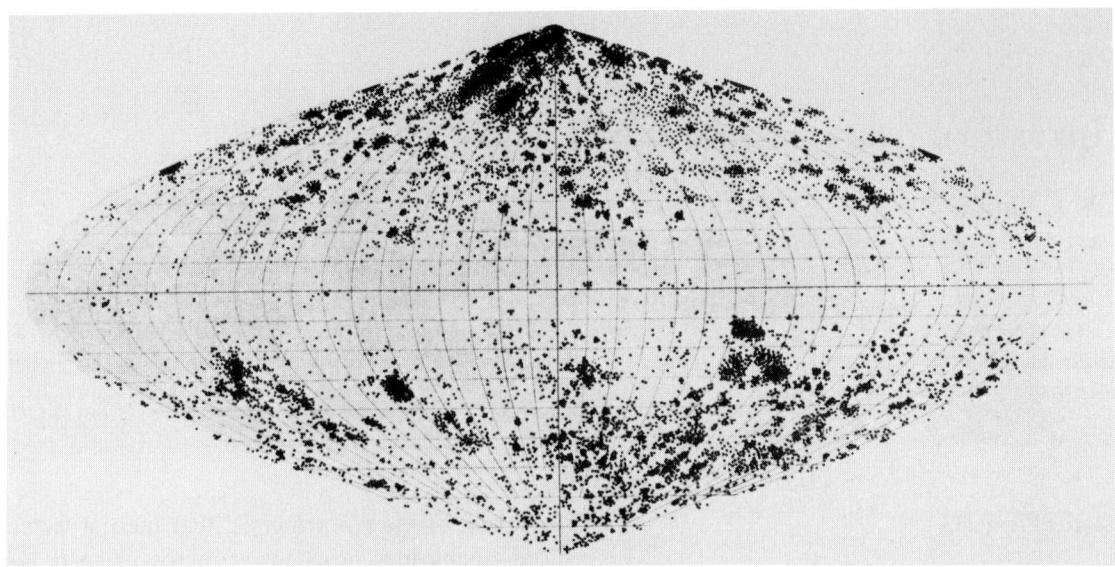

Fig. 1. Most of the points in this all sky map represent nearby galaxies. The low density along the plane is caused by obscuration by the dust in our galaxy. This map suggested to Charlier [2] that the galaxy distribution is a clustering hierarchy, or fractal.

former are given by Limber [7] and Totsuji and Kihara [8]; of the latter by Yu and me [9]. For a point process, the autocorrelation function becomes the two-point position correlation function, $\xi(r)$. Assuming a spatially homogeneous process, $\xi(r)$ is defined by the joint probability of finding galaxies in each of the volume elements δV_1 and δV_2 at separation r:

$$\delta P = n^2 [1 + \xi(r)] \delta V_1 \delta V_2, \qquad (1)$$

where n is the mean galaxy number density. A catalog of angular positions of galaxies with fixed apparent brightness or angular size defines a two-dimensional process for which we can similarly define an angular two-point correlation function, $w(\theta)$. The expectation value of $w(\theta)$, which is a good approximation to what is estimated from a large catalog, is a linear integral over $\xi(r)$. This is the key to the program, because galaxy distance measures are so uncertain that one cannot directly and reliably estimate $\xi(r)$ on small scales. Instead, one measures $w(\theta)$, and then works back to ξ. This approach reveals that, at small r, $\xi(r)$ is close to a power law [8,10],

$$\xi(r_0/r)^\gamma, \quad \gamma = 1.77 \pm 0.04. \qquad (2)$$

One believes that this is a meaningful result, because it is reproduced from catalogs of galaxy distributions at different depths [10]. The significance of this point is discussed further below.

The clustering length r_0 in eq. (2) is about 0.002 times the Hubble length (at which the cosmological expansion velocity extrapolates to the velocity of light, and on the order of the particle horizon beyond which we lose causal connection subsequent to any inflation epoch). At $r \sim 2r_0$, $\xi(r)$ shows a break down from the power law [11–13]. We have only bounds on ξ at larger separations.

The two-point function characterizes the galaxy distribution but certainly does not fix it. One way to proceed is to go to higher-order correlation functions, which Groth and I did in 1973. The reduced three-point function, $\zeta(r_{12}, r_{23}, r_{31})$, is defined as in eq. (1): the joint probability of finding galaxies in each of the volume elements δV_1, δV_2, and δV_3 that define the triangle with sides r_{12}, r_{23}, and r_{31} is

$$\delta P = n^3 \delta V_1 \delta V_2 \delta V_3$$
$$\times [1 + \xi(r_{12}) + \xi(r_{23}) + \xi(r_{31})$$
$$+ \zeta(r_{12}, r_{23}, r_{31})]. \qquad (3)$$

We can similarly define a three-point function $z(\theta_{12}, \theta_{23}, \theta_{31})$ for a projected angular distribution of galaxies. As for the two-point function, z is a linear integral over the spatial function, ζ. A formal inversion of this equation to get ζ is impractical; one models ζ as a function with parameters adjusted to fit z. I chose to try first the exceedingly simple form,

$$\zeta = Q[\xi(r_{12})\,\xi(r_{23}) + \xi(r_{23})\,\xi(r_{31}) + \xi(r_{31})\,\xi(r_{12})]. \quad (4)$$

This form has some good properties; two particularly recommended it to me.

The first is based on a picture of how the clumpy small-scale mass distribution might develop by gravitational instability in an expanding universe. In the standard general relativity cosmology, a small density fluctuation $\delta\rho/\rho$ grows as a power of time reckoned from the start of expansion at the Big Bang. That is not surprising, because there is no characteristic length or time in gravity physics or the world model to permit any other functional form. Where $\delta\rho/\rho$ is positive and has grown to a value of about unity, the material in this dense region breaks away from the general expansion to form a gravitationally bound clump, perhaps a galaxy, or a cluster of galaxies, or a supercluster. What might be the sequence of creation of these objects? If the primeval mass fluctuations approximated a random Gaussian process with power spectrum that is not too steep, then the rms fluctuation in mass averaged through a window of size r increases with decreasing r, so smaller mass objects would break away from the general expansion earlier, forming objects that later became embodied in larger objects than formed at lower mean density. That is, gravitational instability might be expected to form a clustering hierarchy [10,14]. As is now well known from the properties of fractals, this picture would predict that ζ varies with length scale as the square of ξ, as does eq. (4).

The second and even more compelling argument for eq. (4) was simplicity: the integral that projects ζ to the angular function z projects eq. (4) into the same functional form, a symmetric sum of products of the two-point function, w. This functional form has proved to give an excellent fit to the angular data, so it generally is agreed that eq. (4) is a useful approximation to reality.

One can model the reduced four-point function in a similar way as products of ξ, again guided by the clustering hierarchy picture. Fry and I [15] found that this successfully fits the data. I have not found a graduate student with the strength to work through the five-point function.

It was about the time we were working on the four-point function that I became aware of a case of parallel evolution. Mandelbrot gave me a copy of his book, Les Objects Fractals [16], in which he shows models of the galaxy distribution as the stepping points in a Rayleigh–Lévy random walk. I remember looking up his paper [1] on the subject, and seeing the functional form in eq. (4), with an easy generalization to forms for the higher-order functions close to what the observations indicate. So I started referring to the galaxy distribution as a fractal rather than a clustering hierarchy.

2. Bounded and unbounded fractals

Given the success of the fractal picture for the galaxy distribution on small scales, it is natural to ask whether the same fractal character might extend to the large-scale structure of the universe [1,6,16]. My esteemed colleague Mandelbrot and I have been debating the observational merits of this elegant idea for more than a decade. Here is why I have concluded that it is untenable [17].

Mandelbrot dramatic examples have firmly fixed in our minds the point that a pure fractal has the same appearance under a change of scale. If the galaxy distribution were a pure fractal, we would see that galaxy counts fluctuate across the sky as they do in fig. 1, independent of the depth of the sample. To quantify this, suppose one estimates the angular two-point correlation function for the galaxies seen at distance R and in a fixed area of the sky. One could do this for various values of R, holding the sky area fixed. Under the pure fractal picture, one would find that the result is statistically independent of R,

$$w(\theta, R) = W_f(\theta) . \qquad (5)$$

This has to be so because a pure fractal has no characteristic length scale: W_f cannot depend on R because the fractal cannot balance the length unit. On the other hand, if the fractal is bounded, and the depth of the galaxy sample is large, $R \gg r_0$ (eq. (2)), Limber's equation [7] predicts

$$w(\theta, R) = R^{-1} W(\theta R) . \qquad (6)$$

Here the function W contains R in the dimensionless combination $\theta R / r_0$.

We have two pictures, which give unambiguous and quite different predictions, and we have ample data to test them. The observations are consistent with eq. (6), and clearly violate eq. (5). The evidence is reviewed in refs. [17,18]; here I only present an example of why one concludes that the pure fractal picture fails.

As we have noted, if the galaxy distribution were a pure renormalizable fractal a map of more distant galaxies would look like fig. 1, with a higher density of points but similar fluctuations in the density of points across the sky. Fig. 2A shows the galaxy distribution at a depth ≈ 10 times that of fig. 1. This is from the counts of Shane and Wirtanen [19], as reduced at Princeton [20]. (The reduction has been criticized [21], and this led to a lengthy reanalysis [12,22], which left me convinced of the reliability of the results.) It is clear to the eye that this deeper galaxy map does not look like fig. 1.

The fine structure in the deeper galaxy map in fig. 2A is consistent with eq. (6): it has the character one would expect if we were seeing many incoherent patches of the sort seen in fig. 1, with the patches uniformly spread through space. This is illustrated in fig. 2B, which shows a clustering model Soneira and I put together [23]. It has a statistically uniform space distribution of bounded fractal clumps, each clump having diameter $\sim 4r_0$. (This is roughly equivalent to many Rayleigh–Lévy walks started at randomly chosen places.) Most people prefer the real universe over this artificial one; our map has the flavor of a student's copy of a master's painting. But it does illustrate my point that the galaxy distribution looks like a fractal on small scales, while at $r \gg r_0$ it is inconsistent with a pure fractal, consistent with a homogeneous random process.

A related point might be mentioned. I have shown angular galaxy distributions, which are relatively easy to obtain. One can estimate distances from red-shifts (the shift in wavelength of spectral features). The accuracy is a matter of debate (it is complicated by the motions of the galaxies), but clearly is good enough for an exploratory survey. This has only now become feasible, with improvements in detector efficiency and data handling. It has revealed that galaxies tend to define sheets [24], some of which are a good deal broader than r_0. The latter effect has led some to ask whether there might be something wrong with the correlation function statistics, and therefore some merit in the pure fractal picture. A simple example helps one to see why neither follows.

To understand how the statistic $\xi(r)$ responds to sheet-like distributions, imagine a model universe in which galaxies are uniformly distributed on flat sheets, with the sheets placed at separation L to form a cubic lattice. We might call this the "egg crate" model universe. It has two-point correlation function

$$\xi(r) = \tfrac{1}{6} [L/r + 2(L/r) \operatorname{Int}(r/L) - 2] . \qquad (7)$$

This is shallower than the galaxy function at small r; we could fix that by clustering the points on the sheets and adding clusters of points off the sheets, both of which would make the model more realistic. However, the point the model is meant to illustrate is the large-scale behavior. Eq. (7) defines a clustering length where $\xi(r_0) = 1$:

$$r_0 = \tfrac{1}{8} L . \qquad (8)$$

If r_0 were taken from the galaxy distribution, L would be equal to the width of the largest known hole in the galaxy distribution [25]. That is, this example shows that the existence of empty regions and of sheets of galaxies an order of magnitude broader than the clustering length r_0 is not contradictory.

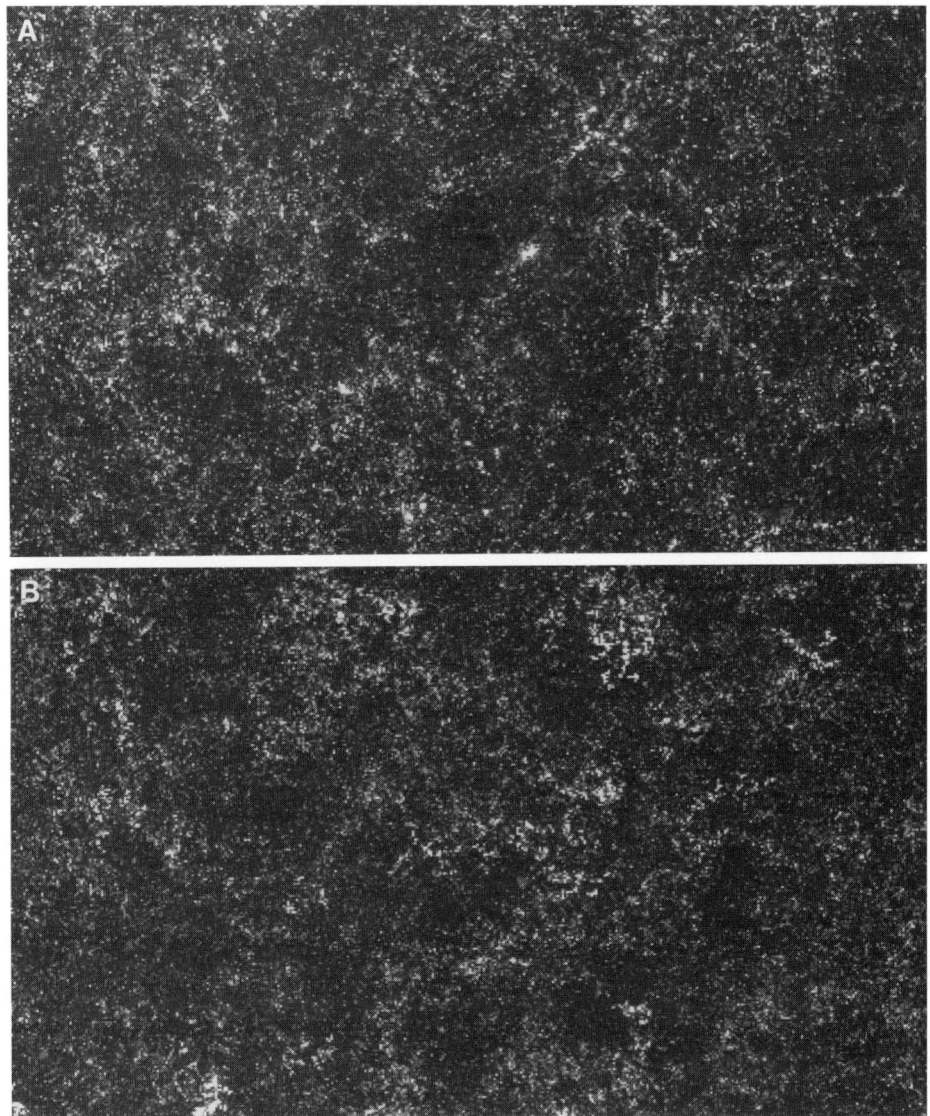

Fig. 2. (A) Map of the angular positions of galaxies at about 10 times the distance of those in fig. 1. The map covers a 40° by 70° area of the sky centered on the top of fig. 1. For details see ref. [20]. The relatively smooth appearance shows that the large-scale galaxy distribution is not a direct extension of the fractal observed on the smaller scales of fig. 1. (B) Simulated galaxy distribution constructed from bounded fractals placed in a statistically homogeneous way [23].

3. Comments

I conclude that the character of the galaxy distribution changes from a pure fractal with dimension $3-\gamma=1.23$ on scales $r<r_0$ to something consistent with homogeneity on much larger scales. To those who remain unconvinced, I renew my offer of a bottle of fine New Jersey wine for the first observationally reasonable example of a pure fractal galaxy space distribution.

As so often happens in science, the discovery of the character of the galaxy space distribution leads to a host of new puzzles. Can we check whether the smaller-scale fractal developed out of the gravitational instability of the expanding universe? One clue, yet to be tested in detail, is that the small-scale galaxy velocity field would be expected to approximate a fractal [26]. Another test might be found in the character of the galaxy space distribution on scales ≈ 1 to $10r_0$, at the transition from the fractal sector. In numerical solutions for the growth of mass clustering in an expanding universe, the mass autocorrelation function at $r \approx r_0$ is steeper than the galaxy two-point function [27]. That may be because galaxies do not trace mass [28], or it may be because the solution assumes Gaussian initial fluctuations from homogeneity. Yet to be explored are the possibilities offered by fractal primeval density fluctuations. Finally, as indicated above, the fact that one sees empty regions larger than r_0 is to be expected given the existence of strong clustering on smaller scales; the surprise was the tendency for the bounding walls of these empty regions to be smooth. This is an example of the tendency of galaxies to define linear structures. The exploration of this phenomenon is just beginning, as is the discussion of whether it might be understood within the gravitational instability picture.

Acknowledgement

This research was supported in part by the US National Science Foundation.

References

[1] B.B. Mandelbrot, C.R. Acad. Sci. (Paris) A 280 (1975) 1551.
[2] C.V.L. Charlier, Ark. Mat. Astron. Fys. 16, No. 16 (1922).
[3] G. de Vaucouleurs, Astron. J. 58 (1953) 30.
[4] A. Einstein, Ann. Phys. 69 (1922) 436.
[5] E. Hubble, Astrophys. J. 79 (1934) 8.
[6] G. de Vaucouleurs, Science 167 (1970) 1203.
[7] D.N. Limber, Astrophys. J. 117 (1953) 134.
[8] H. Totsuji and T. Kihara, Publ. Astron. Soc. Japan 21 (1969) 221.
[9] J.T. Yu and P.J.E. Peebles, Astrophys. J. 158 (1969) 103.
[10] P.J.E. Peebles, Astrophys. J. 189 (1974) L51.
[11] E.J. Groth and P.J.E. Peebles, Astrophys. J. 217 (1977) 385.
[12] E.J. Groth and P.J.E. Peebles, Astrophys. J. 310 (1986) 507.
[13] S.J. Maddox, G. Efstathiou and J. Loveday, in: Evolution of Large-Scale Structure in the Universe, J. Audouze, M.-C. Pelletan and A. Szalay, eds. (Kluwer, Dordrecht, 1989).
[14] D. Layzer, Astrophys. J. 59 (1954) 170.
[15] J.N. Fry and P.J.E. Peebles, Astron. J. 221 (1978) 19.
[16] B.B. Mandelbrot, Les Objects Fractals (Flammarion, Paris, 1975), p. 73.
[17] P.J.E. Peebles, Large-Scale Structure of the Universe (Princeton Univ. Press, Princeton, 1980), p. 243.
[18] P.J.E. Peebles, in: Cosmology and Particle Physics, J.F. Nieves and D.R. Altschuler, eds. (World Scientific, Singapore, 1989).
[19] C.D. Shane and C.A. Wirtanen, Publ. Lick Obs. 22 (1967).
[20] M. Seldner, B. Siebers, E.J. Groth and P.J.E. Peebles, Astron. J. 82 (1977) 249.
[21] M.J. Geller, V. de Lapparent and M.J. Kurz, Astrophys J. 287 (1984) L55.
[22] M.E. Brown and E.J. Groth, Astrophys. J. 338 (1989) 605.
[23] R.M. Soneira and P.J.E. Peebles, Astron. J. 83 (1978) 845.
[24] V. de Lapparent, M.J. Geller and J.P. Huchra, Astrophys. J. 302 (1986) L1.
[25] R.P. Kirschner, O. Oemler, P.L. Schechter and S.A. Shechtman, Astrophys J. 314 (1987) 493.
[26] P.J.E. Peebles, Astron. Astrophys. 68 (1978) 345.
[27] P.J.E. Peebles, Astrophys. J. 297 (1985) 350.
[28] S.D.M. White, C.S. Frenk, M. Davis and G. Efstathiou, Astrophys. J. 313 (1987) 505.

THEORETICAL CONCEPTS FOR FRACTAL GROWTH

L. PIETRONERO

Dipartimento di Fisica, Università di Roma "La Sapienza", Piazzale A. Moro 2, 00185 Rome, Italy

After the introduction of fractal geometry by Benoit Mandelbrot the key problem is to understand *why* nature gives rise to fractal structures. This implies the formulation of models of fractal growth based on physical phenomena and the subsequent understanding of their mathematical structure in the same sense as the renormalization group has allowed to understand Ising-type models. The models of diffusion-limited aggregation and the more general dielectric breakdown model, based on iterative processes governed by the Laplace equation and a stochastic field, have a clear physical meaning and they spontaneously evolve into random fractal structures of great complexity. From a theoretical point of view however it is not possible to describe them within usual concepts. Recently we have introduced a new theoretical framework for this class of problems. This clarifies the origin of fractal structures in these models and provides a systematic method for the calculation of the fractal dimension and the multifractal properties. Here we summarize the basic ideas of this new approach and report about recent developments.

1. Introduction

"Fractal geometry is one of those concepts which at first sight invites disbelief but on a second thought becomes so natural that one wonders why it has only recently been developed". These words by Berry [1] from his review of Mandelbrot's book [2] explain why fractal geometry is having an important influence in all scientific disciplines and in particular in physics. This concept was clearly lacking for the description of complex structures in nature and, by introducing it, Mandelbrot has provided a playground of new problems concerning basic properties of natural phenomena. These problems were up to now left at the margins of scientific activity because it was not possible to cast them within the framework of mathematical methods based on analyticity. From the point of view of fractal geometry it is now possible to pose these problems correctly.

This fact, by itself, has far reaching consequences also on problems that have been object of extensive study along traditional lines. An interesting example is the problem of the statistical properties of the large scale distribution of matter in the universe. Since a few years the complete three-dimensional distribution of galaxies in a certain luminosity range is available for appreciably large volumes. It is possible therefore to perform statistical analysis of these distributions [3]. This has been done extensively using mathematical methods that assume a priori homogeneity at large scale. The reasons for this assumption are basically historical and due to the following argument. The cosmological principle implies local isotropy, this together with analyticity leads to homogeneity [4]. In the absence of an alternative theoretical framework, analyticity was not considered as a property to be checked from experimental data but it was somehow included in the cosmological principle itself. Fractal geometry makes clear instead that local isotropy is not necessarily associated with homogeneity [2]. It was natural therefore to reconsider the statistical analysis of galaxy distribution from a more general point of view that does not assume homogeneity a priori. This has given rise to a surprising result [5]. Contrary to the previous conclusions [3], the galaxy distribution does not show any tendency to homogenize, so the implicit assumption of analyticity is actually incorrect. This new analysis points therefore to the possibility that the large scale distribution of matter in the inverse is fractal to all ob-

Essays in honour of Benoit B. Mandelbrot
Fractals in Physics – A. Aharony and J. Feder (editors)

served scales [5]. This is in agreement with the observation of large voids and superclusters that were up to now in sharp disagreement with the previous statistical analysis. If this conclusion will be confirmed by further studies it will lead to a major change in the foundations of cosmology because all present models are based on a homogeneous metric. This example shows how the same experimental data lead to a different conclusion once they are looked in the broader perspective of fractal geometry.

The next step is now to understand *why* nature gives rise to fractal structures. A deepening of interrelations between fractal geometry and physical phenomena is what may be called a theory of fractals and forms the objective of the present activity in this field [6]. The rest of this lecture will describe some recent developments in this direction [7,8].

It should be noted that the concept of self-similarity is not new in physics. It has been well known since the studies of critical properties of phase transitions, and has been instrumental in the formulation of the renormalization group (RG) method that has essentially solved this problem [9]. In that case, however, self-similarity was considered as a peculiar property of the critical point, characteristic of the competition between order and disorder in equilibrium phase transitions. We can now see that this property is much more common and it appears in several complex structures that are apparently unrelated to phase transitions. In fact the RG theory developed for phase transitions, or variations of it, has not been successful in describing irreversible fractal growth [10]. In the following we are going to see that, despite the common feature of universality, there are basic differences between phase transitions and irreversible fractal growth and new theoretical concepts have to be introduced for their description.

The key problem for fractals is to understand the essential elements for the spontaneous development of these structures. This implies the formulation of models for fractal growth based on physical phenomena and the subsequent understanding of their mathematical structure in the same sense as the renormalization group has allowed to understand Ising-type models.

The first part of this program is reasonably understood since a few years with the formulation of the models of diffusion-limited aggregation (DLA) [11] and the dielectric breakdown model (DBM) [12]. These models consist of simple iterative processes based on the Laplace equation and a stochastic field and their dynamical evolution leads spontaneously to complex structures with a well defined fractal dimension. The relation to physical processes is quite clear and for this reason they have acquired a crucial role in the understanding of fractal growth, similar to that of the Ising model for phase transitions [13].

Until recently most of the work has been based on computer simulations [6]. From a theoretical point of view these models, despite their apparently simple formulation, have eluded all approaches based on known methods, like the RG theory [10]. Recently we have introduced a new theoretical framework that appears especially suitable for irreversible fractal growth. It clarifies the origin of fractal structures in these models and provides a systematic method for the calculation of the fractal dimension [7] (for a detailed description, see ref. [8]).

In this paper we are going to summarize the basic ideas of this new method and to report about recent developments along these lines. The work described refers to collaborations with A. Erzan, C. Evertsz, A.P. Siebesma, R.R. Tremblay, M. Marsili and A. Vespignani.

2. Physical models for fractals

As already mentioned the basic models to understand the physical origin of fractals are DLA and DBM [11,12]. From the point of view of DBM one assumes that the already grown pattern at a given time is equipotential. It is possible then to compute the local field around this structure by solving the Laplace equation

$$\nabla^2 \Phi = 0 \qquad (1)$$

with the boundary condition of constant potential on the grown structure and a different value of the potential at infinity. The growth probability p_j for each bond (j) connected to the structure is then related to the local field,

$$p_j = |\nabla\phi|_j^\eta / \sum_j |\nabla\phi|_j^\eta, \qquad (2)$$

where η is a parameter that modulates the randomness of the process. This probability distribution is then used to select one bond that becomes part of the pattern. The process then continues by iterating this procedure and the result is a spontaneous generation of a random fractal structure. Laplacian fields are common ingredients in diverse physical processes and realistic models for these processes often incorporate stochasticity. For these reasons these models are now believed to capture the essential features of pattern formation in seemingly different phenomena like electrochemical deposition, dendritic growth, dielectric breakdown, viscous fingering in fluids, fracture propagation and others [6].

From computer simulations we have learned that these models generate structures of great complexity whose fractal dimension $D(\eta, d)$ depends continuously on the parameter η and on the Euclidean dimension d of the space in which the structure is embedded [6]. With respect to the formulation of an analytical theory one should note that the process consists of an irreversible dynamics with long-range couplings both in space and in time. None of the known mathematical methods seems to be able to apply to this situation. The absence of a Hamiltonian, in the sense that the statistical weight of a configuration depends on its entire history, gives rise to serious conceptual difficulties if one tries to adapt to this problem the RG methods developed for phase transitions [9]. In addition these growth models do not show an upper critical dimension and therefore ϵ-expansion methods cannot even be formulated.

3. Theoretical framework

3.1. Search for an iterative equation and its fixed point

The growth dynamics is irreversible and it leads spontaneously to an asymptotic structure with well-defined statistical fractal properties that are independent of the initial conditions. The fractal structure can be seen therefore as the *attractor* of a dynamical system and this suggests to look for iterative equations whose *fixed point* should correspond to the *fractal* structure. The first important problem is therefore to define an appropriate *subject* that should appear in these iterative equations.

It is useful in this respect to consider the two-dimensional case with boundary conditions consisting of two parallel lines [7,8,14]. The discussion is easily generalizable to more dimensions. This geometry has the conceptual advantage of defining a unique growth direction. The intersection of a given fractal structure of dimension D with a line perpendicular to the growth direction gives rise to a set of points of dimension $D' = D - 1$. By analyzing this set of points with a procedure of box covering we assign a black dot to a box if this contains some point of the set and a white dot otherwise. The elementary process by which a black box is subdivided into two leads to two possible configurations indicated as type 1 (one black and one white sub-box) and type 2 (both sub-boxes black). The corresponding probabilities in this process of fine graining are indicated by C_1 and C_2 respectively. The average number of black sub-boxes that appear at the next level of fine graining from one black box is

$$\langle n \rangle = \sum_i n_i C_i = C_1 + 2C_2. \qquad (3)$$

It is easy to show that the fractal dimension of the whole structure is related to the values of C_1 and C_2 by [7,8]

$$D = 1 + \frac{\ln\langle n \rangle}{\ln 2}. \qquad (4)$$

It is clear therefore that the natural subject for the

iterative fixed point problem should be the *distribution of elementary configurations* $\{C_i\}$ that appear from one scale to the next in the process of fine graining. Note that this concept is easily generalizable to other fractal problems and even to fully developed turbulence [15,16].

3.2. Fixed scale transformation

We have now to define the appropriate iterative equations for the distribution $\{C_i\}$. From the point of view of the analogy with critical phenomena one would attempt to define the iteration via a renormalization scheme. The present problem however has the peculiarity of being *intrinsically critical*; namely there is no order parameter that leads to scale invariance only for a particular value of it. The growth rules are scale invariant by construction. This leads to a higher degree of symmetry:

(i) The structure is invariant by *scale transformation* in the sense that the values of (C_1, C_2) that are obtained via a fine graining from scale l to scale $\frac{1}{2}l$ are the same as those from scale l' to scale $\frac{1}{2}l'$.

(ii) In addition there is also an invariance with respect to the *dynamical evolution at the same scale*. Namely if one considers two different sections of the original structure and performs a fine graining analysis at the same scale for the two different sets one obtains the same distribution (C_1, C_2) that should also be identical to case (i).

Note that this discussion holds for a homogeneous as well as a self-affine structure [14]. We have now a choice for the formulation of the iterative process: One can use a renormalization transformation based on the invariance property (i), or one can define the iteration from the dynamical evolution at the same scale (ii) and then use the fixed point value of the (C_1, C_2) distribution at all other scales via eqs. (3) and (4). For reasons that will become clear later this second possibility is more convenient for this problem [7,8].

We define a *fixed scale transformation* (FST) that corresponds to the dynamical evolution at the same scale. This gives rise to an iterative equation of type

$$\begin{pmatrix} C_1^{(k+1)} \\ C_2^{(k+1)} \end{pmatrix} = \begin{pmatrix} M_{1,1} & M_{2,1} \\ M_{1,2} & M_{2,2} \end{pmatrix} \begin{pmatrix} C_1^{(k)} \\ C_2^{(k)} \end{pmatrix}, \qquad (5)$$

where the matrix element $M_{i,j}$ defines the conditional probability to have a configuration of type i followed by one of type j in the growth direction. Note that this refers to a "frozen structure" that has already grown to its asymptotic state and in which no more growth will occur in the future.

The practical calculation of $M_{i,j}$ can be done by starting with a configuration of type i and consider the growth processes that can lead to a subsequent configuration j. For example

$$M_{1,2} = \sum \text{ (growth processes leading from 1 to 2)} \qquad (6)$$
$$= p_\alpha + (1-p_\alpha)p_\beta + ... ,$$

where p_α represents in this case the probability that the first nontrivial growth process leads to a configuration of type 2, while p_β corresponds to the second-order term and so on. More details about the explicit construction of eq. (6) can be found in refs. [7,8] but for the boundary condition problem see section 3.3.

Once the matrix elements are defined and considering that

$$M_{1,1} + M_{1,2} = 1 , \quad M_{1,2} = M_{2,2} = 1 , \qquad (7)$$

the fixed point condition gives

$$C_1 = \left(1 + \frac{M_{1,2}}{M_{2,1}}\right)^{-1} \qquad (8)$$

and from this, via eqs. (3) and (4) one obtains the fractal dimension D. The *mathematical condition for the generation of fractal structures* can now be given by studying the value to which the series of eq. (6) will converge. If this value is less than one this implies that there is a finite probability that the growth process will leave sites asymptotically empty. In view of the scale invariance of the Laplace equation and of the whole growth process, this conclusion holds at any scale and therefore holes of all scales will be generated.

In view of the screening properties of the Laplace equation one can see that the convergence of this series is quite fast and indeed to a number smaller than one. In the limit $\eta \to 0$ the screening is eliminated, the convergence is then slower and the series converges to one, leading to a compact structure with dimension 2 [7,8].

3.3. Fluctuations of boundary conditions and void distribution

In order to compute explicitly the matrix elements one has to specify the boundary conditions for the process of conditional growth. This is a new type of problem that usually does not appear in the renormalization group method in which one integrates over the internal degrees of freedom and boundary conditions play no role. In our problem the boundary conditions instead have an important effect. In fact the growth of a structure will be very different if it is isolated or surrounded by other growing structures. In principle therefore one has to take into account all these possibilities, each with its own probability. Given the probability distribution (C_1, C_2) it is possible to derive explicitly the void distribution $P(\lambda)$ which is the conditional probability that, given an occupied box, it is neighboured at the right by a void of size λ. In refs. [7,8] we have presented an approximate calculation but one can actually do it exactly for all the values of the void size [17]. The result is given in an iterative form

$$P(\lambda=0) = \frac{C_2}{(1+\tfrac{1}{2}C_1)(C_2+\tfrac{1}{2}C_1)},$$

$$P(\lambda=2l+1) = \tfrac{1}{2}(1-C_2)P(\lambda=l),\qquad(9)$$

$$P(\lambda=2l) = \tfrac{1}{4}(1+C_2)P(\lambda=l)$$
$$+ \frac{(1-C_2)^2}{4(1+C_2)}P(\lambda=l-1).$$

It is easy to check that the integrated void distribution has the expected behaviour [2,17]

$$P(\lambda \geq \Lambda) = F\Lambda^{-D'},\qquad(10)$$

where F is the lacunarity and D' the fractal dimension of the intersection set.

With respect to the use of this distribution as defining the boundary conditions of our elementary growth processes there is an additional fact to consider. Our basic configurations correspond to pairs of sites of which at least one is black, so at the next level of coarse graining they correspond necessarily to a black site. To this site we can directly apply the above void distribution. However if we ask for the probability that a pair of sites is immediately followed by a black (occupied) site the answer is slightly more complicated. In fact with probability $P(\lambda=0)$ this pair is followed by another pair in which at least one of the two subsites is black. So if we consider the probability that the site immediately following the considered pair is black this will require the extra condition that the pair configuration on the right is of type 2 or of type 1 but with the black site on the left. This gives rise to an additional factor $(C_2+\tfrac{1}{2}C_1)$ and therefore the corrected probability (for example for $\lambda=0$) to be used in the calculation of the matrix elements is

$$P'(\lambda=0) = \frac{C_2}{1+\tfrac{1}{2}C_1}.\qquad(11)$$

Note that in refs. [7,8] the discussion of this distribution is not very clear but the actual calculation is finally correct and in agreement with the present discussion.

The probability distribution for the size of voids fixes the probability distribution for having a certain boundary condition in the elementary growth process. This implies that the generic matrix element $M_{i,j}$ should be interpreted as the convolution over all possible boundary conditions

$$M_{i,j} \Rightarrow \sum_n P'(\lambda_n; C_1, C_2) M_{i,j}(\lambda_n).\qquad(12)$$

Therefore these matrix elements become nonlinear functions of the variables (C_1, C_2). The iterative equation has then the following structure:

$$\begin{pmatrix} C_1^{(k+1)} \\ C_2^{(k+1)} \end{pmatrix} = \sum_n P'(\lambda_n; C_1, C_2)$$
$$\times \begin{pmatrix} M_{1,1}(\lambda_n) & M_{2,1}(\lambda_n) \\ M_{1,2}(\lambda_n) & M_{2,2}(\lambda_n) \end{pmatrix} \begin{pmatrix} C_1^{(k)} \\ C_2^{(k)} \end{pmatrix}\qquad(13)$$

and the corresponding fixed point equation

$$C_1 = \left(1 + \frac{\sum_n P(\lambda_n; C_1, C_2) M_{1,2}(\lambda_n)}{\sum_n P(\lambda_n; C_1, C_2) M_{2,1}(\lambda_n)}\right)^{-1} \quad (14)$$

is now a nonlinear system of equations of infinite order. This can be solved by suitable truncation schemes because the higher orders give exponentially decreasing contributions. The simplest nontrivial method to include the boundary condition fluctuations consists in assuming that as soon as λ_n is different from zero the boundary condition is essentially open (in the sense that the next branch can be considered as infinitely far). This approximation is based on the fact that the convergence of the series that define the matrix elements is rather fast so one has to include only a few orders in the calculation. This implies that the considered structure does not change its size appreciably and therefore the relative probabilities within the structure itself are not too sensitive to the distance of the next branch as soon as this is not very close. This we have called the open–closed approximation and it is possible to treat it analytically [7,8].

The results of various approximation schemes are shown in table 1 and they show an appreciable degree of systematicity. In particular one can see that the self-consistent treatment of boundary conditions has an important effect. It can also be noted that the agreement with the computer simulations is very good for large values of η and less accurate for smaller values. This can be understood in terms of the screening properties of the Laplace equation that give rise to a faster convergence of the series for large values of η.

Finally it should be remarked that the spirit of the present calculation is rather different from real space renormalization not only because one uses a transformation at fixed scale. In fact the size and the number of basic configurations is *strictly fixed* and the calculation is improvable in a systematic way by adding more terms in the series that define the matrix elements of the transformation. In addition growth processes outside the considered cell must be included up the desired level of convergence and the fluctuations of boundary conditions play an important role. These concepts appear natural and necessary in all problems of irreversible growth.

4. Recent developments

4.1. *Empty configurations*

From the way we have defined the basic configurations that appear in the iterative equation it is clear that they should not contain the completely empty configuration. In fact this cannot be generated in the process of fine graining. For this reason we have not included this possibility in the growth dynamics that

Table 1
Fractal dimension $D(\eta)$ [a]

Present theory	$\eta=0$	$\eta=0.5$	$\eta=1$ (DLA)	$\eta=2$
(λ_0); first order	1.7885	1.6465	1.4747	1.1885
(λ_0); second order	1.8990	1.7515	1.5418	1.1997
$(\lambda_0, \lambda_\infty)$; second order	1.8896	1.7549	1.6080	1.3956
$(\lambda_0, \lambda_\infty)$; third order	1.9039	1.7830	1.6406	1.4190
(any λ); ∞ order	2	–	–	–
computer simulations	2	1.92	1.70	1.43

[a] Values of the fractal dimension as a function of the parameter η for DBM model in two dimensions computed with the various schemes of the theory discussed in this paper and compared with the results of computer simulations. The case $\eta=1$ corresponds to DLA while $\eta=0$ gives one type of Eden model.

gives rise to the matrix elements. The idea is that the empty sites of the configurations of type 1 already generate voids of all sizes. However the conditional growth we consider for the definition of the matrix elements can, in principle, lead to the direct generation of a pair of empty sites. This is due to the fact that nearby branches could grow faster than the considered one and it is therefore possible that the pair of sites above the initial configuration remains empty. In order to study this effect one has to include this possibility explicitly in the fixed scale transformation but then consider only the ratio at the fixed point of the two usual configurations. For DLA in two dimensions the effect is to reduce the value of the fractal dimension by about 1% [18], justifying therefore the previous conjecture that this is a higher-order effect and, in general, it can be neglected.

4.2. Results for three dimensions

We have recently completed an analytical calculation of the properties of DLA and DBM in three dimensions. The intersection in this case is done with a plane and the basic configurations are five. The calculation can be performed along lines similar to those we have discussed here for two dimensions. The only exception is that the inclusion of boundary condition fluctuations is in this case more complex. The preliminary result for the fractal dimension of DLA is $D=2.4$ [18] to be compared with the numerical value $D=2.5$ [6].

4.3. Multifractal spectrum of growth probability

There have been several numerical studies concerning the multifractal spectrum of the growth probability for these systems. After the initial enthusiasm it is now becoming clear that there are serious conceptual problems with this approach. The problem is that the present formulation of multifractals assumes a homogeneous fractal as the support of the measure while in the present case the support consists of a self-affine structure. It is difficult therefore to judge the results of this studies. It seems clear that the negative moments are not very meaningful while the question is open as far as the positive moments are concerned. The only quantity that appears reasonably well defined and stable seems to be the strongest singularity strength that has also a direct relation to the fractal dimension [6].

From the point of view of the present theory it is clear that the growth probabilities we use are those that would define the generators of the multiplicative process that would lead to the multifractal spectrum. But here we use them directly as they appear in the growth process and within a framework that is fully compatible also with a self-affine structure. It is not clear that casing them in a multifractal scheme would be of any advantage. However the multifractal spectrum can also be computed analytically as a byproduct of the present theory. In practice one defines suitable square configurations that contain the probabilities used in the process of conditional growth [7,8,19] and consider these as the generators of a multiplicative process. The results for the strongest singularity strength of DLA in two dimensions give values of 0.6–0.7 depending on the particular scheme used [19].

References

[1] M.V. Berry, New Sci. (27 Jan, 1983).
[2] B.B. Mandelbrot, The Fractal Geometry of Nature (Freeman, San Francisco, 1983).
[3] P.J.E. Peebles, The Large Scale Structure of the Universe (Princeton Univ. Press, Princeton, 1980);
M. Davis and P.J.E. Peebles, Astrophys. J. 267 (1983) 465.
[4] S. Weinberg, Gravitation and Cosmology (Wiley, New York, 1972).
[5] L. Pietronero, Physica A 144 (1987) 257;
P.H. Coleman, L. Pietronero and R.H. Sanders, Astron. Astrophys. 200 (1988) L 32.
[6] L. Pietronero and E. Tosatti, eds., Fractals in Physics (North-Holland, Amsterdam, 1986);
H.E. Stanley and N. Ostrowsky, eds., On Growth and Form (Nijhoff, Dordrecht, 1986).
[7] L. Pietronero, A. Erzan and C. Evertsz, Phys. Rev. Lett. 61 (1988) 861.
[8] L. Pietronero, A. Erzan and C. Evertsz, Physica A 151 (1988) 207.

[9] D.J. Amit, Field Theory, the Renormalization Group and Critical Phenomena (McGraw-Hill, New York, 1978).
[10] L.P. Kadanoff, Phys. Today (Feb. 1986) 6.
[11] T.A. Witten and L.M. Sander, Phys. Rev. Lett. 47 (1981) 1400.
[12] L. Niemeyer, L. Pietronero and H.J. Wiesmann, Phys. Rev. Lett. 52 (1984) 1033;
L. Pietronero and H.J. Wiesmann, J. Stat. Phys. 36 (1984) 909.
[13] H.E. Stanley, Phil. Mag. B 56 (1987) 665.
[14] C. Evertsz, Laplacian Fractals, Thesis, University of Groningen (1989), unpublished.
[15] G. Paladin and A. Vulpiani, Phys. Rep. 156 (1987) 147.
[16] A.P. Siebesma, R.R. Tremblay, A. Erzan and L. Pietronero, Physica A 156 (1989) 613.
[17] A.P. Siebesma and R.R. Tremblay, Phys. Rev. B, in press.
[18] A. Vespignani and L. Pietronero, to be published.
[19] M. Marsili and L. Pietronero, to be published.

SUPERDIFFUSIVE TRANSPORT DUE TO RANDOM VELOCITY FIELDS ☆

Sidney REDNER

Center for Polymer Studies and Department of Physics, Boston University, Boston, MA 02215, USA

The motion of a random walk in a medium containing random, but spatially correlated velocity fields is discussed. This type of disorder generally leads to *superdiffusive* behavior in which the mean-square displacement, $\langle x^2(t) \rangle$, grows faster than linearly with time. For a two-dimensional medium with layers of random velocities in the x-direction, $\langle x^2(t) \rangle$ is found to increase as $t^{2\nu}$ with $2\nu = 3/2$. The probability distribution of displacements appears to fit the form $P(x, t) \sim t^{-3/4} \exp[-(x/t^{3/4})^\delta]$, with $\delta \lesssim 1.7$. However, a Lifshitz argument for the tail of the distribution suggests that $\delta \leq 4/3$. This discrepancy is yet to be fully resolved. We also discuss some intriguing properties in a model with isotropic velocity fields. In two dimensions, we find $\nu = 4/3$, while $\delta = 3$. These values obey the general scaling relation $\delta = (1 - \nu)^{-1}$.

1. Introduction

Stochastic transport in spatially heterogeneous media often leads to anomalous, or subdiffusive, behavior in which the mean-square displacement, $\langle x^2(t) \rangle$, grows more slowly than linearly with time. This has been a topic of enormous interest in the recent past (see e.g. refs. [1–3]). An essential mechanism for this anomalous behavior is the presence of disorder on all length scales. As a function of time, the random walk explores regions of progressively higher "resistance" to transport. This leads to a diffusion coefficient which is a decreasing function of length scale, or time. The vanishing of the diffusion coefficient at large scales can be viewed as the mechanism that leads to anomalous diffusion. Corresponding to this anomalously slow transport, the probability distribution is generally non-Gaussian in nature.

In this article, I discuss a very simple mechanism that gives rise to the complementary situation of *superdiffusive* behavior, in which $\langle x^2(t) \rangle$ grows faster than linearly in time. Owing to the relative simplicity of the models for which this behavior can be realized, this general phenomenon should prove to be a productive area for future work. The models that I will discuss are based on the coupling between diffusion and random velocity fields. The physical motivation for this class of models arises in attempting to describe ground water transport in macroscopically heterogeneous rocks [4]. Consider, e.g., a two-dimensional stratified medium consisting of strips which are infinitely long in the x direction and of random widths in the y direction. Each layer is homogeneous but distinct, so that transport properties may vary from layer to layer. This might describe a two-dimensional section of a typical sedimentary rock.

Suppose that fluid is flowing in the x direction (along the strata) and that diffusive mixing takes place between layers. Owing to the differences in each layer, the fluid velocity correspondingly varies from layer to layer. In a center-of-mass frame of reference, then, there are random velocities in the x direction, and pure diffusion in the y direction. If a random walk is released into the flow, then these two driving forces lead to an anomalously

---

☆ This work has been performed in collaboration with J. Koplik, F. Leyvraz, and A. Provata.

large, superdiffusive spread of the probability distribution. Some of the intriguing features associated with this class of transport processes are outlined below.

2. Unidirectional random velocities

For the case of random layers of differing velocities, consider the following simple lattice model on the square lattice. Each line in the x direction is randomly assigned a velocity $\pm$. At each vertex of the lattice, this means that a random walk moves either in the $+y$ or $-y$ direction with probability 1/4, and moves along the direction of the pre-assigned velocity in the x direction with probability 1/2. This corresponds to an infinite velocity bias in the longitudinal direction. For this system, the mean-square displacement of a random walk at time t increases as $t^{3/2}$, as Matheron and de Marsily [4] were apparently the first to recognize. This remarkable result arises because in a time t a random walker explores of the order of $D_\perp t$ horizontal layers, where $D_\perp$ denotes the (microscopic) diffusion coefficient in the transverse direction. Although the longitudinal velocity averaged over an infinite number of layers is zero, the average over a finite number of layers will generally be fluctuating and a vanishing function of the number of layers that the random walk visits. This non-vanishing bias underlies superdiffusive transport.

The average velocity at time t is given by

$$\langle v \rangle_t = (D_\perp t)^{-1/2} \sum_{i=1}^{(D_\perp t)^{1/2}} v_i \sim (D_\perp t)^{-1/4}. \tag{1}$$

The average has been taken only over the $D_\perp^{1/2}$ layers that a typical random walk visits. Correspondingly, the rms longitudinal displacement at time t, $x_{\mathrm{rms}}(t) \equiv \langle x^2(t) \rangle^{1/2}$, may be estimated as

$$x_{\mathrm{rms}}(t) \sim \langle v \rangle_t t \sim D_\perp^{-1/4} t^{3/4}. \tag{2}$$

In addition to investigating the moments of the probability distribution, it is important to study the probability distribution itself [5]. Our preliminary work suggests that the tail of this distribution function is governed by a Lifshitz-type singularity whose manifestations are extremely difficult to observe numerically, while the peak of the distribution shows markedly different behavior.

As a natural first hypothesis, the probability distribution for the longitudinal displacement, averaged over all configurations of random velocities, is taken to have the form

$$\langle P(x, t) \rangle \propto t^{-\nu} \exp[-(x/t^\nu)^\delta], \tag{3}$$

with $\nu = 3/4$. Our goal is to find the shape exponent δ. A Gaussian distribution corresponds to $\delta = 2$, while for many stochastic walk models, an argument by Fisher [6,7] generally gives $\delta = (1 - \nu)^{-1}$. For random velocity layers, this would yield $\delta = 4$. In order to test the hypothesis of eq. (3), we have performed three independent calculational schemes, all of which yield the same result. In these methods, we first compute the dimensionless moment ratios,

$$m_{2k}(t) \equiv \langle x^{2k}(t) \rangle / \langle x^2(t) \rangle^k. \tag{4}$$

The m_{2k} are found to approach constants as $t \to \infty$ whose values depend on δ. By attempting to match these moments to those that arise by directly computing the moments from eq. (3), we infer a value $\delta \lesssim 1.7$.

One way to compute the moments is by a direct evaluation of the stochastic integral over the differing paths that the random walk takes in the transverse direction. By the definition of the displacement, we have

$$\langle\langle x(t)\rangle_w\rangle_c = \int_0^t dt' \langle\langle u(y(t'))\rangle_w\rangle_c, \tag{5}$$

where we have explicitly denoted that the average is first performed over all random walk trajectories, and then over all configurations of random velocities. The moments of the displacement can therefore be written as

$$\begin{aligned}\langle\langle x^n(t)\rangle_w\rangle_c &= \int_0^t dt_1 ... \int_0^t dt_n \langle\langle u(y(t_1))...u(y(t_n))\rangle_w\rangle_c \\ &= n! \int_0^t dt_1 \int_0^{t_1} dt_w ... \int_0^{t_{n-1}} dt_n \langle\langle u(y(t_1))...u(y(t_n))\rangle_w\rangle_c.\end{aligned} \tag{6}$$

The time-ordered product of the velocity correlation function is

$$\begin{aligned}&\langle\langle u(y(t_1))...u(y(t_n))\rangle_w\rangle_c \\ &= \int_{-\infty}^{+\infty} dy_1 dy_2...dy_n \langle u(y_1)...u(y_n)\rangle_c p(y_n, t_n) p(y_{n-1}-y_n, t_{n-1}-t_n)...p(y_1-y_2, t_1-t_2),\end{aligned} \tag{7}$$

where

$$p(x, t) = (4\pi Dt)^{-1/2} \exp(-x^2/4Dt) \tag{8}$$

is the Gaussian probability distribution for the motion in the transverse direction. The evaluation of the integral in eq. (6) is straightforward, but quite tedious. Up to the sixth order (only even powers of velocities within a given layer give non-zero contributions), we find the moment ratios $m_4 \simeq 3.3$ and $m_6 \simeq 19.1$. These are reasonably consistent with the value $\delta \simeq 1.7$ quoted above. These results are also corroborated by direct Monte Carlo simulations, and by a direct enumeration of the occupancy correlation function of the one-dimensional random walk problem in the transverse direction.

On the other hand, a Lifshitz-type argument based on considering the walks which contribute to the extreme large-distance tail of the distribution suggests that $\delta \leq 4/3$. Consider the average probability of a "stretched out" walk, namely, $\langle P(x \sim t, t) \rangle$. According to eq. (3), this should vary as $\exp(-t^{\delta/4})$. On the other hand, the probability of a walk being stretched out can be bounded from below by the probability of remaining confined to one "avenue" in which the velocity bias is in the same direction. In direct analogy with the survival probability of a one-dimensional random walk in the presence of randomly distributed traps [8], the confining probability, averaged over all configurations of the environment, varies as $\exp(-at^{1/3})$. This implies the exponent inequality, $\delta \leq 4/3$. As suggested by our various calculations, the asymptotic behavior is likely to be masked by very slow crossover effects.

3. Generalizations to higher dimensions and to isotropic velocities

Consider the quasi-two-dimensional problem in which the y-z plane consists of random "filaments" in the $\pm x$ direction. (This can be generalized to d'-dimensional strata in a d-dimensional system.) For this system, the arguments leading to eqs. (1) and (2) can again be invoked. Now a random walk visits of the order of $t/\ln t$ different filaments in a time t. Thus $\langle v \rangle_t \sim (\ln t/t)^{1/2}$. Consequently, the rms displacement should vary as $(t \ln t)^{1/2}$. The logarithmic correction to otherwise diffusive behavior suggests that the upper critical dimension for this system is 3.

The behavior of the probability distribution of the displacements also appears to be quite interesting. On the basis of the diffusive behavior of the $x_{\rm rms}$, one might expect that $\langle P(x,t) \rangle$ would be much closer to a Gaussian than in the two-dimensional layered model. On the other hand, a strict (and naive) application of the Lifshitz-tail argument suggests that the exponent inequality $\delta \leq 1$, even further from Gaussian behavior than in two dimensions. The mechanism by which the Lifshitz-tail argument breaks down appears to be quite subtle and interesting. No reliable numerical results for three dimensions are as yet available. The resolution of these conflicting results should prove to be quite interesting.

A second general class of models is the motion of a random walk in the presence of random, but *isotropic* random velocities. For example, consider a random walk on a random "Manhattan" grid, in which the directionality along any Avenue or Street is fixed along its entire length, but whose orientation is random. For this system, we generalize the arguments of eqs. (1) and (2) by decomposing the motion into transverse and longitudinal components, even though the motion is isotropic. Assuming that $x_{\rm rms} \sim t^\nu$, then from eq. (1), the mean velocity, averaged over these t^ν layers, vanishes as $t^{-\nu/2}$. From eq. (2), one then concludes that $x_{\rm rms} \sim t^{1-\nu/2}$. Since the motion is isotropic, however, we must have $\nu = 1 - \nu/2$, or $\nu = 2/3$. Generalizing these arguments to higher dimensions suggests that $\nu = 2/(d+1)$ for spatial dimension $d \leq d_c = 3$.

For the probability distribution of displacements in two dimensions, relatively modest simulations indicate that eq. (3) holds with the exponents $\nu = 2/3$ and with $\delta = 3$. Interestingly, these values are in accord with the general Fisher argument between the shape and size exponent, $\delta = (1-\nu)^{-1}$. It is slightly mysterious that the Fisher argument appears to work for isotropic random velocities but fails for layered random velocities.

4. Summary and discussion

We have discussed some intriguing aspects of superdiffusive transport that occur in media which possess random, but correlated velocity fields. By elementary arguments, one can understand how superdiffusive transport arises from the interplay between diffusion and the geometry of the random velocity fields. For the layered system, it is possible to develop formal calculational schemes to obtain the positive integer moments of the probability distribution. However, the utility of these approaches appears to be limited by the existence of Lifshitz-type singularities which control the asymptotic behavior of the distribution. For an isotropic two-dimensional "Manhattan" system, the probability distribution appears to exhibit conventional scaling in which $\delta = (1-\nu)^{-1}$.

While the work thus far has focused on the spatial moments of the probability distribution, it should also prove fruitful to study the first passage probabilities in superdiffusive transport processes. The mechanisms that yield the basic features of the distribution of first passage times between an input and absorber may well be very different from those that describe the spatial moments of the distribution. In short, there are still a wide variety of fundamental, puzzling questions for which satisfactory, first-principles explanations are still lacking.

References

[1] S. Alexander, J. Bernasconi, W.R. Schneider and R. Orbach, Rev. Mod. Phys. 53 (1981) 175.
[2] Y. Gefen, A. Aharony and S. Alexander, Phys. Rev. Lett. 50 (1983) 77.
[3] S. Havlin and D. Ben-Avraham, Adv. Phys. 36 (1987) 695.
[4] G. Matheron and G. de Marsily, Water Resources Res. 16 (1980) 901.
[5] A. Georges, Ph.D. Thesis, Université de Paris-Sud, Paris (1988).
[6] M.E. Fisher, J. Chem. Phys. 44 (1966) 616.
[7] P.G. de Gennes, Scaling Concepts in Polymer Physics (Cornell Univ. Press, Ithaca, 1979).
[8] P. Grassberger and I. Procaccia, J. Chem. Phys. 77 (1982) 6281.

LUMINESCENCE DECAY IN CHAIN-LIKE POLYMERS USING FRACTAL CONCEPTS

A.K. ROY[1] and A. BLUMEN

Physikalisches Institut and BIMF, University of Bayreuth, D-8580 Bayreuth, Fed. Rep. Germany

We study the direct, incoherent energy transfer from excited donors to acceptor molecules, which are attached to chain-like polymers. Fractal ideas are very fruitful in this study, the polymers being describable through Gaussian random walks or self-avoiding walks, depending on the solvent. Using the end-to-end distribution function of the walks, we determine analytically the approximate expressions for the ensemble-averaged decay forms of the donor excitation, both for multipolar and for exchange-type interactions. Here the fractal dimension plays a central role. We show how the decay form follows from the conformational distribution of polymer chains and how this form allows to determine the chains' fractal dimension.

1. Introduction

Here we study the direct, incoherent energy transfer from energetically excited donors to acceptor molecules, which are all attached to chain-like polymers. The problem is of central importance in the determination of the end-to-end conformations of biopolymers. The experimental means of choice consists in monitoring the fluorescence decay of donor–acceptor pairs due to the Förster mechanism [1]. Such energy transfer studies also help to unravel the internal morphology of polymer Latex particles [2] where the decay laws of the excited donors directly reflect the fractal dimension of the Latex.

We will centre on a single polymer chain which is quite isolated from the others (i.e. the polymer density is in the very dilute limit). We model the chain through quasi-linear fractals [3], such as random walks or self-avoiding walks, depending on the solvent. It will turn out that for multipolar interactions the fluorescence decay obeys Kohlrausch–Williams–Watts (KWW) stretched exponential laws:

$$\Phi(t) \approx \exp[-C(t/\tau)^\alpha] \quad (0 < \alpha < 1, t > \tau). \quad (1)$$

Here α depends on the geometry of the polymer and on the character of the multipolar interaction and C is a constant. On the other hand, for exchange-type interactions the decay law follows exponential–logarithmic patterns:

$$\Phi(t) \approx \exp[-C \ln^\beta(t/\tau)] \quad (\beta \geq 1, t > \tau), \quad (2)$$

where again β is geometry-dependent. Such decay forms obtain when one donor and one acceptor molecule are placed at the ends of the isolated polymer chain, and also when the chain contains a small concentration of randomly placed acceptors. Furthermore, the decay through the fastest channel also leads to equations of the form (1) or (2). These distinct decay mechanisms lead in general, however, to different α and β values, as we show in this work. This fact allows to determine experimentally the microscopic situation prevailing.

Eqs. (1) or (2) hold when the donor–acceptor transmission is fast compared to the relative end-to-end motion of the chain. Experimentally this condition is satisfied in highly viscous solvents. Furthermore the transfer should be quick on the scale of the donor-specific radiative and radiationless relaxation. If this is not the case, then the donor-specific relaxation (rate τ_R^{-1}) can be accounted for by multiplying all decay forms which now follow by $\exp(-t/\tau_R)$.

The paper is organised as follows: First we study the exact form of the ensemble-averaged decay function. Then we present approximate expressions for the decay laws in the case of multipolar and ex-

---

[1] On leave from Santipur College, Nadia 741 404, India.

Essays in honour of Benoit B. Mandelbrot
Fractals in Physics – A. Aharony and J. Feder (editors)

2. The ensemble-averaged decay laws

We start by considering one donor and one acceptor, located at opposite ends of a polymer chain. We let r be the donor–acceptor Euclidean distance and we centre the co-ordinate system at the donor. The time evolution for the decay of the donor excitation $\Phi(c, t)$, for a particular configuration c of the polymer chain, is exponential:

$$\Phi(t) = \exp[-tw(r)]. \tag{3}$$

Here $w(r)$ denotes the energy transfer rate to the acceptor at position r. We neglect the back transfer to the donor. Generally accepted isotropic forms for $w(r)$ are

$$w(r) = \tau^{-1}(l/r)^s \tag{4}$$

for multipolar interactions [4] and

$$w(r) = \tau^{-1} \exp[\xi(l-r)] \tag{5}$$

for exchange-type interactions [5]. The parameter s in eq. (4) depends on the multipolar interaction ($s=6$ for dipole–dipole, $s=8$ for dipole–quadrupole), the quantity ξ in eq. (5) measures the range of the exchange and τ is the energy transfer rate at the minimal donor–acceptor distance l.

The quantity of interest experimentally is the ensemble average of $\Phi(c, t)$ over all the chain configurations. In the continuum version in d dimensions and assuming isotropic interactions we have

$$\Phi(t) = \langle \Phi(c, t) \rangle$$

$$= dV_d \int_0^\infty r^{d-1} p_N(r) \exp[-tw(r)] \, dr, \tag{6}$$

where $V_d = \pi^{d/2}/\Gamma(1+\tfrac{1}{2}d)$, $\Gamma(x)$ is the Gamma function and $p_N(r)$ is the probability density for finding in a polymer consisting of N monomers, the end-to-end distance to be r.

We recall now that chain-like polymers with N monomers can be modelled through fractal objects, e.g. random walks for ideal, and self-avoiding walks for non-ideal chain polymers.

For Gaussian walks one has [6]

$$p_N(r) = A_1 \exp(-B_1 r^2), \tag{7}$$

where $A_1 = (d/2\pi \langle R_N^2 \rangle)^{d/2}$ and $B_1 = d/2 \langle R_N^2 \rangle$ (expressed in terms of the mean-squared end-to-end distance $\langle R_N^2 \rangle$ of a walk of N steps).

For self-avoiding walks (SAWs) [7]:

$$p_N(r) = A_2 r^\theta \exp(-B_2 r^\delta), \tag{8}$$

where

$$B_2 = \{\Gamma[(d+\theta+2)/\delta]\}^{\delta/2}$$
$$\times \{\Gamma[(d+\theta)/\delta] \langle R_N^2 \rangle\}^{-\delta/2} \tag{9}$$

and

$$A_2 = \frac{\delta B_2^{(d+\theta)/\delta}}{dV_d \Gamma[(d+\theta)/\delta]}. \tag{10}$$

In eq. (8) θ and δ are two dimension-dependent critical exponents. The important parameter in the following discussion is δ. δ equals 4 in $d=2$ and 2.44 in $d=3$. The critical exponents θ and δ may be expressed as $\theta = (\gamma - 1)\nu^{-1}$ and $\delta = (1-\nu)^{-1}$, where γ and ν are two universal critical exponents defined via the relations [8]

$$\langle R_N^2 \rangle \approx N^{2\nu} \tag{11}$$

and

$$G_N \approx \mu^N N^{\gamma - 1}. \tag{12}$$

Here G_N is the total number of configurations of SAWs with N steps and μ is the average connectivity constant, which depends both on the dimension and on the particular lattice involved. For higher dimensions the effect of the excluded volume diminishes and thus for $d \geq 4$ the SAW turns into a Gaussian walk. Then one readily verifies for $d \geq 4$ that $\delta = 2$ and $\theta = 0$ (i.e. $\gamma = 1$, $\nu = 1/2$).

In the second part of this section, we remove the restriction of a single donor–acceptor pair and consider one donor surrounded by a small concentration of acceptors, all of which are attached to the chain.

For a small concentration p of acceptors in d dimensions, the decay takes to a good approximation the following closed form (see ref. [9] for details):

$$\Phi(t) = \exp\left(-pdV_d \int_0^\infty r^{d-1}\{1-\exp[-tw(r)]\}\,dr\right). \tag{13}$$

The last expression can be extended to fractals [10]:

$$\Phi(t) = \exp\left(-\bar{C}p \int_0^\infty r^{\bar{d}-1}\{1-\exp[-tw(r)]\}\,dr\right), \tag{14}$$

where $\bar{d}$ is the fractal dimension and the constant $\bar{C}$ may be written as $\bar{C}=\bar{d}V_{\bar{d}}$. Turning now to polymers we remark that by modelling them as quasi-linear fractals, the conformational probability of the polymer chain can be accounted for by performing the integration in (14) over the fractal volume of the chain. For ideal polymer chains $\bar{d}=2$ (superuniversal), whereas for non-ideal polymer chains $\bar{d}=1/\nu$, which is d-dependent for $1<d<4$.

In the last part of this section, we again assume a low concentration of acceptors attached to the chain; however, now the energy transfer occurs only through the fastest channel [11], i.e. only the nearest acceptor (in the Euclidean sense) to the donor is affected. Then

$$\Phi(t) = \int_0^\infty \Omega(r) \exp[-tw(r)]\,dr, \tag{15}$$

where

$$\Omega(r) = pV_d dr^{d-1} \exp(-pV_d r^d). \tag{16}$$

Here $\Omega(r)$ is the probability of having the nearest acceptor at a distance r (Hertz distribution for fractal systems). Comparison of eqs. (6) and (15) with eq. (16) shows that they are formally equivalent, if we set

$$p(r) = p(\bar{d}V_{\bar{d}}/dV_d)r^{\bar{d}-d} \exp(-pV_d r^d). \tag{17}$$

Eq. (17) has the structure of eq. (8), with $\delta=\bar{d}$, $\theta=\bar{d}-d$ and $B_2=pV_d$.

Using the exact expressions, eqs. (6), (14) and (15), we present now approximate expressions for microscopic transfer laws, eqs. (4) and (5).

3. Approximate expressions of the decay laws

We calculate first the decay law when the donor and the acceptor are at the ends of the chain for a non-ideal chain model (SAW). Inserting (4) and (8) into (6) leads to

$$\Phi(t) = dV_d A_2$$

$$\times \int_0^\infty r^{d-1+\theta} \exp[-B_2 r^\delta - (t/\tau)l^s r^{-s}]\,dr. \tag{18}$$

We determine the long-time decay behaviour by the method of steepest descents. For $t/\tau \gg 1$ we obtain

$$\Phi(t) \approx C'(t/\tau)^{[2(d+\theta)-\delta]/2(\delta+s)} \\ \times \exp[-C(t/\tau)^{\delta/(\delta+s)}], \tag{19}$$

where C and C' are time-independent. Evidently, the dominant factor of eq. (19) for $t/\tau \gg 1$ is

$$\Phi(t) \approx \exp[-C(t/\tau)^{\delta/(\delta+s)}]. \tag{20}$$

Eq. (20) corresponds to the KWW stretched exponential law, eq. (1) with $\alpha=\delta/(\delta+s)$. We may also rewrite α, by using the expression $\delta=(1-\nu)^{-1}$, see eq. (10) and further. Setting $1/\nu=\bar{d}$, we have

$$\alpha = \bar{d}/[(s+1)\bar{d}-s]. \tag{21}$$

Let us now consider ideal polymer chains (donor–acceptor pair at the ends of the chain). The decay law is obtained by simply noting that eq. (7) is a special case of eq. (8) for $\theta=0$ and $\delta=2$. Reading off from eq. (19) it follows

$$\Phi(t) \approx C'(t/\tau)^{(d-1)/(2+s)} \exp[-C(t/\tau)^{2/(2+s)}]. \tag{22}$$

The dominant factor for $t/\tau \gg 1$ is

$$\Phi(t) \approx \exp[-C(t/\tau)^{2/(2+s)}]. \tag{23}$$

Therefore, we again obtain the KWW stretched exponential decay law, eq. (1), now with

$$\alpha = 2/(2+s). \qquad (24)$$

Next we come to the situation where we allow for several acceptor molecules to be distributed randomly along the chain. Then inserting eq. (4) into (14) and calculating in the same way as in refs. [9,10], we get

$$\Phi(t) \approx \exp[-C(t/\tau)^{\bar{d}/s}], \qquad (25)$$

and when compared with eq. (1) we have

$$\alpha = \bar{d}/s. \qquad (26)$$

One can see that this decay behaviour differs from that which is obtained for a single donor–acceptor pair both for ideal and also for non-ideal polymer chains, compare eq. (26) with eqs. (21) and (24).

Now, we turn our attention to the fastest channel mechanism according to Klafter and Shlesinger [11]. Taking again $w(r)$ as in eq. (4), we obtain from eq. (19), by observing now that $\theta = \bar{d} - d$ and $\delta = \bar{d}$, see eq. (17),

$$\Phi(t) \approx C'(t/\tau)^{d/2(\bar{d}+s)} \exp[-C(t/\tau)^{\bar{d}/(\bar{d}+s)}]. \qquad (27)$$

For $t/\tau \gg 1$ the dominant factor has the form

$$\Phi(t) \approx \exp[-C(t/\tau)^{\bar{d}/(\bar{d}+s)}] \qquad (28)$$

with

$$\alpha = \bar{d}/(\bar{d}+s). \qquad (29)$$

One can easily see that for ideal polymer chains this α (with $\bar{d}=2$) is exactly equal to the form which is obtained when donor and acceptor are at the ends of the chain, compare with eq. (24). This is so only because then there is exactly one acceptor available and because the fractal dimension of the ideal polymer chain is superuniversal and equals 2. On the other hand, for non-ideal polymer chains (SAW), α in eq. (29) depends on the fractal dimension $\bar{d}$. Compared to the previous study, where the interaction was with all the acceptors on the chain and where we found $\alpha = \bar{d}/s$, see eq. (26), where the decay proceeds more slowly.

Finally, for exchange-type interactions, inserting eqs. (5) and (8) into eq. (6) we obtain for non-ideal chains (SAW)

$$\Phi(t) = dV_d A_2$$

$$\times \int_0^\infty r^{d-1+\theta} \exp\{-B_2 r^\delta - (t/\tau) \exp[\xi(l-r)]\} \, dr. \qquad (30)$$

Again eq. (30) can be evaluated approximately by the method of steepest descent. As leading term to $\Phi(t)$ follows

$$\Phi(t) \approx \exp\{-B_2 \xi^{-\delta} \ln^\delta[(t/\tau) \exp(\xi l)]\}. \qquad (31)$$

So this time evolution corresponds to the exponential–logarithmic law, eq. (2), where now $\beta = \delta$. For SAW chains we have $\delta = (1 - 1/\bar{d})^{-1}$, and thus

$$\beta = \delta = \bar{d}/(\bar{d}-1). \qquad (32)$$

In eq. (31) one should keep in mind that the next correction factor to the exponential term is of the order of $\ln^{\delta-1}[(t/\tau)\exp(\xi l)]$. For ideal chains, $\beta = 2$ as $\bar{d} = 2$.

In a similar fashion one obtains exponential–logarithmic decays (for the exchange-type interaction) when several acceptors are on the chain and also when the decay occurs through the fastest channel. In both cases $\beta = \bar{d}$ is found.

In conclusion, under several distinct conditions multipolar interactions on fractal objects lead to stretched exponential decays, whereas for exchange interactions exponential–logarithmic patterns are found. From the corresponding α and β exponents, eqs. (1) and (2), one may, however, distinguish between the models and between the fractal patterns involved. We expect our findings to be useful for the characterisation of chain-like polymers through luminescence-decay methods.

Acknowledgement

One of us (A.K.R.) wishes to thank the Deutscher Akademischer Austauschdienst (DAAD) for financial support through the award of a research fellowship. The support of the Deutsche Forschungsgemeinschaft (SFB 213) and of the Fonds der Chemischen Industrie are gratefully acknowledged.

References

[1] A. Grinvald, E. Haas and I.Z. Steinberg, Proc. Natl. Acad. Sci. US 69 (1972) 2273;
E. Haas, M. Wilchek, E. Katchalskii-Katzir and I.Z. Steinberg, Proc. Natl. Acad. Sci. US 72 (1975) 1807.

[2] O. Pekcan, M.A. Winnik and M.D. Croucher, Phys. Rev. Lett. 61 (1988) 641.

[3] B.B. Mandelbrot, The Fractal Geometry of Nature (Freeman, San Francisco, 1982).

[4] T. Förster, Z. Naturforsch. A 4 (1949) 321.

[5] D.L. Dexter, J. Chem. Phys. 21 (1953) 836.

[6] G.H. Weiss and R.J. Rubin, Adv. Chem. Phys. 52 (1983) 363.

[7] M.E. Fisher, J. Chem. Phys. 44 (1966) 616;
A. Baumgärtner, J. Chem. Phys. 76 (1982) 4275.

[8] P.G. de Gennes, Scaling Concepts in Polymer Physics (Cornell Univ. Press, Ithaca, 1979);
J. des Cloizeaux and G. Jannink, Les Polymères en Solution: Leur Modelisation et Leur Structure (Les Editions de Physique, Les Ulis, 1987).

[9] A. Blumen and J. Manz, J. Chem. Phys. 71 (1979) 4694.

[10] J. Klafter and A. Blumen, J. Chem. Phys. 80 (1984) 875.

[11] J. Klafter and M.F. Shlesinger, Proc. Natl. Acad. Sci. US 83 (1986) 848.

EXPERIMENTAL OBSERVATION OF LOCAL MODES IN FRACTAL DRUMS

B. SAPOVAL

Laboratoire de Physique de la Matière Condensée, Ecole Polytechnique [1], 91128 Palaiseau Cedex, France

It is shown that fractal boundary conditions drastically alter the properties of wave excitations in space. The low-frequency vibration modes of soap bubbles deposited on a fractal contour are observed. There exist localized modes. The vibrations have a singular behavior along the boundary. Consequently fractal resonators are necessarily non-linear resonators and have special damping properties.

Fractal objects [1] have no translational invariance. We know that when a physical system has complete translational invariance it can support wave excitations of any wavelength. When the system has limited translational invariance it can support only a restricted set of wave excitations. This is true of lattice waves in crystals and of microwave cavities. Another question is what happens when a system, or more precisely its boundary, has dilation invariance instead of translational invariance. This is the case of a fractal bounded resonator [2]. It has already been shown that the excitations of a fractal lattice are localized [3,4]. We consider here the excitations of a fractal drum [2] and discuss an experiment which shows that there exist also in that case localized modes. These fractal modes exhibit infinite derivatives near the re-entrant points of the fractal boundary.

The experiment that we discuss consists of the visual observation of low-frequency vibrations of a "soap bubble" deposited on a fractal contour. Low-frequency vibrations, above 20 Hz, are excited by a loudspeaker situated above the fractal drum shown in fig. 1. The modes are observed visually.

The lower-frequency fundamental mode has no node. The amplitude of the vibration is maximum at the center of the drum and decays smoothly when entering the side region A in fig. 1.

[1] Unité associée du CNRS No. 1254.

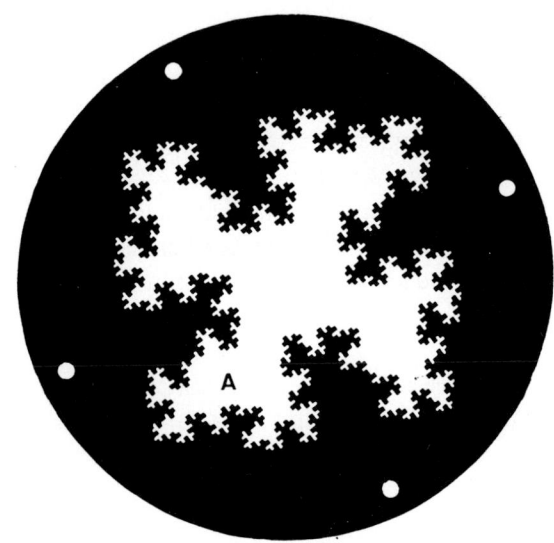

Fig. 1. Picture of the fractal drum. A soap bubble placed on this object constitutes a membrane whose acoustic vibrations are studied. The extreme dimension of the fractal object is 14.3 cm. Its fractal dimension is $D: \ln 8/\ln 4 = 3/2$. A soap bubble (water + 20% detergent + equal volume glycerol) [5] is deposited on that contour. The system is placed horizontally above a loudspeaker fed with an audio signal generator. Different resonances are observed as a function of the excitation frequency.

At next higher frequencies more complex resonances are observed. The most striking fact is the observation of modes which are localized in a finite region of the membrane like part A in fig. 1. Such a

Essays in honour of Benoit B. Mandelbrot
Fractals in Physics – A. Aharony and J. Feder (editors)

mode is confined in an A "bay" and decays slowly toward the center of the drum.

Experiments on soap bubbles are difficult to make quantitatively reproducible because the bubble, even when stabilised with glycerol, has a finite life time. The thickness and consequently the specific mass of the membrane decreases progressively with time due to water evaporation. Generally speaking the possibility of experimental observation of a resonator mode depends on several factors: Frequency of excitation compared to the mode frequency, damping of the mode and coupling between the excitation and the mode under study. This last difficulty is real when using acoustic excitation transmitted through air as we do because it is difficult to obtain a "local" excitation at low frequency.

Strictly speaking, in the rather symmetrical fractal structure that we study there should exist single modes made of combinations of vibrations localized in the four regions of the membrane equivalent to A. (This would correspond for instance to the symmetric and antisymmetric eigenstates of quantum tunneling.) In other words there exist four degenerated A modes which respond to the same excitation frequency. The observation of localized modes can be explained if one remarks that the aperture of the A bay towards the center constitutes a guide below "cut-off" for the frequency of the A mode. In that case the coupling between the equivalent A regions is very weak. Without care one observes a combination of these vibrations but we were able to observe each single one of these modes by displacing carefully the excitation source above the membrane. For a less symmetrical boundary or for a random boundary there should exist no degeneracy and "strict" local modes should exist.

Higher-frequency vibrations were also observed but we were not able to study high-frequency individual modes for the above reasons and because the number of modes for a given frequency range increases at higher frequencies.

We now show that these modes have a particular behavior near the fractal boundary: They are singular in the sense that their derivatives are infinite at particular points on the frontier. To discuss that property, we consider for example fig. 2 in the region around a re-entrant corner and we look for solutions of the wave equation

$$\Delta\psi = K\psi,\qquad(1)$$

where ψ is the wave amplitude. The wave amplitude is 0 along the boundary.

In the immediate vicinity of a re-entrant point we use the polar coordinates (ρ, φ) as shown in fig. 2. Very near the boundary the amplitude of the vibration is very small so that near a point P we look for solutions of

$$\Delta\psi = \partial^2\psi/\partial\rho^2 + \rho^{-1}\partial\psi/\partial\rho + \rho^{-2}\partial^2\psi/\partial\varphi^2 = 0.\qquad(2)$$

The solution is of the form

$$\psi \sim \rho^{2/3}\sin(\tfrac{2}{3}\varphi).\qquad(3)$$

This solution is singular. The derivative $\partial\psi/\partial\rho$ goes to infinity when ρ goes to 0. Such a property should be true near equivalent re-entrant points in the structure.

Visual observation of the low-frequency modes morphology indeed confirms these predictions. Singularities of the amplitude are easily observed around re-entrant points. Would these points possess an infinitely sharp edge, then the singular modes would have infinite derivatives near the fractal boundaries.

Fig. 2. Choice of local polar coordinates near a re-entrant point P.

They would then exhibit infinite stress of the elastic membrane and rupture around the re-entrant points even for a finite amplitude of excitation. These facts are confirmed by experiment. One can observe the mode maximum amplitude near the bursting of the membrane. If instead of the fractal contour of fig. 1 one uses a normal square drum we find that it is possible to achieve much higher vibration amplitudes before the bursting of the membrane with a normal drum than with the fractal drum. Here again quantitative studies are difficult but there can be an order of magnitude difference between the threshold for rupture in the two cases.

These conclusions are compatible with current knowledge about waves and resonators. Waveguides and microwaves cavities are polished because irregularities in the geometry of the boundary cause singularities in the electric field and dielectric breakdown at high power. Perturbation of a resonator mode by a small change of the boundary is widely used in musical instrument manufacturing and microwave technology.

A more general comment should be made following the above observation of singularities. The properties of matter become non-linear beyond a certain threshold. When the membrane is deformed too much, it becomes non-linear and will naturally generate harmonics. These harmonic modes in the same frequency range will be localized themselves and possess their own singularities inducing a kind of cascade. There will exist partially destructive interferences and dissipation at all scales will appear. In that sense fractal structures should possess a strong damping power.

This comment can also explain the appearance of fractal structure, like that of seacoasts, in nature. We know that the seacoast is subject to erosion from currents and waves. But we also know that indented coasts have since ancient times provided shelter to the sailor. Turbulence occurs only above a certain fluid velocity. As water moves along a coast line, the critical velocities above which turbulence starts are very low because the coastline possesses very small features.

We can then suggest [6] that the reason why fractal coastlines exist is precisely because they are best at damping down the currents – even when these are weak – and the waves. As the coast damps down the waves, then the erosion to which it is subjected is reduced. They are thus stabilised by their fractal structure and this is the reason why they exist and we can observe them. It is worth noting that coastline engineering uses empirically porous breakwaters made up of disordered heaps of rocks of various sizes to "absorb energy" when designing breakwaters [7]. The irregular geometry of anechoid chambers is certainly related to the same type of considerations. In the same way, it is clear that the fractal structure of the human circulatory system damps out the hammer blows that our heart generates.

In conclusion fractal boundary conditions indeed alter drastically the wave excitations in space. The singularity at the surface impose infinite value to certain physical quantities. In real physical systems bounded by fractals non-linear effects should be effective in damping vibrations and waves.

The author gratefully acknowledges valuable discussions with J.N. Chazalviel, B.B. Mandelbrot and J. Peyrière.

References

[1] B.B. Mandelbrot, The Fractal Geometry of Nature (Freeman, San Francisco, 1982).
[2] R. Rammal and G. Toulouse, J. Phys. (Paris) 44 (1983) L13.
[3] S. Alexander, in: Annals of the Israel Physical Society, Vol. 5. Percolation Structures and Processes, eds. G. Deutscher, R. Zallen and J. Adler (Israel Phys. Soc., Jerusalem, 1983), p. 149.
[4] M.V. Berry, in: Proc. Tübingen Symp.-Honouring R. Thom, ed. Guttinger (Springer, Berlin, 1979), p. 51.
[5] A.L. Kuehner, J. Chem. Educ. 35 (1958) 337.
[6] B. Sapoval, Fractals (Aditech, Paris, 1989).
[7] B. Le Méhaué, in: The Encyclopedia of Applied Geology, ed. C.W. Finkl Jr. (Van Nostrand Reinhold, New York, 1984), p. 62.

GROWTH VELOCITY OF ELECTROCHEMICAL DEPOSITION AND ITS CONCENTRATION DEPENDENCE

Yasuji SAWADA and Haruhiko HYOSU

Research Institute of Electrical Communication, Tohoku University, Sendai 980, Japan

The structure and stational growth velocity of electrochemical deposits grown from one-dimensional seeds were studied as a function of electrolyte concentration. The growth velocity of zinc metal leaves was found to be independent of the concentration of zinc ions respectively from 0.003 to 0.008 mol/ℓ and from 0.010 to 0.018 mol/ℓ for a fixed potential of 15 V. In this concentration range the local structure remains fractal, but the width of trees and the separation between the trees depends strongly on the concentration. The local structure suddenly changes from fractal to dendritic structure with backbones at 0.020 mol/ℓ. The discrepancy between the experimental results and the velocity selection mechanism based on DLA is discussed.

1. Introduction

Fractal structure is robust in nature [1]. Particularly, fractal growing shapes, such as found in dielectric breakdown [2], crystal growth [3] and viscous fingering [4] were found to be surprisingly similar to each other, and are now one of the most fascinating subjects of nonlinear physics far from equilibrium [5]. Particularly, the problem of crystal growth has been experimentally studied for a long time, primarily by metallurgists. Only a decade ago the growth form of crystals, especially those of dendritic crystals, which are driven far from equilibrium, became a subject for theoretical physicists [6].

The selection mechanism of growth speed and the tip radius of a dendrite was theoretically found to be strongly related to the existence of surface tension and to the anisotropy [7], and experimental results are in accord with the theory [8]. The parabolic tip of a dendrite is unstable without surface tension and even with surface tension if no anisotropy in the microscopic growth mechanism is involved. In such a case, the growth form repeats tip-splitting and often becomes irregular. When the diffusion length is long enough, there is no spatial measure of length in the system and the growth form is fractal up to the diffusion length.

The dendritic growth form was also studied in the electrochemical deposit called metal leaf [3,9,10]. In fact, the zinc metal leaf is now classical as a typical example of fractal growth form. The dependence of the growth form was examined for a wide range of concentrations and applied potentials. With high concentration and low voltage, the growth forms an open fractal. With medium concentration and high voltage, the growth form is dendritic, with a straight backbone. With low concentration and low voltage, the growth form is fractal up to a small length scale and homogeneous above it. The mechanism of the transition of the growth form is not yet well understood.

Recently, Uwaha and Saito [11] proposed a velocity selection mechanism for a DLA grown one-dimensionally from gas phase of a finite particle density. According to them, the velocity of the front is proportional to the $1/(d-D)$ power of the gas density, where d is the Euclidean dimension of the space and D is the fractal dimension of the object. This is an interesting proposal, as it unites the fractal dimension, which is a geometrical index, and the velocity, which is a dynamical quantity. In this paper we report results of experiments on the relation of growth speed and the electrolyte concentration of zinc metal leaves, and compare them with this proposal.

2. Experimental

The experimental setup is almost the same as the one previously reported, except that the present one is rectangular, 24 cm wide and 12 cm long. The depth is 0.25 mm. The applied potential was fixed at 15 V. The concentration of zinc sulphate was varied with 0.001 mol/ℓ steps, in the range from 0.003 to 0.030 mol/ℓ. The growth form was video-recorded and the growth speed of the front was analyzed by a digital image analyzer. The current was measured simultaneously during the growth of each run to measure the amount of deposition in unit time.

3. Results

A series of figures, from fig. 1 to fig. 5, shows the growth form of zinc metal leaves for different concentrations of zinc ions: 0.006 (fig. 1), 0.008 (fig. 2), 0.010 (fig. 3), 0.018 (fig. 4) and 0.020 mol/ℓ (fig. 5). Each picture covers roughly one-fifth of the total width of the cell. The top and the bottom figures correspond respectively to the patterns after 10 and 20 min. The growth forms were similar to the one previously reported in a circular cell [9,10]. They are fractal up to some scale length and homogeneous above it. One immediately notices that the local structure remains almost identical when the concentration is increased up to 0.018 mol/ℓ. The spacings between the trees are widely separated when the concentration is extremely low, but are reduced with increasing concentration. One cannot see any vacant space among trees at the concentration of 0.018 mol/ℓ. This shape corresponds to the one called "homogeneous" [9] or "dense radial" [10]. At the higher concentration, the local structure suddenly changes from fractal to dendritic, with almost straight backbones.

The growth velocity for each concentration remains fixed except for a short duration just after the start and for the later period when the leaves are close to the counter electrode. The growth velocities thus obtained are summarized in fig. 6. Surprisingly, the

Fig. 1. Growth form of electrochemical deposits from a solution including 0.006 mol/ℓ of Zn ions. The upper picture and the lower pictures were taken at 10 and at 20 min, respectively. The electric potential was fixed at 15 V throughout the present work.

growth velocity was found to depend only weakly on the concentration in this range. It is almost constant from 0.003 to 0.008 mol/ℓ. One finds a step-wise change around 0.009 mol/ℓ and the value of the growth velocity again takes a constant value until the local structure changes at around 0.02 mol/ℓ. There seem to exist two distinct regions where the growth mechanisms are different from each other, although the patterns look alike.

Fig. 7 shows the concentration dependence of the current during the growth, which is directly related to the deposition rate. The apparent linearity in the log–log plot indicates that the deposition rate increases with the concentration in a power law, with exponent 0.9 ± 0.1.

Fig. 2. The same as fig. 1 except that the concentration is 0.008 mol/ℓ.

Fig. 3. The same as fig. 1 except that the concentration is 0.010 mol/ℓ.

4. Discussion

According to the present experimental results, the linear growth velocity in the direction parallel to the applied field takes a constant value from 0.003 to 0.008 mol/ℓ and another constant value from 0.010 to 0.018 mol/ℓ, and the total deposition rate is roughly proportional to the concentration over the range measured. The increment of concentration is somehow absorbed by increasing the lateral density. This is in accord with the direct observation that the spacing between the trees decreases with increasing concentration. By a rough estimate one finds that typical spacings decrease with increasing concentration in this concentration range by a power law with exponent -1.7 ± 0.2.

One cannot, however, identify this length with the diffusion length, since the growth speed v should be inversely proportional to the diffusion length, which cannot have a strong concentration dependence from the present experimental results. Instead, the diffusion length should be considered to take two constant values in the two concentration regions, as the growth speed is almost strictly constant in the two concentration ranges. In fact, one observes two length scales in addition to the diffusion length, in the lateral direction to the growth direction. One is the spacing previously discussed, and the other is the width of the trees. When the concentration is low, the width is narrow. The width increases with increasing concentration up to 0.009 mol/ℓ. For the higher concentration the width seems to decrease with increasing con-

Fig. 4. The same as fig. 1 except that the concentration is 0.018 mol/ℓ.

Fig. 5. The same as fig. 1 except that the concentration is 0.020 mol/ℓ.

centration. It is interesting to notice that this value coincides with the concentration for which one observes a stepwise change of the growth velocity. Since the spacing between the trees also decreases very rapidly in this concentration region, the whole structure becomes more and more dense and homogeneous until 0.018 mol/ℓ, still maintaining the local fractal structure (fig. 4). With further increasing of the concentration the local fractal structure cannot be maintained, and the dendritic growth solid backbones appear (fig. 5).

The results described above are very different from the theoretical proposal based on the DLA growth from a finite particle density [11]. For a completely diffusion-limited process, the growth velocity is inversely proportional to the diffusion length. On the

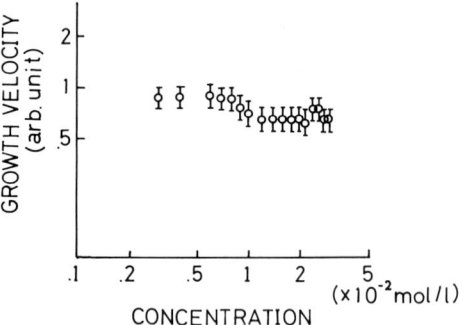

Fig. 6. log–log plot of the growth velocity versus the concentration of Zn ions.

other hand the average density in the solid phase, which is equal to the gas density in the stationary growth condition, should be proportional to the $(D-$

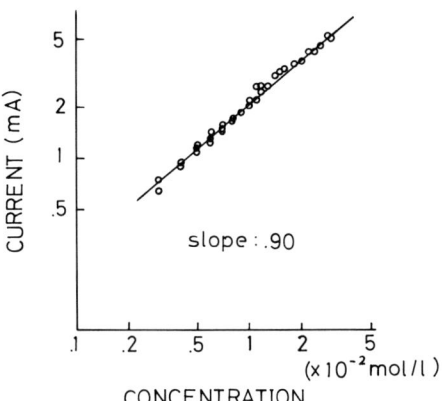

Fig. 7. log–log plot of the current through the anode and cathode versus the concentration of Zn ions.

width of the trees and the average spacing between the trees vary with concentration in the two concentration regions while the growth velocity stays constant in the two regions. Understanding of the underlying mechanism for the present observation may also help to understand other interesting aspects of real world, such as the transition from the isotropic fractal structure to the anisotropic dendrite structure with backbones by a slight change of the concentration.

d) power of the diffusion length, since the diffusion length is the only length scale in the system. By combining two observations one expects that the growth speed increases with the $1/(d-D)$ power of the concentration. For $d=2$ and $D=1.7$ the growth speed is expected to increase with concentration with an exponent of 3.3. The above consideration proposed by Uwaha and Saito was verified by themselves with computer simulation [11]. The present experiments, however, show that the growth velocity stays constant instead of increasing with concentration with a large exponent.

The electrochemical system should certainly be different from the simple DLA in various respects. In the present study it was observed that the average

References

[1] B.B. Mandelbrot, The Fractal Geometry of Nature (Freeman, San Francisco, 1982).
[2] L. Niemeyer, L. Pietronero and H.J. Wiesmann, Phys. Rev. Lett. 52 (1984) 1033.
[3] M. Matsushita, M. Sano, Y. Hayakawa, H. Honjo and Y. Sawada, Phys. Rev. Lett. 53 (1984) 286.
[4] J. Nittmann, G. Daccord and H.E. Stanley, Nature 314 (1985) 314.
[5] E.H. Stanley, ed., STATPHYS 16, Proceedings of the 16th International Conference on Thermodynamics and Statistical Mechanics (North-Holland, Amsterdam 1986).
[6] J.S. Langer, Rev. Mod. Phys. 52 (1980) 1.
[7] D.A. Kessler and H. Levine, Phys. Rev. Lett. 57 (1986) 3069;
D.I. Meiron, Phys. Rev. A 33 (1986) 2704.
[8] H. Honjo, S. Ohta and M. Matsushita, J. Phys. Soc. Japan 55 (1986) 2487.
[9] Y. Sawada, A. Dougherty and J.P. Gollub, Phys. Rev. Lett. 56 (1986) 1260.
[10] D. Grier, E. Ben-Jacob, R. Clarke and L.M. Sander, Phys. Rev. Lett. 56 (1986) 1264.
[11] M. Uwaha and Y. Saito, J. Phys. Soc. Japan 57 (1988) 3285.

LEVY FLIGHTS: VARIATIONS ON A THEME

Michael F. SHLESINGER

Physics Division, Office of Naval Research, 800 North Quincy Street, Arlington, VA 22217-5000, USA

Presented in honor of Benoit Mandelbrot in celebration of his 65th birthday

Levy's original ideas on limit distributions for sums of identically distributed random variables with infinite second moments are applied to a variety of topics including generating nondifferentiable lacunary series, the Riemann hypothesis, and turbulent diffusion.

1. First memories

The year was 1972 and I was a young graduate student of Elliott Montroll at the University of Rochester. I had recently completed my Ph.D. qualifying exams and was casting about for research problems. The world of physics seemed to be quiescent in this era. One week, for the first and only time, Montroll told all his students to attend the weekly thursday colloquium. The speaker was Benoit Mandelbrot, and Montroll promised a lecture that would not only be different, but exciting, while at the same time admitting that he did not know what the topic would be. Montroll, a former Vice President for Research at IBM, knew Benoit well, and felt that he was on the trail of something powerful and beautiful. The day was full of excitement, and even back then Benoit was treated with awe. A crowd of Full Professors completely surrounded him at the tea preceding the colloquium preventing any contact with lowly graduate students. The lecture itself was everything that was promised. Fractal forgeries of galaxies were shown, side by side, with actual pictures of galaxies. The audience was repeatedly told that galactic structures and clusters possessed noninteger dimensions – a startling revelation! The entire day fired my imagination. The next morning I further discussed Levy flights (Benoit's method for constructing his galaxies) with Montroll. He knew all about these strange probability distributions with infinite moments, and was certain that they would play a vital role in physics. All the other Professors in the department were mystified. That day I wrote down the Fourier transform of a probability distribution on a lattice which I thought would generate discrete Levy flights. I did not realize that I had rediscovered the Weierstrass function, and in any event I abandoned this line of thought as I convinced myself erroneously that my random process would only lead to normal Gaussian behavior. It was only in 1981 that I came back to this problem (now at the University of Maryland and again with Elliott Montroll) and used it in various guises (reported here) successfully in several areas of physics. It was Barry Hughes, then a Montroll postdoc from Australia, who told me my random flight structure function was the Weierstrass function. We later generalized this lacunary cosines series to a lacunary Bessel series. When I dropped the Levy flight problem in 1972, it was to develop a theory of scale invariant temporal distributions governing the motion of charges in disordered materials, such as those used in xerography. I met Benoit again in 1976. It was my job to pick up the big shots at the Rochester airport and bring them to the Montroll 60th Birthday Celebration, and I was fortunate to get both Benoit and Mark Kac in the same trip. During the Montroll meeting, Benoit, in a discussion with Harvey Scher (a mentor of mine on the disordered materials problems) said that our work was really on the nature of "fractal time", a notion he introduced in the early 60s to describe error intervals in communication networks. Like the gentleman who was speaking prose, it turned out that I had been working on fractals all along. All through the years, I have found Benoit through his papers, his lectures, and his friendship to be a constant source of inspiration and am honored to have this opportunity to present this work on Levy flights to him in celebration of his 65th birthday.

2. Levy flights [1,2]

2.1. The Fourier representation

Consider a random walker each of whose steps is an identically distributed random variable x with probability density $p(x)$. Levy asked the question of when can the probability density $P_n(x)$ of the position of the walker after n steps be the same as $p(x)$, except for scale factors. This is essentially the question of when does the whole (the sum of the steps) resemble the parts (any sequence of steps). In other words, do such $p(x)$ exist which lead to the random process possessing a self-similar trajectory. If the steps $p(x)$ are Gaussian with zero mean and unit variance, then $P_n(x)$ is Gaussian with variance n. In general,

$$P_n(x) = \int P_{n-m}(x-x') P_m(x') \, dx'$$

$$(\text{for } 0 \leq m \leq n), \tag{1}$$

where $P_1(x) = p(x)$, and $P_0(x)$ is the initial condition. In Fourier space, eq. (1) becomes

$$P_n(k) = P_{n-m}(k) P_m(k), \tag{2}$$

which has the solution

$$P_n(k) = \int \exp(ikx) P_n(x) \, dx = \exp(-n|k|^\beta). \tag{3}$$

The case $\beta = 2$ is the Gaussian. If $\beta < 2$, then the second moment of $P_n(x)$ diverges since

$$\langle x_n^2 \rangle = \sum x^2 P_n(x) = \partial^2 P_n(k)/\partial k^2 |_{k=0}. \tag{4}$$

Expanding the exponential in eq. (3) we find for small k that $P_n(k) \sim 1 - \frac{1}{2}\langle x_n^2 \rangle k^2$ if $\langle x_n^2 \rangle$ is finite, and $\sim 1 - \text{const.} \times |k|^\beta$ if $\langle x^2 \rangle$ is infinite. We must have 2 as the upper critical value of β, or else from eq. (4) $\langle x_n^2 \rangle$ would be zero, a nonsensical result. The above is a simplified version of Levy's discovery that limit distributions, other than the Gaussian, exist for sums of independent identically distributed random variables. The non-Gaussian cases are now called Levy flights and they are composed of self-similar (fractal) trajectories (see fig. 2).

2.2. Weierstrassian Levy flights [3]

Let us study as a specific case of a Levy flight a random walk on a 1D lattice. Let $p(R)$ be the probability that the walker makes a jump of displacement R. Choose

$$p(R) = \frac{N-1}{2N} \sum_{j=0}^{\infty} N^{-j} (\delta_{R,b^j} + \delta_{R,-b^j}). \tag{5}$$

This allows for jumps over an infinite number of orders of magnitude, with an order of magnitude longer jump (in base b) occurring an order of magnitude less often (in base N). Roughly, the random walker makes about N jumps of length one (i.e. visiting a cluster of N points) before a jump of length b occurs. About N jumps are made around the point b before another jump of length b occurs. After about N clusters (each with about N points) are formed from the set of sites visited, a jump of length b^2 occurs. In this manner, one generates clusters within a cluster ad infinitum. One would expect in two or more dimensions that the set of sites visited would have a pointwise fractal dimension of $\beta = \log N/\log b = \log$ (number of subclusters/cluster)/log(scale factor). To make contact with the previous section, let us Fourier transform $p(R)$ to find

$$p(k) = \frac{N-1}{N} \sum_{j=0}^{\infty} N^{-j} \cos(b^j k), \tag{6}$$

which is Weierstrass's famous example of a function which is everywhere continuous, but nowhere differentiable, when $b > N$. Note $p(k)$ satisfies the following scaling equation:

$$p(k) = N^{-1}p(bk) + \frac{N-1}{N}\cos(k),$$

whose homogeneous part $p_h(k) = N^{-1}p_h(bk)$ has the solution

$$p_h(k) = |k|^{\log N/\log b} Q(k), \qquad (7)$$

where $Q(k)$ is periodic in $\log k$ with period $\log b$. The full solution for $p(k)$ is somewhat complicated but for small k, $p(k)$ behaves essentially as $1-|k|^\beta$ or $\exp(-|k|^\beta)$, as expected for a Levy flight. Note that the mean square displacement per step for this random walk diverges as $\sum_j (b^2/N)^j$.

2.3. Fractal Rayleigh–Pearson random walk [4]

To re-do the Weierstrass random walk on higher-dimensional orthogonal lattices will add no new information. Let us however, examine a D-dimensional random walk in the continuum. We choose a spherically symmetric jump distribution,

$$p(\mathbf{x}) = (S_D |\mathbf{x}|^{D-1})^{-1} p_1(x), \qquad (8)$$

where $p_1(x)$ is the probability of jumping a distance $|\mathbf{x}|$, and for normalization S_D is the surface area of a hypersphere of radius unity. Fourier transforming eq. (8) yields

$$p(\mathbf{k}) = \Gamma(\tfrac{1}{2}D) \int_0^\infty (\tfrac{1}{2}|\mathbf{k}|q)^{1-D/2}$$
$$\times J_{D/2-1}(|\mathbf{k}|q)\, p_1(q)\, \mathrm{d}q. \qquad (9)$$

To generate a fractal clustered trajectory we choose

$$p_1(|\mathbf{x}|) = \frac{N-1}{N} \sum_{j=0}^\infty N^{-j}\delta(|\mathbf{x}|-b^j), \qquad (10)$$

which leads to

$$p(\mathbf{k}; N, b) = \frac{N-1}{N} \sum_{j=0}^\infty N^{-1}\Gamma(\tfrac{1}{2}D)$$
$$\times (\tfrac{1}{2}|\mathbf{k}|b^j)^{1-D/2} J_{D/2-1}(|\mathbf{k}|b^j). \qquad (11)$$

This function provides a natural generalization of the Weierstrass function (when $D=1$ it reduces to it). The full expansion of eq. (11) in powers of k is some-

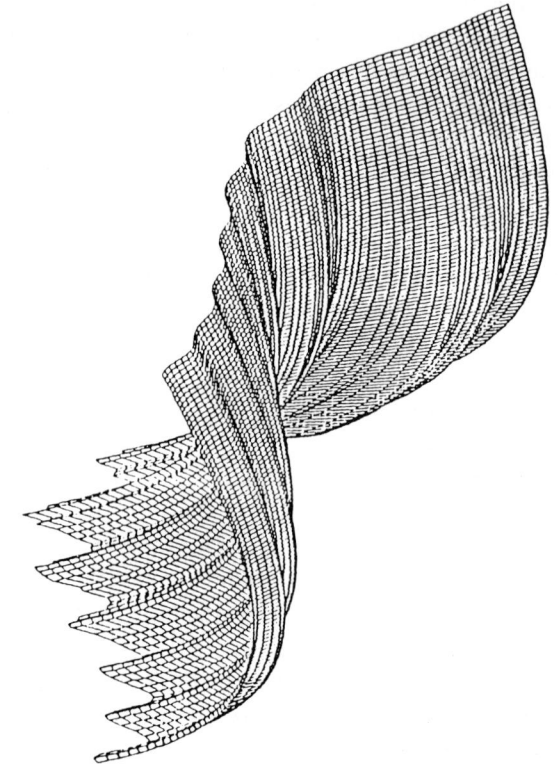

Fig. 1. With k fixed, the first six terms of the Bessel series in eq. (11) is plotted as a function of N and b, for $2<(N, b)<7$. In the regime $(b<N)$ where the function is known to be differentiable the manifold appears smooth. For $b>N$, where we hypothesize nondifferentiability, the manifold is wild. Pictures [4] of $p(k)$ versus k also show a transition from a smooth to a wiggly function as b becomes greater than N.

what complicated, but it is not difficult to see that since

$$p(\mathbf{k}) = N^{-1}p(b\mathbf{k}) + \frac{N-1}{N}\Gamma(\tfrac{1}{2}D)$$
$$\times (\tfrac{1}{2}|\mathbf{k}|)^{1-D/2}J_{D/2-1}(b|\mathbf{k}|),$$

a term of the form $|\mathbf{k}|^{\log N/\log b} Q(|\mathbf{k}|)$ (with Q periodic in $\log k$ with period $\log b$) is generated by the homogeneous part of the equation. If $\log N/\log b > \tfrac{1}{2}(3-D)$, then $\mathrm{d}p(k)/\mathrm{d}k$ exists, which sets boundaries on where the function might be nondifferentiable. It is an open question if higher derivatives of $p(\mathbf{k})$ are nondifferentiable.

2.4. Riemann–Möbius random walk [5]

Let us now consider a Levy flight process on a lattice where allowed jump lengths correspond to integers of number theoretic interest. Choose the jump probability as

$$p(R) = \tfrac{1}{2} C \sum_{|R|=1}^{\infty} \frac{1 \pm \mu(n)}{n^{1+\beta}} (\delta_{R,n} + \delta_{R,-n}), \quad (12)$$

where $\beta > 0$, and C is a normalization constant, and $\mu(n)$, called the Möbius function, is defined as, (-1^R) if n is a product of R distinct primes, and it is zero otherwise. Note that this walk is in the form of a Levy flight when the mean square displacement per jump is infinite. Fourier transforming, we find

$$p(k) = C \sum_{n=1}^{\infty} \frac{1 \pm \mu(n)}{n^{1+\beta}} \cos(nk). \quad (13)$$

Replacing $\cos(nk)$ by its own inverse Mellin transform

$$\cos k = \frac{1}{2\pi i} \oint k^{-s} \cos(\pi s/2) \, \Gamma(s) \, \mathrm{d}s \quad (14)$$

and switching the sum and the integral in eq. (13) yields,

$$p(k) = \frac{C}{2\pi i} \oint k^{-s} [\zeta(1+\beta+s) \pm \zeta^{-}(1+\beta+s)]$$
$$\times \cos(\pi s/2) \, \Gamma(s) \, \mathrm{d}s. \quad (15)$$

We have used the fact that

$$\zeta^{-1}(s) = \prod_{\text{primes}} (1 - p^{-s})$$

$$= (1 - 2^{-s})(1 - 3^{-s})(1 - 5^{-s})\ldots = \sum_{n=1}^{\infty} \frac{\mu(n)}{n^s}.$$

In eq. (15) there are poles at $s = 0, -2, -4, \ldots$, from the $\cos(\pi s/2) \, \Gamma(s)$ term, and a simple pole at $s = -\beta$ from the $\zeta(1+\beta+s)$ term. All other singularities arise from zeros of the zeta function. Examining the integral

$$\int_0^{\infty} \exp(-nx) \, x^{-s-1} \, \mathrm{d}x = \Gamma(s)/n^s,$$

and summing over n on both sides of the equation leads to an expression for $\zeta(s)$ in terms of the Bernoulli numbers which immediately leads to $\zeta(-2n) = 0$ ($n = 1, 2, 3, \ldots$). The first of this set of zeros occurs at $s = -\beta - 3$.

The remaining question is where are the complex zeros of $\zeta(s)$. First, we observe that they must occur in complex conjugate pairs because $p(k)$ is real. Next, we see that they cannot have a real part greater than $-\beta$ or else the singularity from the $\pm \zeta^{-1}(1+\beta+s)$ term would dominate the pole from the $\zeta(1+\beta+s)$ term in the small k expansion of $p(k)$. This cannot occur because the maximum value of $p(k)$ is unity. Thus, the most singular term in the integrand should subtract from unity in the small k expansion of $p(k)$, and the $\pm \zeta^{-1}$ term can add. No double zeros of $\zeta(1+\beta+s)$ can lie on the line Re $s = -\beta$ because this would generate a $\pm k^{\beta} \ln k$ term which again would spoil the normalization of $p(k)$. The functional equation

$$\Gamma(s/2) \pi^{-s/2} \zeta(s) = \Gamma((1-s)/2) \pi^{-(1-s)/2} \zeta(1-s)$$

forces all complex zeros of $\zeta(s)$ to lie symmetrically about the line Re $s = 1/2$. We can conclude from our Levy flight approach that the complex conjugate zeros of the ζ function have a real part bounded by $0 \leq \text{Re } s \leq 1$, and that there are no double poles on the boundary lines. If the random walk were not fractal then the pole at $s = -2$ would dominate and the above arguments on restricting the positions of the zeros of $\zeta(s)$ would have no force. The reader is welcome to try to extract more information from this probabilistic approach. The question of the location of all the complex zeros of $\zeta(s)$ is perhaps the most celebrated problem in mathematics. Riemann hypothesized in 1859 that all of these zeros lie on the line Re $s = 1/2$. The Riemann hypothesis has not been resolved to date.

3. Turbulent diffusion [6]

We have only considered a sum of random variables with each new term in the sum added all at once, not continuously. Let us now explicitly take into account, for a Levy flight (see fig. 2) the time spent in a trajectory between sites. As before we choose the asymptotic behavior $p(R) \sim R^{-1-\beta}$. Next, introduce a new probability function $\Psi(R, t)$ which is the probability density for a jump of displacement R to occur and take a time t. We write

$$\Psi(R, t) = \psi(t|R) p(R), \qquad (16)$$

where $\psi(t|R)$ is the probability density for the transition taking a time t, given that a distance R was covered. For the sake of simplicity, we choose

$$\psi(t|R) = \delta(t - |R|/V(R)), \qquad (17)$$

where $V(R)$ is the velocity with which a jump of distance R occurs. For motion in a turbulent flow one would choose (ignoring the fractal nature of the dissipation field)

$$V(R) = R^{1/3} \qquad (18)$$

so as to be in accord with Kolmogorov's $-5/3$ law, i.e. consider the following scaling relations $E_R \sim V(R)^2$, so $\bar{\epsilon} \sim V^2/T_R \sim V^3/R$, where E_R is the energy associated with a scale R, $\bar{\epsilon}$ is the average dissipation (assumed to be a constant), and $T_R = R/V$

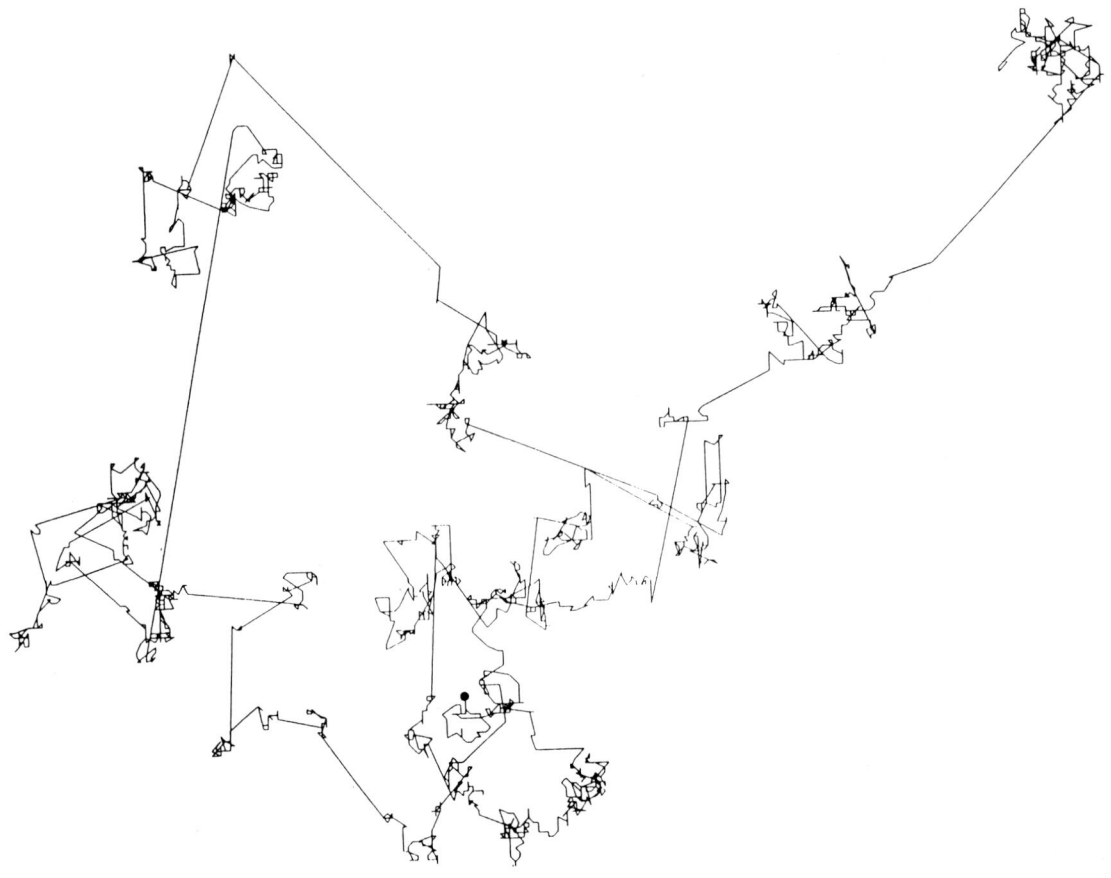

Fig. 2. A realization of a Levy flight [7]. In our model of turbulent diffusion the trajectory can represent the relative separation of two fluid particles. The straight line segments would be traversed with a velocity depending on the jump size, i.e. bigger vortices have more energy and induce bigger velocities.

is a time scale. This leads to eq. (18) and E_k (the Fourier transform of E_R) varying as $k^{-5/3}$. Using all of the above it can be shown that when $\beta \leq 1/3$ (i.e. the mean time spent in a flight is infinite) then

$$\langle R^2(t) \rangle \sim t^3, \qquad (19)$$

which is Richardson's law of turbulent diffusion. Our derivation involves an underlying fractal Levy flight structure, mixed with Kolmogorov scaling, all in a nonlocal integral differential equation of motion.

References

[1] P. Levy, Théorie de l'Addition des Variables Aléatoires (Gauthier-Villars, Paris, 1937).
[2] B.B. Mandelbrot, The Fractal Nature of Geometry (Freeman, San Francisco, 1982).
[3] B.H. Hughes, M.F. Shlesinger and E.W. Montroll, Proc. Nat. Acad. Sci USA 78 (1981) 3287.
[4] B.D. Hughes, E.W. Montroll and M.F. Shlesinger, J. Stat. Phys. 28 (1982) 111.
[5] M.F. Shlesinger, Physica A 138 (1986) 310.
[6] M.F. Shlesinger, B.W. West and J. Klafter, Phys. Rev. Lett. 58 (1987) 1100.
[7] G. Zumofen, A. Blumen, J. Klafter and M.F. Shlesinger, J. Stat. Phys. 54 (1989) 1519.

SCATTERING FROM FRACTAL STRUCTURES

S.K. SINHA

Exxon Research and Engineering Company, Annandale, NJ 08801, USA

We briefly discuss how the property of self-similarity found in nature manifests itself in various types of scaling laws for the scattering of radiation from fractal structures. The cases of scattering from mass and surface fractals are discussed, as is the generalization to the case where an object may possess both mass and surface fractal dimensions. The important case of scattering from self-affine fractal surfaces is also illustrated.

The recognition of the property of self-similarity [1] as a hidden symmetry in a wide variety of apparently random and disordered structures appearing in nature is a remarkable achievement. It amounts to a recognition of a ubiquitous static scaling property of these diverse systems similar to what we have hitherto been familiar with in the study of critical phenomena at phase transitions. The reasons as to why such scaling properties manifest themselves in a variety of growth and formation processes still seems deep and mysterious and are only gradually being understood. However, it was rapidly realized that scattering experiments could provide a nice experimental method of verifying the existence of such scaling properties on microscopic and mesoscopic length scales.

Consider the definition of a mass fractal, namely that the total mass inside a radius R scales as

$$M \sim R^{\bar{d}}, \tag{1}$$

where $\bar{d}$ is the mass fractal dimension. Eq. (1) is a necessary consequence of the property of self-similarity. Differentiating eq. (1), it is easy to show that, given a particle at the origin, the probability of finding a particle in an infinitesimal volume dV at a distance r from the origin is $\sim r^{\bar{d}-3} dV$. We may thus formally write the pair-correlation function as

$$\langle \rho(\boldsymbol{r}) \rho(\boldsymbol{r}') \rangle \equiv g(\boldsymbol{r}-\boldsymbol{r}') \sim |\boldsymbol{r}-\boldsymbol{r}'|^{\bar{d}-3}. \tag{2}$$

We have assumed that, statistically, the system is translationally invariant, as would be true for an extended system without a particular origin site. We assume also that the scattering length density is proportional to the particle density as would be true for a monatomic system or for length scales larger than those on which the atomic composition is inhomogeneous. (The latter is generally true for small angle scattering experiments.) Then the scattering of X-rays, neutrons, or light is proportional to the structure factor

$$S(\boldsymbol{q}) = \iint d\boldsymbol{r} \, d\boldsymbol{r}' \, g(\boldsymbol{r}-\boldsymbol{r}') \exp[i\boldsymbol{q}(\boldsymbol{r}-\boldsymbol{r}')]. \tag{3}$$

Eqs. (2) and (3) yield the simple result for mass fractals that

$$S(q) \sim q^{-\bar{d}}. \tag{4}$$

The observation of such power laws in small-angle scattering experiments provides both a verification of the fractal paradigm for the structure and a convenient way of measuring the fractal dimension $\bar{d}$. By now, a variety of materials have been investigated, which include systems of aggregated colloids, gels, biological systems, polymeric systems, rocks, ceramic materials and metallic systems [2–11], where such power laws have been observed in the scattering over typically a decade in length scales. At very small q, $S(q)$ must saturate, as required by general principles, and we have a transition to the so-called "Guinier regime" [12]. This is due to the fact that there must be an upper length scale cutoff to the fractal correlation function given in eq. (2). This is usually represented by an exponential cutoff [5], which works reasonably well in analyzing scattering data, although it does not always work in detail for specific systems [13]. In this approximation, eq. (2) is modified to

$$g(r) \sim r^{\bar{d}-3} \exp(-r/\xi), \quad (5)$$

where ξ is the "fractal correlation length". The modification to $S(q)$ from eq. (3) is then given by

$$S(q) \sim \Gamma(\bar{d}-1)\xi^{\bar{d}}(1+q^2\xi^2)^{-\bar{d}/2}$$
$$\times \left(\frac{1+q^2\xi^2}{q\xi}\right)^{1/2} \sin[(\bar{d}-1)\arctan(q\xi)], \quad (6)$$

which reduces to the Guinier form as $q \to 0$, with the effective radius of gyration r_g given by

$$r_g^2 = \tfrac{1}{2}(\bar{d}+1)\bar{d}(\bar{d}-1)\xi^2. \quad (7)$$

For large q, the asymptotic form of eq. (4) is regained, *except* in the case were $\bar{d} \to 3$, where instead we obtain the asymptotic form [11]

$$S(q) \sim q^{-4}, \quad (8)$$

which is Porod's law. This is because eq. (6) now describes the correlation function for a uniform solid material with *smooth* but random internal surfaces [14].

For fractal aggregates, the large-q behavior of $S(q)$ will be strongly modified by short-range deviations from true fractal correlations, as must be the case if the aggregating particles are of finite size.

Figs. 1a and 1b show $S(q)$ obtained by small-angle X-ray scattering for one of the best-studied fractally aggregating systems, namely gold colloid particles of diameter 7.5 nm aggregated according to (a) diffusion-limited cluster aggregation (DLCA) [15,16] conditions and (b) reaction-limited cluster aggregation (RLCA) [17] conditions. In these systems, ξ is so large that the Guinier regime is not observed, even at the smallest q for the measurements (3×10^{-3} nm^{-1}). The large q range obtained in this experiment enables a fairly stringent test to be made of short-range and intermediate-range order in this sys-

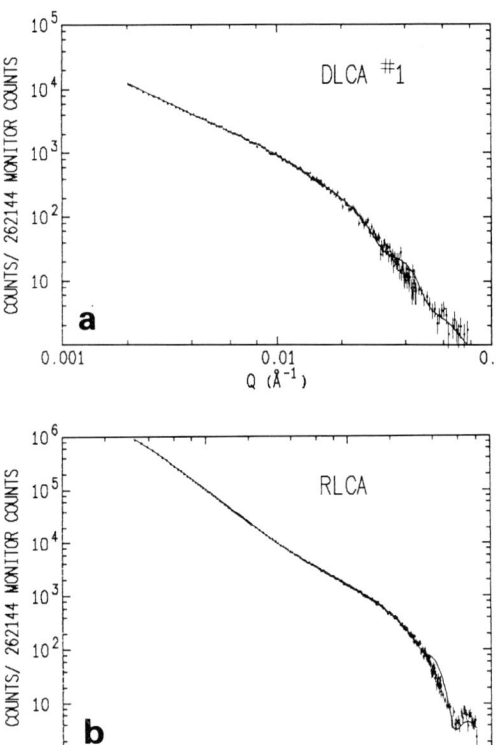

Fig. 1. (a) $S(q)$ for Au colloids aggregated according to DLCA measured with small-angle X-ray scattering. The solid curve is a fit based on short-range random sphere packing and long-range fractal correlations (figurge from ref. [3]). (b) The same curves for Au colloids aggregated according to RLCA (figure from ref. [3]).

tem. Ref. [3] discusses in detail the structural models for these aggregates which were used to fit the data. For most of the q range, the power law behavior characteristic of a mass fractal is evident, and the measurements gave values for $\bar{d}$ of 1.73 for the DLCA clusters and 2.2 for the RLCA clusters, in reasonable agreement with computer simulations of cluster–cluster aggregation processes in these different limits [15–18]. The short-range order is not fractal, but corresponds to random packing of hard spheres, and fig. 2a shows the corresponding $g(r)$ function obtained from the fitting model. Fig. 2b shows the corresponding $g(r)$ obtained by Meakin [18] from a computer simulation of DLCA, and it shows striking

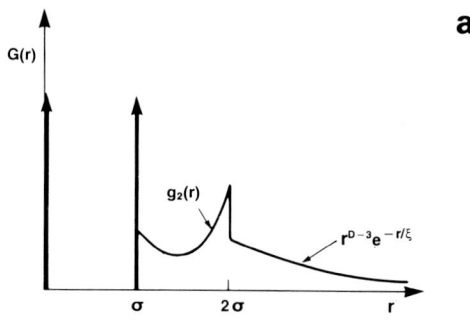

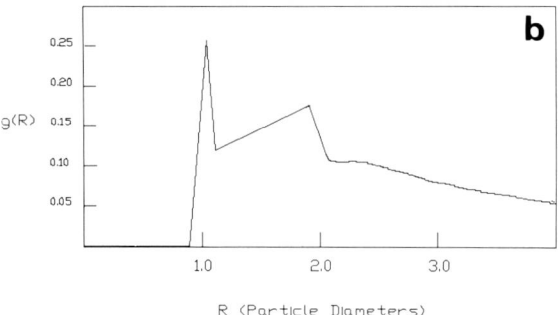

Fig. 2. (a) Model $g(r)$ used to fit $S(q)$ shown in fig. 1. The pair correlation function is that for *centers* of spheres of diameter σ. The spike at σ represents the nearest-neighbor ring of spheres. (b) $g(r)$ obtained from computer simulations by Meakin [18] for DLCA growth.

similarities. The particles appear to be fractally correlated after roughly second-neighbor distances.

Mass fractals are, of course, not the only kind of fractal structures which can be found in nature – one can also have fractal surfaces. For these, the most satisfactory definition of the surface fractal dimension d_s is that given by Mandelbrot [1] in terms of "tiling". If l is the linear dimension of squares used to cover the fractal surface, the number of such tiles required for complete coverage varies as

$$N \sim l^{-d_s}. \tag{9}$$

Bale and Schmidt [19] were the first to show that this leads to an analytic behavior of $g(r)$ for small r such that the asymptotic form of $S(q)$ for large q is given by

$$S(q) \sim q^{d_s - 6}. \tag{10}$$

Since $2 < d_s < 3$, the observation of $q^{-\alpha}$ power laws in $S(q)$, where $3 < \alpha < 4$ has been taken to indicate the existence of fractally rough surfaces [7,19,20], and eq. (10) represents the generalization of Porod's law (eq. (8)) for such surfaces.

There are several reasons as to why the above description of scattering from fractal structures is incomplete. Firstly, we note that a mass fractal usually also has a fractal surface, e.g., a fractal aggregate of particles of dimension a has *both* a mass and surface fractal dimension (which are both equal to $\bar{d}$) for length scales $l \gg a$. For $l < a$, the mass fractal dimension crosses over to 3 for solid particles, while d_s crosses over to 2 for smooth surfaces or a non-trivial d_s between 2 and 3 for rough surfaces. Derivations of eq. (10) such as ref. [19] usually assume that there exists a uniform solid "interior" of the solid away from the surface over the same length scales, whereas this may not be the case (see figs. 3a and 3b). A more general treatment of the scattering from fractal structures, such as in fig. 3b, without such assumptions [21] shows that the general asymptotic form for $S(q)$ is

$$S(q) \sim q^{d_s - 2\bar{d}}, \tag{11}$$

which yields eq. (4) (where $d_s = \bar{d}$, as for fractal aggregates) or eq. (10) (where $\bar{d} = 3$), as particular cases. Whether structures exist with non-integral and different $\bar{d}$ and d_s is not yet clear. Certainly, one can imagine structures where the volumes and surface scale according to different exponents as the length-

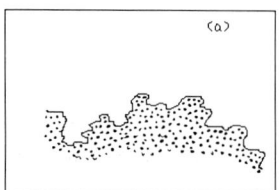 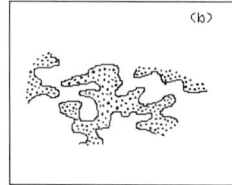

Fig. 3. (a) Schematic of a fractal surface with a uniform solid "interior". (b) Schematic of an object which has no uniform interior volume.

scale of the yardstick is changed. A trivial example would be a dilute (non-overlapping) random set of spheres with a power-law size distribution or dendritic growth with successive branches scaling in length and diameter according to different exponents. (Such a structure would not be strictly self-similar, however.) An interesting case is when d_s, $\bar{d}$ both $\to 3$. Eq. (11) shows that in such a case $S(q) \sim q^{-3}$. (This is to be distinguished from the case where $\bar{d}$ alone $\to 3$, but the internal surfaces stay smooth, where as discussed above, Porod's law, eq. (8) holds.) In fact, a porous solid which is sintered to the point where there is a lot of internal surface but the pore volume is about to disappear, probably approaches the situation where d_s, $\bar{d}$ both $\to 3$. Fig. 4 shows scattering from a heavily sintered aerogel [22] where asymptotic q^{-3} scattering is observed. Exponents close to -3 have been observed for a number of porous solids [20,22].

Finally, we note that a more general (and probably more prevalent) kind of rough surface is the self-affine fractal surface [1,23], which can be mathematically generated by the algorithm of fractional Brownian motion. Such a surface has the property that

$$\langle [z(\mathbf{R}) - z(\mathbf{R}')]^2 \rangle \sim |\mathbf{R} - \mathbf{R}'|^{2h}, \quad (12)$$

where $z(\mathbf{R})$ is the height above a flat reference surface at lateral position $\mathbf{R}$. Most readers are familiar with the beauty of the "fractal landscapes" generated by Voss and Mandelbrot [23] using such algorithms and their striking resemblance to surfaces found in nature. The texture of such landscapes is controlled by the exponent h. Mandelbrot has shown [1] that for such surfaces, a surface fractal exponent can be defined by

$$d_s = 3 - h. \quad (13)$$

It is interesting that the diffuse scattering of radiation from such a surface can also be calculated exactly in the Born approximation [24,25]. For a single self-affine surface, the diffuse scattering for $\mathbf{q}$ vectors normal to the reference surface has the simple asymptotic form

$$S(q_z) \sim q_z^{-(2+2/h)}, \quad (14)$$

while an average over randomly oriented such surfaces yields the asymptotic form

$$S(q) \sim q^{-(3+h)}, \quad (15)$$

which is consistent with the result in eq. (10) for a self-similar fractal surface via the relation given in eq. (13). Fig. 5 shows X-ray diffuse scattering from the surface of a block of polished Pyrex glass for $\mathbf{q}$ normal to the surface (a) and for $\mathbf{q}$ parallel to the surface (b). The observed scattering was fitted to the functional form for diffuse scattering from a self-affine surface given in ref. [24] and yielded for this surface a value of $h = 0.2 \pm 0.02$.

The fractal paradigm has thus provided us with an invaluable working hypothesis to explain various

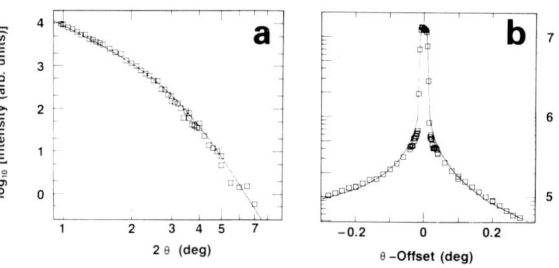

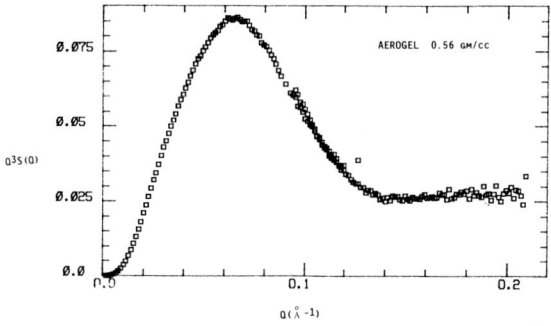

Fig. 4. $q^3 S(q)$ for a sintered aerogel (density=0.56 g/cm³), showing the q^{-3} asymptotic behavior at large q (figure from ref. [22]).

Fig. 5. (a) Diffuse scattering from a piece of polished Pyrex glass measured with 1.5 Å X-rays for $\mathbf{q}$ normal to the surface. (b) A rocking curve at $2\theta = 2°$ corresponding to diffuse scattering with $\mathbf{q}$ parallel to the surface, for $q_z = 0.14$ Å^{-1}. The spike at the center corresponds to the specular reflection. The solid curves represent a fit to a self-affine fractal surface model with $h = 0.2 \pm 0.02$ (figure from ref. [24]).

asymptotic power laws observed in scattering from a large variety of disordered materials. It is, however, a method of analyzing scattering data which must be used with care and discrimination. Where it has been successfully applied, it has brought a new richness and excitement to structural studies of mesoscopic systems.

I am indebted to Tom Witten for introducing me to the possibilities of scattering from fractals, and to David Weitz, Mike Drake, Robin Ball, Yossi Klafter, Jorgen Kjems, Torsten Freltoft, Shlomo Alexander, Po-zen Wong, and Peter Dimon for invaluable collaborations, discussions, and enlightenment.

References

[1] B.B. Mandelbrot, The Fractal Geometry of Nature (Freeman, San Francisco, 1982).
[2] D.A. Weitz, M.Y. Lin, J.S. Huang, T.A. Witten, S.K. Sinha, J.S. Gethner and R.C. Ball, in: Scaling Phenomena in Disordered Systems, R. Pynn and A. Skjeltorp, eds. (Plenum, New York, 1985), p. 171;
H.M. Lindsay, D.A. Weitz, R.C. Ball, R. Klein, P. Meakin and M.Y. Lin, Nature, in press.
[3] P. Dimon, S.K. Sinha, D.A. Weitz, C.R. Safinya, G.S. Smith, W.A. Varady and H.M. Lindsay, Phys. Rev. Lett. 57 (1986) 595.
[4] D.W. Schaefer, J.E. Martin, P. Wiltzius and D.S. Cannell, Phys. Rev. Lett. 52 (1984) 2371;
G. Dietler, G. Aubert, D.S. Cannell and P. Wiltzius, Phys. Rev. Lett. 57 (1986) 3117.
[5] S.K. Sinha, T. Freltoft and J. Kjems, in: Kinetics of Aggregation and Gelation, F. Family and D.P. Landau, eds. (North-Holland, Amsterdam, 1969), p. 23.
[6] T. Freltoft, J.K. Kjems and S.K. Sinha, Phys. Rev. B 33 (1986) 269.
[7] K.D. Keefer and D.W. Schaefer, Phys. Rev. Lett. 56 (1986) 2376.
[8] J. Teixeira, in: On Growth and Form, H.E. Stanley and N. Ostrowsky, eds. (Nijhoff, Dordrecht, 1986), p. 145;
S.H. Chen and J. Teixeira, to be published.
[9] R. Vacher, T. Woignier, J. Pelous and E. Courtens, Phys. Rev. B 37 (1988) 6500.
[10] J.E. Martin and A.J. Hurd, J. Appl. Cryst. 20 (1987) 61.
[11] P. Mangin, B. Rodmacq and A. Chamberod, Phys. Rev. Lett. 55 (1985) 2899.
[12] A. Guinier and G. Fournet, Small Angle Scattering of X-Rays (Wiley, New York, 1955).
[13] D.A. Weitz, M.Y. Lin and H.M. Lindsay, private communication.
[14] P. Debye, H.R. Anderson and J. Brumberger, J. Appl. Phys. 28 (1957) 679;
G. Porod, in: Small Angle X-Ray Scattering, O. Glatter and O. Kratky, eds. (Academic Press, New York, 1982), p. 17.
[15] P. Meakin, Phys. Rev. Lett. 51 (1983) 1119.
[16] M. Kolb, R. Botet and R. Jullien, Phys. Rev. Lett. 51 (1983) 1123;
R. Jullien, M. Kolb and R. Botet, J. Phys. (Paris) 45 (1984) L211.
[17] M. Kolb and R. Jullien, J. Phys. (Paris) 45 (1984) L977.
[18] P. Meakin, private communication;
H.M. Lindsay, R. Klein, D.A. Weitz, M.Y. Lin and P. Meakin, Phys. Rev. A 38 (1988) 2614;
M.Y. Lin, H.M. Lindsay, D.A. Weitz, R.C. Ball, R. Klein and P. Meakin, Proc. Soc. A (1989), in press.
[19] H.D. Bale and P.W. Schmidt, Phys. Rev. Lett. 53 (1984) 596;
J.K. Kjems and P. Schofield in: Scaling Phenomena in Disordered Systems, R. Pynn and A. Skjeltorp, eds., NATO ASI Ser. B133 (Plenum, New York, 1985), p. 141.
[20] P.Z. Wong, J. Howard and J.S. Lin, Phys. Rev. Lett. 57 (1986) 637.
[21] S.K. Sinha and R.C. Ball, to be published.
[22] Z. Djordjevic, S.K. Sinha and C. Glinka, to be published.
[23] R.F. Voss, in: Scaling Phenomena in Disordered Systems, R. Pynn and A. Skjeltorp, eds. (Plenum, New York, 1985), p. 1.
[24] S.K. Sinha, E.B. Sirota, S. Garoff and H.B. Stanley, Phys. Rev. B 38 (1988) 2297.
[25] P.Z. Wong, Phys. Rev. B 32 (1985) 7417;
P.Z. Wong and A.J. Bray, Phys. Rev. B 37 (1988) 7751.

GEOMETRICAL SCALING OF MICROSPHERE-DEPOSITED MONOLAYERS WITH HOLES

A.T. SKJELTORP

Institute for Energy Technology, N-2007 Kjeller, Norway

Uniformly sized microspheres dispersed in water are spread as a thin film on a flat surface. During the drying process, random capillary forces between the spheres produce a random packing resulting in a broad distribution of holes in an otherwise compact, grainy structure. The patterns are shown to have a fractal geometry and the cumulative numbers of holes, $n(s)$, with size larger than s scale as $n(s) \propto s^{-\tau}$.

1. Introduction

The availability of uniformly sized polystyrene microspheres with diameter in the 1–100 μm range [1] has proven to be very useful as a means to construct experimental many-body model systems to study non-equilibrium processes like aggregation [2,3], fracturing [4], and deposition [5]. The resulting, often fractal [6] structures have been studied using optical microscopy and have been compared with computer simulations. They have provided a basis for obtaining a better understanding of growth processes far from equilibrium.

Here, we will show how thin films of such particles dispersed in water can be used to study deposition onto a substrate. The attractive capillary force between the spheres during drying is much stronger than the bonding to the substrate. This non-equilibrium process produces a variety of structures with holes depending on the sphere concentration. The purpose of this work is to show that the two-dimensional "porous" system has fractal properties and that the cumulative numbers of holes, $n(s)$, with sizes larger than s scale as

$$n(s) \propto s^{-\tau}. \qquad (1)$$

Here, the scaling exponent τ characterizes the hole formation. So far, there is no analytical means to predict the values of τ based on the complicated dynamics in the formation of the present system.

It may be noted that there have been earlier attempts to use eq. (1) to express the cumulative numbers of objects for other natural processes like: pieces of drift ice [7], lunar craters [8], and dust particles in atmospheres aerosols [9]. Computer simulations have also been used to show that the cluster-size distribution in diffusion-limited cluster–cluster aggregation follows this power-law dependence [10].

2. Experimental results

The most suitable spheres for the experiments were of size 2–5 μm. For the results reported here 4.8 μm spheres were used. The spheres are dispersed in water and spread to a thin film on a microscope slide, fig. 1a. During the drying process, capillary and surface

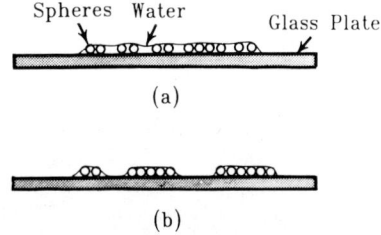

Fig. 1. Schematic, thin-film deposition experiments: (a) monolayer of microspheres dispersed in water; (b) partly dried monolayer.

tension forces produce random attractive forces between the spheres, compacting certain regions of the sample, fig. 1b. The samples were observed directly in a light microscope. A video-camera attachment and frame grabber with 512×512 pixels resolution allowed digital analysis of the structures.

Fig. 2 shows snapshots of the deposition process. As may be seen, initially small holes are formed in random locations. These holes grow in size and new holes are formed. The resulting structure has a wide distribution of hole sizes, as is shown in fig. 3, for increasing magnification in the same region of the sample. It may be noted in passing that this texture has a similar appearance to Mandelbrot's random slices of "Emmenthaler Swiss cheese" displayed as plate 307 in his book [6] with fractal dimension $D=1.90$.

Fig. 4 shows other pattern formations for decreas-

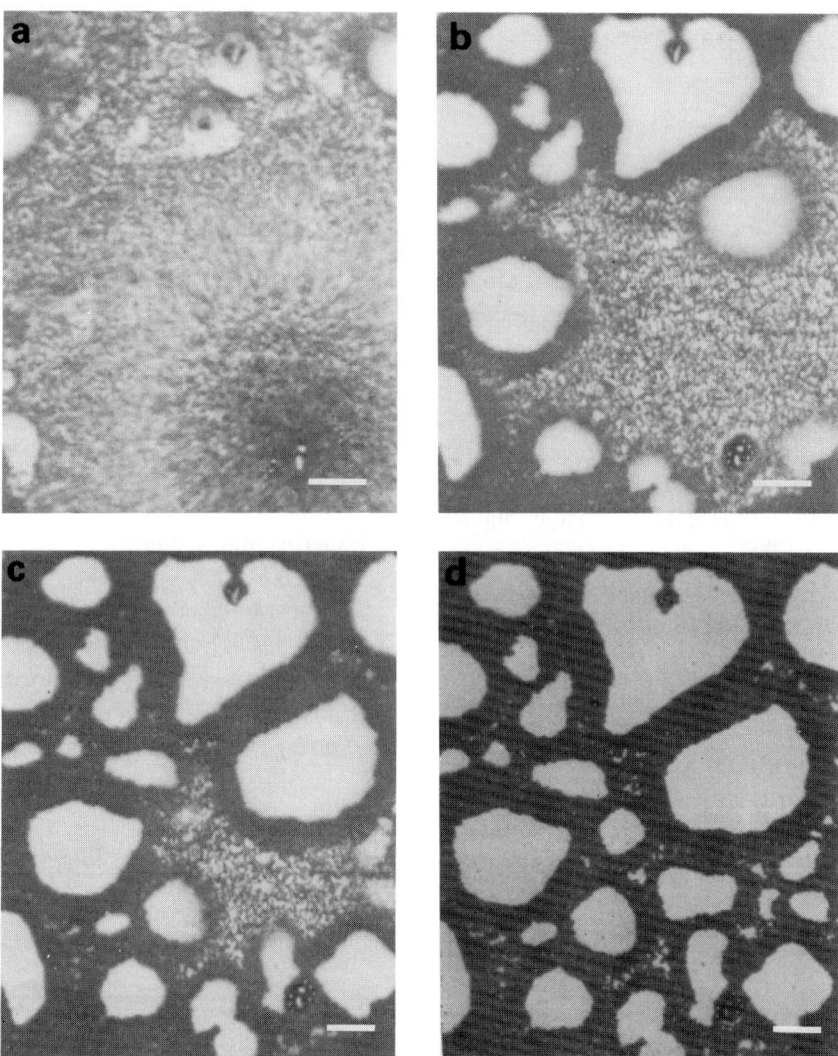

Fig. 2. Snap-shots of the hole-formation during the drying process of a monolayer of 4.8 μm spheres. The time-steps between (a)–(d) are about 0.2 s. The length of the marker is 100 μm.

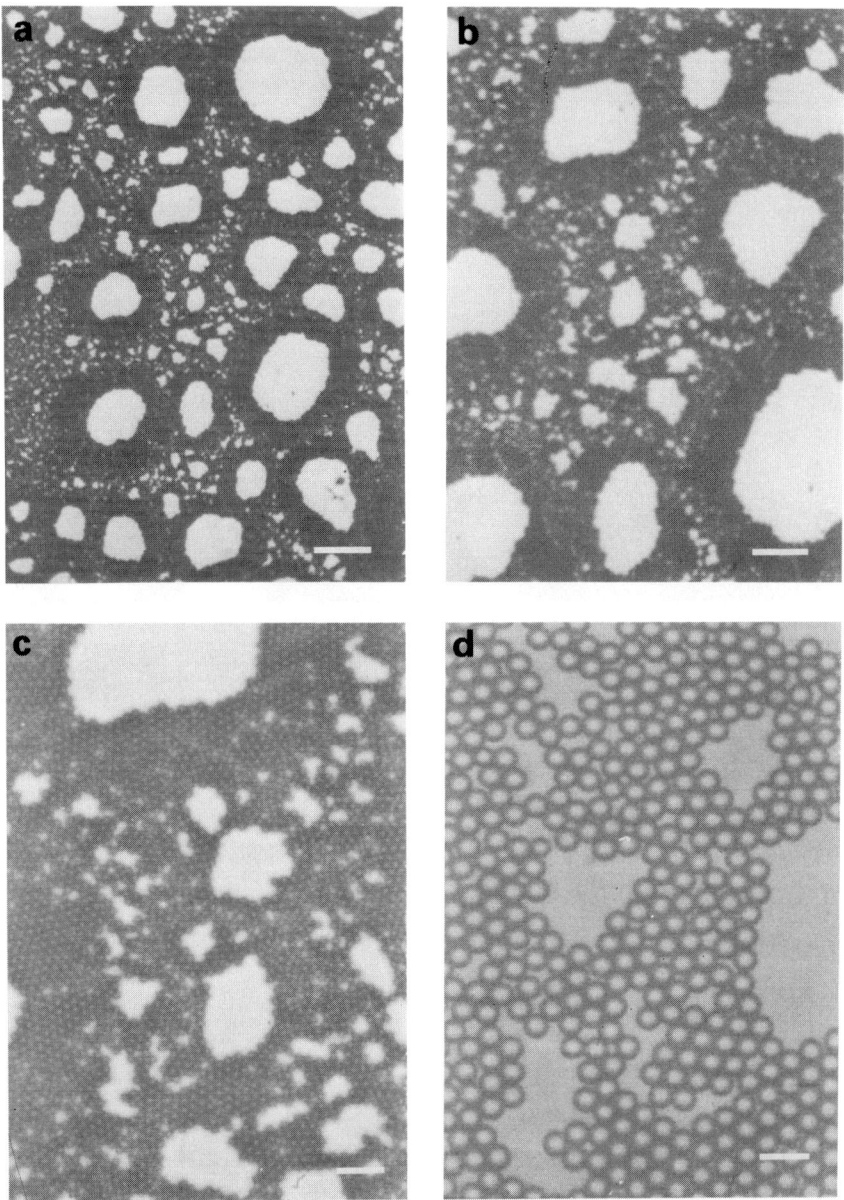

Fig. 3. Typical hole-distribution in a monolayer of 4.8 μm spheres for increasing magnification (a)–(d) from the same region of the sample with packing fraction $\eta \simeq 0.7$. The length of the marker is 100 μm in (a), 50 μm in (b), 20 μm in (c) and 10 μm in (d).

ing sphere concentration with the following typical stages: (a) isolated holes with a broad size-distribution; (b) very "porous" material with large holes; (c) a meandric, continuous path of packed spheres throughout the sample; (d) coexistence of isolated islands of packed spheres and a very stringy path of packed spheres.

In our subsequent geometrical analysis we shall

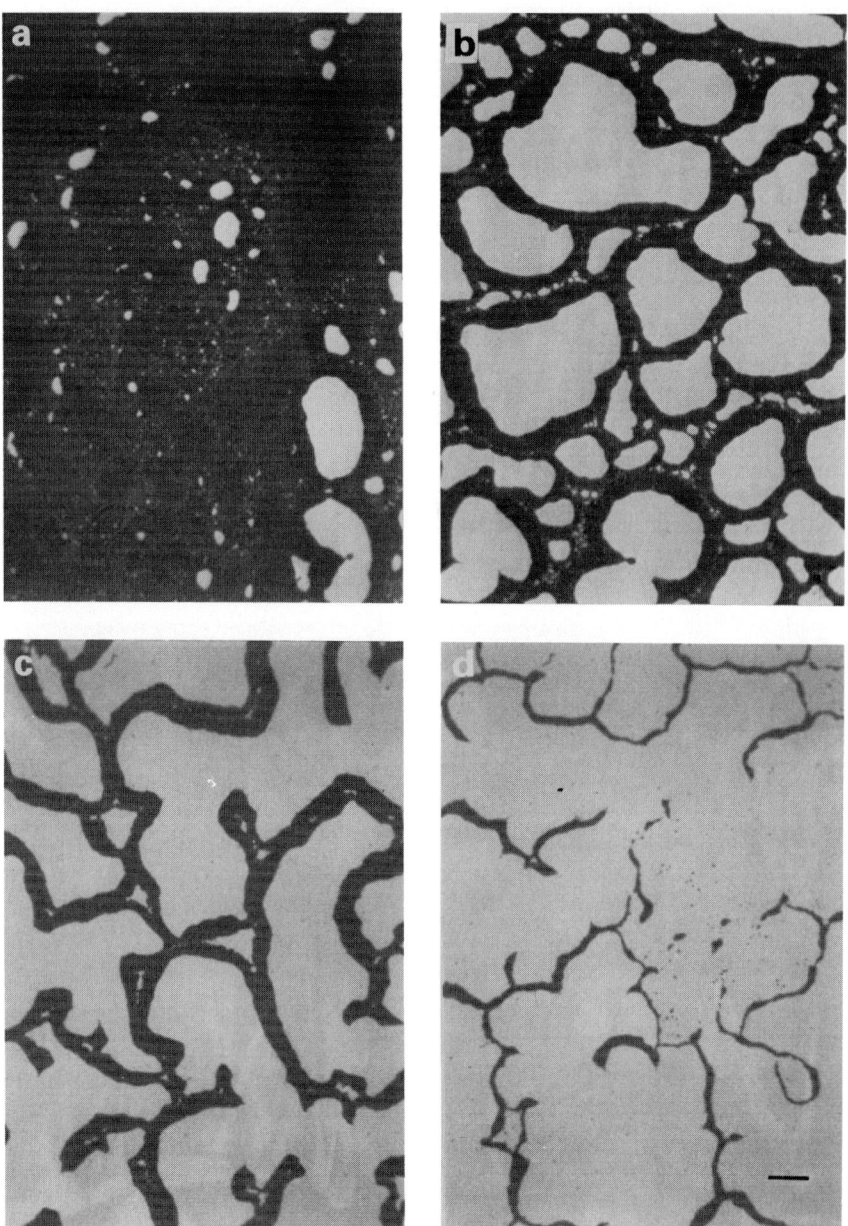

Fig. 4. Variation in the pattern of packed monolayers of 4.8 μm spheres for decreasing initial concentration (a)–(d). The length of the marker is 200 μm.

mainly concentrate upon the results shown in fig. 3. As may be seen for this case, there is an upper length scale in hole-size or separation between holes of about 200 μm and a lower length scale limit of about 5 μm corresponding to the sphere diameter. We observe fractal scaling to be approximately valid in a finite range between these two limits as discussed in the following.

3. Geometrical description

For the analysis of the experimental results, the pictures were digitized, low-pass filtered and thresholded to produce a white and black representation of holes and packed regions, respectively. The filtering was used to eliminate "noise" due to features below the resolution at a particular magnification. The resulting binary images for figs. 3a–3c to be analyzed are shown in figs. 5a–5c.

The first method to characterize the patterns is to search for an experimental fractal size distribution. This may conveniently be performed using the box counting method [11]. The procedure starts by covering the digitized picture with an imaginary net of square boxes of side L. The number of boxes, $N_b(L)$, containing any part of the objects to be analyzed is counted. In order to reduce statistical fluctuations, the net of boxes is shifted to different positions (typically 50), and average values of $N_b(L)$ versus L are obtained. For a fractal structure it is expected that

$$N_b(L) \propto L^{-D}. \qquad (2)$$

A log–log plot of $N_b(L)$ versus L should thus produce a straight line with slope D.

Fig. 6 shows the results for the packed regions in the digitized images of figs. 5a–5c. As may be seen, the data for the three different magnifications from the same region of the sample appear to fall on the same straight line with slope $D = 1.85 \pm 0.05$. This similar behaviour for different magnifications thus signifies a random fractal scaling on length scales between 5 and 200 μm. However, closer inspection shows a slight curvature in the data-sets which implies that a uniform fractal description is only a first approximation.

It is also of interest to find D for different packing fractions (or coverage) defined as η = packed area/total area. The porosity φ is then given by $\varphi = 1 - \beta$. Pictures like those in fig. 4 were analyzed as discussed above and the results are shown in fig. 7. As expected, D approaches 2 as η increases towards 1 (compact structure with no holes).

Finally, we shall analyze the size distribution of the

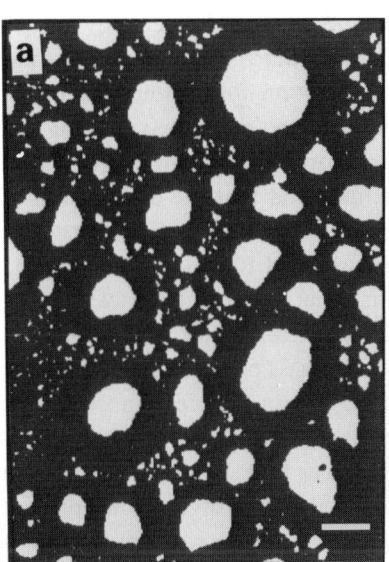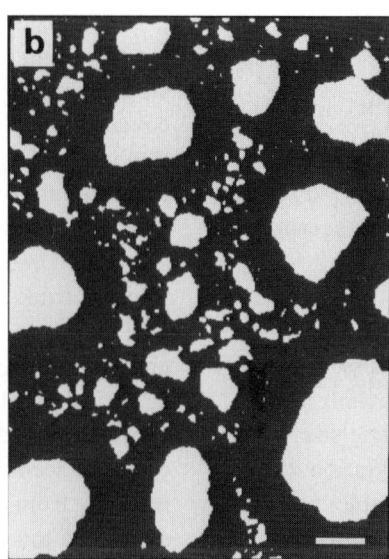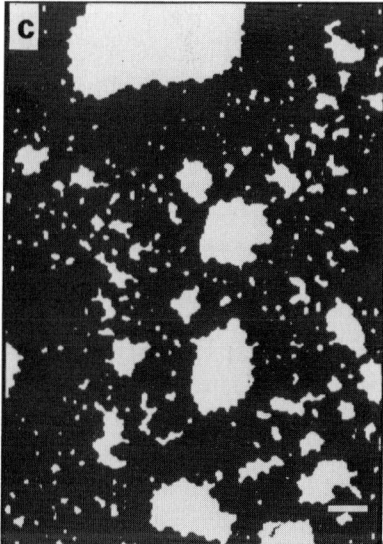

Fig. 5. Binary pictures of packed regions (black) and holes (white) for increasing magnifications (a)–(c) as in fig. 3. The length of the marker is 100 μm in (a), 50 μm in (b) and 20 μm in (c).

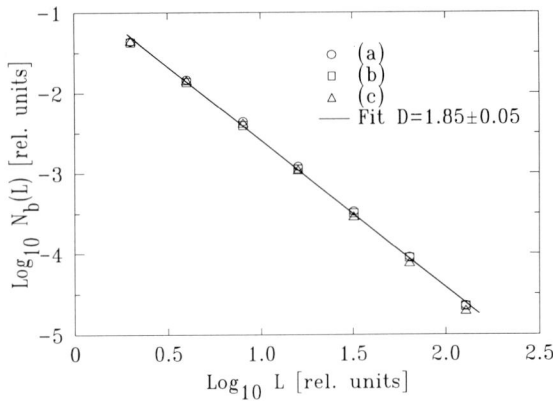

Fig. 6. The fractal dimension D of the packed regions (black) in figs. 5a–5c determined from the log–log plot of the relative number of boxes, $N_b(L)$, needed to cover the objects versus box size L.

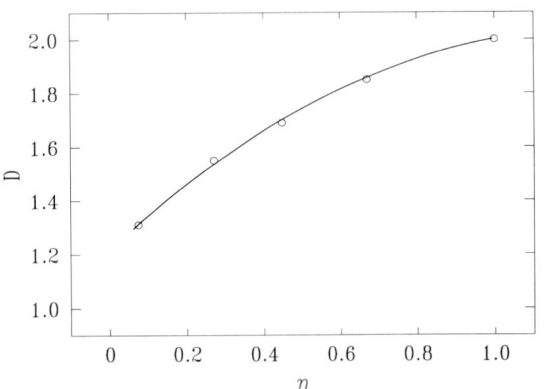

Fig. 7. Fractal dimension D versus packing fraction η for the packed regions (black portions for structures like those in fig. 4). The line is only a guide to the eye.

holes. For this we again choose the structures in fig. 5 with packing fraction $\eta \simeq 0.7$. As discussed in section 1, it is of interest to check out the possible power-law scaling in eq. (1). The characteristic size, s, of the holes was chosen to be the radius of gyration. (One could equally well have chosen the simple mean of the short and long sides of the circumscribing rectangle as a measure of the characteristic size.)

Fig. 8 thus shows the cumulative numbers of holes with sizes larger than s. As seen from the figure the

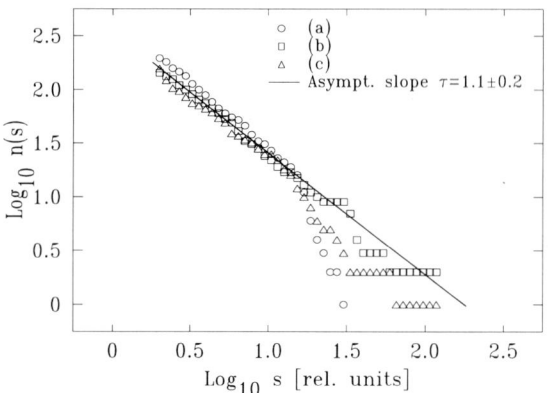

Fig. 8. The cumulative numbers of holes, $n(s)$, with sizes larger than s versus s. (a)–(c) refer to the digitized pictures in fig. 5. τ is the asymptotic scaling exponent used in eq. (1).

three sets of data from the same region of the sample, asymptotically exhibit the proposed power-law dependence in eq. (1) with $\tau = 1.1 \pm 0.2$. (The deviations and large fluctuations in the data for large s are due to finite-size effects: There are few "big holes" around if one views only a tiny portion of the sample.) The present case is thus yet another example of non-equilibrium processes like those noted in section 1 producing objects with this type of scaling.

In conclusion, it has been possible to employ a novel model system of randomly attractive microspheres to study monolayer deposition for various degrees of coverage. It is found that the non-linear dynamics in the pattern formation to a first approximation produces fractal structures and that the distribution of hole sizes follows a simple power law. This illustrates one more time the power of fractal concepts for extracting order from apparent irregularity of high degree in nature. But again, this is also yet another example of how little progress has been made in using empirical numbers to obtain a more comprehensive theoretical understanding of non-equilibrium processes. It seems also clear that the pattern formation resulting from non-linear processes as for the present system, is too irregular to be approximated by uniform (i.e. geometrically self-similar) fractals. An obvious extension would be to attempt to analyze such "frozen" structures as shown here using multifractal concepts.

Acknowledgement

The research was supported in part by Dyno Industrier A.S. and by the Norwegian Council for Science and Humanities (NAVF). I would also like to thank John Ugelstad and collaborators at the Foundation of Scientific and Industrial Research (SINTEF) at the Norwegian Institute of Technology (NTH) for supplying the samples used in the experiments.

References

[1] J. Ugelstad, P.C. Mørk, K.H. Kaggerud, T. Ellingsen and A. Berge, Advan. Colloid Interface Sci. 13 (1980) 101.
[2] A.T. Skjeltorp, Phys. Rev. Lett. 58 (1987) 1444.
[3] G. Helgesen, A.T. Skjeltorp, P.M. More, R. Botet and R. Jullien, Phys. Rev. Lett. 61 (1988) 1736.
[4] A.T. Skjeltorp and P. Meakin, Nature 335, No. 6189 (1988) 424.
[5] A.T. Skjeltorp and P. Meakin, in: Proceedings of the 4th EPS Liquid State Conference on the Hydrodynamics of Dispersed Media, in press.
[6] B.B. Mandelbrot, The Fractal Geometry of Nature (Freeman, San Francisco, 1983).
[7] M. Matsushita, J. Phys. Soc. Japan 54 (1985) 857.
[8] R.B. Baldwin, Astron. J. 69 (1964) 377.
[9] S.K. Friedlander, Smoke, Dust and Haze (Wiley, New York, 1977).
[10] T. Vicsek and F. Family, Phys. Rev. Lett. 52 (1984) 1669.
[11] R. Voss, in: Scaling Phenomena in Disordered Systems, R. Pynn and A. Skjeltorp, eds. (Plenum Press, New York, 1985), p.1.

NEW RESULTS ON THE FRACTAL AND MULTIFRACTAL STRUCTURE OF THE LARGE SCHMIDT NUMBER PASSIVE SCALARS IN FULLY TURBULENT FLOWS

K.R. SREENIVASAN and Rahul R. PRASAD

Mason Laboratory, Yale University, New Haven, CT 06520, USA

By measuring concentration fluctuations of a dye with very fine spatial and temporal resolution in typical unconfined turbulent water flows, we obtain the fractal dimension characteristic of the scalar interface in the range between Kolmogorov and Batchelor scales. We use one-dimensional intersection methods and invoke Taylor's hypothesis, but both of them are amply justified. We obtain a theoretical estimate for the fractal dimension by modifying our earlier arguments for finite (though large) Schmidt number effects. Finally, the multifractal characteristics of the scalar dissipation rate in the same scale range are also presented.

1. Introduction

A trace of dye or smoke, or a suspension of the fine particles of a metal, is considered a passive scalar if it does not affect the dynamics of the flow into which it is introduced. The behavior of passive scalars in turbulent flows is interesting in its own right, and the understanding of its mixing is practically useful in several contexts including combustion; since their evolution is determined by the velocity field, passive scalars can be studied profitably as a diagnostic even if the primary focus is on the dynamics of turbulent motion.

It is now well known [1] that an unbounded turbulent flow such as a jet develops at high Reynolds numbers "fronts" across which vorticity changes are rather sharp on scales larger than the characteristic thickness of the fronts. Such a front, called the vorticity interface, retains its sharpness in spite of the natural tendency of vorticity to diffuse: The nonlinear stretching inherent in the quadratic terms of the fluid equations provides the balancing action. A passive scalar introduced in fully turbulent flows gets dispersed by turbulence, and itself displays a sharp front across which the scalar concentration shows similar large jumps. In analogy with the vorticity interface, this front is called the scalar interface. This is the object of our interest here.

The scalar interface is a complex surface residing in three-dimensional physical space (see fig. 1); it is quite convoluted over a range of scales which are statistically self-similar. At high enough Reynolds numbers, there is a large separation between the largest and smallest scales on which the interface appears convoluted, and this allows the use of fractals [2] in characterizing the interface [3–6]. Unlike a mathematical fractal, the scale-similar regime of the interface is bounded on both sides by physical effects; the upper cutoff occurs at a (fraction of) the integral scale of motion, this being comparable to (but distinctly less than) the gross size of the flow such as the width of the jet, whereas the inner cutoff occurs at a scale where the effects of scalar diffusivity are felt directly. When the Schmidt number σ (that is the ratio of the fluid viscosity to scalar diffusivity) is unity, this scale is the Kolmogorov scale η equal to the smallest dynamical scale of the vorticity interface. If σ is much smaller than unity (as in the steller atmosphere) the smallest scalar scale is the so-called Batchelor scale $\eta_b = \eta = \sigma^{-1/2}$ [7]. This is typically the case of a dye mixed in water; the molecular structure of water is such that the colliding dye molecules transfer mo-

Essays in honour of Benoit B. Mandelbrot
Fractals in Physics – A. Aharony and J. Feder (editors)

0167-2789/89/$03.50 © Elsevier Science Publishers B.V.
(North-Holland Physics Publishing Division)

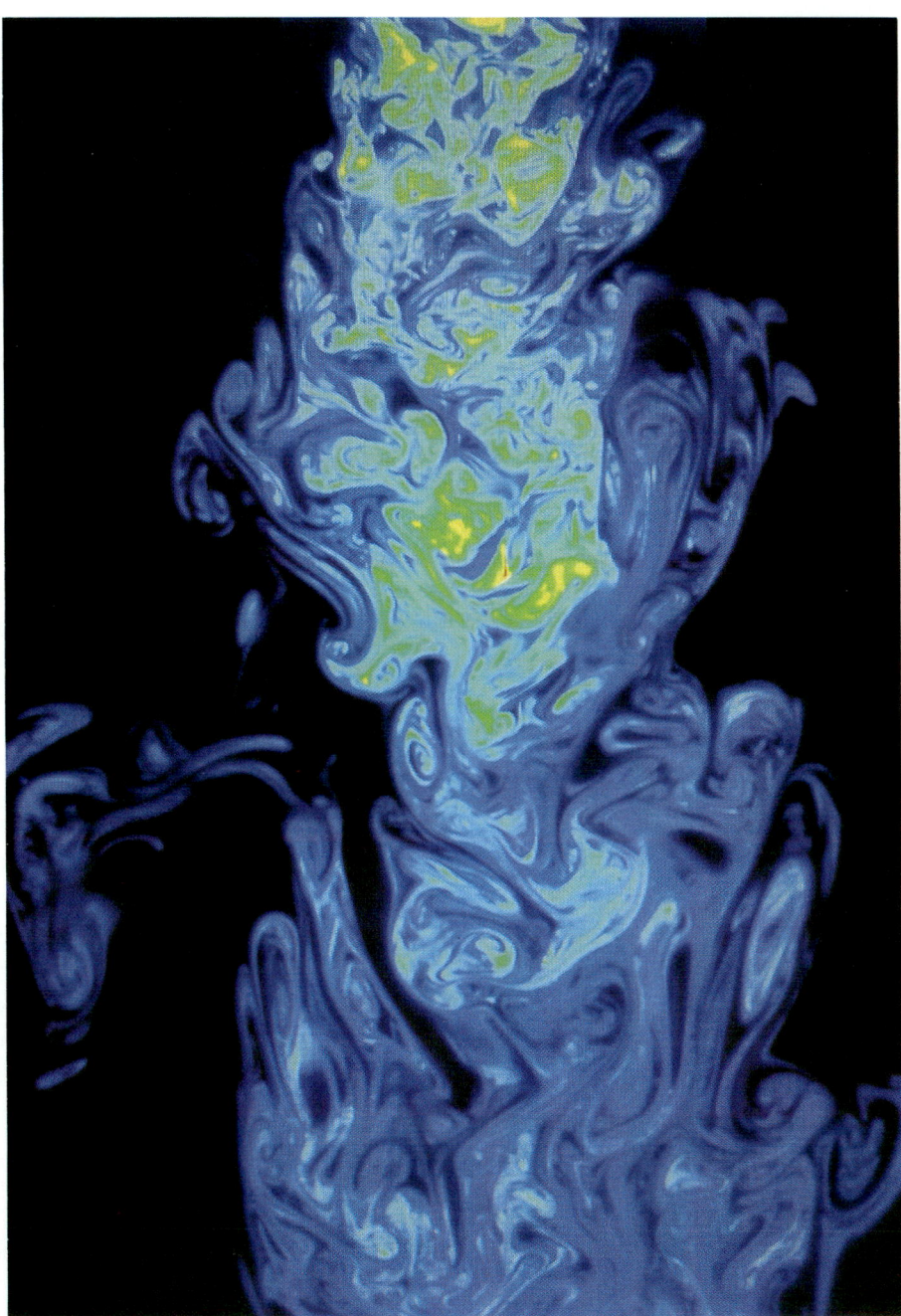

Fig. 1. To demonstrate the complexity of the scalar interface, we show a two-dimensional section of a turbulent jet at a nozzle Reynolds number of about 4000, obtained by the laser-induced fluorescence technique. Only scales coarser than the Kolmogorov scale are resolved. The digital camera used to obtain this image has an array size of about 1300×1000 pixels. The region imaged extends from 8 to 24 diameters downstream of the nozzle. A Nd:YAG laser beam shaped into a sheet of 200–250 µm thickness using suitable lenses was directed into a water tank into which the nozzle fluid containing small amounts of a fluorescing dye was emerging in the form of a jet. The laser had a power density of 2×10^7 J s^{-1} per pulse and a pulse duration of about 10 ns. The flow is thus frozen in this picture to an excellent approximation.

mentum much more efficiently than their own mass.

In our previous work, we used a dye mixed in water flows and resolved all scales above η, which, in a typical experiment, was around 200 µm. The Schmidt number was [8] of the order of 2000, yielding a Batchelor scale of about 4–5 µm. It is easy to argue [7,9] that a different scaling regime should exist between η_b and η. It is therefore interesting to resolve these scales and determine their scaling properties. This is the first purpose of the paper.

In the flow interior far from the boundary, the scalar concentration fluctuates in both space and time. The square of the gradient of these fluctuations represents (to within a constant) the rate at which the fluctuation intensity is being smeared by molecular diffusivity. This quantity is called the scalar dissipation rate, χ. We have shown earlier [10] that χ possesses a multifractal distribution – again with the qualification that cutoffs are present. As before, the spatial resolution was limited to Kolmogorov scale. Our second purpose is to determine the multifractal scaling properties of χ between η_b and η.

2. The method

As remarked earlier, the scalar interface is a fractal-like surface embedded in three-dimensional space, and we want to determine its fractal dimension. Following [3,4], we shall use box-counting methods which involve covering the volume by three-dimensional boxes of varying sizes, and counting the number of boxes containing the interface. The exponent characterizing the variation of this number with respect to the box size will give the fractal (i.e. box) dimension of the interface. The current limitations of instrumentation technology permit this direct method to be used only as long as the volume to be scanned is not too large and the resolution required is not too demanding. Such measurements have been made by Prasad and Sreenivasan [6], who resolved a volume of the order of $25\eta \times 300\eta \times 300\eta$ with resolution of between 2η and 3η. In general, fluid dynamical constraints are much stronger, and the more feasible way of obtaining the fractal dimension is to use the method of intersections. Here, one intersects the interface by a thin plane or a line – thin meaning that the finest scales of interest are resolved – and obtaining the fractal dimension of the intersections. The fractal dimension of the surface itself is then obtained by the so-called additive law for co-dimensions (see ref. [2] and references cited here), according to which the intersection by a plane results in a set whose dimension is one less than the dimension of the original set; when intersected by a line, the fractal dimension is two less than the dimension of the original set.

The requirement that the Batchelor scale be resolved allows only trivial extents of the flow to be mapped even in two-dimensional intersections; one therefore has to resort only to one-dimensional intersections. These can be obtained rather easily by invoking Taylor's frozen flow hypothesis according to which turbulence convects undistorted with the mean motion. This is reasonably accurate, especially for small scales of motion, if the mean convection velocity of the flow is large compared to its fluctuations. The relevant ratio is about 60 for the wake behind a cylinder and is large enough, but is only of the order of 4 for jets. It turns out that *geometric* aspects such as the fractal dimension are quite insensitive to details such as Taylor's hypothesis; in ref. [3], we showed that even for jets the fractal dimension results can be obtained quite accurately in this way. On the other hand, *dynamical* aspects such as the spectral distribution of the scalar variance are much more sensitive to Taylor's hypothesis [11].

3. The flows and the measurement technique

A turbulent wake behind a circular cylinder was produced by lowering a tank of water past a rigidly mounted cylinder. The cylinder was 1 cm in diameter and had an aspect ratio of 58. The tank was lowered at a constant speed of 15 cm/s by means of a hydraulic lift. The fluorescent dye (sodium fluorescein) that seeped into the wake from a narrow channel cut along the length of the cylinder – either at the front or the back stagnation regions – was mixed by the turbulence in the wake. The flow Reynolds number of 1500 (based on the cylinder diameter and the free stream relative speed) is moderate. During data acquisition, the position of measurement varied between 60 and 70 diameters behind the cylinder. The Kolmogorov and Batchelor scales were estimated to be about 160 and 4 µm, respectively.

A jet was produced by allowing water to flow from a settling chamber through a nozzle of circular cross-section (diameter 1.2 cm) into a tank of still water at a constant speed of about 35 cm/s. The nozzle was contoured according a fifth-order polynomial to have zero slopes and curvature at the entrance and the exit. The contraction ratio was about 10. The jet Reynolds number based on nozzle diameter and exit velocity was about 4000. During data acquisition, the position of measurement varied between 20 and 37 nozzle diameters downstream. The estimated Kolmogorov and Batchelor scales in the measurement region are about 200 and 5 µm, respectively.

The optical setup is shown in fig. 2. By various combinations of lenses described in the caption, the beam is focused to a spot of about 4 µm at the desired location in the flow. Concentration fluctuations are detected as fluctuations in fluorescence intensity, the two being in linear proportion to each other. The optical signal from the photomultiplier tube is passed through a current amplifier before being digitized by the 12-bit A/D converter on the MASSCOMP 5000 computer. The digitizing frequency is set at 320 kHz, which is well below the limiting digitization rate of 1 MHz of the A/D converter. The photomultiplier tube is quoted by the manufacturer as having good frequency response up to 50 MHz. So the temporal response of the instrumentation is believed to be much better than is required for present purposes.

From the highly resolved concentration fluctuation signal c, we obtain the fractal dimension of the interface as well as the multifractal aspects of χ. We have demonstrated in ref. [11] that the signal possesses the expected classical properties, for example the correct power law behaviors.

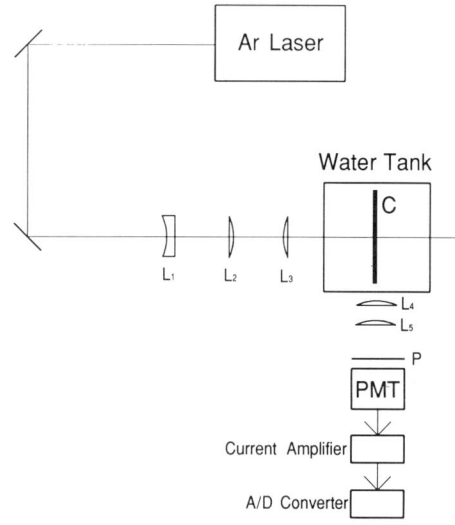

Fig. 2. Schematic of the optical setup. Shown is the orientation of the cylinder (C), whose wake is the flow of interest here. A 5 mm diameter light beam from the continuous argon laser (power output about 7 W) is first expanded into a thicker beam of 60 mm diameter by the combination of spherical lenses L_1 (focal length 25 mm) and L_2 (focal length 300 mm), and then focused to a spot of 5.5 µm diameter by means of a convex lens L_3 of focal length 500 mm. The optical signal is collected by a photomultiplier tube (PMT). In the optical path upstream of the photomultiplier tube is a combination of lenses L_4 and L_5 (focal lengths 400 and 1000 mm, respectively) that give an image enlargement by a factor of 2.5. This combination enlarges the 5.5 µm focal spot in the flow to a size of about 13 µm. Ahead of the photomultiplier tube, a 10 µm diameter pinhole (P) is located. This effectively reduces the size of the spot imaged onto the phototube to 4 µm, this being the spatial resolution of measurement.

4. Results

Fig. 3 shows a typical plot of the logarithm of the number of boxes containing the intersection points of the interface as a function of box size. There are two distinct power law regimes, one of which occurs (roughly) between η_b and η, and the other to the right of η. As expected from earlier measurements [3–6], the negative slope in the latter region is around 0.36 – giving a fractal dimension of 2.36. (The scatter in that region is relatively large because the limited duration of the signal did not contain too many inter-

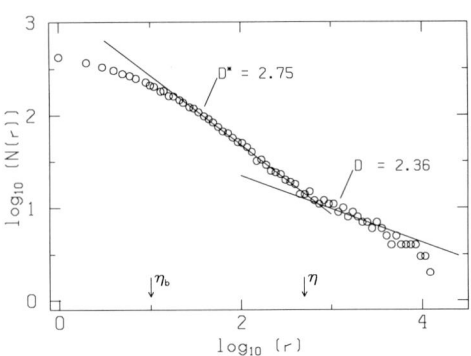

Fig. 3. A typical log–log plot of the number $N(r)$ of length elements or "boxes"' of size r containing the interface versus the box size r. The flow is the wake of a circular cylinder. The negative slopes of the straight parts give, in the respective scaling regime, the fractal dimensions of one-dimensional intersections of the boundary. The dimension, corresponding to the slope of the line drawn in the region between η and L, is about 2.36. That in the range between η and η_b is about 2.75.

sections comparable to the bigger boxes.) The region between η_b and η has a slope of about -0.75, giving the fractal dimension to be about 2.75. The average slope from several realizations is 2.7 ± 0.03. Data from jets confirm this conclusion.

This is our first main result, and we should like to explain it. This is done by considering the dye mixing at infinitely large Schmidt numbers [9], and then providing corrections for the finite (but large) Schmidt numbers.

The basic idea is that the properties of the scalar interface and the mixing of the scalar with the ambient fluid are related. Since the amount of mixing is governed by large eddies in the flow, the actual process of mixing (by which we mean molecular mixing) is accomplished by diffusion across the surface whose geometry is determined by the requirement that it accomplish the exact amount of mixing set by the large scales. Thus, even though the process is initiated by large scales, one can legitimately concentrate on the diffusion end. This approach has a much better likelihood of yielding results of some "universality", simply because the small-scale features of the flow are, to a first approximation, independent of configurational aspects of the flow. The fractal dimension of the interface is but one example. This approach neither minimizes the role of large eddies nor resorts to gradient transport models usually discredited in turbulence theory.

Motivated by this thinking, Sreenivasan et al. [9] concentrated on the last stages of the mixing process by working with diffusion across the fractal-like interface. They proceeded from Fick's law of diffusion, which can be expected to hold accurately in spite of the high degree of convolutedness of the surface (because the scales of convolutions are significantly larger than the molecular mean free path), and showed that the flux of the scalar is given by

$$\beta \, \mathrm{Re}^{3(D-7/3)/4} \sigma^{(D^*-3)/2} , \qquad (1)$$

where β (which in ref. [9] has been written down explicitly) consists only of quantities depending on the large-scale features of the flow and are independent of Reynolds number. D is the fractal dimension in the scale range between η and L, and D^* in the range between η and η_b. Re is the flow Reynolds number given by $u'L/\nu$; u' is the root-mean-square fluctuation velocity and ν is the kinematic viscosity of the fluid.

One can then invoke [9] the so-called Reynolds number similarity, which is merely a statement of the observed fact that all fluxes (mass, momentum, energy) must be independent of Reynolds number in fully turbulent flows. According to (1), Reynolds number similarity requires that

$$D = 7/3 , \qquad (2)$$

in rough agreement with experiments [3–7,9]. Multifractal corrections [9,12] change this value slightly to $D=2.36$, bringing it identically equal to the measured average [4,9].

Similarly, Schmidt number similarity requires that

$$D^* = 3 . \qquad (3)$$

This means that convolutions of the interface on scales between η_b and η are space-filling. The physical picture corresponding to this situation was described in ref. [9].

This last result, of specific interest here, gets modified when the Schmidt number remains finite (though large). To quantify the effect, we recapitulate that an essential argument used in ref. [9] is that the concentration gradient across the interface is of the order of c'/η_b, where c' is the root-mean-square of the concentration fluctuation c – a large-scale feature. It turns out [7] that the time taken by the scalar to diffuse down to the Batchelor scale increases logarithmically with the Schmidt number. There is also a corresponding pile up of fluctuation intensity in the scalar patches as the straining by the velocity field continues unabated. The effective concentration gradient is then given by $c' (\ln \sqrt{\sigma})/\eta_b$, and the expression (1) for the flux gets multiplied by the factor $\ln \sqrt{\sigma}$. It is then easy to show that the Schmidt number similarity requires that

$$D^* = 3 - 2 \ln(\ln \sqrt{\sigma})/\ln \sigma. \qquad (4)$$

In the limit of infinite Schmidt numbers (4) reduces to (3). For a Schmidt number of 1930, as for the fluorescing dye [8], (4) yields the result that $D^* = 2.65$, quite close to the measured value of 2.7.

We reiterate that the present arguments hold in circumstances where the amount of mixing is determined by the large scale, and the surface adjusts itself accordingly. For large eddies to be the controlling factor at infinitely large Schmidt numbers, it is necessary that the Reynolds number must be correspondingly large, the precise condition being that $(\ln \sigma)/\text{Re}^{1/2} \ll 1$. As expected on physical grounds, this condition never lets the characteristic gradient across the interface exceed $\Delta c/\eta_b$, where Δc is the maximum concentration difference in the flow.

The result that the interface has space-filling characteristics in the Batchelor regime (scale sizes between η_b and η) suggests that other aspects of the scalar in this scaling regime might also be space-filling in the limit of infinite σ. In particular, the scalar dissipation rate χ might be space-filling also. If so, all the generalized dimensions [13] will all be unity, and the multifractal spectrum, or the $f(\alpha)$ curve [14], trivially reduces to the point (3, 3) in three dimensions and to the point (1, 1) in one-dimensional intersections. Finite Schmidt number effects may alter this result, and it would therefore be useful to obtain from experiment the generalized dimensions. We follow the procedure described in ref. [12].

The generalized dimensions D_q are obtained by dividing a record of scalar dissipation into smaller boxes of size r, and identifying power laws of the type

$$\sum (X_r)^q \sim r^{(q-1)D_q}, \qquad (5)$$

where X_r is the total dissipation over a box of size r, and the sum is taken over all boxes of size r; q is any real number. It is clear that if q is positive and large, only the large intensity regions will be picked by the summation in (5) while the least intense regions correspond to large negative q's. According to (5), if log–log plots of $[\sum (X_r)^q]^{1/(q-1)}$ versus r present linear regions within the scaling range $\eta_b < r < \eta$, the slopes correspond to the D_q's.

The D_q and the $f(\alpha)$ curves for one-dimensional sections of the dissipation of turbulent kinetic energy were measured in ref. [12] and shown to be universal features of fully developed turbulence. Similar measurements in refs. [6,10] for the scalar dissipation were made from two- and three-dimensional images in the scale range between η and L. Since we have measured – as already explained, by the application of Taylor's hypothesis – one component of χ with resolution of the order of the Batchelor scale, our purpose here is to measure the D_q curve in the Batchelor regime. We note the earlier result [6,10] that the multifractal properties of a single component of χ are the same as those of χ itself, and that Taylor's hypothesis is adequate for the purpose.

Typical log–log plots of $[\sum (X_r)^q]^{1/(q-1)}$ versus r are shown in fig. 4 for some representative q values. For clarity and convenience, only the scaling in the Batchelor regime is shown. The straight line regions yielding the D_q's are quite unambiguous. The $f(\alpha)$ curve can be computed from Legendre transforms, but these results are not presented here.

Fig. 5 shows the curve of D_q versus q for the Batchelor regime. We have invoked the additive law and added 2 to the results obtained from one-dimensional intersections. As expected, all the generalized dimensions are quite close to the box dimension D_0

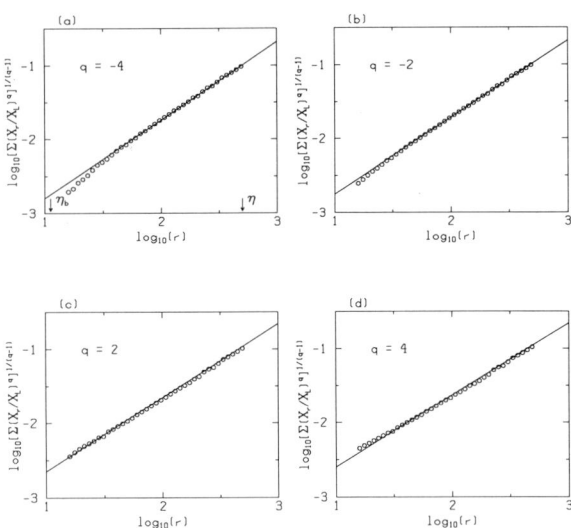

Fig. 4. Typical log–log plots of $[\sum (X_r)^q]^{1/(q-1)}$ versus r from the dissipation field of the jet for four different values of q; (a) $q = -4$, (b) -2, (c) 2, (d) 4. Power law regions are seen for each q, extending approximately between η and η_b.

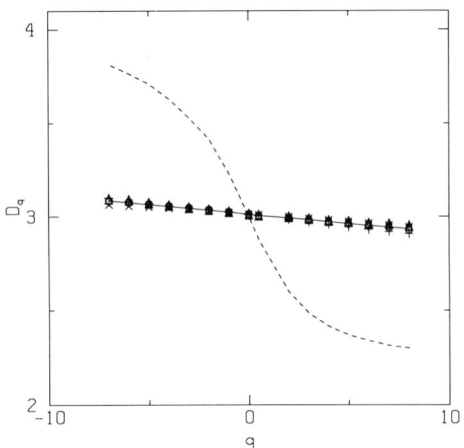

Fig. 5. The generalized dimensions for the scalar dissipation. Different symbols correspond to different realizations, and the solid line represents the mean. The dashed line shows results for the range between η and L [10]. The present results for the range between η and η_b show that all the D_q's are much closer to D_0. The expectation is that they will identically be equal to D_0 in the limit $\sigma = \infty$. The additive law has been used in presenting the results.

of the support. Fig. 5 also compares the present results to those previously obtained [10] in the scaling range between η and L. Unlike the interface dimension, it is difficult to interpret the generalized dimensions physically and obtain theoretical estimates.

5. Summary of results

Two distinct scaling regimes, and therefore two distinct fractal dimensions, exist for the scalar interface in the high Schmidt number case. The two separate scaling regimes reflect the fact that the dominant physical effects are different in the two regimes. For example, the Kolmogorov scale plays no role in the Batchelor regime except that it acts as a cutoff scale analogous to the integral length scale in the regime between η and L. In the scaling range between L and η, the fractal dimension is 2.36 ± 0.05; this result also holds for the vorticity interface [3]. The Batchelor regime possesses another fractal dimension, which is 3 for infinitely large Schmidt numbers – assuming, of course, that mixing is still controlled by large eddies. As remarked already, the condition for the latter is that the square root of the Reynolds number must be large compared to the logarithm of the Schmidt number. Unlike finite Reynolds number effects, finite Schmidt number corrections are significant; even if σ is about 2000, both experiments and a simple theory of mixing show that the fractal dimension is only as high as 2.7.

Generalized dimensions of χ in the flow interior are quite close to D_0 in the Batchelor regime. Among other things, it means that the intermittency corrections in that regime are quite negligible. If we extrapolate our experience with finite Schmidt number corrections for the interface dimension, we may speculate that all the D_q's for $\sigma = \infty$ will equal D_0, and that the intermittency corrections are identically zero.

By his own work and through his influence on others, Benoit Mandelbrot has had a vigorous and long-lasting influence in charting frontiers of science in many areas including turbulence. This paper is a modest expression of our intellectual indebtedness to him. It is a pleasure to dedicate it to Benoit on the occasion of his 65th birthday. The work was financially supported by DARPA (URI) and AFOSR.

References

[1] S. Corrsin and A.L. Kistler, NACA Rep. (1955) 1244.
[2] B.B. Mandelbrot, The Fractal Geometry of Nature (Freeman, San Francisco, 1982).
[3] K.R. Sreenivasan and C. Meneveau, J. Fluid Mech. 173 (1986) 357.
[4] K.R. Sreenivasan, R.R. Prasad, C. Meneveau and R. Ramshankar, J. Pure Applied Geophys. 131 (1989) 297.
[5] R.R. Prasad and K.R. Sreenivasan, Exp. Fluids 7 (1989) 259.
[6] R.R. Prasad and K.R. Sreenivasan, in: Proceedings of Turbulent Shear Flows (Seventh Symposium), Stanford (August 1989), also submitted to J. Fluid Mech.
[7] G.K. Batchelor, J. Fluid Mech. 5 (1959) 113.
[8] B.R. Ware et al., in: Measurement of Suspended Particles by Quasi-elastic Light Scattering, B.E. Dahneke, ed. (Wiley, New York, 1983), p. 435.
[9] K.R. Sreenivasan, C. Meneveau and R. Ramshankar, Proc. Roy. Soc. (London) A 421 (1989) 79.
[10] R.R. Prasad, C. Meneveau and K.R. Sreenivasan, Phys. Rev. Lett. 61 (1988) 74.
[11] K.R. Sreenivasan and R.R. Prasad, in preparation.
[12] C. Meneveau and K.R. Sreenivasan, Nucl. Phys. B. Proc. Suppl. 2 (1987) 49.
[13] H.G.E. Hentschel and I. Procaccia, Physica D 8 (1983) 435.
[14] T.C. Halsey, M.H. Jensen, L.P. Kadanoff, I. Procaccia and B.I. Shraiman, Phys. Rev. A 33 (1986) 1141.

LEARNING CONCEPTS OF FRACTALS AND PROBABILITY BY "DOING SCIENCE"

H. Eugene STANLEY

Center for Polymer Studies and Department of Physics, Boston University, Boston, MA 02215, USA

Dedicated to Benoit B. Mandelbrot on his 65th birthday

Very recent advances in computer technology provide the power of mainframe systems in relatively compact and inexpensive personal computers; soon the computing power of even a supercomputer will be available on a desktop at a price comparable to today's personal computers. Over the next decade this tremendous computing power can and probably will become available in schools throughout the world. Here we discuss the possibility of harnessing this new technological resource as a teaching tool for specific topics in mathematics and science, focusing on random processes in nature and their deep connection to concepts in probability and fractal geometry. Such natural phenomena as the growth of snowflakes via random aggregation and the disordered geometric configurations of polymer chains demonstrate that fundamentally random microscopic processes can give rise to predictable macroscopic behaviors. They also give rise to *random* fractal structures of inherent interest and great beauty. Because it is impossible to view the underlying processes directly, computer simulation and visualization is an indispensable tool for understanding and studying these phenomena. In the process of "doing science" with both hands-on experiments and computer simulations, students would learn abstract mathematical concepts in a context which is at once concrete and inherently motivating. Furthermore, the techniques they could employ would mirror in most respects those in current use by researchers, thus forging an unprecedented link between this curriculum and the professional worlds of science and mathematics.

I ascribe to nature neither beauty, deformity, order nor confusion. It is only from the viewpoint of our imagination that we say that things are beautiful or unsightly, orderly or chaotic.

– Baruch Spinoza, 1665

In fact, all epistemological value of the theory of probability is based on this: that large-scale random phenomena in their collective action create strict, nonrandom regularities.

– B.V. Gnedenko and A.N. Kolmogorov, 1954

Benoit Mandelbrot is the first great mathematician of the "rough edges" of existence. Plato sought to explain nature with five regular solids; Newton and Kepler bent Plato's circle to an ellipse; modern science analyzed Plato's shapes into particles and waves, and generalized the curves of Newton and Kepler to relative probabilities – still without a single rough edge.

Now, more than two thousand years after Plato, nearly three hundred years after Newton, and after thirty strenuous years of wily insinuation, calculated argument, and stunning demonstration, Mandelbrot has established a discovery that ranks with the laws of *regular* motion. Bespeaking the knowledge possessed by every child and every great painter, Mandelbrot has said, "Clouds are not spheres, mountains are not cones, coastlines are not circles, bark is not smooth, nor does lightning travel in a straight line."

What Mandelbrot has named fractal geometry describes not only the zigzag of Zeus's thunderbolt, or the branching and the varying densities of Pan's forests. It describes as well the Mercurial irregularities of the commodities market, the heretofore unaccountable fits of Poseidon the earthshaker, and a myriad of phenomena in the realm of lesser deities – snowflakes, shale, lava, gels, the rise and fall of rivers, fibrillations of the heart, the surging of electronic noise. Fractal geometry points to a symmetry of pattern within each of the meldings, branchings, and shatterings of nature.

Mandelbrot's work has also produced a new aesthetic. Strange phenomena once considered the monsters of mathematics are now seen as beauties of nature. In the inhuman flourishing of diffusion-limited aggregation (DLA), which from a simple seed and an iterative process develop structures fractured and yet orderly beyond imagining is a beauty evident to all scientists and mathematicians.

Mandelbrot's prolonged and energetic refusal to confine himself within accepted rigidities, the dedication with which he has pursued the geometry of the irregular, the scientific, mathematical, and aesthetic insights by which he has have brought us closer to the character of the universe – all these innovative aspects are lacking in today's curriculum, but almost certainly *must* be in place by the end of the century, as Mandelbrot's geometry becomes the cornerstone of a larger and larger part of the edifice of science.

A book that preceded by more than half a century Mandelbrot's 1982 classic The Fractal Geometry of Nature [1] and was known by every scientist and mathematician at that time is On Growth and Form by D'Arcy Wentworth Thompson [2]. On Growth and Form called attention to the fact that a large part of science was based on structures and processes that on a microscopic level are completely random, despite the fact that on the macroscopic level one can perceive patterns and structure. This classic has become popular again, in large part due to the fact that in the past few years the advent of advanced computing and sophisticated experimental techniques have led to remarkable progress in our understanding of the connection between the structure of a variety of random "forms" and the fashion in which these forms "grow". Examples range from fractal viscous fingering (a topic from fluid mechanics) to dielectric breakdown (a topic from electromagnetism). Photographs of these two phenomena are shown in fig. 1.

The striking similarity of these two structures suggests that there might be some unifying elementary physical principle underlying both. This unifying principle is remarkable: These and a vast number of other phenomena can be related to the properties of a simple random walk. For example, the equations describing fluid flow and electrical breakdown processes can be formally related to the equations describing the diffusion of random walks. This discov-

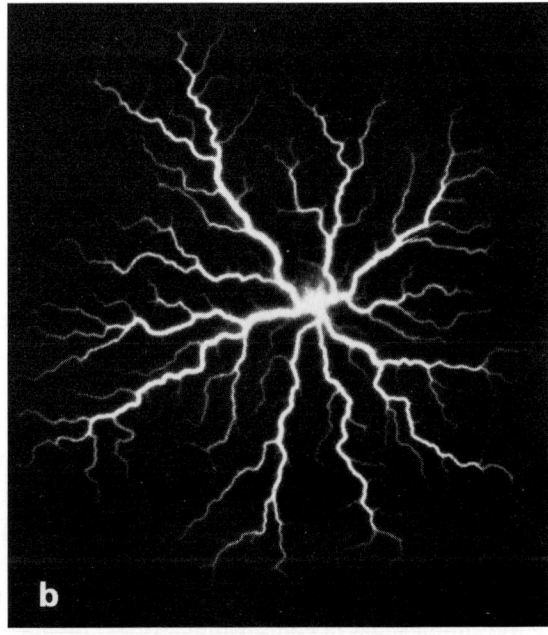

Fig. 1. Two of the many naturally occurring fractal structures discussed in this article, (a) viscous fingers and (b) dielectric breakdown. Experimental details are given in the original references [3,4].

ery forms the basis of much of the recent interest of scientists in fractals.

Within the scientific community there has been a tremendous upsurge of interest in this opportunity to unify a large number of diverse phenomena, ranging from chemistry and biology to physics and materials science. Now the time has come for this discovery to make the leap from the research laboratory to the introductory science curriculum. The lack of computing power in the high-school classroom has unfortunately been a barrier to this transfer in the past, but within a decade this situation will change.

There is an intimate connection between *fractal* concepts and the laws of *probability*; indeed, Benoit Mandelbrot was himself a student of probability before he did his seminal work on fractals. Although the general public is customarily introduced to that subset of fractal objects that are deterministic (such as the Sierpiński gasket, or the Mandelbrot set), it is the broader class of random fractals that has the greater impact on the sciences. Thus a study of random fractals in nature provides a unique opportunity for students to master key concepts in probability and statistics by focusing on their application to concrete scientific problems.

The great scientist and educator Victor Weisskopf has frequently stated that while the first half of the twentieth century will be known for the development and appreciation of quantum mechanics, the second half will be known for the development and appreciation of random materials and processes. Probability concepts form the cornerstone of our understanding of such random phenomena, yet almost no student emerges from school with an understanding of how to use probabilistic reasoning as a tool for understanding how things work.

Not only can the students learn about the laws of probability from the study of real fractal phenomena in nature, but the converse is also true. By learning about probability, students will be prepared to learn those aspects of modern biology, chemistry and physics that *depend crucially on a working knowledge of probability concepts*. Indeed, to most of the natural world simple ideas of equilibrium "order" do not apply. At the forefront of current scientific interest are

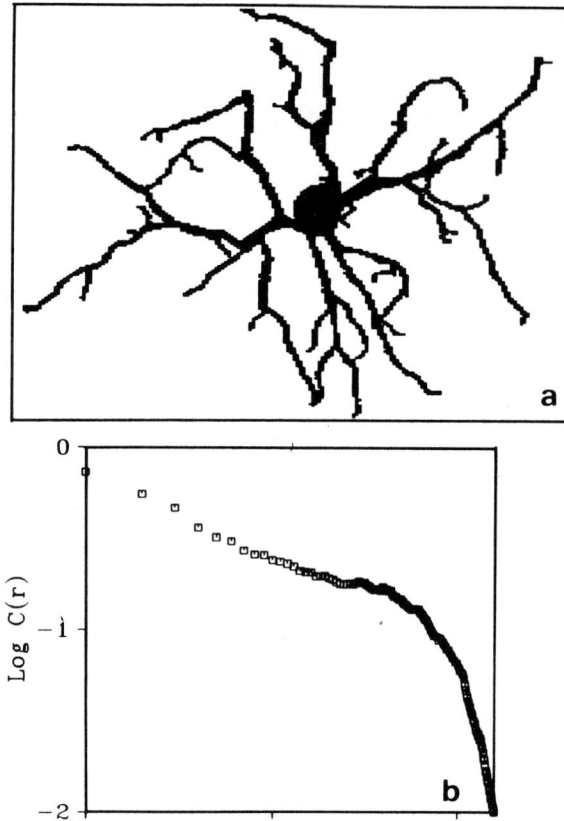

Fig. 2. The interested student can discover that the shapes of many quasi-two-dimensional retinal neurons are fractal objects, and hence may be quantitatively characterized by their fractal dimension d_f. The student can also understand a tentative explanation of neuronal shape in terms of diffusion-limited aggregation (DLA) model, which predicts $d_f \approx 1.70$. Part (a) is a digitization of a Golgi-stained horizontal cell in a turtle retina, taken from ref. [5]. Part (b) shows an example of the corresponding fractal analysis; from the slope of the straight-line portion the student can estimate $d_f = 1.63 \pm 0.05$. This figure is discussed in further detail in ref. [6], while analogous results on the retinal vasculature are presented in this volume by Family [7].

a large number of disordered, non-equilibrium structures. These range from neurons and neural networks in biology (fig. 2) to chemical oscillations in the atmosphere and superclusters in the cosmos. At each level of randomness, the same questions are asked: "*What are the mechanisms leading to the initial break with complete homogeneity and order?*" and "*What *determines the limits of growth and the final form of the pattern?*" It is a remarkable fact that the introduction of extremely simplified models has directly led to understanding of the morphology of a vast range of disordered structures.

Despite the ubiquity of these new ideas, there appears almost no discussion of them in the school curriculum. This is unfortunate for many reasons, among which are:

(1) Students are not learning enough *new* ideas of science. For many students, the term "modern science" connotes developments such as quantum mechanics, nuclear science and atomic theory, when in fact modern science includes a vast array of topics based on probability and statistical mathematics.

(2) Students have difficulty becoming attracted to science and mathematics in part because the day-to-day developments that we read about in the newspaper are rarely taught in the school. A key rationale of the topics related to fractals is that they are sufficiently simple that students can truly understand them.

(3) Students also have difficulty becoming attracted to the opportunity of learning the laws of probability. Indeed, a *major* problem is how to educate high-school students in probability and statistics! The motivational barrier for these topics of abstract mathematics is significantly lowered by focusing on enticing applications from biology, chemistry and physics.

Thus the members of our community are in a unique position to construct an innovative approach to this problem that serves to draw together probability and various disciplines of science, capitalizing on a student's natural curiosity about nature to drive his or her desire to learn basic mathematics.

The student can utilize the computer in much the same fashion as a researcher: to test a microscopic model of what is going on in the world, and to visualize its result! Very useful in this context is the extensive use of a multi-window mode to view simultaneously the simulation of a *microscopic* process in one window and the macroscopic "*emergent phenomena*" in the other. Thus one can convey a set of abstract concepts that have the following *two* meanings:

(i) the microscopic process (which is unpredictable), and

(ii) the macroscopic result (which is completely predictable).

One example is particle diffusion. In the laboratory, one cannot possibly trace the random motion of individual particles (e.g. individual molecules in the atmosphere). Using a mainframe or other extremely powerful computer, however, we can view *in real time* the motion of a random walk. One can also continuously update appropriate "running averages" – such as the mean square displacement. One can thereby appreciate how the individual *unpredictable* microscopic events – such as the random walk's erratic motion – can lead to surprisingly *predictable* macroscopic averages. The powerful computer permits tracking complex motions and the multi-window presentation trains the student to learn to appreciate *alternative representations of the same thing*: the lower level is the underlying phenomenon, while the higher level is the build-up of the probabilistic analysis (the running averages, for example). Students will be able to control interactively parameters of the simulation, and track its progress in both windows simultaneously. Rather than being simply entertaining video games or "cartoons," the simulations are actually self-contained "simulation experiments" from which students can collect data.

In the remainder of this article, we offer some specific examples of the innovations we are attempting, and of the advanced technologies that we shall bring to bear. The key fact is that very recent advances in technology, which can essentially put the power of a mainframe computer in a high-school environment, have the potential to revolutionize the teaching of certain topics in science and mathematics to students of all ages. The key innovations that the new technology makes possible are the following:

(1) Students can make tangible the macroscopic effects of a given probability concept (e.g. a specific dendritic "form" arises from a specific probability rule).

(2) Students can make their own structures by inventing their own probability rules.

(3) Students can perform their own calculations in

much the same spirit that scientists perform calculations – with state-of-the-art computational tools.

(4) By actually "doing science" (which is fun!) students can learn about fundamentals of mathematics (which is sometimes boring!).

(5) Because neither teachers nor students know the outcome of each experiment, scientific curiosity draws the entire class together.

Our own experiences have been divided between students in the 8–12 year old group and in the college age group. Surprisingly, the advanced technology permits ideas to be conveyed with equal ease to both groups. With both groups, we place a great deal of emphasis on first acquiring a "gut" feeling for the key ideas. Games often help in this regard. For example, to demonstrate a one-dimensional random walk, a student can toss a coin, moving to the right each time the coin is "heads" and to the left each time the coin is "tails". The idea of a *probability distribution* arises when the entire class performs the same experiment, each student in his own "racetrack." Even a class of 30 students is not great statistics, so the students can appreciate the power of the computer in transforming from a sample of 30 classroom students (and perhaps 20 time steps each) to a sample of thousand of random walkers (each stepping perhaps a million time steps). After becoming extremely familiar with the statistical properties of one-dimensional random walks, we introduce two-dimensional random walks. Again, we begin with classroom experiments to get the students to appreciate the feeling for the problem. The student realizes that although the nature of the walk changes from being one-dimensional (along a line in the classroom) to being two-dimensional (requiring, probably, a gymnasium or a football field). Although the coin thrown must now have four sides instead of two, the essential statistical properties of the walk are quite similar. The intriguing connection between the microscopic probabilities and the macroscopic distribution functions tend to excite the imagination of the students and teachers alike. Particularly satisfying to both groups is the fact that the outcome of each experiment is different. This is in marked contrast to the outcome of many classroom experiments, which everyone knows in advance from

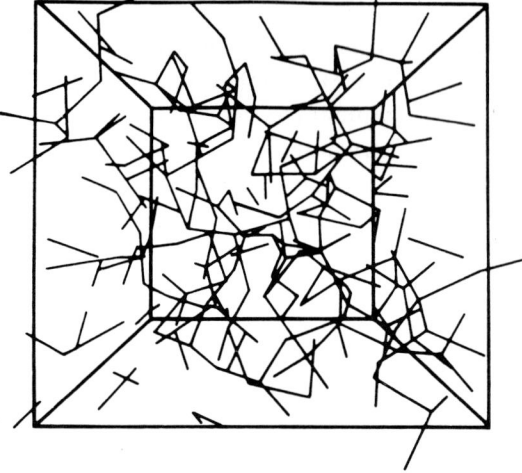

Fig. 3. This computer simulation makes possible the visualization of the details of the hydrogen-bonded network in liquid water. The students can actually construct this network in a high-power computer, and recognize that it is identical to a bond percolation network. From ref. [8].

the students who took the same class during the previous period (or previous year!).

Finally, we can move to a class of problems associated with percolation connectivity (fig. 3). In the simplest form, one throws a coin and if the outcome is heads one places a circle on the vertex of a piece of graph paper ruled into a triangular lattice. After every vertex has been examined (and covered with probability 0.5), the students draw lines connecting all nearest-neighbor pairs of vertices which are covered with circles. The students can understand that the circles are like molecules and the lines joining neighboring circles are like chemical bonds. The entire system can undergo a gelation transition in which there appears, in addition to many tiny ramified "molecules", a single "macromolecule" that spans the entire sheet of paper. This gelation is not altogether unlike what happens in a classroom experiment in which real gels are prepared using recipes resembling that for making jello, except that the strength of the resulting gel is sufficiently great that one can play games with it. Thus again the student appreciates the connection between a simple probabilistic process (placing circles on the vertices of triangular graph paper) and a macroscopic phenomenon (the formation of a magical material termed a gel). Students can then be taught that hydrogen-bond networks in liquid water are quite analogous to gels, except that the lifetime of the hydrogen bonds in water are so short (picoseconds) that water "flows" while jello does not flow. Thus children can learn that materials in their everyday world that initially appear to be quite different are at their foundation rather similar, and that probability theory is the basis of this similarity.

The preceding examples have been used with 8-year olds to stimulate their curiosity in mathematics, and they have also been used with college students. Other topics from polymer chemistry also fascinate students of all ages, including adults. For example, we have many rather dramatic *hands-on* demonstrations that utilize very recent technological developments in the polymer chemistry area. After the students become familiar with the experimental phenomenon, they are motivated to understand its underlying basis in applied probability theory. In this regard, the ability to have the computing power of a mainframe computer in a classroom makes a significant difference, since with this tremendous degree of computational power it becomes possible to make the connection between observed phenomena and microscopic models.

We have found that computer graphics, in particular those with appropriate use of color, can significantly aid in exciting the interest and involvement of children, as well as in teaching subtleties of the subject. First we can connect to students' natural interest in biology by simulating the simplest "cancer growth model" (fig. 4). We then proceed to the diffusion-limited aggregation (DLA) model. In DLA one releases particles from the edge of the computer screen. The particles diffuse randomly until they touch a perimeter site of the growing cluster, whereupon they stick irreversibly to the cluster. Eventually a large ramified cluster results (fig. 5). Students will learn to appreciate an important concept of probability: the newly arriving particles are much more likely to stick on the tips of the cluster than deep in the fjords.

Much of the excitement is watching this structure grow, in real time, on a personal computer. Children enjoy predicting where the new particle will stick: "*Which tip will grow next?*" After one watches such a beautiful structure grow, one can appreciate how it is that completely unpredictable microscopic events can nonetheless lead to a recognizable pattern. Each time the computer experiment is repeated with a different random number seed, a different cluster is produced – no two are alike. Nonetheless, the same overall "form" arises.

We then proceed to alter the rules slightly: we grow the DLA on a surface instead of in free space. The pattern we find (fig. 6) has the identical form to that grown in free space! However, if we change abruptly from DLA rules to Eden rules, then the cluster changes its form abruptly from fractal to non-fractal (fig. 7).

Students usually love the ballistic deposition model. Here particles fall in straight-line trajectories on a surface like hail falling upon a pavement. They form clusters that resemble tall poplar trees (fig. 8).

Having excited the students' interest in simple random models, we can proceed to do some experiments. Every child knows that acid eats its way into materials, but most do not know that it can do so in a "worm-holing" fashion that lends itself to a beautiful visualization. Our colleagues at the Dowell–Schlumberger research laboratory (a multi-national oil company) have perfected a technique that permits these wormholes to be visualized (fig. 9), and computer simulations can be done by the students.

The concept of multifractals can also be introduced here (fig. 10). Multifractals are an important generalization of the concept of fractals that have very recently found a tremendously wide range of applications in chemistry and physics. The key idea is that each element of an object can have a different probabilistic weight attached to it, and the distribution of these weights can be described by a multifractal distribution function, a concept first proposed by Mandelbrot – see, e.g., the recent article by Mandelbrot [10]. Their study by high-school students serves to place the students in touch with the latest and most exciting developments in chemistry and physics, and the availability of a multi-tasking machine (such as the NeXT personal computer) has the effect that each night students can "put on a computer run" and come in the next morning to see results such as shown in fig. 10.

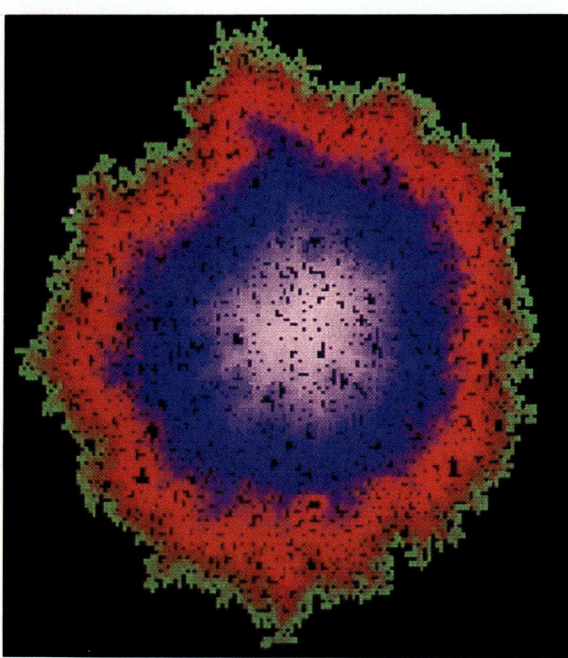

Fig. 4. A typical cluster grown according to the rules of the Eden cancer growth model. The color indicates the order in which particles are added to the cluster. Courtesy of P. Meakin.

Fig. 5. A typical cluster grown according to the diffusion-limited aggregation (DLA) model. A total of 50 000 particles are added, one at a time, to form the aggregate. The color indicates the successive order in which particles have been added. Courtesy of P. Meakin.

Fig. 6. A DLA cluster where the seed particle is placed on a surface instead of in free space. The similarity between this disorderly growth pattern and that of fig. 5 exemplifies the "universality" of the cluster rules. Courtesy of P. Meakin.

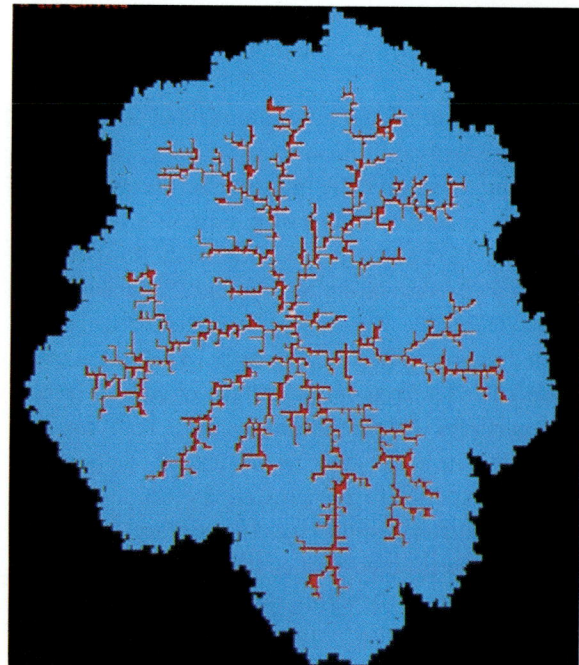

Fig. 7. A hybrid cluster, grown first with DLA rules and then with Eden rules. Courtesy of P. Meakin.

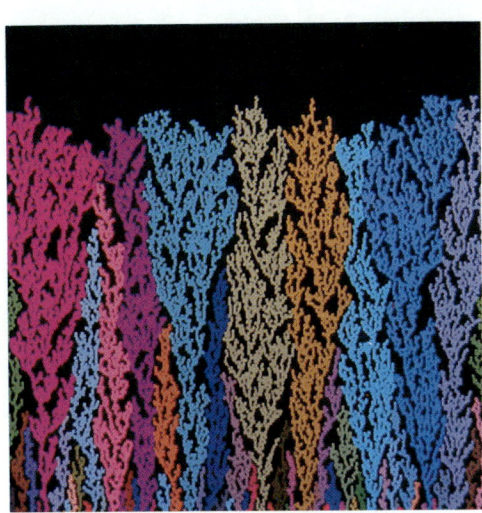

Fig. 8. The results of a "ballistic deposition" of particles on a substrate. The colors serve to distinguish individual clusters of connected particles. Courtesy of P. Meakin.

Fig. 9. The results of an experiment in which plaster of Paris is dissolved by water in much the same fashion as acid can dissolve rock when poured in vast quantities down the bore hole of an oil well. Students can reproduce this experiment in the classroom laboratory and can also perform a computer simulation that gives a quite similar pattern. Courtesy of G. Daccord.

Fig. 10. Computer simulation representing a non-viscous fluid being forced under pressure into a viscous fluid. The pressure in the box is color coded from orange (zero at the edge of the box) to black (highest pressure in the non-viscous fluid). Multifractal concepts are used to describe this distribution of colors. Reprinted by permission from Nature, Vol. 335, p. 405, Copyright © 1988, MacMillan Magazines Ltd. [9].

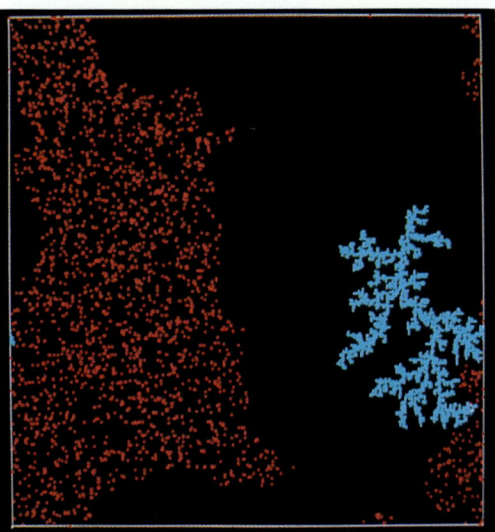

Fig. 11. The results of a cluster–cluster aggregation computer simulation showing how a fast moving large cluster (in blue) can serve to "eat up" all the smaller clusters (in red). Courtesy of P. Meakin.

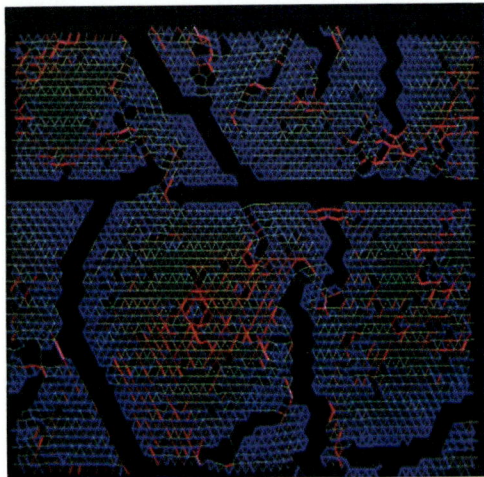

Fig. 12. The result of a computer simulation of a fracture experiment in which a large homogeneous plate is allowed to crack in much the same fashion that a dried up salt flat cracks. Courtesy of P. Meakin.

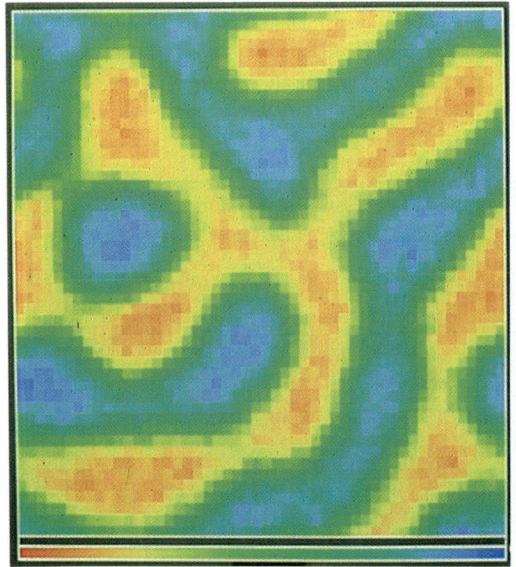

Fig. 13. Computer simulation of "spinodal decomposition" whereby two immiscible fluids separate when the system is cooled below the consolute temperature. Courtesy of P. Meakin.

Fig. 14. Comparison between a photograph of a real snowflake (a) and a student-generated computer model of a snowflake (b); this ability of kids to "make their own snowflakes" on a computer by using the microscopic rules of probability was the germ behind a recent 30 min *3-2-1 Contact* segment on snowflake growth. After Nittmann and Stanley [11].

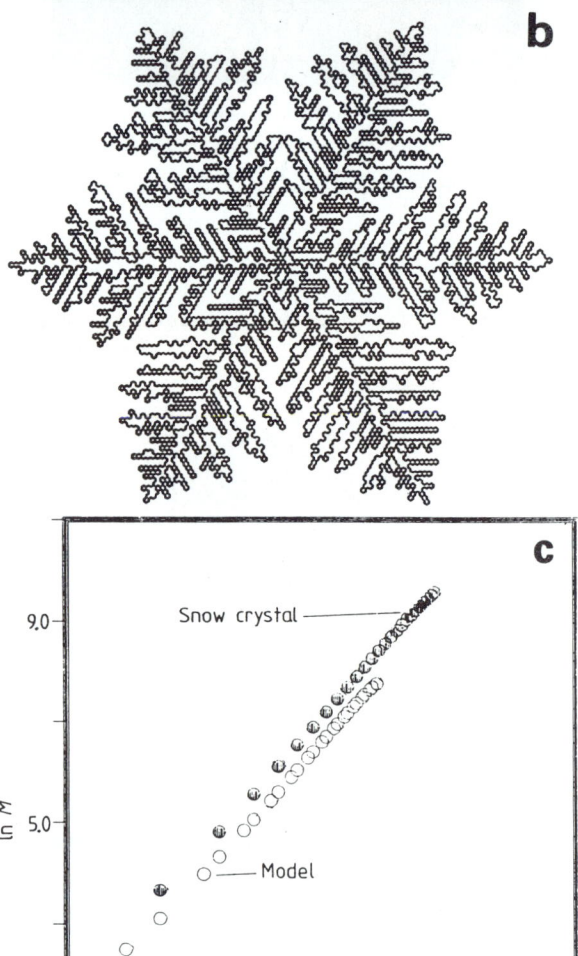

Suppose our clusters are free to move about? Then we have what is called *cluster–cluster aggregation*. One can watch on a computer screen the wild dance of individual clusters, inexorably colliding and linking together forever on collision. An example from a rather late stage of this growth process is shown in fig. 11.

Students are fascinated with fracture – fracture of dropped dishes, fracture of bones in athletics. Fracture involves simple concepts of probability, and fractal models can be used to get a good feeling for the key scientific ideas underlying this phenomenon. Fig. 12 shows a simulation of fracture corresponding to a dried out mud flat.

Another application of fractal concepts to real systems occurs when two immiscible fluids separate when the system is cooled below its consolute temperature. Fig. 13 shows a simulation of this system.

Next we teach students about snowflakes. This topic is always an exciting one: students have heard that no two snowflakes are identical, and are always fascinated by this concept. By using the computer, we can allow each student to build his or her own personal snowflake, and they can recognize (fig. 14) that although every snowflake has the same general form, indeed, no two are identical!

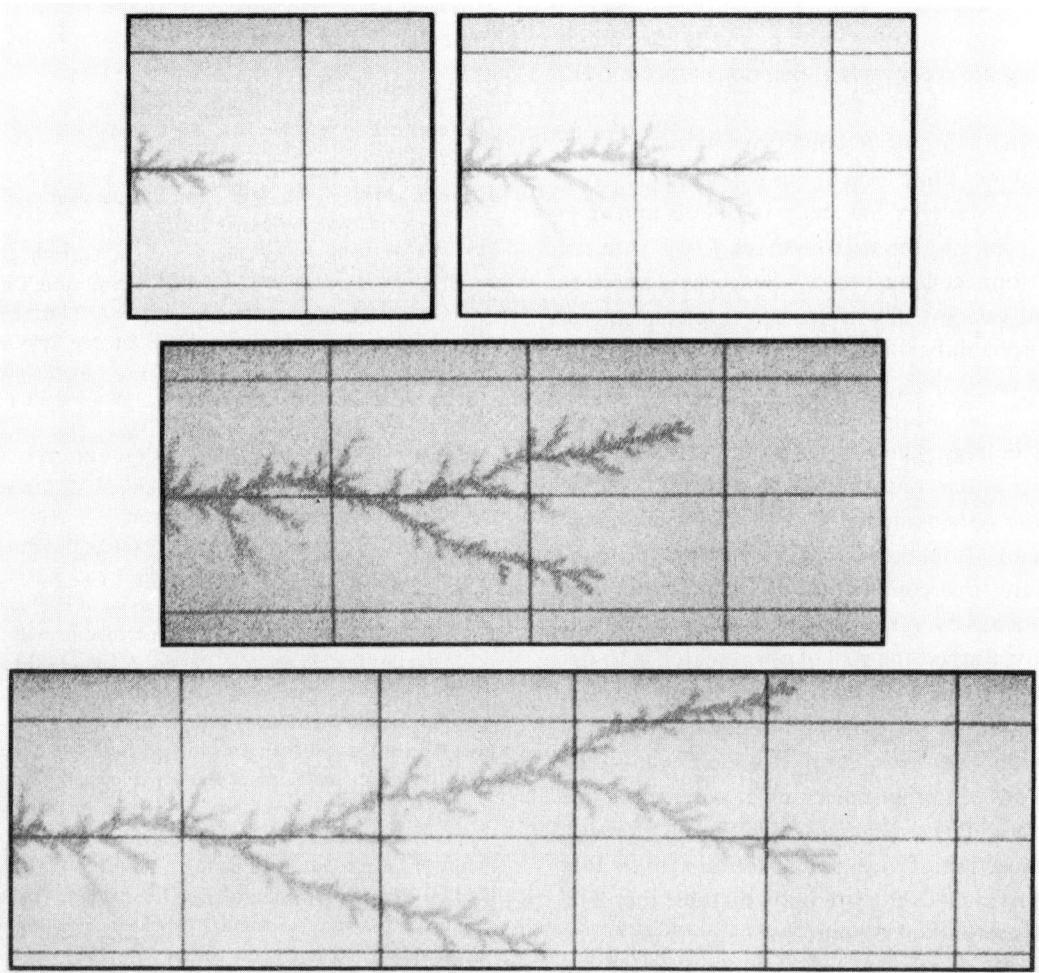

Fig. 15. Experiments on viscous fingers: The first demonstration that this simple experiment in fluid flow leads to fractal forms. From ref. [12]. A random porous medium can produce fractal fingers for immiscible fluids; see, e.g., ref. [13].

A final example concerns fluid instabilities. The key phenomenon is that when a low-viscosity fluid like water is forced under pressure into a high-viscosity fluid, one has a classic instability termed viscous fingering. It is possible to reproduce the classic experiments in this field using materials available to high-school-science laboratories (fig. 15), in order that students acquire a feel for this phenomenon. Then the students can simulate the underlying motion of the fluid particles, and so appreciate that at each instant of time a fundamentally *random* choice is made by each water molecule. Although the *microscopic* events are random, the resulting *macroscopic* form (fig. 15) is recognizable as being essentially the same as the DLA cluster studied previously. Thus the student can connect diverse fields of science, as well as learn about topics that are very much the object of current research.

Under the appropriate conditions, we can relate fluid phenomena to a class of familiar geologic structures, which students can study in details and then simulate – see, e.g., the analysis in ref. [14]. Thus the students connect branches of science as diverse as fluids and geology, all with underlying concepts of applied probability.

In summary, then, concepts of probability and fractals underlie a huge fraction of recent developments in biology, chemistry, physics and geology. Probability is the mathematics of random systems, and random systems are found at the forefront of most technologies – from the integrated circuit to the high-temperature superconductor. Of course, we do not teach students everything there is to know about probability. Rather, the goal of our research is to figure out how to best use advanced technology to teach the important but extremely subtle idea of how it is that individual *unpredictable* microscopic events – be they motions of atoms, molecules, or polymer chains – correspond to *predictable* macroscopic averages. The first reaction of most people, of all ages, is that "if the individual events are unpredictable then one can say nothing about the outcome of averages". The dramatic discovery which students will make for themselves is that "*even if the individual events are unpredictable, one can nonetheless say a great deal about the averages*".

The author wishes to thank all who have made possible this author's appreciation of fractal objects and their significance in science. First and foremost, he wishes to thank Professor Benoit Mandelbrot for introducing him to the subject in 1975 – in a long discussion following an MIT Colloquium on current theories of turbulence (by P.C. Martin) – and for having patience with his own feeble attempts to interpret the geometry of percolation clusters in terms of fractal dimensions. Secondly, he wishes to thank the many visitors to Boston who have given so generously of their time and of their ideas. Last but certainly not least, he wishes to thank his students from which he has learned so much.

The author also wishes to thank Brian Jorgensen for collaboration on the opening paragraphs.

References

[1] B.B. Mandelbrot, The Fractal Geometry of Nature (Freeman, San Francisco, 1982).
[2] D'Arcy Wentworth Thompson, On Growth and Form (1917), abridged edition, 1st Ed. (Cambridge Univ. Press, Cambridge, 1961), paperback edition, 1st Ed. (1966).
[3] G. Daccord, J. Nittmann and H.E. Stanley, Phys. Rev. Lett. 56 (1986) 336.
[4] L. Niemeyer, L. Pietronero and A.J. Wiesmann, Phys. Rev. Lett. 52 (1984) 1033.
[5] H.F. Leeper, J. Comp. Neurol. 182 (1978) 777.
[6] F. Caserta, H.E. Stanely, W.D. Eldred, G. Daccord, R.E. Hausman and J. Nittmann, preprint.
[7] F. Family, Physica D 38 (1989) 98–103, this Proceedings.
[8] H.E. Stanley, J. Teixeira, A. Geiger and R.L. Blumberg, Interpretation of the unusual behavior of H_2O and D_2O at low temperature: are concepts of percolation relevant to the "puzzle of liquid water"?, Physica A 106 (1981) 260.
[9] H.E. Stanley and P. Meakin, Multifractal phenomena in physics and chemistry, Nature 335 (1988) 405.
[10] B.B. Mandelbrot, An introduction to multifractal distribution functions, in: Random Fluctuations and Pattern Growth: Experiments and Models, H.E. Stanley and N. Ostrowsky, eds. (Kluwer, Dordrecht, 1988).
[11] J. Nittmann and H.E. Stanley, J. Phys. A 20 (1987) L1185.
[12] J. Nittmann, G. Daccord and H.E. Stanley, Fractal growth of viscous fingers: quantitative characterisation of a fluid instability phenomenon, Nature 314 (1985) 141.
[13] U. Oxaal, M. Murat, F. Boger, A. Aharony, J. Feder and T. Jøssang, Nature 329 (1987) 32.

HUNTING FOR THE FRACTAL DIMENSION OF THE KAUFFMAN MODEL

D. STAUFFER

HLRZ, c/o KFA Julich, D-5170 Julich 1, Fed. Rep. Germany

The Kauffman model is a random mixture of all possible cellular automata. On the square lattice with nearest-neighbor interactions simulations give a fractal dimension D above 1.8, compatible with the usual scaling law $D = d - \beta/\nu$, for damage spreading.

When the fractal dimension D is related to the usual critical exponents at a second-order phase transition, then usually one takes

$$D = d - \beta/\nu$$

in d dimensions, where β is the exponent with which the order parameter vanishes, and ν is the exponent for the divergence of the correlation length. A different choice is

$$D = d - 2\beta/\nu .$$

This controversy even had the honor to enter The Book [1] in connection with percolation theory [2,3]. Thus the present author was deeply disturbed when recent determinations of critical exponents [4] in the two-dimensional Kauffman model gave better agreement for the codimension $d - D$ with $2\beta/\nu$ than with β/ν. The present work thus re-examines the evaluation of D to find out the truth (?).

What is the Kauffman model [5]? It is a random mixture of all possible cellular automata. For cellular automata on the square lattice [6], each site carries a spin which is up or down. The orientation of the spin at the next time step $t+1$ is determined completely by the orientation of its neighbor spins at time t. With nearest neighbors and no memory effects, we have four nearest neighbors, thus $2^4 = 16$ neighbor configurations, and therefore $2^{16} = 65536$ different possible rules, since for each neighbor configuration the rule can force the center spin up or down. For normal cellular automata, each site obeys the same rule, and then one can investigate 65536 different cellular automata [7].

In one dimension, we have only a left and a right neighbor, and thus four possible neighbor configurations. Thus we have $2^4 = 16$ possible rules. Among them is the logical AND which gives up if both neighbors are up. The logical OR gives up if at least one neighbor spin points up; tautology gives always up, and contradiction never. 12 other rules of a similar kind complete the list of 16 possible cellular automata. However, usually one-dimensional systems with short-range interactions do not show the kind of phase transition observed for higher dimension, and this is also true for the Kauffman model. Thus we ignore here the simple one-dimensional case.

For the Kauffman model, each site selects at the beginning randomly the rule it wants to obey, and sticks with that rule for the rest of the evolution. Originally this model was invented [5] to describe genetic interactions and for that purpose had an infinite range of interaction. For example, each site selected randomly at the beginning four other sites from anywhere in the system, and treated them as its neighbors. This case is similar to the mean-field limit and ignores geometry; thus while biologically more relevant it does not have a fractal mass dimension. Instead, we look at it as some critical phenomenon with nearest-neighbor interactions only [8].

To have a critical point we need a continuously varying parameter p such that at a certain value of p something special is happening. Thus during the ini-

Essays in honour of Benoit B. Mandelbrot
Fractals in Physics – A. Aharony and J. Feder (editors)

tial selection of rules, each site picks, for each neighbor configuration separately, the rule resulting in spin up with probability p, and the rule resulting in spin down with probability $1-p$. In a computer program, we initially go through all sites to set their rules. For this purpose, after having selected the next site, we go through all 16 possible configurations of its four neighbors. For each such configuration we produce a random number between zero and unity. If that random number is smaller than p we fix the rule as resulting in an up spin, whereas for larger random numbers we fix it as giving a down spin. After 16 such random numbers, the rule for this site is determined and we go to the next site. When all sites are treated, the rules of the lattice are fixed, we orient the spins randomly up or down, and start the simulation.

To define critical exponents we need an order parameter such that for p on one side of a threshold p_c the order parameter is zero, and on the other it is nonzero. Following Kauffman [5] we look at the stability of the spin configuration (i.e. the genetic setup) against small perturbations (i.e. mutations) of the spins. Does the flipping of one spin against the rules (the initial error or *damage*) affect only a small neighborhood of the spin, or does it affect in the later time development a finite fraction of all spins even in the thermodynamic limit? In the latter case we call the system chaotic [8]. Thus we simulate two lattices, identical in the size and rules, and also identical in the initial spin distribution except that a small number of spins in one replica is different from the corresponding spins in the other replica. The number of spins which at some later time are different in a site-by-site comparison of the two replicas is the Hamming distance; we call it the damage if initially the different spins are restricted to a geometrically well localized region, as in the present study. We thus study the spreading of that damage throughout the system, and its fractal dimension right at the critical point $p=p_c$ of the transition to chaos.

In the past, the initial damage in such studies [9] was often taken as one single site. Then accidentally the damage often vanishes even in the chaotic phase since the actual configuration could prevent it from spreading; for example the initially selected site could be a spin which has selected the rule "always point down" and thus is insensitive to flipping. Efficiency is improved [10,11] if a whole line of spins is damaged initially; it is improved even more drastically if along that line we take the spins always up, for all times, as in refs. [12,13]. Log–log plots of damage versus system size gave compatible slopes for these three initial damages; but only with the second method, keeping one line of spins always up in one of the two replicas, did we achieve nearly 100% efficiency, i.e. the damage restricted initially to one lattice line spread over the whole lattice for nearly all samples considered.

We took $p=p_c$, the threshold for the transition to chaos. We initially damaged the center line of the $L\times L$ square lattice, and watched if the damage touches a lattice boundary parallel to the initially damaged line. We stop the simulation at the time of touching and then count the number of sites (the unnormalized Hamming distance) damaged at the moment of touching. This set of sites need not be a geometrically connected cluster and might better be called a cloud. In averaging over different runs (different rules and different initial spin configuration) we ignore those few runs where the damage did not touch the boundary. We define the mass fractal dimension D through $M\propto L^D$ and the time fractal dimension D' through $t\propto L^{D'}$, where t is the average time (number of sweeps through the lattice) after which the damage cloud hits the boundary of the $L\times L$ lattice, and M is the number of sites in the damage cloud at the moment of touching. Ref. [4] gave $D=1.6$, 1.8 and 1.8 from simulations in 2, 3 and 4 dimensions, whereas D' was 1.6, 2.2 and 2.1 there.

Our simulations used the da Silva–Herrmann method [11] with modifications pushing its speed to about 150 million sites updated per second by one Cray-XMP processor. Up to 4 megawords of memory were used, since each site occupied one bit of the 64-bit computer words.

We first used $p=0.298$, the threshold given in ref. [11] with an error bar of 0.005. With up to one million sites in a square lattice, nearly all samples lead to a damage cloud touching the boundary. For bigger lattices this success fraction, i.e. the fraction of samples with damage hitting the boundary, decayed rapidly. Fig. 1 explains why: The true (?) threshold seems closer to 0.303 than to 0.298. We plot in fig. 1 the effective threshold $p_c(L)$ for which the success fraction is about 1/2. For smaller lattices, the effective threshold is smaller than for bigger lattices; for example it is 0.266 for $L=64$ (close to the first [8] threshold estimate of 0.26) and it is 0.302 for $L=3392$, our biggest lattice. Thus with a fixed p close to the asymptotic threshold, small lattices will be relatively far in the chaotic region and the success ratio is above 99%. Only for bigger lattices do we see possible inaccuracies in p_c. (With the gradient method of ref. [11] the finite-size effect is smaller and less accurate.) Had the finite-size effect had the opposite sign, our method would have been very inefficient.

In ref. [4] we studied the fraction of sites damaged after a sufficiently long time in the chaotic phase; this fraction was going to zero as $(p-p_c)^\beta$. On the non-chaotic side of the transition, the ratio of the final number of damaged sites to the initial number of damaged sites (in the limit where both are a small fraction of the lattice size) varied as $(p_c-p)^{-\gamma}$. These quantities are perhaps the analogs [8] of the order parameter and its susceptibility near other critical points. Our numerical estimates were $\beta=0.34$, $\gamma=2.5$ for the square lattice. Assuming hyperscaling, $d\nu=\gamma+2\beta$, we get $\nu=1.6$, which explains why fig. 1 plots the thresholds versus $L^{-5/8}=L^{-1/\nu}$. Thus the relation $D=d-\beta/\nu$ predicts $D=1.79$, whereas $D=d-2\beta/\nu$ predicts $D=1.58$. The latter, not the first prediction, agrees with the old $D=1.6$ from ref. [9]. (With our new p_c near 0.303, the estimates for γ, ν and D increase whereas β decreases.)

To see if this contradiction to some theories is a numerical error or real, we recalculated damage mass M and touching time t for L between 64 and 3392 using the estimate $p_c=0.303$ for the threshold. Fig. 2 suggests $D=1.8$ to 1.9 (instead of the old $D=1.6$) for the mass fractal dimension, and a somewhat smaller value 1.3 to 1.4 for the time fractal dimension, D'. Thus agreement has been restored with the traditional prediction $D=d-\beta/\nu$. Also the new result $D>D'$ explains why right at $p=p_c$ in very large systems, the Hamming distance increases slightly stronger than linearly [4] with the number of sweeps through the lattice, if we damage initially a single site only; the exponent for this power law should be D/D', and ref. [4] gives about 1.2 for this ratio. We can no longer exclude the possibility that D agrees with the fractal dimension 91/48 of two-dimensional percolation, though the other exponents still are different.

The slight curvature for the largest systems

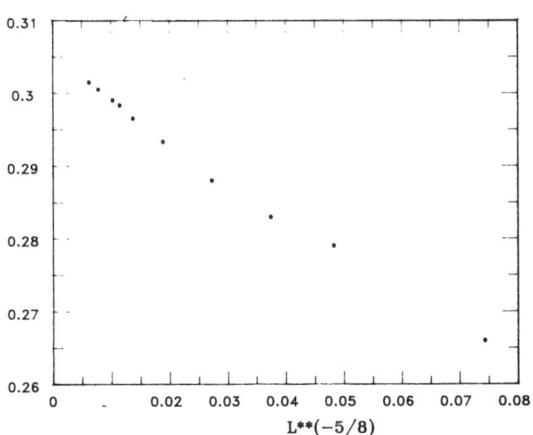

Fig. 1. Variation of effective threshold with power of inverse system size. Typical error bar is 0.001.

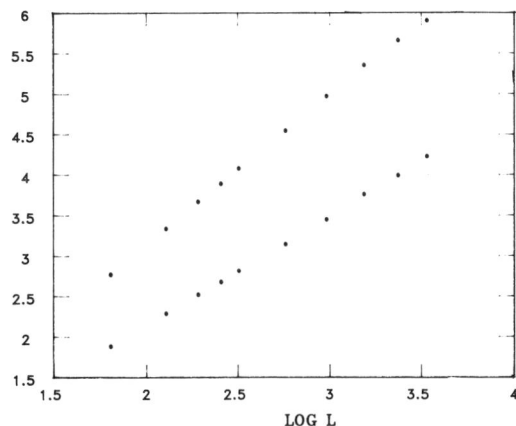

Fig. 2. log–log plot of touching damage (upper data) and touching time (lower data) versus system size L; decadic logarithms. The errors are a few percent.

($L=2368$ and 3392) may suggest that even 0.303 is slightly too small an estimate for the threshold, a suspicion compatible with fig. 1. We get about the same exponents (without this curvature for the largest systems) if we look at damage and time for the effective, size-dependent thresholds of fig. 1, as a function of system size: the damage is then about half as large and the time about twice as large as in fig. 2.

Thus everything seems to be in order again: scaling relations are valid. Of course, the reader may accuse this writer of wishful thinking and of pushing p_c upward to 0.30 from the very first estimate 0.26 in order to get data in agreement with his prejudices. Another explanation is the rather high value of the correlation exponent $\nu = 1.6$, as a result of which finite-size effects were unusually strong and required particularly large lattices. In both interpretations, there is no guarantee that the present exponents are the final word on the subject. Once one believes a threshold near 0.303 and uses the above trick to damage a whole line permanently, a nice numerical result is obtained with much less computational efforts than our previous less successful (?) attempts.

I thank H.J. Herrmann and N. Jan for suggesting this work.

References

[1] B.B. Mandelbrot, The Fractal Geometry of Nature (Freeman, San Francisco, 1982).
[2] D. Stauffer, Z. Physik B 25 (1976) 391.
[3] H.E. Stanley, R.J. Birgeneau, P.J. Reynolds and J.F. Nicoll, J. Phys. C 9 (1976) L553.
[4] D. Stauffer, in: Computer Simulation Studies in Condensed Matter Physics, D.P. Landau, K.K. Mon and H.B. Schuttler, eds. (Springer, Berlin, 1989).
[5] S.A. Kauffman, J. Theor. Biol. 22 (1969) 437.
[6] S. Wolfram, Theory and Applications of Cellular Automata (World Scientific, Singapore, 1986).
[7] D. Stauffer, Physica A 157 (1989) 645.
[8] B. Derrida and D. Stauffer, Europhys. Lett. 2 (1986) 739.
[9] L. de Arcangelis, J. Phys. A 20 (1987) L369;
A. Hansen, J. Phys. A 21 (1988) 2481;
D. Stauffer, Phil. Mag. B 56 (1987) 901;
R. Jullien, L. Peliti, R. Rammal and N. Boccara, eds., Universalities in Condensed Matter (Springer, Berlin, 1988), p. 246.
[10] P.M. Lam, J. Stat. Phys. 50 (1988) 1263.
[11] L.R. da Silva and H.J. Herrmann, J. Stat. Phys. 52 (1988) 463.
[12] M. Corsten and P. Poole, J. Stat. Phys. 50 (1988) 461.
[13] A. Coniglio, L. de Arcangelis, H.J. Herrmann and N. Jan, Europhys. Lett. 8 (1989) 315.

FRUSTRATION AND CORRELATIONS IN FRACTALS

R.B. STINCHCOMBE

Department of Theoretical Physics, Oxford University, 1 Keble Road, Oxford OX1 3NP, UK

A technique for obtaining exact thermodynamic and correlation functions of regular fractals is illustrated by a development for the Ising system on the triangular Sierpińsky gasket. A generalised hyperscaling statement is obtained. The technique is then applied to the fully frustrated antiferromagnetic case, where the exact ground state entropy per spin is shown to be $S(0) = 0.4930...$. This result shows that the introduction of (regularly disposed) holes into the triangular lattice does not relieve frustration, as is commonly supposed. Relations completely determining the correlation functions are obtained. That for the zero field pair correlation function of vertex spins at separation r is $G^{-1}(2r) = G^{-2}(r) - G^{-1}(r) + 1$, together with a relation which sets the correlation-length scale. Exact asymptotic expressions are derived for ferromagnetic correlations in both zero and non-zero field. The zero-field, zero-temperature correlation length of the fully frustrated system is shown to be $\xi = 0.79$, representing extremely rapid decay of correlations.

1. Introduction

Mandelbrot [1] and many subsequent workers have emphasised the wide range of situations in which fractal characteristics occur and can be exploited. In the field of statistical physics, especially critical phenomena, fractal viewpoints have been particularly useful, both in describing the real fractals that occur (e.g. in dilute systems) and in providing models which have helped our understanding of scaling phenomena.

In particular, Mandelbrot and coworkers [2] were among the first to realise and exploit the fact that some of the famous models of statistical physics can be solved on certain fractals.

That provided the stimulus for the present paper, which is presented as a small tribute to Mandelbrot's pioneering work. We follow a similar philosophy, and provide exact solutions for the fractal versions of some important problems in the statistical physics of lattice spin systems.

The problems addressed are the calculation of free energy and correlation functions, hyperscaling, and the properties of fully frustrated lattices.

A method is outlined here by which free energy and all thermodynamic functions, and more importantly correlation functions, of Ising and other classical discrete spin models can be obtained exactly for non-zero as well as zero field on typical fractal lattices. The correlation functions provide the detailed microscopic characterisation of the properties of such models, yet little is known about them. So it is hoped that the results presented here will lead to further understanding of such models. One aim in the investigation was the interpretation of recent work using conformal field theory techniques [3], which implies that in non-zero field the Ising two spin correlation function has eight correlation lengths.

The best known fully frustrated system is the Ising antiferromagnet on the triangular lattice. Its exact solution [4] exhibits disorder down to zero temperature and a macroscopic ground state entropy. The absence of other exact solutions of fully frustrated systems has left open a number of interesting questions. One, important in its own right but especially for the understanding of spin glasses [5–7], is whether introduction of holes or other disorder relieves the frustration and reduces the huge ground state degeneracy and, with it, the ground state entropy. Another concerns the nature of correlations in frustrated lattices.

This paper gives an exact solution for thermodynamic functions and in particular entropy, and correlation functions on a fully frustrated fractal lattice which has the property of holes (on all scales). It turns out that this has a greater ground state entropy per spin than in the triangular lattice, in contrast to the usual assumption of holes relieving frustration; and the correlations, even at absolute zero, decay extremely rapidly.

The technique which provides both the exact free energy (hence entropy, etc.) and the correlation functions will be illustrated with the spin-1/2 Ising model on the triangular Sierpińsky gasket fractal since that will be our model for the fully frustrated lattice. But the procedure applies very generally, as will be remarked later.

2. Transformation of parameters, partition function, and correlations

The techniques introduced here for the exact partition function and correlation functions rely on the hierarchical nature of regular fractals. For the specific case we consider, the triangular Sierpińsky gasket, the composition of three nth generation gaskets, each sharing a vertex with each of the others, produces an $(n+1)$th generation triangular gasket of twice the side length, whose vertices are the unshared vertices of the earlier three. Let $Z_n^{\alpha\beta\gamma}$ denote the "constrained" partition function of the spin-1/2 Ising model on the nth generation gasket, but with the three vertex spins in arbitrarily specified states α, β, γ respectively (with $\alpha = \pm 1$ denoting spin up or down, etc.). Then

$$Z_{n+1}^{\alpha\beta\gamma} = \sum_{\alpha'\beta'\gamma'} Z_n^{\alpha\beta'\gamma'} Z_n^{\beta\gamma'\alpha'} Z_n^{\gamma\alpha'\beta'}. \qquad (1)$$

It can be shown that under this recurrence procedure $Z_n^{\alpha\beta\gamma}$ retains the same form as it has for $n=0$ (when it is just the Boltzmann weight for a specified configuration α, β, γ of three interacting spins); however, the parameters renormalise, and the constrained partition function becomes at the nth generation

$$Z_n^{\alpha\beta\gamma} = u_n(1 + x_n(\alpha+\beta+\gamma) \\ + y_n(\alpha\beta+\beta\gamma+\gamma\alpha) + z_n\alpha\beta\gamma), \qquad (2)$$

where

$$u_{n+1} = 8u_n^3[3x_n^2(1+y_n) + 1 + y_n^3] \\ \equiv 8u_n^3 D_n, \qquad (3)$$

$$x_{n+1} = [x_n^2(x_n+z_n) + 2x_ny_n(1+y_n) \\ + x_n + y_n^2 z_n]D_n^{-1}, \qquad (4)$$

$$y_{n+1} = [y_n^2(1+y_n) + 2x_ny_n(x_n+z_n) \\ + x_n^2 + y_nz_n^2]D_n^{-1}, \qquad (5)$$

$$z_{n+1} = [3y_n^2(x_n+z_n) + x_n^3 + z_n^3]D_n^{-1}. \qquad (6)$$

The initial values of the parameters are

$$u_0 = c^3C^3(1+t^3)(1+3v\tau^2), \qquad (7)$$

$$x_0 = \frac{\tau(1+2v+v\tau^2)}{1+3v\tau^2}, \qquad (8)$$

$$y_0 = \frac{\tau^2+v+2v\tau^2}{1+3v\tau^2}, \qquad (9)$$

$$z_0 = \frac{\tau(\tau^2+3v)}{1+3v\tau^2}, \qquad (10)$$

where

$$c \equiv \cosh K, \quad t \equiv \tanh K,$$

$$v \equiv [t-1+1/t]^{-1}, \qquad (11)$$

$$C \equiv \cosh h, \quad \tau \equiv \tanh h,$$

where K and h are the exchange interaction and field, each divided by k_BT.

Iteration of (3)–(6) using the initial conditions (7)–(10) provides $Z_n^{\alpha\beta\gamma}$, from which the usual partition function Z_n (and hence the free energy and all thermodynamic functions) is obtained using

$$Z_n = \sum_{\alpha\beta\gamma} Z_n^{\alpha\beta\gamma} = 8u_n. \qquad (12)$$

Correlation functions such as $\langle \sigma_i\sigma_j \rangle$ can also be obtained as follows. If i is a vertex joining two triangles corresponding to mth generation Sierpińsky gaskets, the spin σ_i will correspond to one (α', say) of the variables α', β', γ' being summed over at stage $n=m$. So allowing for an extra factor α' in carrying out that sum correctly incorporates the effect of σ_i. Similarly incorporating σ_j leads to an explicit result for the correlation function for arbitrary i, j. The technique generalises to correlation functions of arbitrary order.

For the particular case of spins at the three corner vertices α, β, γ of the nth generation gasket, the correlation functions are just

$$\langle \sigma^\alpha \rangle_{(n)} = x_n, \tag{13}$$

$$\langle \sigma^\alpha \sigma^\beta \rangle_{(n)} = y_n, \tag{14}$$

$$\langle \sigma^\alpha \sigma^\beta \sigma^\gamma \rangle_{(n)} = z_n. \tag{15}$$

A similar method, both for the partition function and for correlation functions, applies for other hierarchically constructed regular fractals (such as the Berker lattice) and for other models (e.g. higher-spin Ising, Potts models, classical XY, etc.). But we continue here with the Sierpiński gasket.

The zero-field limit is an important special case of the above. In this limit x_n, z_n vanish and the transformations of u_n, y_n simplify to

$$u_{n+1} = 8u_n^3(1+y_n^3), \tag{16}$$

$$y_{n+1} = \frac{y_n^2}{1-y_n+y_n^2}. \tag{17}$$

Using (12), eq. (16) can be converted into the following equation for the transformation of the free energy per spin, $F_n \equiv (\ln Z_n)/N_n$:

$$F_{n+1} = 3F_n \frac{N_n}{N_{n+1}} + \frac{1}{N_{n+1}} \ln[\tfrac{1}{8}(1+y_n^3)]. \tag{18}$$

Here, $N_n = \tfrac{1}{2}(3^{n+1}+3)$ is the number of spins on the nth generation gasket. Eq. (18) is the form arising in standard scaling approaches to free energies [8].

Eq. (17) is equivalent to the thermal scaling from which the following relationship between thermal correlation length and temperature has been earlier deduced [2,9–11] for the ferromagnetic case at low temperatures:

$$\xi \sim \exp[\tfrac{1}{4} \exp(4K) \ln 2]. \tag{19}$$

In the light of this anomalous relationship between ξ and temperature, and the anomalous fractal dimension $d_f = \ln 3/\ln 2$ it is interesting to enquire whether hyperscaling ("$2-\alpha = d\nu$") applies. The answer provided by considering the free energy result (18) is that hyperscaling is satisfied in the generalised sense that the total free energy scales like ξ^{-d_f}.

3. Fully frustrated fractal

We now turn to the fully frustrated system provided by the antiferromagnetic Ising model on the Sierpiński gasket, in zero field. This is like an Ising antiferromagnet on the triangular lattice but with holes on all scales. The above procedures allow its exact solution to the same degree as is provided by the famous solution [4] for the triangular lattice.

In this case K is negative. To obtain the ground state entropy per spin, $S(0)$, it is only necessary to use low-temperature forms for F_n, y_n in (18), leading to the exact result

$$S(0) = \tfrac{2}{3} \ln 6 + \tfrac{2}{9} \sum_{p=0}^{\infty} 3^{-p} \ln[\tfrac{1}{8}(1+c_p^{-3})], \tag{20}$$

where c_p is the zero-temperature limit of $1/y_p$. This is obtained by iterating (17). Since $c_0 = -3$ goes into $c_1 = 13$, $c_2 = 157$, ..., etc., the series in (20) is extremely rapidly convergent and yields

$$S(0) = 0.493006107\ldots. \tag{21}$$

This agrees, to the accuracy of that work, with recent numerical calculations [12] on the same system. $S(0)$ is higher than that for the triangular lattice ($S(0) = 0.3383\ldots$), showing that introduction of holes has not relieved frustration but instead has increased the degeneracy of the ground state. The ground state entropy per spin is actually $\approx 70\%$ of that ($\ln 2$) at infinite temperature, and the ground state degeneracy of the N-spin fractal is $\exp(0.49\ldots N)$.

4. Correlations

Turning next to the correlation functions, we first

continue with the triangular Sierpińsky gasket in zero field.

Let ABC denote the vertex sites of the nth generation triangular gasket, and let (AB) denote the site midway between A and B, etc. Then using the procedure explained before, in the zero-field case some representative pair correlation functions are found to be

$$\langle \sigma_A \sigma_B \rangle_{(n)} = y_n, \tag{22}$$

$$\langle \sigma_A \sigma_{(AB)} \rangle_{(n)} = \frac{y_{n-1} + y_{n-1}^3}{1 + y_{n-1}^3}, \tag{23}$$

$$\langle \sigma_A \sigma_{(BC)} \rangle_{(n)} = \frac{2 y_{n-1}^2}{1 + y_{n-1}^3}, \tag{24}$$

$$\langle \sigma_{(AB)} \sigma_{(BC)} \rangle_{(n)} = \frac{y_{n-1}(1 + y_{n-1})}{1 + y_{n-1}^3}, \tag{25}$$

$$\langle \sigma_{(A(AB))} \sigma_{(B(BC))} \rangle_{(n)} = \frac{2 y_n (1 + y_{n-1}^2)}{y_{n-2}(1 + y_{n-1}^2)}. \tag{26}$$

For separations much larger than the correlation length (19), all pair correlation functions fall off with a factor $\exp(-r/\xi)$, where r is the length of the shortest path joining the two sites.

To illustrate how this and more detailed results can be obtained for the correlations, consider the special case of $\langle \sigma_A \sigma_B \rangle_{(n)} = y_n$. In going from n to $n+1$, y_n transforms according to (17), while the separation r of sites A, B changes by a factor 2. So, writing this correlation function as $G(r)$, eq. (17) implies

$$G^{-1}(2r) = G^{-2}(r) - G^{-1}(r) + 1. \tag{27}$$

Also, measuring r in units of the primitive bond length,

$$G(1) = y_0 = (t - 1 + 1/t)^{-1}. \tag{28}$$

The solution of (27) subject to the boundary condition (28) completely determines this pair correlation function.

For small G, the first term on the right-hand side of (27) dominates, and the functional equation is then solved by

$$G(r) = \exp(-r/\xi), \quad r \gg \xi. \tag{29}$$

The "arbitrary constant" ξ is determined by (28), e.g. by iterating (27) from $r=1$ until r becomes large enough to match $G(r)$ to (29). Only a few iterations (or none, at high temperatures where $\xi = \ln^{-1}(1/y_0) \ll 1$) are required, except for the ferromagnetic case at low temperatures where, by (28), $G(1)$ is close to 1. Then, for $1 - G \equiv \delta$ small, (27) becomes

$$\delta(2r) = \delta(r) + \delta^2(r) + \ldots, \tag{30}$$

which implies

$$G(r) = 1 - \frac{\ln 2}{\ln(\Lambda/r)},$$

$$r \ll \Lambda \equiv \exp[\ln 2/\delta(1)]. \tag{31}$$

A few iterations of (27) join the regime where (31) applies to that where (29) does, and the matching requires

$$\xi = c\Lambda = c \exp[\tfrac{1}{4}(\ln 2) \exp(4K)], \tag{32}$$

where c is a constant of order 1. The resulting ξ is, of course, the same correlation length as given in (19). The analysis gives explicit asymptotic forms (29) and (31) for this correlation function. We stress, however, that the basic equation (17), with (22)–(26) etc. determines all the pair correlation functions completely, in zero field.

For the fully frustrated antiferromagnetic case similar procedures show that even at absolute zero the correlation functions die away very rapidly. Eqs. (27) and (28) again apply, and in this case give

$$G(1) = -1/3, \quad G(2) = 1/13, \quad G(3) = 1/157, \ldots. \tag{33}$$

The first correlation has antiferromagnetic character, and all the others ferromagnetic. G is always small enough to make the exponential decay result (29) apply quite accurately, and matching at $r=4$ gives $\xi = 0.79$. This is a very short correlation length, particularly for a system at absolute zero, but it is the frustration which gives rise to this striking result.

Non-zero field introduces many qualitatively new features into the correlations. Firstly, the odd correlations (e.g. x_n, z_n in (13), (15)) now become non-zero. Secondly, correlations like those defined in (14) and (15) and (22)–(26) do not go to zero as spin separation increases, because of the non-zero value of $x_n \equiv \langle \sigma \rangle_{(n)}$. However, the cumulant correlations, for example

$$\langle \sigma^\alpha \sigma^\beta \rangle_{(n)}^{(c)} \equiv \langle \sigma^\alpha \sigma^\beta \rangle_{(n)} - \langle \sigma^\alpha \rangle_{(n)} \langle \sigma^\beta \rangle_{(n)}$$
$$= y_n - x_n^2 \equiv g_n, \qquad (34)$$

$$\langle \sigma^\alpha \sigma^\beta \sigma^\varphi \rangle_{(n)}^{(c)} \equiv z_n - x_n^3 \equiv j_n \qquad (35)$$

do die away, and their associated correlation lengths depend on field as well as temperature. The exponential decay is related to the fact that for small g_n, j_n the transformation equations (4), (5), (6) take the form

$$x_{n+1} = x_n \left(1 + \frac{g_n(2 - x_n^2) + j_n x_n}{(1 + x_n^2)^2} + \ldots \right), \qquad (36)$$

$$g_{n+1} = \left(\frac{g_n(1 - 2x_n^2) + j_n x_n}{(1 + x_n^2)^2} \right)^2 + \ldots, \qquad (37)$$

$$j_{n+1} = 3 x_n g_{n+1} + \ldots . \qquad (38)$$

The neglected terms are of higher order in g, j. These equations imply that as n increases (i.e. with increasing spin separation) g and j vanish and x goes to a constant. The exponential vanishing of the correlations arises, as in the derivation of (29), from the quadratic dependence of the right-hand side of (37) on g and j, which is proportional to g by (38). More detailed consideration of (36)–(38) shows that they give rise to a hierarchy of smaller and smaller correlation lengths in the correlation functions g, j. For the ferromagnetic case at the critical temperature (zero), these lengths all depend in the same way on field, and that dependence can be got for the critical regime by linearising the scaling relations (3), (4), (5) about the zero-temperature, zero-field fixed point, or as a by-product of matchings (cf. (27)–(32)) being carried out to arrive at the correlation functions. The new field-determined dominant correlation length is

$$\xi_h \propto h^{-\nu_h}, \quad \nu_h = \frac{\ln 2}{\ln 3} \left(= \frac{1}{d_f} \right). \qquad (39)$$

At zero temperature and small field the cumulant pair correlation function for the ferromagnetic case has asymptotic forms,

$$g(r) \sim 1 - \frac{1}{2} \left(\frac{r}{\xi_h \ln 2} \right)^{2/\nu_h}, \quad r \ll \xi_h, \qquad (40)$$

$$\sim \exp(-r/\xi_h), \qquad r \gg \xi_h. \qquad (41)$$

Only the dominant correlation length shows up in (41). The appearance of many correlation lengths (for somewhat smaller r) is reminiscent of the recent conformal algebra results [3] for the Ising model in a field. However, it seems that in the case of the fractal they are a consequence of the discrete scale invariance it possesses, since many subdominant correlation lengths also appear in the zero-field correlations.

5. Concluding remarks

As mentioned earlier, the procedures developed here generalise readily to higher spin, or to Potts models, etc., and to other hierarchically constructed fractals such as the Berker lattice of tetrahedral Sierpiński gasket. For such fractals an equation like (1) again applies, but with different numbers of vertex labels (2,4 respectively for the two fractals mentioned). The detailed development then proceeds along similar lines up to and including the calculation of correlation functions.

For the example of the Berker lattice, the equations analogous to (27), (28) for the Ising vertex spin pair correlation function in zero field are

$$G(2r) = \frac{2G^2(r)}{1 + G^4(r)}, \quad G(1) = y_0 = t. \qquad (42)$$

The consequent $G(r)$ for temperatures just above the transition (where $t = t^*$) is rather like that for the Sierpiński gasket in a field (eqs. (40), (41)), but with $\frac{1}{2}\nu_h$, ξ_h replaced by the thermal eigenvalue ν and $\xi_T = (t^* - t)^{-\nu}$.

For all these fractals above their transition in zero field, it can be shown that $1/\xi$ (as a function of the thermal parameter y_0) and the vertex spin pair correlation function G (as a function of r at fixed y_0) are inverse functions of each other.

In interpreting the exact results available from such treatments it is necessary to distinguish the features resulting from special characteristics of the fractals. For the Sierpińsky gasket these characteristics include, as is well known, the disconnecting nodes, which cause the criticality to appear at zero temperature and give rise to the anomalous form (19) for the thermal correlation length. Because of the inverse function relationship referred to above, that anomalous form is reflected in a corresponding anomalous behaviour (eq. (31)) in the low-temperature correlation function, which therefore also has to be seen as a consequence of the disconnecting nodes.

Equally important for our present considerations is the discrete scale invariance of the fractal, which as remarked above is probably the cause of the hierarchy of correlation lengths since it would be expected to give rise to periodic dependences on $\ln r$ in position-dependent quantities like correlation functions. The other special characteristic of relevance here is the special disposition of holes on the fractal. This feature may mean that the increased degeneracy of the ground state found to be caused by the holes may not be shared by randomly diluted frustrated systems.

References

[1] B.B. Mandelbrot, Fractals, Form, Chance and Dimension (Freeman, San Francisco, 1977).
[2] Y. Gefen, A. Aharony, B.B. Mandelbrot and S. Kirkpatrick, Phys. Rev. Lett. 45 (1980) 855.
[3] A.B. Zamolodchikov, Integrals of motion and the S-matrix of the (scaled) $T=T_c$ Ising model with magnetic field, Rutherford preprint RAL-89-001.
[4] G.H. Wannier, Phys. Rev. 79 (1950) 357.
[5] L. De Seze, J. Phys. C 10 (1977) L353.
[6] J. Villain, Z. Phys. B 33 (1979) 31.
[7] E.G. Gabl and G. Grest, Phys. Rev. Lett. 43 (1979) 1182.
[8] M. Nauenberg and B. Nienhuis, Phys. Rev. Lett. 33 (1974) 1593.
[9] Y. Gefen, A. Aharony and B.B. Mandelbrot, J. Phys. A 16 (1983) 1267.
[10] Y. Gefen, A. Aharony, Y. Shapir and B.B. Mandelbrot, J. Phys. A 17 (1984) 435.
[11] Y. Gefen, A. Aharony and B.B. Mandelbrot, J. Phys. A 17 (1984) 1277.
[12] M. Grillon and F.G. Brady Moreira, private communication.

PHASE TRANSITIONS FOR POLYMERS ON FRACTAL LATTICES

J. VANNIMENUS

*Laboratoire de Physique Statistique, Ecole Normale Supérieure, 24 rue Lhomond, 75231 Paris Cedex 05, France
and Laboratoire Louis-Néel, CNRS, 38042 Grenoble Cedex, France*

Models of linear and branched polymers on finitely ramified fractal lattices are briefly reviewed. Their foremost interest is to provide exactly solvable systems with phase transitions at finite values of the fugacity or temperature. This gives insight into the behaviour of polymers in inhomogeneous media, their collapse transition and their adsorption at surfaces. Some unexpected results are also found, such as the possible existence of essential singularities in generating functions, or the non-convergence of critical exponents towards their Euclidean limit.

1. Historical background

This Conference offers a chance to reflect on how and why one got involved in the physics of fractals, and I will give an account of the motivations and general thread of our work, rather than insist on technical details.

My first contact with the world of fractals occurred as early as 1974, while a post-doctoral fellow at the IBM laboratory in Yorktown Heights. There, Benoit Mandelbrot showed me the first chapters of his manuscript "Les Objets Fractals" [1], asking for the reactions of a "typical reader". In fact he needed advice on French literary style rather than scientific contents, and my wife was better qualified to make suggestions in that respect – so she, not me, is thanked in the foreword to the book. It was amusing anyway to learn that the coast of Brittany has infinite length and to discover the concept of self-similarity, but frankly all this looked so far from my research interests – "serious" calculations on the electronic structure of metallic surfaces – that I did not imagine to work some day on these questions. In hindsight, I missed a unique opportunity to enter a new field as a pioneer, but then very few people around proved more far-sighted!

Eight years later, fractals had become fashionable and provided very useful models for the random media we were studying at Ecole Normale. Also, they appeared to open a new approach to phase transitions by offering well-defined realizations of spaces of non-integral dimensionality [2]. In particular one could hope to gain a deeper insight into the ϵ-expansion, scaling laws, universality, ... from the solution of new non-trivial models. Many spin models (Ising, Potts, XY) have been subsequently considered in that spirit, but either they have transitions only at $T=0$ [3], or they are not exactly solvable and one has to resort to various approximations [4–7], a limitation which lessens their usefulness as testing grounds.

Polymer models turn out to be the most notable exception to that disappointing situation. In their case a well-defined thermodynamic transition exists as a function of monomer fugacity, and exact results can be obtained on fractal lattices of finite ramification such as the Sierpinski gasket [8,9]. Various critical properties and their relationships can thus be studied in detail, and most aspects are found to be qualitatively similar to their counterparts on regular Euclidean lattices, with just different values of critical exponents. But the real reward is that polymers on fractals have yielded several surprises which open the way to new questions and possible generalizations, and are the best justification for my lasting interest in that class of systems.

2. Linear polymers

The simplest "polymer" system on a lattice is the standard self-avoiding walk (SAW), where the links do not interact except for the non-crossing constraint. This models a linear polymer in a good solvent and one expects that the gyration radius of an N-monomer chain behaves for large N as

$$\langle R \rangle \approx N^\nu, \tag{1}$$

where the average is taken over all configurations with equal weight.

We considered this problem on fractals with Rammal and Toulouse [9], as a natural generalization of their work on random walks [10]. One first question was to understand on which properties of the lattice the exponent ν depends: Is a simple Flory-type formula based on the fractal dimension D [11],

$$\nu = 3/(2+D), \tag{2}$$

valid, at least as a first-order approximation? Also, we hoped to gain insight in the controversial and deep problem of polymers in random media [12–14]. SAW models on fractals had in fact already been studied by Dhar [8], although he used the term "pseudo-lattices" and his work was not very widely known. Our main contribution was to point out that the product $D\nu$ must be an intrinsic quantity, i.e., it is invariant if the lattice is crumpled or distorted but its topology is fixed. An "improved" Flory expression was proposed, introducing the spectral dimension $\tilde{d}$ as a new ingredient, and it was in reasonable agreement with the known exact cases – but clearly our expression was still very crude.

Recently that question has been taken up again from a deeper point of view by Bouchaud and Georges [15] and by Aharony and Harris [16], who independently obtained the formula

$$\nu = \frac{1}{D} \frac{4d_c - \tilde{d}}{2 + 2d_c - \tilde{d}}, \tag{3}$$

where d_c is the chemical or "spreading" dimension, which describes the average number of accessible sites on the lattice [17]. Expression (3) is based on reasonable assumptions and correctly reproduces the trend among some families of fractals, but in spite of its dependence on three distinct properties it is not yet in very close agreement with all known values – for instance it predicts $\nu = 0.825$ and 0.725, respectively for the 2-d and 3-d gaskets, while the exact values are 0.798 and 0.674.

Anyhow, it is very useful to test in that way the validity of arguments proposed to explain the success of the Flory approximation and of its generalizations to inhomogeneous media. One also sees clearly that universality on fractals is much weaker than on regular lattices [18]: it is possible to determine some properties on which critical exponents must at least depend, but not to list a finite set of properties that uniquely define a universality class.

3. The collapse transition

On regular lattices a collapse transition of the polymer may occur when a monomer–monomer short-range attractive interaction is introduced to model the effect of a bad solvent. At the transition temperature (θ point) the polymer behaves on large scales essentially as an ideal Gaussian chain, with $\nu_\theta = 1/2$ for $d \geq 3$, and at lower temperatures it forms a compact globule, with $\nu_c = 1/d$. The idea to look for a similar transition on fractal lattices is due to Klein and Seitz [19], who concluded that none occurs for finite values of the interaction on the $d=2$ Sierpinski gasket (SG).

With Dhar, we realized that this negative result might be due to a topological constraint: a SAW can only cross once a given kth-order triangle of the gasket, so self-interactions only occur at the vertices and they become rarer and rarer as the size N increases. This effect does not occur on the 3-d gasket, and indeed we readily discovered the existence in that case of a new fixed point corresponding to the usual θ point, with $\nu_\theta = 0.5294$. In fact the transition was implicitly contained in the recurrence equations written down eight years earlier by Dhar!

A whole family of fractals depending on one (integer) scale factor p, the "modified rectangular lattices", were then studied along these lines [20]. The resulting equations can even be continued formally for $p \to 1$, enabling us to mimic a "quasi-Euclidean" system in the sense of Gefen et al. [2]. This trick provides a new Migdal–Kadanoff type renormalization scheme for the collapse transition problem in two dimensions, which gives an estimate $\nu_\theta = 0.546 \pm 0.010$, in very good agreement with the best numerical determination $\nu_\theta = 0.55 \pm 0.01$. It also shows how the deep connection between fractal models and real-space renormalization group equations may be exploited in a quantitative manner.

4. Adsorption

In the presence of an impenetrable surface such as a solid wall, a polymer in dilute solution feels an effective repulsion due to the loss of configurational entropy with respect to the bulk, so the solution is depleted near the surface. An attractive surface potential is necessary to balance that effect. Then there exists a transition temperature T_a, below which a finite fraction of the monomers remains adsorbed. The transition is analogous to a tricritical point and there is a cross-over region in which the fraction M of adsorbed monomers scales as

$$M \approx N^\phi \approx |T - T_a|^{-1}. \quad (4)$$

The value of the cross-over exponent ϕ is $1/2$ for an ideal chain and ≈ 0.6 for a SAW in $d=3$ [21].

With Bouchaud, we have recently considered that problem when the polymer is restricted to reside on a fractal lattice and interacts with a fractal surface of dimension d_s [22]. By making reasonable physical assumptions and assuming a scaling form for the density profile, as for the Euclidean case, simple bounds on the exponent ϕ may be derived:

$$d_s/D \geq \phi \geq 1 - (D - d_s)\nu, \quad (5)$$

where ν is the gyration radius exponent in the bulk solution.

On the other hand exact results may be obtained for the gaskets, if the natural boundary is chosen as adsorbing surface. We find that an adsorption transition takes place at a finite T_a and close to T_a the polymer behaviour is correctly described by eq. (4), with $\phi = 0.59152$ and 0.7481 for the 2-d and 3-d SG, respectively, in agreement with the stringent bounds given in eq. (5). It is also possible to study adsorption in the presence of attractive self-interactions on the 3-d SG, and one observes a multicritical point, where a collapse and an adsorption transition coexist.

5. Branched polymers

After linear polymers, it sounds natural to study their branched cousins on fractals, but the generalization is far from straightforward. It took the strong motivation of Knezevic to overcome my pessimism and to enumerate the possible configurations on the SG and get their recurrence relations. Even then, their behaviour was confusing and quite different from the linear polymer case. We had to find ways to extract the dominant terms for the asymptotic behaviour before we could understand what was happening and locate the relevant fixed points: they correspond to finite values for specific combinations of the basic variables, which themselves go to zero or to infinity. Armed with that experience we could also solve the 3-d gasket, where it is necessary to keep track of about 10^7 configurations and to generate an 11-equation recursion system – about the limit for an "exact" solution, even by computer!

That effort yielded several very interesting results [23]:

(1) Loops are irrelevant on large scales, so branched polymers indeed belong to the same universality class as lattice animals.

(2) One can calculate exactly ν as well as the other basic exponent θ, which appears in the singular part of the generating function near the critical fugacity $1/\mu$,

$$G(x) \approx (1 - \mu x)^{\theta - 1}. \quad (6)$$

These exponents have "reasonable" values:

$\nu=0.71655$ and $\theta=0.5328$ for the 2-d SG, to be compared to $\nu=0.641$ and $\theta=1$ for the square lattice. The new feature is that the two exponents are independent, while on Euclidean lattices they obey the Parisi–Sourlas relation: $(\theta-1)/\nu=d-2$. On fractal lattices one can write by analogy $(\theta-1)/\nu=D-\delta$, where δ is very close to 2 for some quasi-linear lattices, while $\delta=2.237$ for the 2-d SG. For all lattices studied so far, $\delta \geq 2$, but there is no interpretation of that finding yet.

(3) A collapse transition exists for the gaskets, with an exponent ν_t at the transition temperature extremely close to the value in the compact phase $\nu_c = 1/D$. For instance, $\nu_t = 0.6325$, $\nu_c = 0.6309$ for the 2-d SG, and $\nu_t = 0.5055$, $\nu_c = 1/2$ for the 3-d SG. This is very similar to the situation in $d=2$, where the best numerical determination is $\nu_t = 0.509 \pm 0.003$, and it suggests the possibility for a deeper explanation in terms of some small parameter.

These results confirm that there is a deep similarity with the case of regular lattices, and that many features of critical phenomena survive the loss of translational invariance, but further studies revealed more intriguing behaviour in other systems.

6. Surprises

A first type of unexpected behaviour was discovered for branched polymers on a generalized gasket, the $b=3$ member of the GM family introduced by Given and Mandelbrot [24]. As expected intuitively, the exponent $\nu=0.7068$ is a little closer to the Euclidean value 0.664 than for the 2-d SG. But the generating function has an essential singularity close to the critical fugacity [25]

$$G(x) \approx \exp[c(1-\mu x)^{-\psi}] \quad (7)$$

with $\psi = \ln(3-\sqrt{3})/\ln(3+\sqrt{3})$, instead of the standard power law, eq. (6). The argument for the power law singularity relies on translation invariance, and it does not apply to inhomogeneous systems, so the behaviour discovered on one particular fractal might well be much more general. At least, one should keep that possibility in mind, for instance while analyzing numerical data for polymers in random media.

Another counter-intuitive effect concerns linear polymers on the GM gaskets. There is no essential singularity in that case, and the critical exponent γ of the generating function is well defined for all scale factors b. When $b \to \infty$, the system becomes more and more similar to an Euclidean triangular lattice, and one would expect the exponents to converge to their $d=2$ values. This is indeed the case for ν, but Elezovic et al. [26] found numerically that the convergence of γ was at best very slow and non-monotonic. Through a deep analysis of the scaling properties of SAW, Dhar was finally able to show [27] that $\lim \gamma(b \to \infty) = 133/32$ is completely different from $\gamma(d=2) = 43/32$.

The collapse transition of self-interacting polymers itself is not always as simple as expected. On the 3-d GM gasket a "topological frustration" effect prevents a liner polymer from filling densely the lattice, but contrarily to the 2-d SG gasket (where no transition occurs at finite T), we discovered a novel "quasi-compact" phase [28], where the fractal dimension of the polymer is slightly less than the lattice D. For lattice trails, i.e. walks which may self-cross at a site but not on a bond, Chang and Shapir [29] found a θ point on the 2-d SG, while none exists for SAW [19], thus suggesting that the two problems belong to different universality classes.

In a similar vein, Maritan pointed out recently [30] that the random walk and the ideal chain have very different asymptotic behaviour on many fractals, contrarily to the intuition built from experience with random walks on percolation clusters, and that in some cases one also finds essential singularities linked to localization effects.

7. Conclusion

In many cases the equations describing polymers on fractal lattices are complex enough to display a rich phase diagram, with several fixed points, and to reveal unexpected features. The main challenge raised by the wealth of exact results so obtained is now to understand in depth when the critical behaviour will be qualitatively similar to the standard one on regular lattices, and when it may be radically different. Also, it would be very useful to develop the connections with real-space renormalization [20,27], in order to achieve a systematic approach applicable to more general systems.

References

[1] B.B. Mandelbrot, Les Objets Fractals: Forme, Hasard et Dimension (Flammarion, Paris, 1975).
[2] Y. Gefen, Y. Meir, B.B. Mandelbrot and A. Aharony, Phys. Rev. Lett. 50 (1983) 145.
[3] Y. Gefen, A. Aharony, Y. Shapir and B.B. Mandelbrot, J. Phys. A 17 (1984) 435.
[4] Y. Gefen, A. Aharony and B.B. Mandelbrot, J. Phys. A 17 (1984) 1277.
[5] G. Bhanot, H. Neuberger and J.A. Shapiro, Phys. Rev. Lett. 53 (1984) 2277.
[6] P.Y. Lai and Y.Y. Goldschmidt, J. Phys. A 20 (1987) 2159.
[7] B. Bonnier, Y. Leroyer and C. Meyers, Phys. Rev. B 37 (1988) 5205.
[8] D. Dhar, J. Math. Phys. 19 (1978) 5.
[9] R. Rammal, G. Toulouse and J. Vannimenus, J. Phys. (Paris) 45 (1984) 389.
[10] R. Rammal and G. Toulouse, J. Phys. (Paris) 44 (1983) L13.
[11] K. Kremer, Z. Phys. B 45 (1981) 148.
[12] B. Derrida, J. Phys. A 15 (1982) L119.
[13] A.B. Harris, Z. Phys. B 49 (1983) 347.
[14] A.K. Roy, B.K. Chakrabarti, J. Phys. A 20 (1987) 215.
[15] J.-P. Bouchaud and A. Georges, Phys. Rev. B 39 (1989) 2846.
[16] A. Aharony and A. Brooks Harris, J. Stat. Phys. 54 (1989) 1091.
[17] J. Vannimenus, J.-P. Nadal and H. Martin, J. Phys. A 17 (1984) L351.
[18] B. Hu, Phys. Rev. B 33 (1986) 6503.
[19] D.J. Klein and W.A. Seitz, J. Phys. Lett. 45 (1984) L241.
[20] D. Dhar and J. Vannimenus, J. Phys. A 20 (1987) 199.
[21] E. Eisenriegler, K. Kremer and K. Binder, J. Chem. Phys. 77 (1982) 6296.
[22] E. Bouchaud and J. Vannimenus, Ecole Normale Supérieure preprint (1989).
[23] M. Knezevic and J. Vannimenus, Phys. Rev. Lett. 56 (1986) 1591; Phys. Rev. B 35 (1987) 4988.
[24] J.A. Given and B.B. Mandelbrot, J. Phys. A 16 (1983) L565.
[25] J. Vannimenus and M. Knezevic, Europhys. Lett. 3 (1987) 21.
[26] S. Elezovic, M. Knezevic and S. Milosevic, J. Phys. A 20 (1987) 1215.
[27] D. Dhar, J. Phys. A 49 (1988) 397.
[28] M. Knezevic and J. Vannimenus, J. Phys. A 20 (1987) L969.
[29] I.S. Chang and Y. Shapir, J. Phys. A 21 (1988) L903.
[30] A. Maritan, University of Padova preprint (1989).

DETERMINISTIC MODELS OF FRACTAL AND MULTIFRACTAL GROWTH

Tamás VICSEK [1]

Department of Physics, Emory University, Atlanta, GA 30322, USA

Four deterministic models are presented in order to get more insight into the geometry and multifractal behavior associated with fractal growth phenomena. These models allow for exact treatment and are used to demonstrate such properties as directed self-affinity and self-similarity and multifractal growth probability and mass distribution. It is pointed out that in some cases the various fractal dimension definitions may lead to inconsistent results.

One of the standard approaches to the understanding of fractal objects observed in nature is the construction of various deterministic fractals using recursion procedures. Although these models are not stochastic by definition (unlike most of the real fractals), they have been successfully used by Mandelbrot [1–3] to describe a wide class of natural structures including examples ranging from microscopic percolating networks to clusters of galaxies.

Deterministic models based on constructing a fractal starting from an initial configuration and substituting its units with the initial configuration itself in a recursive manner represent a useful tool in studying the behavior of systems with fractal geometry [1–4] because they allow for exact treatment of a number of properties. The early examples were designed to describe the physics of inhomogeneous networks and include a model for the backbone at the percolation threshold [5] and a Sierpinski carpet type construction to treat the Ising model in $D = 1 + \epsilon$ dimension [6].

More recently, a number of recursive models have been utilized to characterise fractal growth phenomena [7] as well. The goal of the related calculations is not to build a model which is nearly identical to the structure which is to be described, instead, the models are used to carry out exact calculations. Then the related results can be used to check theoretical predictions and are helpful in many other ways in revealing the complicated behavior of the original process. For example, the scaling of the cluster size distribution in diffusion-limited deposits observed in the simulations holds exactly in a deterministic model [8] for diffusion-limited aggregation (DLA) [9]. A similar construction was used to describe structures with multifractal geometry [10]. As a rather different approach, branching Julia sets were proposed as deterministic structures with a harmonic measure on them analogous to the growth probability distribution of Laplacian patterns [11]. Finally, two very recent deterministic models aimed at the description of two-phase fluid displacement in inhomogeneous media. Aharony et al. [12] used the Mandelbrot–Given fractal [13] to calculate the fractal dimension of viscous fingering patterns on a percolation network. The self-similarity of DLA and viscous fingers was described by Hinrichsen et al. [14] in terms of branch orders within a fractal which is similar to one of the fractals to be described later.

The main goal of this paper is to present a few deterministic recursive models for growing fractals. The growth of objects with fractal geometry [1–4, 7] is a common phenomenon in nature. Typically, the reason for the fractal properties of the interface is the instability of the process due to the presence of a non-local field satisfying the Laplace equation (Laplacian

---

[1] On leave from: Institute for Technical Physics, Budapest, P.O. Box. 76, 1325 Hungary.

Essays in honour of Benoit B. Mandelbrot
Fractals in Physics – A. Aharony and J. Feder (editors)

growth). As a rule, this type of growth leads to branching fractal structures with deep fjords between the branches. The most studied examples include electrodeposition [15,16], crystallization [17,18], dielectric breakdown [19] and viscous fingering [20,21]. Although considerable amount of information has accumulated about these processes, the basic phenomenon still lacks theoretical description based on first principles.

In the process of designing a model for fractal growth one takes into account the already known features of the corresponding real structures. Studies of the diffusion-limited aggregation model of Witten and Sander [8] and the related experimental systems [15–21] have greatly contributed to our knowledge of these properties. The structure of DLA clusters grown on a square lattice has been shown to exhibit a crossover: The overall shape of clusters is initially approximately circular, but for very large sizes it becomes a cross-like pattern [22,23]. As far as concerning the multifractal nature [24–27] of the growth probability distribution of Laplacian patterns numerical [28–30] and experimental [31,32] evidence supports that the growth velocity of the interface can be described in terms of a fractal measure.

Four models will be described in the paper. The first two are based on growing clusters from a seed, while the two remaining ones are constructed by subsequent division of the initial configuration (generator) made of intervals. Each model is used for different purposes which are the following: (i) to demonstrate how one can obtain fractal structures imitating intricate realistic growth patterns, (ii) to give an example of a cluster growing in a self-affine, directed way, (iii) to study the structure and the growth probability distribution of directed, but self-similar objects, and (iv) to describe the multifractal distribution of mass in some fractals.

In general, when constructing a growing fractal cluster, one starts with a particle of linear size a. In the first step ($k=1$) $n-1$ particles are added to the original one so that the linear size of the resulting cluster becomes ra. We call this configuration as generator. Next ($k=2$) each particle in the first stage is substituted by the whole $k=1$ configuration itself. In the kth step the same rule is applied: each particle is replaced by the $k=1$ configuration. In this way the linear size and the number of particles in the structure at the kth stage become r^k and n^k, respectively.

(i) To obtain a *snowflake-like* fractal one can use the generator shown at the upper left corner of fig. 1. It is remarkable how the simplest generator imitating the six arms of a snowflake gives rise to a complicated realistic pattern. The fractal dimension D corresponding to this model can be obtained using the relationship $N(R) \sim R^D$, where $N(R)$ is the number of particles within a region of radius R. Thus, we have

$$D = \ln n / \ln r = \ln 13 / \ln 5 \simeq 1.59 . \quad (1)$$

The fractal shown in fig. 1 is not consistent with one of the characteristic features of Laplacian growth. In growth phenomena the motion of the interface usually starts from a small bounded region or from a hyperplane and correspondingly, the growth takes place in a direction pointing away from the seed con-

Fig. 1. The generator (upper left corner) and the third stage of constructing a snowflake-like fractal.

figuration or from the initial surface. This behavior is obvious from simply looking at the computer-generated or experimental structures [7]. The branches tend to grow outward, and practically there are no branches developing in a direction of the seed. Thus, there exists a *directedness* inherent to Laplacian patterns due to this simple geometrical constraint.

(ii) In some cases the higher chance of growing into a given direction leads to clusters whose linear size diverges with a smaller exponent in the direction perpendicular to that of the preferred growth. As a result the clusters become self-affine instead of self-similar which means that the clusters of very different sizes can be scaled onto each other only by using direction-dependent rescaling factors. Next we consider a recursion process which generates tree-like structures which are *both directed and self-affine*. The construction is demonstrated in fig. 2. Using the generator shown in this figure ($k=1$) produces a structure which can be looked at as a set of trees growing out from a horizontal interval. Interestingly, the question of attributing a unique fractal dimensionality to such clusters has some unclarified aspects. It is clear, that in the asymptotic limit (when $k \to \infty$) the length of the trees L becomes infinitely larger than their width W. Therefore, covering these objects with boxes of size l and determining the box counting dimension D_b from $N(l) \sim (l/L)^{-D_b}$ (where $N(l)$ is the number of boxes needed to cover the cluster) one obtains

$$D_b = 1 . \qquad (2)$$

However, the scaling of the number of particles within a region of radius L (see above) would give

$$D = \ln 7 / \ln 4 \simeq 1.4 . \qquad (3)$$

Not all of the directed structures are self-affine. Analysis of very large off-lattice DLA clusters indicates that they are made of a few major branches with a length to width ratio remaining constant in the asymptotic limit. During the growth of diffusion-limited aggregates the distribution of growth probabilities satisfies multifractal scaling, and in the following we shall describe a deterministic model which is particularly helpful in demonstrating the above properties.

(iii) In the construction described here the units of the previous stage are replaced with an appropriately *rotated and reflected* (mirror image) version of the generating configuration [33]. The model is demonstrated in fig. 3. The first stage ($k=1$) is the generator: a simple branching structure made of three units (intervals of the same length). At the next stages each of the units obtained at the previous steps are replaced by the $k=1$ configuration while simultaneously obeying the following rules: (a) none of the branches should point in a direction below the horizontal, (b) no branches are allowed to overlap or touch each other. These rules can be satisfied uniquely by replacing a unit with either the generating configuration or with its mirror image.

The fractal dimension of the above construction is

$$D = \ln 3 / \ln 2 \simeq 1.585 . \qquad (4)$$

In the present approach the angle θ between the vertical branch and the branch growing out to the right in the generating configuration is an adjustable parameter. Varying the angle θ generates analogous, but differently looking configurations. It is important to

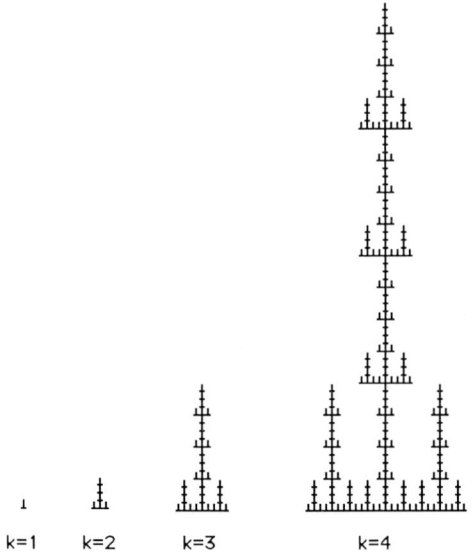

Fig. 2. Four stages of generating a fractal which consists of directed and self-affine trees growing out from a horizontal interval.

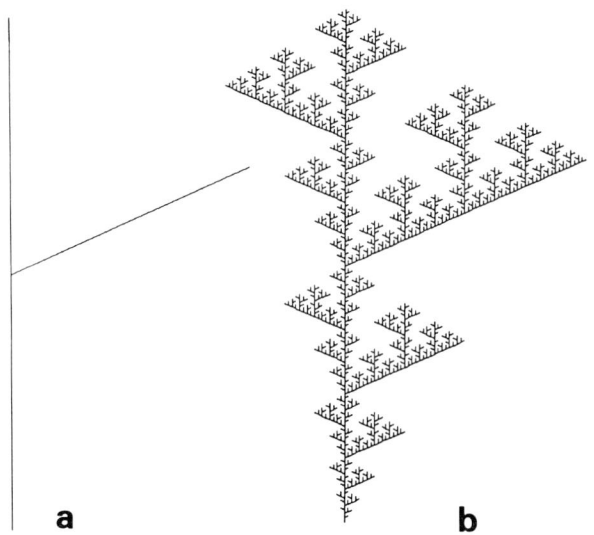

Fig. 3. The generator (a) and the seventh stage (b) of the recursion process resulting in a directed but self-similar fractal.

note that since this fractal is constructed using a recursion procedure which does not result in differently scaling lengths, it is bound to be self-similar.

Let us next analyze the model displayed in fig. 3 from the point of view of its multifractal behaviour. Imagine that the structure is made of a perfect conductor and is electrically charged. Then the electric field (or the density of charge carriers) at the surface is called the harmonic measure corresponding to the given configuration. It is the gradient of a scalar field satisfying the Laplace equation with appropriate boundary conditions, and as such is in complete analogy with the growth probability or interface velocity distribution of Laplacian patterns.

The structure in fig. 3b is made of two kinds of loops being open to a different degree. This observation makes it easier to understand what is the mechanism which produces the multifractal nature of the harmonic measure of DLA clusters. Each time we get deeper into the pattern and enter a loop, the field is decreased (screened) by a given factor λ_1 or λ_2 (depending which type of loop we enter). As a result, the strength of the field is determined by a *multiplicative process* with different weights λ_1 and λ_2 which is well known to lead to fractal measures [24,25].

To make the above picture more quantitative [33] we note that the openings of the consecutive loops of the same type produce two characteristic cones with the corresponding angles φ_1 and φ_2 as demonstrated in fig. 4a. Next we assume that the ragged interface between the fractal and the cone-shaped empty regions can be approximated by a smooth surface like in the probability scaling theory of DLA [34,35]. However, in the present case instead of the convex shape with a tip we treat an incision having the shape of a cone. Nevertheless, we shall use the term tip for the place where the smoothed-out interface has a sharp turn.

The potential along the above described cones is constant and we expect that its gradient (the electric field playing the role of the measure defined on the fractal) goes to zero as some power of the distance from the tip. The situation is analogous to the picture used in the probability scaling approach, however, in our case φ_1 and φ_2 are smaller than $\pi/2$. Thus, we can make use of the known solution of the simplified problem shown in fig. 4b

$$\nabla\phi(r,\varphi) = (C\pi/2\varphi)r^{\pi/\varphi-1}. \tag{5}$$

where ϕ denotes the electric potential, C is a constant and r is the distance from the tip. Let us now assume that the structure is covered by boxes of linear size ϵ. Then the amount of measure in a box consisting of a tip of angle φ is given by

$$m(\epsilon) \sim \epsilon^{\pi/\varphi}. \tag{6}$$

This expression is equivalent to a singularity exponent

$$\alpha = \ln m(\epsilon)/\ln \epsilon = \pi/\varphi. \tag{7}$$

As we saw there are two characteristic angles and the smaller one gives the largest α. In the limit $\varphi \to 0$ we have $\alpha_{\max} \to \infty$. One can find $\alpha_{\max}(\min)$, the smallest value (as a function of θ) of $\alpha_{\max}$ from the condition $\varphi_1 = \varphi_2$. The corresponding equation can easily be solved numerically, and gives $\alpha \simeq 1.005$ and $\beta \simeq 0.305$.

(iv) Finally, let us consider another branching structure which is generated in an asymmetric way [10]. To obtain the fractal displayed in fig. 5 one replaces each of the four intervals with the generator.

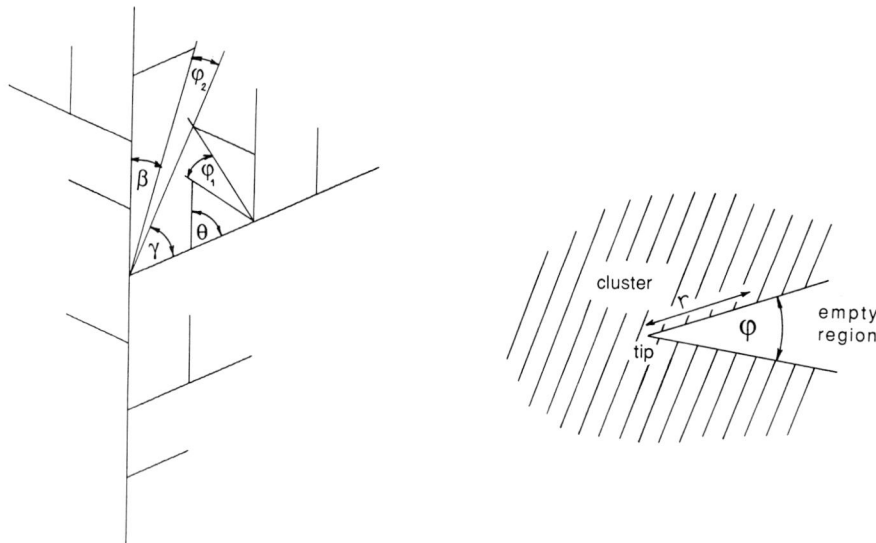

Fig. 4. The angles depicted are used in the text to calculate the largest singularity exponent $\alpha_{\max}$ of the harmonic measure associated with the fractal in fig. 1.

However, this time the lengths of the intervals are not the same, and the generator has to be rotated and scaled accordingly. To simulate growth, the structure is blown up by a trivial factor at each step so that the already existing part does not change its size.

Next we determine the distribution of mass in the above fractal. Suppose that we cover our fractal with boxes of size l. Then the expression

$$\chi_q(l) = \sum_i (M_i/M_0)^q \sim (l/L)^{(q-1)D_q} \qquad (8)$$

defines a non-trivial set of dimensions D_q for each q

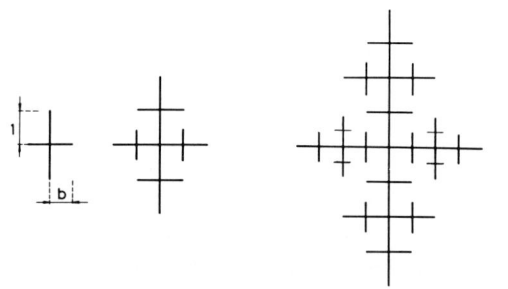

Fig. 5. The recursion process demonstrated here leads to a structure with multifractal mass distribution (see text).

[10]. Here M_i is the mass (the total length of the intervals) within the ith box, M_0 is the total mass of the structure and L is its linear size. Now we note that at each stage the structure is made of four parts and the two smaller parts are copies of the two larger ones (their mass is b times smaller). Thus, because of self-similarity

$$\chi_q(l) = 2[(1+b)]^{-q}\chi_q(l/2) \\ + 2[b(1+b)]^{-q}\chi_q(bl/2) . \qquad (9)$$

Using eq. (8) we get

$$(\tfrac{1}{2})^{(q-1)D_q} + (\tfrac{1}{2}b)^{(q-1)D_q} = \tfrac{1}{2}(1+b)^q . \qquad (10)$$

The main message of the above expression is that the mass distribution within this fractal can be related to a set of dimensions described by an equation which has different solutions depending on the actual values of q and b. The usual fractal dimension corresponds to $q=0$. It can be obtained from the equation

$$(\tfrac{1}{2})^D + (\tfrac{1}{2}b)^D = \tfrac{1}{2} , \qquad (11)$$

which can be solved numerically for any fixed b. For

some of the b values the above implicit equation can be inverted, e.g., if $b=1/2$, $D=1-\ln(\sqrt{3}-1)/\ln 2 \simeq 1.45$. Again, if one defines the fractal dimension through $M(L) \sim L^D$ the corresponding result

$$D = 1 + \ln(1+b)/\ln 2 \qquad (12)$$

is not consistent with (11).

In conclusion, the various deterministic models are useful in getting an insight into the complex geometrical structure and multifractal properties associated with fractal growth phenomena. In some cases the approach based on generating deterministically growing clusters raises interesting questions about the equivalence of various definitions for the fractal dimension.

I am grateful to Benoit B. Mandelbrot for his many valuable suggestions during my visit to the Mathematics Department of Yale University where we were developing together one of the models described in this paper. I would also like to thank Tamás Tél for collaboration and Fereydoon Family and János Kertész for helpful discussions.

References

[1] B.B. Mandelbrot, Les Objets Fractals: Form, Hasard et Dimension (Flammarion, Paris, 1975).
[2] B.B. Mandelbrot, The Fractal Geometry of Nature (Freeman, San Francisco, 1982).
[3] B.B. Mandelbrot, Physica Scripta 32 (1985) 257.
[4] J. Feder, Fractals (Plenum, New York, 1988).
[5] Y. Gefen, A. Aharony, B.B. Mandelbrot and S. Kirkpatrick, Phys. Rev. Lett. 47 (1981) 1771.
[6] Y. Gefen, Y. Meir, B.B. Mandelbrot and A. Aharony, Phys. Rev. Lett. 50 (1983) 145.
[7] T. Vicsek, Fractal Growth Phenomena (World Scientific, Singapore, 1989).
[8] T. Vicsek, J. Phys. A 16 (1983) L647.
[9] T.A. Witten and L.M. Sander, Phys. Rev. Lett. 47 (1981) 1400.
[10] T. Tél and T. Vicsek, J. Phys. A 20 (1987) L835.
[11] I. Procaccia and R. Zeitak, Phys. Rev. Lett. 60 (1988) 2511.
[12] A. Aharony, U. Oxaal, M. Murat, Y. Meir, F. Boger, J. Feder and T. Jossang, in: Random Fluctuations and Pattern Growth, H.E. Stanley and N. Ostrowsky, eds. (Kluwer, Dordrecht, 1988) p. 83.
[13] B.B. Mandelbrot and J. Given, Phys. Rev. Lett. 52 (1984) 1853.
[14] E.L. Hinrichsen, K.J. Maloy, J. Feder and T. Jossang, J. Phys. A 22 (1989) L271.
[15] R.M. Brady and R.C. Ball, Nature 309 (1984) 225.
[16] M. Matsushita, M. Sano, Y. Hayakawa, H. Honjo and Y. Sawada, Phys. Rev. Lett. 53 (1984) 286.
[17] H. Honjo, S. Ohta and M. Matsushita, J. Phys. Soc. Japan 55 (1986) 2487.
[18] Gy. Radnóczy, T. Vicsek, L.M. Sander and D. Grier, Phys. Rev. A 35 (1987) 4012.
[19] L. Niemeyer, L. Pietronero and H.J. Wiesmann, Phys. Rev. Lett. 52 (1984) 1033.
[20] G. Daccord, J. Nittmann and H.E. Stanley, Phys. Rev. Lett. 56 (1986) 336.
[21] K.J. Maloy, J. Feder and J. Jossang, Phys. Rev. Lett. 55 (1985) 2681.
[22] P. Meakin and T. Vicsek, Phys. Rev. A 32 (1985) 685.
[23] P. Meakin, R. Ball, P. Ramanlal and L.M. Sander, Phys. Rev. A 35 (1987) 5233.
[24] B.B. Mandelbrot, J. Fluid Mech. 62 (1974) 331.
[25] U. Frisch and G. Parisi, in: Turbulence and Predictability in Geophysical Fluid Dynamics and Climate Dynamics, M. Ghil, R. Benzi and G. Parisi, eds. (North-Holland, Amsterdam, 1985).
[26] T.C. Halsey, M.H. Jensen, L.P. Kadanoff, I. Procaccia and B. Shraiman, Phys. Rev. A 33 (1986) 1141.
[27] B.B. Mandelbrot, in: Random Fluctuations and Pattern Growth, H.E. Stanley and N. Ostrowsky, eds. (Kluwer, Dordrecht, 1988) p. 333.
[28] T.C. Halsey, P. Meakin and I. Procaccia, Phys. Rev. Lett. 56 (1986) 854.
[29] C. Amitrano, A. Coniglio and F. di Liberto, Phys. Rev. Lett. 57 (1986) 1016.
[30] Y. Hayakawa, S. Sato and M. Matsushita, Phys. Rev. A 36 (1987) 1963.
[31] J. Nittmann, H.E. Stanley, E. Torboul and G. Daccord, Phys. Rev. Lett. 58 (1987) 619.
[32] S. Ohta and H. Honjo, Phys. Rev. Lett. 60 (1988) 611.
[33] B.B. Mandelbrot and T. Vicsek, J. Phys., to be published.
[34] L.A. Turkevich and H. Scher, Phys. Rev. Lett. 55 (1985) 1026.
[35] R.C. Ball, R.M. Brady, G. Rossi and B.R. Thomson, Phys. Rev. Lett. 55 (1985) 1046.

RANDOM FRACTALS:
SELF-AFFINITY IN NOISE, MUSIC, MOUNTAINS, AND CLOUDS

Richard F. VOSS

IBM Thomas J. Watson Research Center, Yorktown Heights, NY 10598, USA

Many of nature's seemingly complex shapes can be effectively characterized and modeled as random fractals based on generalizations of fractional Brownian motion, fBm. As a function of one dimension, t, the trace $V_H(t)$ provides a model for the "$1/f$" noises. Extending fBm's to higher dimensions gives $V_H(x, y)$ as landscapes and $V_H(x, y, z)$ as clouds. Although all such fBm's are statistically self-affine, as characterized by the parameter H or the spectral density exponent β, either zerosets or trails of independent fBm's are statistically self-similar and may be represented by the fractal dimension D.

1. Random fractals: an introduction

Geometry, the mathematical language for the description and manipulation of shapes, is often viewed as dry, dull, and synonymous with the simple triangles, circles, and cones of classical Euclidean geometry. Such basic shapes, moreover, offer a poor vocabulary for nature's seemingly complex forms. Benoit Mandelbrot was the first to both recognize the obvious problem, "mountains are not cones, clouds are not spheres", and to provide the answer in terms of a new mathematical language, the fractal geometry of nature [1]. Since his invention of the name *fractal* in 1975, Mandelbrot's fractal geometry has revolutionized the application of non-Euclidean geometric constructs to the natural sciences. The imagery and language of fractals has also stimulated renewed interest in science, computation, and mathematics by visual designers, artists, movie makers and composers.

Fractals provide a framework for the characterization and modeling of the irregular, seemingly complex shapes found in nature. Some of the building blocks of fractal geometry originated in the deterministic, exactly self-similar mathematical "monsters" (such as the Koch curve and Sierpiński gasket) of the early 1900's. Although such constructs serve to build intuition and a vocabulary for scaling shapes, most of the fractals found in nature possess a statistical rather than exact self-similarity and self-affinity. The following sections present an expository summary [2] of the characterization and modeling of random fractals based on generalizations of *fractional Brownian motion* (fBm).

2. Traces of fractional Brownian motion

Mandelbrot and Walls originally advanced [1,3] the concept of *fractional Brownian motion* as a model for noise or random processes in time. It is an extension of the central concept of *Brownian motion* that has played an important role in both physics and mathematics. Sample *traces* of fBm are shown in fig. 1. The term *trace* evokes views of oscilloscope traces of electrical noise.

A fractional Brownian motion, $V_H(t)$, is a single-valued function of one variable, t (usually time). Its increments $\Delta V_H(\Delta t) = V_H(t_2) - V_H(t_1)$ have a Gaussian distribution with variance

$$\langle \Delta V_H^2(\Delta t) \rangle \propto \Delta t^{2H}, \tag{1}$$

where $\Delta t = |t_2 - t_1|$ and the angular brackets denote averages over many samples of $V_H(t)$. The parameter H has a value $0 < H < 1$. Such a function is stationary. Its mean-square increments depend only on the

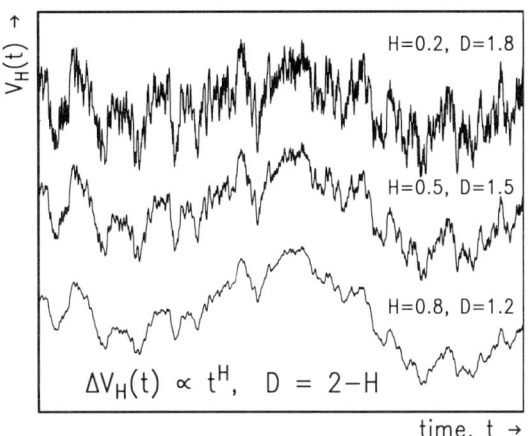

Fig. 1. Sample traces of fBm $V_H(t)$ versus t with different H.

time difference Δt and all t's are statistically equivalent. The value $H=1/2$ gives the familiar Brownian motion with $\Delta V^2 \propto \Delta t$.

Although $V_H(t)$ is continuous, it is nowhere differentiable. Nevertheless, many constructs have been developed to give meaning to "derivative of fractional Brownian motion" as *fractional Gaussian noises* [1,3]. Such constructs are usually based on averages of $V_H(t)$ over decreasing scales. The derivative of normal Brownian motion, $H=1/2$, corresponds to the uncorrelated *Gaussian white noise* of fig. 3, below, and Brownian motion is said to have *independent increments*. For $H>1/2$ there is a positive correlation for the increments of $V_H(t)$. For $H<1/2$ the increments are negatively correlated. Such correlations extend to arbitrarily long time scales.

$V_H(t)$ shows a statistical scaling behavior. If the time scale Δt is changed by the factor r, then the increments ΔV_H change by a factor r^H,

$$\langle \Delta V_H^2(r\,\Delta t) \rangle \propto r^{2H} \langle \Delta V_H^2(\Delta t) \rangle . \qquad (2)$$

Unlike the more familiar statistically self-similar curves, a $V_H(t)$ trace requires *different* scaling factors in the two coordinates (r for t but r^H for V_H) reflecting the special status of the t coordinate. Such non-uniform scaling is known as *self-affinity* rather than self-similarity.

3. Trails of fractional Brownian motion

Consider a particle or animal undergoing a fractional Brownian motion or random walk in which each coordinate is tracing out an independent fBm in time. Although each independent trace of position versus time is self-affine, the corresponding *trail* in space (the set of all points visited) is self-similar.

This duality is illustrated in fig. 2. Fig. 2a shows a sample trail resulting from two independent fBm's, $Y(t)$ versus $X(t)$, characterized by $H=0.6$. A magnification of a low-resolution segment by a factor of 10 yields a high-resolution view that is statistically self-similar. Fig. 2b shows a trace of one of the coordinates, $X(t)$ versus t. Here a magnification of a small portion by a factor of 10 shows the characteristic self-affine scaling. The effective local slope appears to increase and the trace seems to fill more of the plane.

4. Self-similar versus self-affine fractals

The distinction between similarity and affinity illustrated in fig. 2 is important. By way of summary

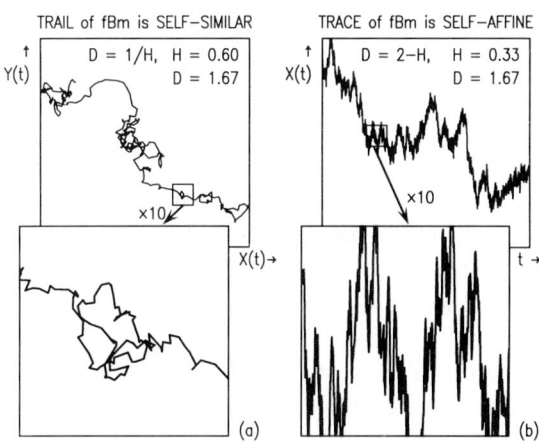

Fig. 2. Comparison of a self-similar trail of fBm $Y(t)$ versus $X(t)$, with the self-affine trace of $X(t)$ versus t.

[1,4], a *self-similar* object is composed of N copies of itself (with possible translations and rotations) each of which is scaled down by the ratio r in all E coordinates from the whole. More formally, consider a set S of points at positions $x = (x_1, ..., x_E)$ in Euclidean space of dimension E. Under a *similarity* transform with real scaling ratio $0 < r < 1$, the set S becomes rS with points at $rx = (rx_1, ..., rx_E)$. A bounded set S is *self-similar* when S is the union of N distinct (non-overlapping) subsets each of which is congruent to rS (identical under translations and rotations). The set S is *statistically self-similar* if it is composed of N distinct subsets each of which is scaled down by the ratio r from the original and is identical in all statistical respects to rS. In either case, the fractal or *similarity dimension* of S is given by

$$1 = Nr^D \quad \text{or} \quad D = \log(N)/\log(1/r) . \qquad (3)$$

This relation leads to several important methods of estimating D for a given set S.

For topologically one-dimensional fractal "curves", with $D \geq 1$ the apparent "length", varies with the measuring ruler size L and the number of steps of size L, $N(L)$, as

$$\langle \text{LENGTH} \rangle \propto L \times N(L) \propto 1/L^{D-1} . \qquad (4)$$

D also characterizes the covering of the set S by E-dimensional "boxes" of linear size L. If the entire S is contained within one box of size $L_{\max}$, then each of the $N = 1/r^D$ subsets will fall within one box of size $L = rL_{\max}$. Thus, the average number of boxes of size L, $N_{\text{box}}(L)$, needed to cover S is given by

$$\langle N_{\text{box}}(L) \rangle = (L_{\max}/L)^D \propto 1/L^D . \qquad (5)$$

This *box dimension* can be conveniently estimated by dividing the E-dimensional Euclidean space containing the set into a grid of boxes of size L^E and counting the number of such boxes $N_{\text{box}}(L)$ that are non-empty.

One can also estimate the average "volume" or "mass" of the set S within a distance L about a given point in S, as

$$\langle M(L) \rangle \propto L^D . \qquad (6)$$

For the trail of fBm in fig. 2a over an interval Δt each coordinate will vary by typically $L = \Delta t^H$. If overlap can be neglected, the "mass" of the trail $M \propto \Delta t \propto L^{1/H}$. In comparison with eq. (6), the trail of fBm has a fractal dimension

$$D = 1/H \quad \text{for a trail of } V_H(t) , \qquad (7)$$

provided $1/H < E$. Normal Brownian motion with $H = 0.5$ has $D = 2$. $1/H$ is known as the *latent* fractal dimension [1,4] of a *trail* of fBm. When $1/H > E$ overlap cannot be neglected and the actual $D = E$.

5. Relation of D to H for self-affine traces of fBm

Under an *affine* transform each of the E coordinates of x may be scaled by a different ratio $(r_1, ..., r_E)$. Thus, the set S is transformed to $r(S)$ with points at $r(x) = (r_1 x_1, ..., r_E x_E)$. A bounded set S is *self-affine* when S is the union of N distinct (non-overlapping) subsets each of which is congruent to $r(S)$. Similarly, S is *statistically self-affine* when S is the union of N distinct subsets each of which is congruent *in distribution* to $r(S)$. The fractal dimension D, however, is not as easily defined as with self-similarity [4].

Consider a trace of $V_H(t)$ covering a time span $\Delta t = 1$ and a vertical range $\Delta V_H = 1$. $V_H(t)$ is statistically self-affine when t is scaled by r and V_H is scaled by r^H. Suppose the time span is divided into N equal intervals each with $\Delta t = 1/N$. Each of these intervals will contain one portion of $V_H(t)$ with vertical range $\Delta V_H = \Delta t^H = 1/N^H$. Since $0 < H < 1$ each of these new sections will exhibit the increasing local slope of fig. 2b and the occupied portion of each interval will be covered by $\Delta V_H / \Delta t = (1/N^H)/(1/N) = N/N^H$ square boxes of linear scale $L = 1/N$. In terms of box dimension, as t is scaled down by a ratio $r = 1/N$ the number of covering boxes increases from 1 to $N(L) = $ number of intervals $\times$ boxes per interval $= N \times N/N^H = N^{2-H} = 1/L^{2-H}$. Thus, by comparison with eq. (5), the box dimension

$$D = 2 - H \quad \text{for a trace of } V_H(t) . \qquad (8)$$

It is important to note that the association of a similarity dimension D with a self-affine fractal such as fBm is implicitly fixing a scaling between the (otherwise independent) coordinates. Moreover, unlike statistical self-similarity, different methods of estimating the fractal dimension may give different answers in different limits relative to this (artificially introduced) characteristic length. The difference is particularly clear when one attempts to estimate D for a trace of fBm from eq. (4). As above, one can divide the t axis into N segments of size $\Delta t = 1/N$. For each segment $\Delta V \simeq \Delta t^H$, and the length L along each segment is given by $L^2 = \Delta t^2 + \Delta V^2 \simeq \Delta t^2 + \Delta t^{2H}$. On small scales as $L \to 0$, $\Delta t \ll 1$, $L \propto \delta t^H$ so $N \propto 1/L^{1/H}$ and LENGTH $\propto 1/L^{1-1/H}$ giving $D = 1/H$. On large scales with $\Delta t \gg 1$, $L \propto \Delta t \propto 1/N$, LENGTH is independent of Δt, and $D = 1$. Thus, different self-similarity based methods of estimating D, can give $D = 2 - H$, $D = 1/H$, or $D = 1$ for the same fBm trace.

The *zeroset* of fBm is the intersection of the trace of $V_H(t)$ with the t axis, the set of all points such that $V_H(t) = 0$. The zeroset is a disconnected set of points with topological dimension zero and a fractal dimension $D_0 = D - 1 = 1 - H$ that is less than 1 but greater than 0. Although the trace of $V_H(t)$ is self-affine, its zeroset is self-similar and its fractal dimension is well defined. Thus, the zeroset is the least ambiguous method of characterizing the fractal dimension for fBm as $D = D_0 + 1 = 2 - H$ in agreement with eq. (8).

6. Spectral density exponent β for fBm

Random functions in time $V(t)$ are often characterized [5] by their *spectral densities* $S_V(f)$ as shown in fig. 3. If $V(t)$ is the input to a narrow bandpass filter at frequency f and bandwidth Δf, then $S_V(f)$ is the mean-square output $V(f)$ divided by Δf,

$$S_V(f) = |V(f)|^2 / \Delta f.$$

$S_V(f)$ gives information about the time correlations of $V(t)$. When $S_V(f)$ increases steeply at low f, $V(t)$ varies more slowly. $S_V(f)$ is often estimated from Fourier transforms of sample traces of $V(t)$.

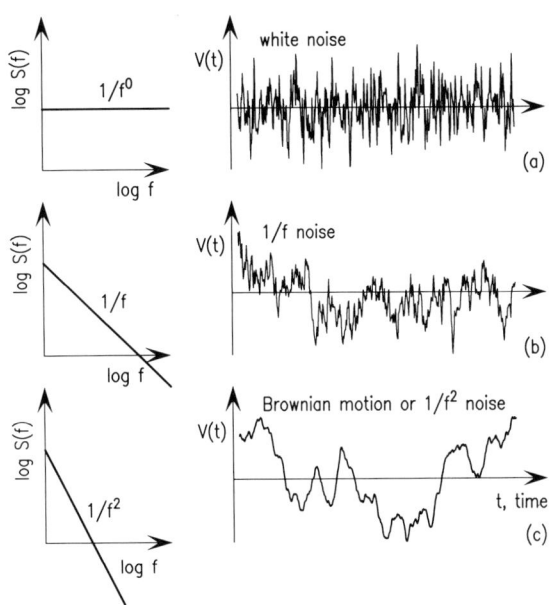

Fig. 3. Typical noises and their spectral densities $S_V(f)$.

The 2-*point autocorrelation function*,

$$G_V(\tau) = \langle V(t) V(t+\tau) \rangle - \langle V(t) \rangle^2,$$

provides a measure of how the fluctuations at two times separated by τ are related. $G_V(\tau)$ and $S_V(f)$ are not independent. In many cases they are related by the Wiener–Khintchine relation [6,7],

$$G_V(f) = \int S_V(f) \cos(2\pi f \tau) \, df.$$

For the Gaussian white noise of fig. 3a, $S_V(f) = $ constant and $G_V(\tau) = \Delta V^2 \delta(\tau)$ is completely uncorrelated. For simple power laws where $S_V(f) \propto 1/f^\beta$ with $1 < \beta < 3$, $G_V(\tau)$ is directly related to the mean-square increments of fBm and

$$\beta = 2H + 1. \tag{9}$$

This result, together with eq. (8) provides an extremely useful connection between D, H and β for finite simulations.

7. Noise as fBm in one dimension

Most experimental observations of fluctuations in

time resemble one of the three typical noise samples in fig. 3. When $1 < \beta < 3$ the noise samples correspond directly to traces of fBm and they may be characterized either by β, H, or $D = 2 - H$. However, when $-1 < \beta < 1$, the latent D exceeds 2 and the noise samples are more conveniently treated as the *increments*

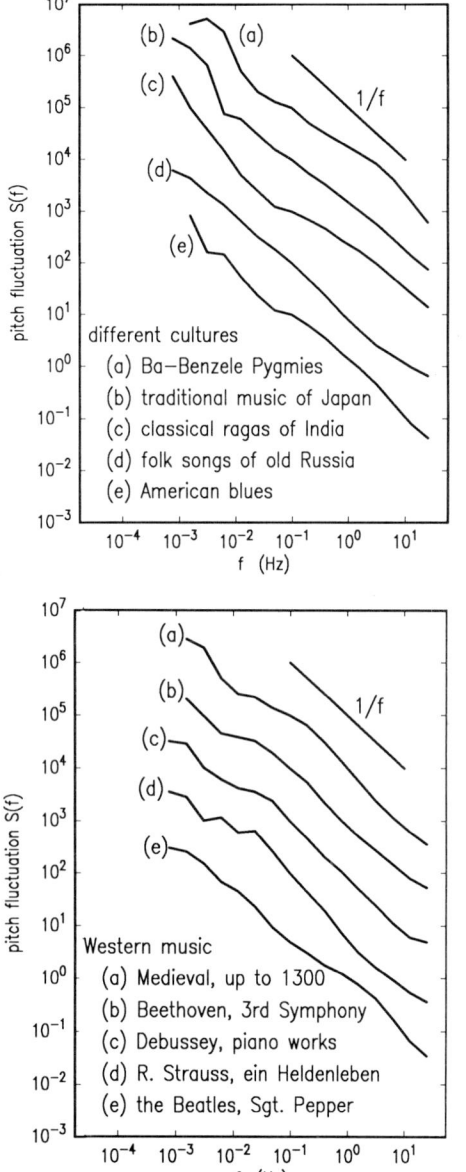

Fig. 4. Melody fluctuation spectral densities.

of fractional Brownian motion. It is useful to consider integration and an appropriate definition of "derivative" as extending the range of H. Thus, integration of a fBm produces a new function with H increased by 1, while "differentiation" reduces H by 1. When $H \to 1$, *the derivative of fBm looks like a fBm with $H \to 0$*. If $V(t)$ has $S_V(f) \propto 1/f^\beta$ then its derivative dV/dt has spectral density $f^2/f^\beta = 1/f^{\beta-2}$. Differentiation decreases β by 2 and decreases H by 1.

The case of $1/f$ noise as in fig. 3b is special. It is at the boundary between fBm's with $H \to 0$ and the increments of fBm with $H \to 1$. Samples of $1/f$ noise look statistically similar on all time scales. According to eq. (2) with $H = 0$, changing the time base of an oscilloscope displaying $1/f$ noise, requires no change in vertical gain. $1/f$ noise, moreover, represents the most common $S_V(f)$ observed for natural systems [6]. It

Fig. 5. Samples of stochastically composed fractal music based on the different types of noises shown in fig. 3. (a) "White" music is too random; (b) "$1/f$" music is the closest to actual music (and most pleasing) and (c) "brown" or $1/f^2$ music is too correlated.

is found in almost all electronic components from simple carbon resistors to vacuum tubes and all semiconducting devices; in all time standards from the most accurate atomic clocks and quartz oscillators to the ancient hourglass; in ocean flows and the changes in yearly flood levels of the river Nile as recorded by the ancient Egyptians; in the small voltages measurable across nerve membranes due to sodium and potassium flow; and even in the flow of automobiles on an expressway. $1/f$ noise is also found in music.

8. Music as $1/f$ noise

One of my most exciting discoveries was that almost all musical melodies mimic $1/f$ noise [7]. Music has the same blend of randomness and predictability that is found in $1/f$ noise. The melody rises and falls with the same time correlations as the $1/f$ noise of fig. 3b. Fig. 4 shows some of the measured melody spectral densities for different types of music. This type of analysis is surprisingly insensitive to the widely different types of music. With the exception of very modern composers, like Stockhausen, Jolas, and Carter (where the melody fluctuations approach white noise at low frequencies), all types of music share this $1/f$ noise base. Such a view of melody fluctuations emphasizes the common element in music and suggests an answer to a question that has long troubled philosophers: "what does music imitate?" The measurements suggest that music is imitating the characteristic way our world changes in time.

It is, of course, possible to use simulations of fBm's as $1/f^\beta$ noises for stochastic music composition. Fig. 5 shows samples of scores generated from the three characteristic *noises* of fig. 3. Although none of these samples corresponds to a sophisticated composition of a specific type of music, that generated from $1/f$ noise is the closest to real music. Such $1/f$ compositions sound recognizably musical, but from a foreign or unknown culture.

9. Self-affine fBm in higher dimensions: Mandelbrot landscapes and clouds

The traces of fBm, particularly fig. 1 with $H=0.8$, bear a striking resemblance to a mountainous horizon. The modeling of the irregular earth's surface as a generalization of traces of fBm was first proposed by Mandelbrot [1]. The single variable t can be replaced by coordinates x and y in the plane to give $V_H(x, y)$ as the surface altitude at position x, y as shown in fig. 6. In this case, the altitude variations of a hiker following any straight line path at constant speed in the xy plane is a fBm. The increments, $\langle \Delta V_H^2(\Delta r) \rangle \propto \Delta r^{2H}$, where $\Delta r^2 = \Delta x^2 + \Delta y^2$ in analogy with eq. (1), and the box dimension

$$D = 3 - H \quad \text{for a landscape } V_H(x, y). \tag{10}$$

Samples of such landscapes with different D and H are shown in fig. 6. The self-affine nature becomes apparent as one approaches (magnifies) these landscapes. As with fig. 2b, the slopes seem steeper to a mountain hiker than to a distant airplane pilot.

The intersection of a vertical plane with the surface $V_H(x, y)$ is a self-affine fBm trace with $D=2-H$, smaller by 1 than the value of eq. (10). The zeroset of $V_H(x, y)$, its intersection with a horizontal plane, also has a fractal dimension $D_0 = 2 - H$. This special intersection, which produces a family of (possibly disconnected) curves as the coastlines of the landscape, is, however, self-similar. Fig. 7 shows successive magnifications of the self-similar coastline of a self-affine landscape.

The extension of fBm can continue to still higher dimensions to produce, for example, a self-affine fractal temperature or density distribution $V_H(x, y, z)$, shown as the cloud in fig. 8. Here, the variations of an observer moving at constant speed along any straight line path in space generate a fBm trace and the box dimension

$$D = 4 - H \quad \text{for a cloud } V_H(x, y, z). \tag{11}$$

The zeroset $V_H(x, y, z) = \text{constant}$ now gives a self-similar fractal with $D_0 = 3 - H$.

Fig. 6. Changing the landscape D from 2.1 to 2.5 to 2.8.

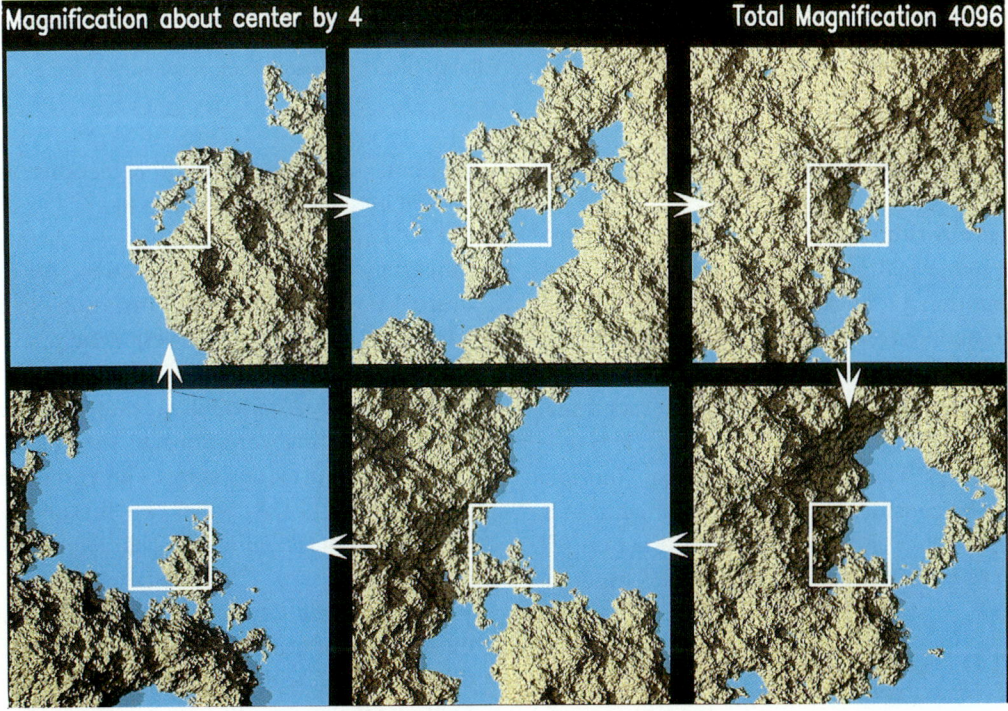
Fig. 7. The self-similar coastline of a self-affine landscape.

Fig. 8. Light scattered from a fractal cloud $V_H(x, y, z)$ in front of a fractal planet generated from random cuts and a portion of the Mandelbrot set boundary.

10. Algorithms for fBm simulation

Numerous algorithms have been developed to provide finite numerical approximations to fractional Brownian motion [1,2]. Many offer a trade off between computational efficiency and accuracy in terms of stationarity or isotropy. Most use the general principle of adding appropriate perturbations at each scale of interest.

Random cuts. The normal Brownian motion $V_{1/2}(t)$ may be considered either as the integral of a white noise $W(t)$ or as the sum of step-function responses $\Theta(t)$ to uncorrelated impulses of random amplitudes A_i,

$$V_{1/2}(t) = \int W(t) \, dt = \sum A_i \Theta(t-t_i).$$

Thus, a surface becomes the sum of independent step-function faults in random directions as demonstrated by the fractal planet in fig. 8.

FFT filtering. White noise traces, surfaces, or volumes may be numerically filtered to give directly the desired power law $1/f^\beta$ for variations in any direction. Figs. 1, 2, 6, and 8 were generated with FFT filtering.

Midpoint displacement. A sample is doubled in stages by adding variations to the calculated midpoints of the previous stage. The variations scale to agree with eq. (1). Although fast, this produces a non-stationary landscape with a characteristic crumpled paper artifact.

Successive random additions. This method reduces the artifacts of midpoint displacement by expanding the sample and adding scaling variations to *all* points (not just the midpoints) at each stage. It was used to generate fig. 7.

Weirstrass–Mandelbrot random function. Whereas the Fourier transform sums a linear progression of frequencies, the Weirstrass–Mandelbrot function sums a geometric progression of ratio r,

$$\sum A_n r^{nH} \sin(2\pi r^{-n} t + \phi_n) ,$$

where the A_n are random amplitudes and the ϕ_n are random phases.

References

[1] B.B. Mandelbrot, The Fractal Geometry of Nature (Freeman, San Francisco, 1982), and references therein; Fractals: Form, Chance, and Dimension (Freeman, San Francisco, 1977).

[2] R.F. Voss, in: The Science of Fractal Images, H.-O. Peitgen and D. Saupe, eds. (Springer, Berlin, 1988); in: Fundamental Algorithms for Computer Graphics, R.A. Earnshaw, ed., NATO ASI Ser. F Vol. 17 (Springer, Berlin, 1985), pp. 805–835; Physica Scripta T 13 (1986) 27–32.

[3] B.B. Mandelbrot and J.R. Wallis, SIAM Rev. 10 (1968) 422–437.

[4] B.B. Mandelbrot, J. Stat. Phys. 34 (1984) 895–930; Physica Scripta 32 (1985) 257–260.

[5] F. Reif, Statistical and Thermal Physics (McGraw-Hill, New York, 1965) ch. 15;
J.J. Freeman, Principles of Noise (Wiley, New York, 1958) ch 1;
F.N.H. Robinson, Noise and Fluctuations (Clarendon Press, Oxford, 1974).

[6] R.F. Voss, Proceedings of the 32rd Annual Symposium on Frequency Control, Atlantic City (1979) pp. 40–46, and references therein.

[7] R.F. Voss and J. Clarke, J. Accous. Soc. Am. 63 (1978) 258–263; Nature 258 (1975) 317–318;
M. Gardner, Mathematical Games (column), Sci. American (April 1978) 16.

LAGRANGIAN CHAOS AND SMALL SCALE STRUCTURE OF PASSIVE SCALARS

Angelo VULPIANI [1]

Dipartimento di Fisica, Università de L'Aquila, Piazza dell'Annunziata 1, 67100 L'Aquila, Italy

We revise the classical theory of Batchelor, which gives a k^{-1} law for the power spectrum of a passive scalar at wavenumbers k, for which the molecular diffusion is unimportant and much smaller than the fluid viscosity. Using some ideas borrowed from the theory of dynamical systems, we show that this power law is related to the chaotic motion of marker particles (Lagrangian chaos) and to the incompressibility constraint. Moreover our approach permits showing that the k^{-1} regime is present in fluids which are not turbulent and it is valid for all dimensionalities $d \geq 2$.

1. Introduction

Mixing and transport properties of fluids present a quite general problem of great theoretical and experimental interest [1]. In common situations this problem can be reduced to the study of the behavior of a passive scalar Θ whose evolution equation is given by:

$$\partial_t \Theta + (\boldsymbol{u} \cdot \nabla) \Theta = D \Delta \Theta, \tag{1}$$

where D is the diffusion coefficient and $\boldsymbol{u}$ is the (Eulerian) velocity field of the fluid.

Eq. (1) is related to the motion of marker particles in the fluid, therefore an obvious conjecture is that Lagrangian chaoticity has a predominant role in determining the statistical properties of the convection in fluids. Since a chaotic Lagrangian behavior typically arises even in regular velocity fields [2-5], we expect that the Θ-behavior on small scales is qualitatively the same either for turbulent or laminar fluids, in presence of Lagrangian chaoticity.

The fluctuations of the passive scalar field can be characterized by the correlation function

$$C(r) = \langle \Theta(x) \Theta(x+r) \rangle, \tag{2}$$

where $\langle ... \rangle$ means spatial average. In three-dimensional turbulent fluids, Batchelor [6] has found that for large Prandtl number ν/D (ν is the kinematic viscosity) the power spectrum

$$\Gamma(k) = \int_{|\boldsymbol{k}|=k} d\boldsymbol{k} \int C(r) e^{-i\boldsymbol{k}\cdot\boldsymbol{r}} d\boldsymbol{r} \tag{3}$$

obeys the scaling law $\Gamma(k) \propto k^{-1}$ in the viscous convective subrange (where the molecular diffusion of the passive scalar is still negligible, but the velocity field behavior is dominated by viscous effects).

We show that the k^{-1} regime is a property which follows directly from Lagrangian chaoticity, and does not depend on the dimensionality d, for $d \geq 2$. Turbulent fluids and very simple flows (for instance periodic in time) exhibit this same property.

2. A dynamical system approach

One can see, by dimensional analysis, that the molecular diffusion becomes dominant only for scales smaller than the Batchelor length $l_B \propto (\nu D^2)^{1/4}$. In the following we assume that our system has a high Prandtl number, so that $l_B \ll l_K$ (where $l_K \propto \nu^{3/4}$ is the Kolmogorov length). In the interval $[l_B, l_K]$ (viscous convective subrange), the passive scalar fluctuations are expected to exhibit universal features. The vis-

---

[1] Also at INFN Sez. di Roma and GNSM–CISM Unità di Roma.

cous convective subrange does exist even in absence of an inertial range, i.e. at moderate Reynolds numbers.

It is easy to realize that neglecting the diffusion term, the behavior of eq. (1) is determined only by the features of the Lagrangian motion of marker particles in the fluid, i.e. by the equation

$$d\mathbf{x}/dt = \mathbf{u}(\mathbf{x}, t) . \qquad (4)$$

Thus we can write the evolution equation for Θ as

$$\Theta(\mathbf{x}, t) = \Theta_0(S^{-t}\mathbf{x}) , \qquad (5)$$

where $\Theta_0(\mathbf{x}) = \Theta(\mathbf{x}, 0)$, and S^t is the evolution operator defined by eq. (4), i.e. $\mathbf{x}(t) = S^t \mathbf{x}(0)$.

Let us consider a smooth initial condition $\Theta_0(\mathbf{x})$ and $N \gg 1$ particles, moving according to eq. (4). The ith particle is initially in $\mathbf{x}^{(i)}(0)$ and transports its own $\Theta^{(i)} = \Theta_0(\mathbf{x}^{(i)}(0))$. At time t, one has

$$\langle |\Theta(\mathbf{x}+\mathbf{r}) - \Theta(\mathbf{x})|^2 \rangle$$
$$\approx N^{-2} \sum_{i,j} (\Theta^{(i)} - \Theta^{(j)})^2 P_r(\mathbf{x}^{(i)}, \mathbf{x}^{(j)}) ,$$

where $P_r(\mathbf{x}^{(i)}, \mathbf{x}^{(j)})$ is the probability density to have $\mathbf{x}^{(i)} - \mathbf{x}^{(j)} = \mathbf{r}$. In presence of Lagrangian chaoticity, two particles with large $\Theta^{(i)} - \Theta^{(j)}$ (and therefore distant at the initial time) can approach up to a distance $\mathcal{O}(r)$ after a time $t \sim \lambda^{-1} |\ln r|$, where λ is the maximal Lyapunov exponent of eq. (4). In other terms, $\lambda^{-1} |\ln r|$ is the typical time necessary to get a "good mixing" on scale r: the $\Theta^{(i)}$ of the particles in a box of edge r, after this time, are distributed in the whole range of admissible values. $P_r(\mathbf{x}^{(i)}, \mathbf{x}^{(j)})$ is therefore approximately constant and

$$\langle |\Theta(\mathbf{x}+\mathbf{r}) - \Theta(\mathbf{x})|^2 \rangle \sim r^\zeta, \quad \text{with } \zeta = 0 . \qquad (6)$$

It follows that

$$\Gamma(k) \propto k^{-\alpha} \quad \text{with } \alpha \leq 1 .$$

The conservation law $\int \Gamma(k) \, dk = \text{constant}$ eventually implies $\alpha = 1$, i.e. the Batchelor result. There is an intuitive interpretation in terms of the fractal structure of the iso-Θ surfaces. The fractal dimension D_Θ of these surfaces and the exponent ζ are related by [7]

$$D_\Theta = d - \tfrac{1}{2}\zeta .$$

The k^{-1} law corresponds to surfaces (lines in 2D) invading the whole space, as $D_\Theta = d$, that is the most chaotic situation. In fig. 1 we show one iso-Θ line in a two-dimensional fluid, with a time periodic velocity field obtained by a five-mode truncation of the Navier–Stokes equations, whose Eulerian [8] and Lagrangian [4] behaviors have been studied. Impressive is the similarity with the result obtained in 2D fully developed turbulent flows with $\approx 10^4$ modes (see fig. 4b of ref. [9]).

Let us briefly discuss some numerical simulations for two- and three-dimensional volume-preserving maps. The maps are used to describe the motion of particles in time-periodic velocity fields $\mathbf{u}(\mathbf{x}, t+T) = \mathbf{u}(\mathbf{x}, t)$. The numerical simulations were done using the so-called "water-bag" method [10].

We present results for the following two maps:
(i) the standard map ($d=2$);
(ii) the ABC map [3,5] ($d=3$).

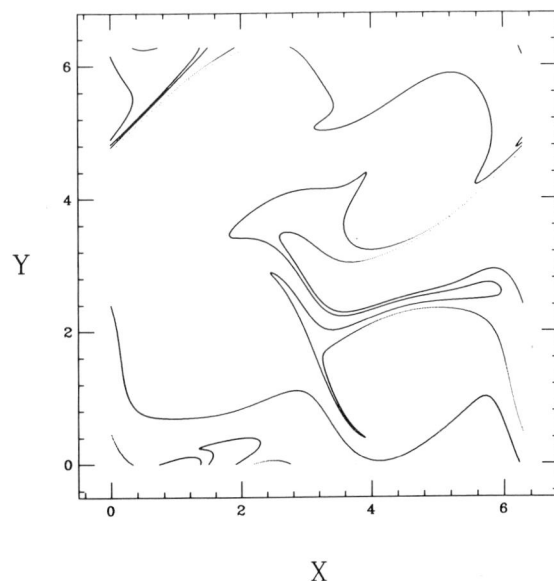

Fig. 1. The shape of the line $(x-3.20)^2 + (y-3.10)^2 = (3.05)^2$, convected by the velocity field obtained by a 5-mode truncation of the 2D Navier–Stokes equations [8] with a velocity field periodic in time, after ≈ 8 characteristic times.

Figs. 2 and 3 show $\Gamma(k)$ versus k at different times for the two maps. In both cases the initial Θ_0 has a spectrum concentrated in a range of small k. As time goes on $\Gamma(k)$ develops a k^{-1} shape. As there is no forcing, the spectrum shifts forward to large k.

We conclude this section with some remarks:

(I) It is possible to show [11] that the k^{-1} law is correct as long as one can linearize the stability equation which describes the growth of the separation between two marker particles in the fluid. This is not the case for 3D turbulence in the inertial range, where the velocity field is highly irregular and one has $|u(x+l)-u(x)|\propto l^h$ with $h<1$ ($h=1/3$ in the Kolmogorov theory). So the k^{-1} law holds only in the viscous convective subrange, where the velocity field is smoothed by the viscous dissipation.

(II) It is well known [12] that in 2D turbulent flows there exist some regularity properties, absent in $d=3$, which lead to $|u(x+l)-u(x)|\propto l$ also in the inertial range. We therefore conclude that in 2D the Batchelor law holds at all length scales, and there are not the two different regimes exhibited by 3D turbulence. Some closure approximations [13] and numerical simulations of the 2D Navier–Stokes equations [9] provide evidence for a k^{-1} power law in the inertial range.

(III) The k^{-1} is an exact result [14]; this is due to the fact that we have considered a Lagrangian dynamics which is volume preserving and chaotic. On the other hand the constant in front of this scaling law is sensitive to details of the Lagrangian chaoticity, e.g. the temporal intermittency.

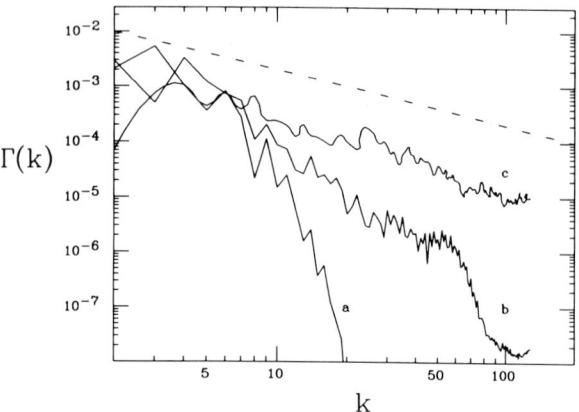

Fig. 2. Power spectrum $\Gamma(k)$ versus k at different times for the standard map (with $K=0.99$). The initial condition is $\Theta_0(x,y) = 1+0.2\cos(x)\sin(x+y)$; the times shown are $n=2$ (a), 4 (b), 25 (c). The line with slope -1 is drawn for comparison.

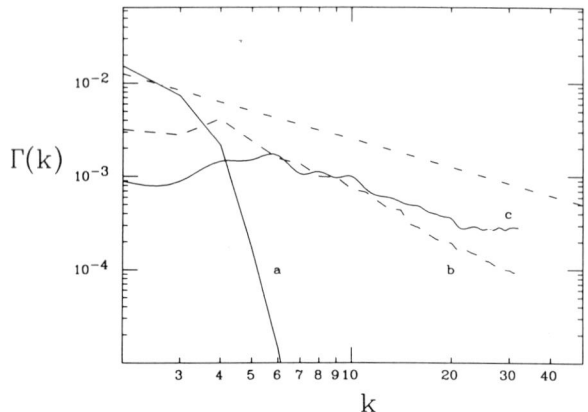

Fig. 3. The same as in fig. 2 for the ABC map [3] (with $A=0.5$, $B=0.08$ and $C=0.16$). The initial condition is $\Theta_0(x,y,z) = 1+0.1\cos(x+z)+0.2\sin(x+y)$, the times shown are $n=3$ (a), 15 (b), 45 (c).

3. Multifractal structure at very small scales

It has been recently shown [15] that the non-uniform stretching of a typical chaotic flow leads to a multifractal structure for the gradients of passive scalars. We want to point out that multifractality does not imply corrections to Batchelor's law on "physical" length scales, although a new regime should hold on very small scales. Let us consider eq. (4). Using the property (5) we have

$$\frac{\partial \Theta(x,t)}{\partial x_k} = \sum_{j=1}^{d} \frac{\partial \Theta_0(y)}{\partial y_j}\frac{\partial y_j}{\partial x_k}, \qquad (7)$$

where $y=S^{-t}x$. The terms $\partial y_j/\partial x_k$ are connected to the time-reversed equation of (4): $dx/dt = -u(x,-t)$.

If the system (4) is chaotic, because of the volume-preserving properties of the dynamics, one has

$$\overline{\left|\frac{\partial(S^{-t}\boldsymbol{x})_j}{\partial x_k}\right|^q} \sim \overline{\left|\frac{\partial(S^{t}\boldsymbol{x})_j}{\partial x_k}\right|^q} \sim \exp[L(q)\,t]\,, \qquad (8)$$

for large t, where $\overline{(...)}$ means time average and $L(q)$ are the generalized Lyapunov exponents [16]. Since the system is chaotic, we have that $\overline{(...)} = \langle(...)\rangle$, so that from eqs. (7) and (8) one obtains

$$\langle[\nabla\Theta(\boldsymbol{x},t)]^q\rangle \sim \exp[L(q)\,t]\,.$$

On finite times, one observes that $|\nabla\Theta| \propto \exp(\gamma t)$, with a local growth rate γ depending on the initial condition. The probability of finding $\gamma \neq \lambda =$ maximal Lyapunov exponent vanishes exponentially with time [17]. The relevance of the fluctuations can be measured by the deviation of $L(q)$ from the linear shape λq. These deviations lead to a multifractal structure when considering the probability measure $d\mu(\boldsymbol{x}) \propto |\nabla\Theta(\boldsymbol{x})|^2 d\boldsymbol{x}$. Basically one has an anomalous scaling for the coarse-grained measure $p_i(l) \propto \int_{\Lambda_i} |\nabla\Theta(\boldsymbol{x})|^2 d^d\boldsymbol{x}$, over a box Λ_i of size l, that is: $\langle p(l)^q\rangle \propto l^{(q-1)d_q+d}$, where d_q are called generalized Renyi dimensions [17]. Nevertheless the multifractality is observable only on length scales $\exp(-\gamma_{\max}t) < l < \exp(-\lambda t)$, where $\gamma_{\max}$ is the maximum local growth rate. On the contrary the onset of the k^{-1} law requires good mixing properties, which hold when initially distant particles of the fluid may approach, as it should happen on length scales $\approx k^{-1} > \exp(-\lambda t)$. The multifractal regime thus seems very difficult to observe in real experiments, since the diffusion coefficient introduces the natural cutoff $k_B \propto D^{-1/2}$, which is much smaller than $\exp(\lambda t)$ for reasonable times.

A very peculiar situation arises near the onset of Lagrangian chaos in incompressible fluids. In 2D the Lagrangian chaoticity firstly appears around tiny regions of the fluids (separatrices of the stream function), embedded into large regions with regular Lagrangian behavior [4]. It follows that the exponential amplification of $\nabla\Theta$ is limited to the tiny chaotic regions, while one has a polynomial growth in the regular regions.

4. Conclusions

We have revised the classical results of Batchelor on small scale fluctuations of passive scalar fields, in the limit of high Prandtl number. Our derivation stresses that the existence of the k^{-1} power law regime requires:

(1) Lagrangian chaoticity, i.e. exponential divergence of the distance between two initially close marker particles;

(2) the possibility of performing a Taylor expansion of the velocity field, at first order in the space of the coordinates.

These two hypotheses imply:

(a) The Eulerian turbulence *is not a necessary condition* for the k^{-1} regime.

(b) There exists a k^{-1} regime in *all* dimensions $d \geq 2$.

(c) In 2D Navier–Stokes equations, $\Gamma(k) \sim k^{-1}$ both in the *inertial* and *viscous convective* ranges.

Acknowledgement

I enjoyed many exchanges of ideas with A. Crisanti, M. Falcioni and G. Paladin.

References

[1] J.M. Ottino, C.W. Leong, H. Rising and P.D. Swanson, Nature 333 (1988) 419.

[2] M. Henon, C.R. Acad. Sci. (Paris) A 262 (1966) 312; H. Aref, J. Fluid Mech. 143 (1984) 1.

[3] T. Dombre, U. Frisch, J.M. Greene, M. Henon, A. Mehr and A.M. Soward, J. Fluid Mech. 167 (1986) 353.

[4] M. Falcioni, G. Paladin and A. Vulpiani, J. Phys. A 21 (1988) 3451.

[5] M. Feingold, L.P. Kadanoff and O. Piro, J. Stat. Phys. 50 (1988) 529.

[6] G.K. Batchelor, J. Fluid Mech. 5 (1959) 113.

[7] B.B. Mandelbrot, J. Fluid Mech. 72 (1975) 401; I. Procaccia, J. Stat. Phys. 36 (1984) 649.

[8] C. Boldrighini and V. Franceschini, Comm. Math. Phys. 64 (1979) 159.

[9] A. Babiano, C. Basdevant, B. Legras and R. Sadourny, J. Fluid Mech. 183 (1987) 379.

[10] H.L Berk and K.V. Roberts, in: Methods in Computational Physics, Vol. 9, B. Adler, S. Fernbach and M. Rotenberg, eds. (Academic Press, New York, 1970), p. 87.
[11] A. Crisanti, M. Falcioni, G. Paladin and A. Vulpiani, to be published.
[12] H.A. Rose and P.L. Sulem, J. Phys. (Paris) 39 (1978) 441.
[13] M. Lesieur and J. Herring, J. Fluid Mech. 161 (1985) 77.
[14] R.H. Kraichnan, Phys. Fluids 11 (1968) 945; J. Fluid Mech. 64 (1974) 737.
[15] E. Ott and T.M. Antonsen, Phys. Rev. Lett. 61 (1988) 2839.
[16] H. Fujsaka, Prog. Theor. Phys. 70 (1983) 1264; R. Benzi, G. Paladin, G. Parisi and A. Vulpiani, J. Phys. A 18 (1985) 2157.
[17] G. Paladin and A. Vulpiani, Phys. Rep. 156 (1987) 147.

HULL-GENERATING WALKS

Robert M. ZIFF

Department of Chemical Engineering, The University of Michigan, Ann Arbor, MI 48109-2136, USA

A hull-generating walk (HGW) is a type of kinetic random walk that generates the hull or perimeter of a percolation cluster, and thus has a fractal dimension of 1.75. Some examples of HGWs for site and bond percolation on a square lattice are described.

1. Introduction

A percolation cluster is a collection of occupied sites connected to each other by paths along nearest-neighbor pairs of sites, and surrounded inside and outside by vacant sites. (For bond percolation, this definition holds by reformulating the problem as site percolation on the covering lattice.) A closed circuit along the boundary of adjacent occupied and vacant sites is called a perimeter or the hull of the cluster. The term "hull" was first used by Mandelbrot [1] to describe the island of points enclosed by the external boundary of a cluster, but it has been generalized to refer to the boundary as well, and that meaning will be used here. One can have both external hulls, in which the occupied sites are on the inside and the vacant sites on the outside, and internal hulls, in which the occupied sites are on the outside and the vacant sites are on the inside.

Mandelbrot's influence on the study of percolation hulls goes far beyond the coining of the name, of course. The invention of fractals and the resulting interest in the study of growth processes and geometric properties has stimulated a great deal of work on percolation clusters and their hulls, which are among the simplest and most elegant of random fractals, and which result from many growth and epidemic models (see for example refs. [2–4]). While perimeters of percolation clusters have been studied for many years in the context of cluster statistics [5,6] or boundary properties [7], the introduction of fractals has led to substantial advances in the understanding of their properties.

Indeed, the fractal nature turned out to be the key to the discovery of expressions for all the critical exponents of percolation hulls. In their investigation of the scaling of percolation-gradient frontiers (hulls), Sapoval, Rosso and Gouyet [8] were led to the conjecture that the fractal dimension of the hull is exactly

$$D = 1 + 1/\nu = 1.75, \qquad (1)$$

where $\nu = 4/3$ is the correlation-length exponent. This conjecture is supported by numerical studies [8–11]. Then, by scaling arguments, (1) was shown [10,12] to give simple values for the other critical exponents, such as $\gamma' = 2$ (for the mean hull-size exponent) and $\beta' = 1/3$. Finally, Saleur and Duplantier [13] derived exact expressions for these exponents from first principles, thus verifying (1) by scaling. Theoretical arguments for (1) have also been given by Bunde and Gouyet [14]. This work has shown that percolation hulls have simpler critical exponents than the clusters themselves.

Many types of kinetic random walks in two dimensions have been found to generate percolation-cluster hulls – often quite unexpectedly. What makes a given path a percolation-cluster hull is that it is generated with the same probability (weight) as it would be found on a lattice that has been randomly populated with occupied and vacant sites. In this paper I am concerned with these walks, which I call hull-generating walks (HGWs).

In general, HGWs have the following properties:

(1) They are generated on a lattice that starts out completely blank (untested), except perhaps for one or two sites to start the walk.

(2) They leave a path of both occupied and vacant sites, or some equivalent representation of a cluster boundary.

(3) Their growth process is local (depending only upon a local set of sites) and kinetic (they grow step by step).

(4) They eventually close to form a completed loop (on an infinite lattice).

(5) Once they close, they are no longer kinetic in nature, meaning that their probability of growing (weight) is no longer dependent upon the starting point.

(6) When closed, their weight is that of the corresponding perimeter of a percolation cluster. If the occupied sites are on the outside, then the path represents an external perimeter, while if the occupied sites are in the inside then the path represents an internal perimeter.

(7) Their fractal dimension is 1.75.

HGWs were first introduced by Ziff, Cummings, and Stell [15,16] for the specific purpose of generating percolation cluster perimeters. They were independently found to result from quite different considerations. Kremer and Lyklema [17] devised an indefinitely growing self-avoiding walk (IGSAW) on a square lattice which satisfies properties (1)–(3) above, but not the rest because the walks never close. However, Weinrib and Trugman [12] studied a similar walk on a honeycomb lattice, which they call the smart kinetic walk (SKW), and found that it is precisely a HGW for site percolation on the dual (triangular) lattice. Gunn and Ortuño [18] considered a random system of sites on a lattice that have the property of rotating the direction of a walk passing through them by given amounts, and found under certain circumstances that the paths are equivalent to a HGW for bond percolation on the square lattice. A similar walk was used by Grassberger [11]. Recently, Roux et al. [19] have introduced a step-by-step tiling process that is equivalent to the Gunn and Ortuño model and also to the representation by Saleur and Duplantier [13], and thus is equivalent to the bond-percolation HGW. These models will be described in more detail below.

HGWs are useful for finding the percolation threshold [10,20]; in fact using the gradient-probability method [20] they appear to be the most efficient Monte Carlo way to find p_c. HGWs allow one to generate the hull of the backbone of a percolation cluster [21], and also the "accessible perimeter" of Grossman and Aharony [22], as discussed below. Coniglio et al. [23] and later Duplantier and Saleur [24] and Bradley [25] have argued that the HGW is appropriate to represent a two-dimensional polymer chain at the θ or θ' point, and so these walks are more than just a mathematical curiosity but have physical significance as well.

In general, a HGW can be constructed for a given system by the following procedure [15]: First devise an algorithm to trace out the perimeter of an existing cluster. Then repeat the same algorithm on a blank (untested) lattice, with the modification that when the state of any site that is still untested is needed, that site is made "occupied" with probability p and "vacant" otherwise, and the algorithm is continued according to that decision. Moreover, the state of the site must be remembered so that if it is ever visited again it will be treated the same way. The perimeter produced by this walk has the same weight as the corresponding perimeter on a populated lattice, because, in random percolation, the state of a site (or bond) is assigned with statistical independence, and it is irrelevant whether the choise of the state is made beforehand or during the walk.

An interesting aspect of the HGW is the behavior when p is increased beyond p_c. For $p < p_c$, external hulls are more likely, while for $p > p_c$ the internal hulls are more likely [15]. However, the existence of the infinite cluster is not evident – there is no hull associated with it. The internal hulls that are produced when $p > p_c$ may be holes within the infinite cluster, or holes within a larger finite cluster. There is a natural symmetry for the behavior of the walks about p_c, which for site percolation on the triangular lattice and bond percolation on the square lattice is perfect because of the identity of matching lattices.

By considering different lattices and definitions of the perimeter, a great variety of HGWs can be constructed, all of which satisfy properties (1)–(7) above. Furthermore, even for the same system one can devise different walks that generate the perimeter. This will be illustrated by some examples of HGWs for site and bond percolation on a square lattice.

2. Site percolation on a square lattice

In fig. 1 a simple cluster of five occupied sites (shaded circles) and nine vacant sites (open circles) for site percolation on a square lattice is shown. First consider the walk that follows the occupied sites at the boundary, which is illustrated in fig. 1a. To follow that boundary, the walker moves from occupied site to occupied site, always keeping vacant sites to its right. The walker "looks" first to the right, then straight, then left, then back, where "straight" is the direction of the previous step. The HGW that results from this process follows the following rule on a initially untested lattice: If the site being looked at is

(1) occupied: the walker moves to it;

(2) vacant: the walker looks to the next site in counter-clockwise order;

(3) untested: the site is made "occupied" with

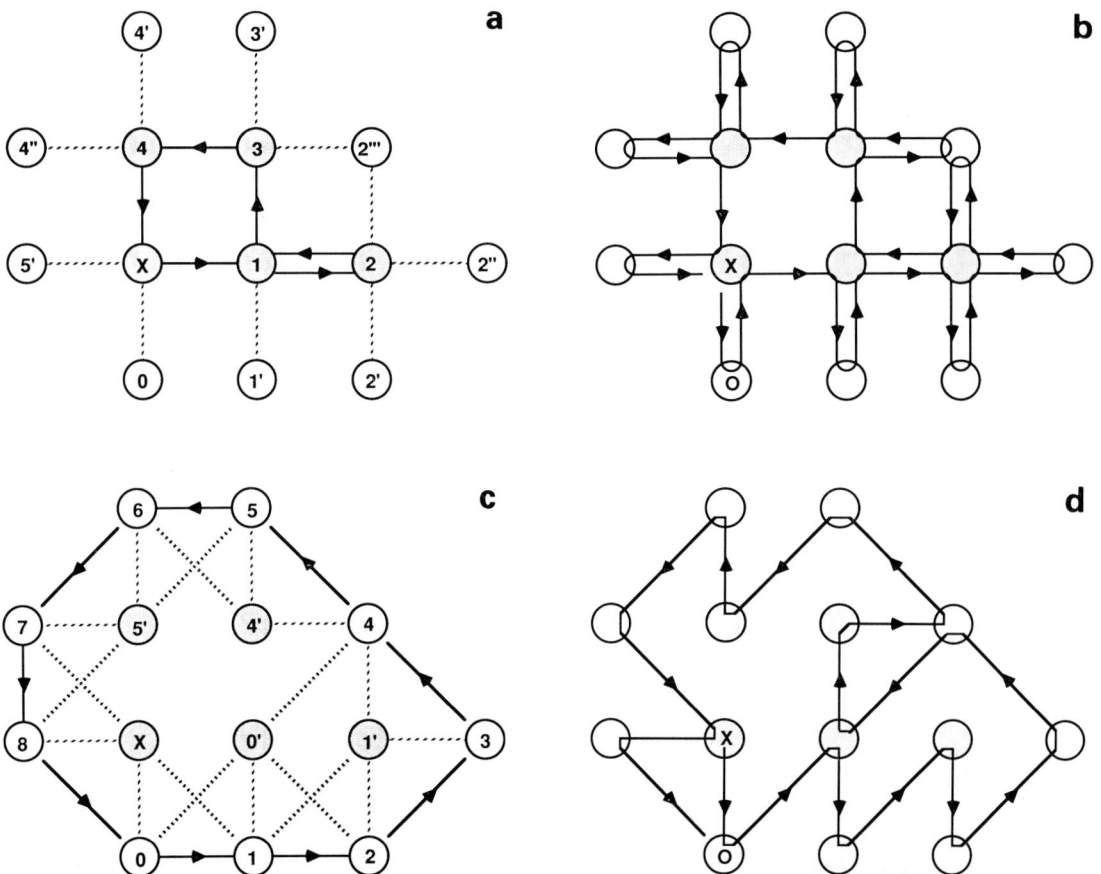

Fig. 1. Site percolation on a square lattice: (a) the walk that connects occupied sites (shaded circles) of the perimeter, (b) the equivalent "blind ant" walk, (c) the generic walk that moves along the vacant sites (open circles) of the perimeter, and (d) a new more efficient walk that goes along diagonals from vacant sites and vertical and horizontal lines from occupied sites.

probability p and (1) is followed, otherwise the site is made "vacant" and (2) is followed.

The walk is started by placing down the occupied–vacant site pair (marked X and O in fig. 1a), and finishes when the walker returns to X and attempts to go in the direction of the first step. Also in fig. 1a, the occupied sites are numbered according to the order in which they are created, and the vacant sites are labelled with a prime, double prime, etc., and a number corresponding to the occupied site where the walker was when the vacant site was created. Thus after the walker reaches occupied site 1, the vacant site 1' is first created, before the occupied site 2 is created. Because of the three vacant sites created around 2, the walker must backtrack to site 1, which is allowed here. When the walker reaches occupied site 3, it first looks to the right and sees 2''', which was already made vacant before, and so goes on to site 3' – and so on. This walk was simulated very extensively in refs. [10,15,20], where the scaling relations and (1) were verified, and the value of p_c(square) was found, all to high accuracy.

In this algorithm it is assumed that the walker is able to "look" at a neighboring site before deciding to move to it. In this case the walker is called a "myopic" ant [4]. If the walker does not have this ability (it is a "blind" ant), then it must move to every site in the perimeter, and the walk of fig. 1b results. Here a vacant site rotates the walk by π, while an occupied site rotates it by $-\pi/4$, and evidently this is a walk of the Gunn–Ortuño type.

The basic idea behind the walk of fig. 1b can be used to define a "generic" walk for any site-percolation problem: the vacant sites send the walker back ($\Delta\theta=\pi$) while an occupied site rotates the walk to the next direction of the lattice. (I arbitrarily use negative θ to define the next direction here.) A similar walk along the vacant sites can be made by going to the dual lattice and reversing the roles of the sites.

In fig. 1c the generic walk that joins the vacant sites of this same cluster is shown. The vacant sites satisfy the connectivity of the dual lattice, which in this case is the square lattice with nearest-neighbor and next-nearest-neighbor communication.

Inspection of figs. 1b and 1c shows that many of the steps are redundant in that a walker sometimes goes to a site that is guarranteed to be of a certain state by virtue of the walker's previous position. In fig. 1d a simpler, more efficient walk that visits both the occupied and vacant sites of a perimeter is shown. In this walk, the occupied sites rotate the walk to the first vertical or horizontal direction to the right, and the vacant sites rotate it to the first diagonal direction to the left. The angles of rotation are thus not fixed but either $\pm\pi/2$ or $\pm 3\pi/4$ depending upon the direction from which the site is approached. The asymmetry between the occupied and vacant sites reflects the different nature of these two sites on this lattice.

3. Bond percolation on a square lattice

A great variety of HGWs for bond percolation on a square lattice have been found, and I will briefly describe them here.

In fig. 2a a bond-percolation cluster with five occupied bonds (solid lines) and ten vacant bonds (shaded lines) is shown. The arrows follow a step-by-step path from bond center to bond center that traces out the boundary of this cluster.

In fig. 2b the same process is shown on the equivalent covering site-percolation lattice, where the sites are placed at the centers of the bonds and each site is connected to six other sites. The path of the connected arrows is precisely the walk of Gunn and Ortuño [18] in a system containing sites that rotate the walk by either $-\pi/2$ or $\pi/2$, corresponding to the occupied bonds and the vacant bonds, respectively. In contrast, the generic walk for this system is shown in fig. 2c, which is evidently more complicated. A generic walk can also be constructed that steps from vacant site to vacant site, analogous to fig. 1c. Notice that in fig. 2b the walker checks only the diagonals of the lattice.

Manna and Guttmann [26] have pointed out that the paths of connected arrows in fig. 2a or 2b are kinetic growth trails (KGTs) [27], also called growing self-avoiding trails (GSATs) [28], on the directed

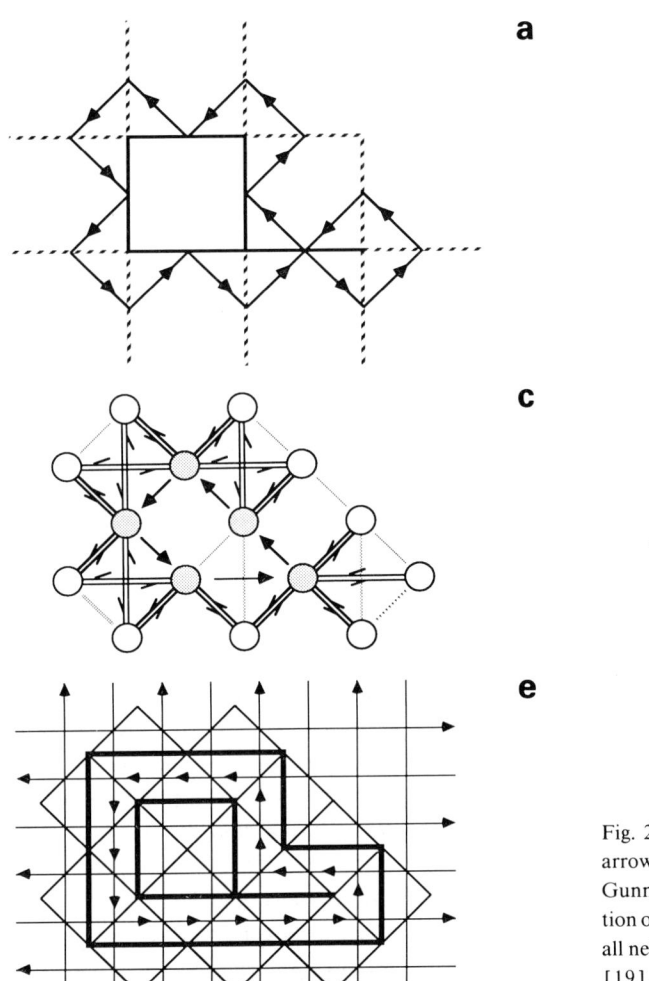
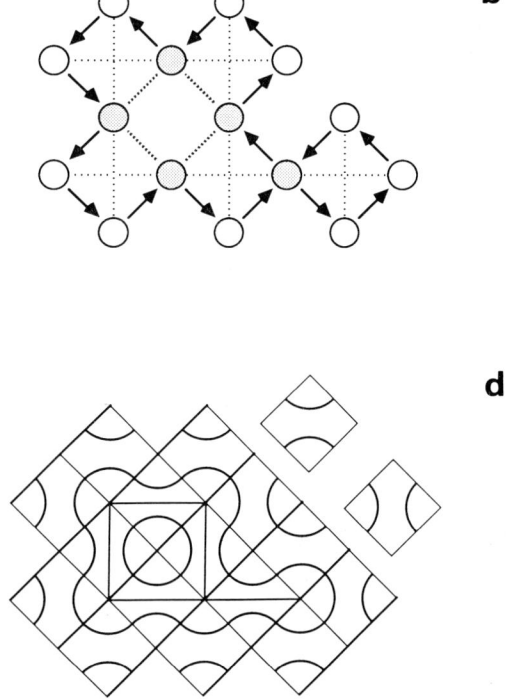

Fig. 2. Bond percolation on a square lattice: (a) a cluster with arrows showing an external perimeter path equivalent to the Gunn–Ortuño [18] or Grassberger [11] walk, (b) site percolation of the covering lattice, (c) the generic walk algorithm in which all neighbors are checked, (d) the equivalent tiling of Roux et al. [19], and (e) the cluster placed on a Manhattan lattice on which the perimeter is simply a KGW [25,26].

L-lattice, in which at each site there is a pair of arrows pointing in and a pair pointing out. A KGT is a kinetic walk on a lattice that can visit each site without restriction but each bond only once. On the L-lattice, the KGT is automatically a SKW because it never gets trapped except to close.

In fig. 2d the random tiling model of Roux et al. [19] is shown. In this model, the two tiles shown in the upper right-hand corner of that figure are randomly placed with equal probability on a square lattice (rotated by $\pi/2$ here) and connected paths are formed. The tiles evidently have the effect of rotating the direction of the walk as in fig. 2b, and thus this process is equivalent to the bond HGW, as shown by Duplantier [29], and Manna and Guttmann [26]. Note that the two tiles do not correspond directly to occupied and vacant sites, however. If one thinks of the lattice as being a checkerboard, then the occupied sites will correspond to tiles of one type on the white squares but the tiles of the opposite type on black squares [29]. Roux et al. [19] always found criticality for any mixture of the two tiles randomly put on all squares, since they thus always created an equal number of vacant and occupied bonds (however, placed with a spatial bias towards different colors on the checkerboard).

In fig. 2e the cluster is placed on an underlying Manhattan lattice with half the lattice spacing of the percolation lattice. The arrows of the walk around the cluster are seen to obey the restrictions of this lattice, as shown by Bradley [25] and by Manna and Guttmann [26]. Each step on this lattice corresponds to either a cut of a vacant bond or a step parallel to an occupied bond. Because of the properties of the Manhattan lattice, at each step there are two possible directions to continue. The walk is a simple kinetic growth walk (KGW). A KGW is a walk which steps with equal probability to any neighboring site that was not previously visited [30–32]. On the Manhattan lattice, the KGW is therefore a SKW [25,26].

In summary, the following walks are HGWs for bond percolation on a square lattice:

(1) The generic walk on the covering site lattice, which can be constructed to step between either the occupied (fig. 2c) or vacant sites.

(2) The paths on the $\pm\pi/2$ model of Gunn and Ortuño [18].

(3) Hull percolation on the random tiling of Roux et al. [19].

(4) The KGT (or GSAT) on an L-lattice [26].

(5) The KGW on a Manhattan lattice [25,26].

4. Discussion

Thus, we have seen that many walk-forming processes, which have mostly arisen independently from a variety of problems, are in fact different forms of HGW. This paper has been mainly a pedagogical review, although the walk of fig. 1d is a new and efficient HGW for site percolation hulls. A tiling procedure to generate these paths can also be given, although it is not as elegant as the tiling for bond percolation. This walk can also be generalized for site–bond percolation.

The perimeters considered here are related, but not identical, to the accessible perimeter introduced by Grossman and Aharony [22]. The accessible perimeter is the external perimeter of a cluster that can be probed by a particle of a given size moving along a path of nearest-neighbor vacant sites from infinity. When this particle is sufficiently large (depending upon the lattice), the invaginations of the cluster are cut off and the remaining hull is found to have a fractal dimension of $\approx 4/3$ rather than the 7/4 of the complete perimeter, and thus of a different universality class. Note that for a perimeter generated by the HGW, we can define the accessible perimeter as all sites that can be reached from infinity without crossing any path of the HGW, which is thus a definition independent of the size of a probe particle and the type of lattice. To generate the accessible perimeter by a walk process, one must first generate the complete perimeter in the usual way with a HGW, and then carry out another scouting walk around the perimeter to identify the "hull" of the hull [21,22]. In fact, this new hull is more in the spirit of Mandelbrot's original definition of the word than the more common usage as any perimeter, and furthermore the fractal dimension (4/3) is exactly identical to the value conjectured by Mandelbrot [1] for the Brown hull (which is the accessible perimeter of the Brown trail), based upon the value for the self-avoiding random walk. There is no local walk that can generate the accessible perimeter from scratch, which is perhaps related to its being of a different universality class.

The connections between HGWs and KGWs, SKWs, IGSAWs, etc. are numerous but their exact nature is dependent upon the specific lattice and system being considered. In many cases, such walks are not precisely HGWs but of the same universality class. One example is the IGSAW of Kremer and Lyklema [17] on a square lattice. While the growing end of the IGSAW never gets trapped, the non-growing end easily does [17]. In contrast, for the IGSAW (or the SKW) on the honeycomb lattice introduced in ref. [12] neither end will get trapped and the walk will always eventually close, because it is a HGW.

It is useful to make this distinction between "being of the same universality class as a HGW" and "being a type of a HGW", which is a stronger statement. As we have seen, there are many random walks that are a type of HGW, which means that any results of their simulation apply equally to percolation hulls. This can lead to ambiguity when refering to such walks – do they represent SKWs, or percolation hulls [23]? The answer depends upon the lattice – for site percolation on the triangular lattice the HGWs are both, while for the square lattice the HGW (fig. 1) is somewhat different than the SKW or IGSAW. I would also like to point out that in ref. [23], the very extensive simulations of ref. [10], which gave $1/D=4/7\pm0.0005$, were misquoted to a much lower precision.

The various considerations given here for the square lattice can be applied to the many other two-dimensional lattices, including directed ones, to yield a great variety of interesting HGWs.

Acknowledgements

I thank S. Manna and A. Guttmann, B. Duplantier and H. Saleur, and R. Bradley for sending preprints of their work. I also wish to acknowledge support from the National Science Foundation grant No. DMR-8619731.

References

[1] B.B. Mandelbrot, The Fractal Geometry of Nature (Freeman, San Francisco 1983) pp. 126–130, 132, 242.
[2] H.J. Herrmann, Phys. Rep. 136 (1986) 153.
[3] T. Vicsek, Fractal Growth Phenomena (World Scientific, Singapore, 1989).
[4] H.E. Stanley, in: On Growth and Form, H.E. Stanley and N. Ostrowsky, eds. (Nijhoff, The Hague, 1986).
[5] P.G. de Gennes, P. Lafore and J.P. Millot, J. Phys. Chem. Solids 11 (1959) 105.
[6] M.E. Fisher and J. Essam, J. Math. Phys. 2 (1961) 609.
[7] G.R. Reich and P.L. Leath, J. Phys. C 11 (1978) 1155, 4017.
[8] B. Sapoval, M. Rosso and J.-F. Gouyet, J. Phys. (Paris) 46 (1985) L149.
[9] R. Voss, J. Phys. A 17 (1984) L373.
[10] R.M. Ziff, Phys. Rev. Lett. 56 (1986) 545.
[11] P. Grassberger, J. Phys. A 19 (1986) 2675.
[12] A. Weinrib and S.A. Trugman, Phys. Rev. B 31 (1985) 2993.
[13] H. Saleur and B. Duplantier, Phys. Rev. Lett. 58 (1987) 2325.
[14] A. Bunde and J.-F. Gouyet, J. Phys. A 18 (1985) L185.
[15] R.M. Ziff, P.T. Cummings and G. Stell, J. Phys. A 17 (1984) 3009.
[16] R.M. Ziff, presented at Statistical Mechanics Meeting at Rutgers, May, 1982, J. Stat. Phys. 28 (1982) 838.
[17] K. Kremer and J.W. Lyklema, Phys. Rev. Lett. 54 (1985) 267; J. Phys. A 18 (1985) 1515.
[18] J.M.F. Gunn and M. Ortuño, J. Phys. A 18 (1985) L1095.
[19] S. Roux, E. Guyon and D. Sornette, J. Phys. A 21 (1988) L475.
[20] R.M. Ziff and B. Sapoval, J. Phys. A 18 (1986) L1169.
[21] S.S. Manna, J. Phys. A 22 (1988) 433.
[22] T. Grossman and A. Aharony, J. Phys. A 19 (1986) L745; 20 (1987) L1193.
[23] A. Coniglio, N. Jan, I. Majid and H.E. Stanley, Phys. Rev. B 35 (1987) 3617.
[24] B. Duplantier and H. Saleur, Phys. Rev. Lett. 59 (1987) 539; 60 (1988) 1204; 61 (1988) 1521.
[25] R.M. Bradley, Phys. Rev. A 39 (1989) 3738.
[26] S.S. Manna and A.J. Guttmann, preprint (1988).
[27] A. Malakis, J. Phys. A 8 (1975) 1885.
[28] J.W. Lyklema, J. Phys. A 18 (1985) L617.
[29] B. Duplantier, J. Phys. A 21 (1989) 3969.
[30] J.W. Lyklema and K. Kremer, J. Phys. A 17 (1984) L691; 19 (1986) 279.
[31] I. Majid, N. Jan, A. Coniglio and H.E. Stanley, Phys. Rev. Lett. 52 (1984) 1257.
[32] S. Hemmer and P.C. Hemmer, J. Chem. Phys. 81 (1984) 584.

VITA AND PUBLICATIONS OF BENOIT B. MANDELBROT

VITA

Born 20 November 1924, Warsaw, Poland.
Ingénieur diplômé, *Ecole Polytechnique*, Paris: Admission class of 1944, graduated in 1947.
Master of Science, then Professional Engineer in Aeronautics:
  California Institute of Technology, Pasadena CA, 1948 and 1949.
Docteur d'Etat ès Sciences Mathématiques: *Faculté des Sciences de Paris*, 1952.

1949–57   Staff member (Attaché, then Chargé, then Maître de Recherches):
          Centre National de la Recherche Scientifique, Paris.
1957–58   Maître de Conférences de Mathématiques Appliquées: *Université*, Lille.
          Maître de Conférences d'Analyse Mathématique: *Ecole Polytechnique*, Paris.
1958→   Research Staff Member until 1974; IBM Fellow since 1974:
          IBM Thomas J. Watson Research Center, Yorktown Heights NY.
1987→   Abraham Robinson Adjunct Professor of Mathematical Sciences:
          Yale University, New Haven CT.

Positions held on long-term leave

1950–53   Ingénieur, Groupe de Télévision en Couleur: *LEP, S.A.* (Groupe Philips), Paris.
1953–54   Member of the School of Mathematics: *Institute for Advanced Study*, Princeton NJ.
1955–57   Chargé de Cours de Mathématiques et Membre du Séminaire Jean Piaget:
          Université, Genève.
1962–63   Visiting Professor of Economics, and Research Fellow in Psychology:
          Harvard University, Cambridge MA.
1963–64   Visiting Professor of Applied Mathematics, and Staff Member of the Joint Committee on
          Biomedical Computer Science: *Harvard University*, Cambridge MA.
1979–80   Visiting Professor, later Professor of the Practice of Mathematics, Mathematics Department:
1984–87   *Harvard University*, Cambridge MA.

Selected part-time or short-term leave activities

1953–71   Research Associate, later Lecturer in Electrical Engineering, most recently Institute Lecturer:
          Massachusetts Institute of Technology, Cambridge MA.
1969–77   Senior Staff Member: *National Bureau of Economic Research*, New York NY.
1970   Visiting Professor of Engineering and Applied Science: *Yale University*, New Haven CT.
1972   Visiting Professor of Physiology: *Albert Einstein College of Medicine*, Bronx NY.

1974   Visiting Professor of Physiology: *SUNY Downstate Medical Center*, Brooklyn NY.
1980   Visiteur: *Institut des Hautes Etudes Scientifiques (IHES)*, Bures-sur-Yvette, France.
1984   Visitor: *Mittag–Leffler Institute*, Djursholm, Sweden.
       Walker-Ames Distinguished Professor: *University of Washington*, Seattle WA.
1987   Regents' Lecturer: *University of California*, Santa Cruz CA.

Honors

Decorations

Chevalier: L'Ordre de la Légion d'Honneur (Nommé en 1989).

Academies

Fellow: *American Academy of Arts and Sciences*, Cambridge MA, since 1982.
Associate (Foreign): *USA National Academy of Sciences*, Washington DC, since 1987.
Member: *European Academy of Arts, Sciences and Humanities*, Paris, since 1987.

Doctorates of Science or Humane Letters, Honoris Causa

D.Sc.:   *Syracuse University*, Syracuse NY, 1986.
D.Sc.:   *Laurentian University*, Sudbury ON, Canada, 1986.
D.Sc.:   *Boston University*, Boston MA, 1987.
D.Sc.:   *State University of New York*, Albany NY, 1988.
D.Sc.:   *Universität Bremen*, Bremen, Fed. Rep. Germany, 1988.
D.H.L.: *Pace University*, New York NY, 1988.
D.Sc.:   *University of Guelph*, Guelph ON, Canada, 1989.

Awards and medals

Scholar: *Rockefeller Foundation*, 1953.
Fellow: *John Simon Guggenheim Memorial Foundation*, 1968 (resigned).
Recipient: Research Division Outstanding Innovation Award in 1983, Corporate Award in 1984:
            IBM Corporation.
Recipient: 1985 Barnard Medal for Meritorious Service to Science, "Magna est Veritas":
            USA National Academy of Sciences and Columbia University.
Recipient: 1986 Franklin Medal for Signal and Eminent Service in Science:
            The Franklin Institute, Philadelphia PA.
Recipient: 1988 Charles Proteus Steinmetz Medal:
            IEEE and Union College, Schenectady NY.
Recipient: 1988 Alumni Distinguished Service Award for Outstanding Achievement:
            California Institute of Technology, Pasadena CA.
Recipient: 1988 Senior Award (Humboldt Preis):
            Alexander-von-Humboldt-Stiftung, Bonn, Fed. Rep. Germany.

Recipient: 1988 "Science for Art" Prize:
Fondation Moet–Hennessy–Louis Vuitton, Paris.
Recipient: 1989 Harvey Prize for Science and Technology:
Technion–Israel Institute of Technology, Haifa, Israel.

Scientific Societies

Fellow: *American Physical Society.*
American Geophysical Union.
Institute of Mathematical Statistics.
American Statistical Association.
Institute of Electrical and Electronics Engineers.
Econometric Society.
American Association for the Advancement of Science.
Member (elected): *International Statistical Institute.*
Member: *Société Mathématique de France.*
American Mathematical Society.
Society for Industrial and Applied Mathematics.

Selected boards and committees

1964–82 Editorial Board of the journal *Information and Control.*
1969–72 Committee on the Applications of Mathematics: *USA National Academy of Sciences.*
1974–78 Editorial Board of the *Journal of Financial Economics.*
1982–88 Editorial Board of the journal *Pure and Applied Geophysics.*
1984→   Editorial Board of the journal *Advances in Applied Mathematics.*

PUBLICATIONS

Books and their translations

A   Logique, langage et théorie de l'information (avec L'Apostel et A. Morf)
    (Presses Universitaires de France, Paris, 1957).
B   Les Objets Fractals: Forme, Hasard et Dimension (Flammarion, Paris, 1975, 1984, 1989).
    Hungarian translation by G. David (Gondolat Konyvkiado, Budapest).
    Gli Oggetti Frattali: Forma, Case e Dimensione,
        Italian translation by R. Pignoni; preface by L. Peliti and A. Vulpiani
        (Giulio Einaudi, Torino, 1987).
    Los Objetos Fractales: Forma, Azar y Dimensión,
        Spanish translation by J.M. Llosa (Tusquets, Barcelona, 1987).
C   Fractals: Form, Chance and Dimension (Freeman, San Francisco, 1977).
D   The Fractal Geometry of Nature (Freeman, New York, 1982).
    Die fraktale Geometrie der Natur, German translation by R. Zähle and U. Zähle
        (Birkhauser/Akademie-Verlag, Basel/Berlin, 1987).
    Fraktal Kikagaku, Japanese translation directed by H. Hironaka
        (Nikkei Science, Tokyo, 1984).
E   La Geometria della Natura
    (Imago (per Montedison Progetto Cultura), Milan, 1987);
    (Edizioni Theoria, Rome, 1989).

Research publications other than books

1951
1   Adaptation d'un message à la ligne de transmission, I and II, Compt. Rend. (Paris) 232 (1951) 1638–1740 and 2003–2005.
1952
2   Sur la notion générale d'information et la durée intrinsèque d'une stratégie, Compt. Rend. (Paris) 234 (1952) 1346–1348.
3   Les démons de Maxwell, Compt. Rend. (Paris) 234 (1952) 1842–1844.
1953
4   Contribution à la théorie mathématique des jeux de communication (Ph.D. Thesis), Publications de l'Institut de Statistique de l'Université de Paris 2 (1953) 1–124.
5   An informational theory of the statistical structure of language, in: Communication Theory, the Second London Symposium, W. Jackson, ed. (Butterworth/Academic Press, London/New York, 1953) pp. 486–504.

1954

6   Structure formelle des textes et communication (deux études), Word 10 (1954) 1–27.

7   Simple games of strategy occurring in communication through natural languages, Trans. IRE Prof. Group Information Theory 3 (1954) 124–137.

1955

8   On recurrent noise limiting coding, in: Information Networks, the Brooklyn Polytechnic Institute Symposium, E. Weber, ed. (Interscience, New York, 1955) pp. 205–221.

9   Diagnostic en l'absence de bruit, Institut de Statistique de Université de Paris (1955) pp. 1–73 (booklet).

10   Théorie de la précorrection des erreurs de transmission, Ann. Télécommun. 10 (1955) 122–134.

1956

11   La distribution de Willis–Yule, relative au nombre d'espèces dans les genres taxonomiques, Compt. Rend. (Paris) 242 (1956) 2223–2225.

12   On the language of taxonomy: an outline of a thermo-statistical theory of systems of categories, with Willis (natural) structure, in: Information Theory, the Third London Symposium, Colin Cherry, ed. (Butterworth/Academic Press, London/New York, 1956) pp. 135–145.

13   Exhaustivité de l'énergie d'un système, pour l'estimation de sa température, Compt. Rend. (Paris) 243 (1956) 1835–1837.

14   A purely phenomenological theory of statistical thermodynamics: canonical ensembles, IRE Trans. Information Theory 112 (1956) 190–203.

1957

15   Note on a law of J. Berry and on insistence stress, Information Control 1 (1957) 76–81.

16   Théorie mathématique de la loi d'Estoup–Zipf, Institut de Statistique de l'Université de Paris (1957) pp. 1–80 (booklet).

17   Application of thermodynamical methods in communication theory and in econometrics, Institut Mathématique de l'Université de Lille (1957).

1958

18   Les lois statistique macroscopiques du comportement (rôle de la loi de Gauss et des lois de Paul Lévy), Psychologie Française 3 (1958) 237–249.

1959

19   A note on a class of skew distribution functions, Information Control 2 (1959) 90–99.

20   Variables et processus stochastiques de Pareto-Lévy et la répartition des revenus, I and II, Compt. Rend. (Paris) 249 (1959) 613–615 and 2153–2155.

21   Ensembles grand canoniques de Gibbs; justification de leur unicité basée sur la divisibilité infinie de leur énergie aléatoire, Compt. Rend. (Paris) 249 (1959) 1464–1466.

1960

22   Processus stochastiques à loi stable positive, permanents, markoviens et stationnaires (non additifs), Compt. Rend. (Paris) 250 (1960) 451–453.

23   The Pareto–Lévy law and the distribution of income, Int. Economic Rev. 1 (1960) 79–106.

1961

24   On the theory of word frequencies and on related markovian models of discourse, in: Structure of Language and its Mathematical Aspects, Symposia in Applied Mathematics XII, R. Jakobsen, ed. (Am. Math. Soc., Providence RI, 1961) pp. 190–219.

25 Final note on a class of skew distribution functions (with a post-script), Information Control 4 (1961) 198–216 and 300–304.
26 Stable Paretian random functions and the multiplicative variation of income, Econometrica 29 (1961) 517–543.

1962
27 Paretian distributions and income maximization, Quart. J. Economics 76 (1962) 57–85.
28 Sur certains prix spéculatifs: faits empiriques et modèle basé sur les processus stables additifs de Paul Lévy, Compt. Rend. (Paris) 254 (1962) 3968–3970.
29 The role of sufficiency and estimation in thermodynamics, Ann. Math. Statistics 33 (1962) 1021–1038.

1963
30 A new model for the clustering of errors on telephone circuits (with J.M. Berger), IBM J. Res. Dev. 7 (1963) 224–236.
31 The stable Paretian income distribution, when the apparent exponent is near two, Int. Economic Rev. 4 (1963) 111–115.
32 New methods in statistical economics, J. Political Economy 71 (1963) 421–440.
33 The variation of certain speculative prices, J. Business Univ. Chicago 36 (1963) 394–419.

1964
34 On the derivation of statistical thermodynamics from purely phenomenological principles, J. Math. Phys. 5 (1964) 164–171.
35 Random walk models for the spike activity of a single neuron (with G.L. Gerstein), Biophys. J. 4 (1964) 41–68.
36 Random walks, fire damage amount, and other Paretian risk phenomena, Operations Res. 12 (1964) 582–585.

1965
37 Self-similar error clusters in communications systems and the concept of conditional stationarity, IEEE Trans. Commun. Technol. COM-13 (1965) 71–90.
38 Une classe de processus stochastiques homothétiques à soi. Application à la loi climatologique de H.E. Hurst, Compt. Rend. (Paris) 260 (1965) 3274–3277.
39 Leo Szilard and unique decipherability, IEEE Trans. Information Theory IT-11 (1965) 455–456.
40 Ensembles de multiplicité aléatoires (with J.-P. Kahane), Compt. Rend. (Paris) 262 (1965) 3931–3933.
41 Very long-tailed probability distributions and the empirical distribution of city sizes, in: Mathematical Explorations in Behavioral Science (Cambria Pines CA, 1964), F. Massarik and Ph. Ratoosh, eds. (Irwin, Homewood IL, 1965) pp. 322–332.
42 Information theory and psycholinguistics, in: Scientific Psychology: Principles and Approaches, B.B. Wolman and E. Nagel, eds. (Basic Books, New York, 1965) pp. 550–562.

1966
43 Forecasts of future prices, unbiased markets and "martingale" models, J. Business Univ. Chicago 39 (1966) 242–255.
44 Nouveaux modèles de la variation des prix (cycles lents et changements instantanés), Cah. Séminaire d'Econométrie 9 (1966) 53–66.

1967

45 Sporadic random functions and conditional spectral analysis; self-similar examples and limits, in: Proceedings of the Fifth (1965) Berkeley Symposium on Mathematical Statistics and Probability, Vol. 3, L. LeCam and J. Neyman, eds. (University of California Press, Berkeley, 1967) pp. 155–179.

46 Some noises with $1/f$ spectrum, a bridge between direct current and white noise, IEEE Trans. Information Theory IT-13 (1967) 289–298.

47 How long is the coast of Britain? Statistical self-similarity and fractional dimension, Science 155 (5 May 1967) 636–638.

48 The variation of some other speculative prices, J. Business Univ. Chicago 40 (1967) 393–413.

49 Sporadic turbulence, in: Proceedings of the International Symposium on Boundary Layers and Turbulence including Geophysical Applicatons, supplement to Phys. Fluids 10 (September 1967) S302–S303.

50 On the distribution of stock price differences (with H.M. Taylor), Operations Res. 15 (1967) 1057–1062.

1968

51 On intermittent free turbulence, in: Turbulence of Fluids and Plasmas (Polytechnic Press of the Polytechnic Institute of Brooklyn, April 1968).
  – The geometry of turbulence, in: Conference on Prospects for Theoretical Turbulence Research (N.C.A.R., Boulder CO, June 14–20, 1974) pp. 9–12.

52 Noah, Joseph and operational hydrology (with J.R. Wallis), Water Resources Res. 4 (1968) 909–918.

53 Fractional Brownian motions, fractional noises and applications (with J.W. Van Ness), SIAM Rev. 10 (1968) 422–437.
  – Critique of a would-be improvement: On an eigenfunction expansion and on fractional Brownian motions, Lett. Nuovo Cimento 33 (1982) 549–550.

1969

54 Long-run linearity, locally Gaussian processes, H-spectra and infinite variances, Int. Economic Rev. 10 (1969) 82–111.

55 Computer experiments with fractional Gaussian noises (with J.R. Wallis), Water Resources Res. 5 (1969) 228–267.

56 Some long-run properties of geophysical records (with J.R. Wallis), Water Resources Res. 5 (1969) 321–340.

57 Robustness of the rescaled range R/S in the measurement of noncyclic long-run statistical dependence (with J.R. Wallis), Water Resources Res. 5 (1969) 967–988.

1970

58 On the secular pole motion and the Chandler wobble (with K. McCamy), Geophys. J. 21 (1970) 217–232.

59 Statistical dependence in prices and interest rates, in: Papers of the Second World Congress of the Econometric Society, Cambridge, England (8–14 September, 1970).

1971

60 A fast fractional Gaussian noise generator, Water Resources Res. 7 (1971) 543–553.

61 When can price be arbitraged efficiently? A limit to the validity of the random walk and martingale models, Rev. Economics Statistics 53 (1971) 225–236.

1972

62 Renewal sets and random cutouts, Z. Wahrscheinlichkeitstheorie 22 (1972) 145–157.

63 On Dvoretzky coverings for the circle, Z. Wahrscheinlichkeitstheorie 22 (1972) 158–160.

64 Possible refinement of the lognormal hypothesis concerning the distribution of energy dissipation in intermittent turbulence, in: Statistical Models and Turbulence (La Jolla, CA), M. Rosenblatt and Ch. Van Atta, eds., Lecture Notes in Physics, Vol. 12 (Springer, New York, 1972) pp. 333–351.

65 Statistical methodology for non-periodic cycles: from the covariance to R/S analysis, Ann. Economic Social Measur. 1 (1972) 257–288.

66 Broken line process derived as an approximation to fractional noise, Water Resources Res. 8 (1972) 1354–1356.

1973

67 Tests of the degree of word clustering in samples of written English (with F.J. Damerau), Linguistics 102 (1973) 58–75.

68 Comments on "A subordinated stochastic process model with finite variance for speculative prices", by P.K. Clark, Econometrica 41 (1973) 157–160.

69 Formes nouvelles du hasard dans les sciences, Economie Appl. 26 (1973) 307–319.

1974

70 Intermittent turbulence in self similar cascades; divergence of high moments and dimension of the carrier, J. Fluid Mech. 62 (1974) 331–358.

71 Multiplications aléatoires itérées et distributions invariantes par moyenne pondérée aléatoire, I and II, Compt. Rend. (Paris) A 278 (1974) 289–292 and 355–358.

72 A population birth and mutation process, I: Explicit distributions for the number of mutants in an old culture of bacteria, J. Appl. Probability 11 (1974) 437–444.

1975

73 Limit theorems on the self-normalized range for weakly and strongly dependent processes, Z. Wahrscheinlichkeitstheorie 31 (1975) 271–285.

74 Fonctions aléatoires pluri-temporelles: approximation poissonienne du cas brownien et généralisations, Compt. Rend. (Paris) A 280 (1975) 1075–1078.

75 On the geometry of homogeneous turbulence, with stress on the fractal dimension of the isosurfaces of scalars, J. Fluid Mech. 72 (1975) 401–416.

76 Stochastic models for the Earth's relief, the shape and the fractal dimension of the coastlines, and the number–area rule for islands, Proc. Natl. Acad. Sci. (USA) 72 (1975) 3825–3828.

77 Sur un modèle décomposable d'Univers hiérarchisé: déduction des corrélations galactiques sur la sphère céleste, Compt. Rend. (Paris) A 280 (1975) 1551–1554.

78 Hasards et tourbillons (quatre contes à clef), Ann. Mines (November 1975) 61–66.

1976

79 Géométrie fractale de la turbulence. Dimension de Hausdorff, dispersion et nature des singularités du mouvement des fluides, Compt. Rend. (Paris) A 282 (1976) 119–120.

80 Intermittent turbulence and fractal dimension: kurtosis and the spectral exponent $5/3+B$, in: Turbulence and Navier Stokes Equations (Orsay, 1975), R. Temam, ed., Lecture Notes in Mathematics, Vol. 565 (Springer, New York, 1976) pp. 121–145.

1977

81 Fractals and turbulence: attractors and dispersion, in: Seminar on Turbulence (Berkeley, 1976), P. Bernard and T. Ratiu, eds., Lecture Notes in Mathematics, Vol. 615 (Springer, New York, 1977) pp. 83–93.

1978

82   The fractal geometry of trees and other natural phenomena, in: Geometrical Probability and Biological Structures: Buffon's 100th Anniversary Conference (Paris, 1977), R. Miles and J. Serra, eds., Lecture Notes in Biomathematics, Vol. 23 (Springer, New York, 1978) pp. 235-249.

83   Geometric facets of statistical physics: scaling and fractals, Stat. Phys. 13, International IUPAP Conference (Haifa, 1977), D. Cabib, C.G. Kuper and I. Riess, eds., Ann. Israel Phys. Soc. 2 (1) (1978) 225-233.

84   Les objets fractals, La Recherche 9, 85 (1978) 1-13.

85   Colliers aléatoires et une alternative aux promenades au hasard sans boucle: les cordonnets discrets et fractals, Compt. Rend. (Paris) 286 (1978) 933-936.

1979

86   Corrélations et texture dans un nouveau modèle d'Univers hiérarchisé, basé sur les ensembles trémas, Compt. Rend. (Paris) 288 (1979) 81-83.

87   Robust R/S analysis of long run serial correlation (with M.S. Taqqu), Bull. Int. Statistical Inst., 42nd Session, Manila, 46 (book 2) (1979) 79-104.

1980

88   Critical phenomena on fractals (with Y. Gefen and A. Aharony), Phys. Rev. Lett. 45 (1980) 855-858.

89   Fractal aspects of the iteration of $z \to \lambda z(1-z)$ for complex λ and z, in: Non-linear Dynamics (New York, 1979), R.H.G. Helleman, ed., Ann. NY Acad. Sci. 357 (1980) 249-259.

1981

90   Scalebound or scaling shapes: A useful distinction in the visual arts and in the natural sciences, Leonardo 14 (1981) 45-47.

91   Solvable fractal family, and its possible relation to the backbone at percolation (with Y. Gefen, A. Aharony and S. Kirkpatrick) Phys. Rev. Lett. 47 (1981) 1771-1774.

1982

92   Comments on computer rendering of fractal stochastic models, Commun. Assoc. Computing Machinery 25 (1982) 581-584 (and cover).

1983

93   Fractal curves osculated by sigma-discs, and construction of self-inverse limit sets, Math. Intelligencer 5 (2) (1983) 9-17 (and front and back covers).

94   Geometric implementation of hypercubic lattices with noninteger dimensionality, using low lacunarity fractal lattices (with Y. Gefen, Y. Meir and A. Aharony), Phys. Rev. Lett. 50 (1983) 145-148.

95   Diffusion on fractal lattices and the fractal Einstein relation (with J.A. Given), J. Phys. A 16 (1983) L565-L569.
    – Elaboration: Comment on transport processes on fractal structures (with J.A. Given) J. Phys. A 17 (1984) 1937-1939.

96   On the quadratic mapping $z \to z^2 - \mu$ for complex μ and z: the fractal structure of its $\mathscr{M}$-set, and scaling, Physica D 7 (1983) 224-239.

97   Fractal nature of software-cache interaction (with J. Voldman, L.W. Hoevel, J. Knight and Ph. Rosenfeld), IBM J. Res. Dev. 27 (1983) 164-170.

98   Phase transitions on fractals: I. Quasi-linear lattice (with Y. Gefen and A. Aharony), J. Phys. A 16 (1983) 1267-1278.

1984

99 Phase transitions on fractals: II. Sierpiński gaskets (with Y. Gefen, A. Aharony and Y. Shapir), J. Phys. A 17 (1984) 435–444.

100 Phase transitions on fractals: III. Infinitely ramified lattices (with Y. Gefen and A. Aharony), J. Phys. A 17 (1984) 1277–1289.

101 The fractal character of the fracture surfaces of metals (with D.E. Passoja and A.J. Paullay), Nature 308 (19 April, 1984) 721–722.

102 Fractals in physics: squig clusters, diffusions, fractal measures and the unicity of fractal dimension, Stat. Phys. 15, International IUPAP Conference (Edinburgh, 1983), D. Wallace and A. Bruce, eds., J. Stat. Phys. 34 (1984) 895–930.
– Illustration: On the aggregative fractals called squigs, which include recursive models of polymers and of percolation clusters, in: Kinetics of Aggregation and Gelation, Proceedings of the International Topical Conference, University of Georgia, Athens (April 1984), F. Family and D.P. Landau, eds. (North-Holland, Amsterdam, 1984) pp. 5–7.

103 Physical properties of a new fractal model of percolation clusters (with J.A. Given), Phys. Rev. Lett. 52 (1984) 1853–1856.

104 Squig sheets and some other squig fractal constructions, followed by Comment on the equivalence between fracton/spectral dimensionality and the dimensionality of recurrence, J. Stat. Phys. 36 (book a) (1984) 519–545.

105 On the dynamics of iterated maps VIII: The map $z \to k\lambda(z+1/z)$, from linear to planar chaos, and the measurement of chaos, in: Chaos and Statistical Methods (Kyoto Summer Institute, 1983), Y. Kuramoto, ed. (Springer, New York, 1984) pp. 32–41.

106 On fractal geometry and a few of the mathematical questions it has raised, in: Proceedings of the International Congress of Mathematicians (Warsaw, 1983), C. Olech and Z. Ciesielski, eds. (PWN/North-Holland, Warsaw/Amsterdam, 1984) pp. 1661–1675.

1985

107 Continuous interpolation of the complex discrete map $z \to \lambda z(1-z)$, and related topics (On the dynamics of iterated maps, IX), Nobel Foundation Symposium 59 on the Physics of Chaos, N.R. Nilsson, ed., Physica Scripta T9 (1985) 59–63.

108 Fractals, their transfer matrices and their eigen-dimensional sequences (with Y. Gefen, A. Aharony and J. Peyrière), J. Phys. A 18 (1985) 335–354.
– Variant: Partial dimensional sequences and percolation (with Y. Gefen, A. Aharony and A. Kapitulnik), J. Stat. Phys. 36 (1984) 827–830.

109 Fractal properties of rain, and a fractal model (with S. Lovejoy), Tellus A 37 (1985) 209–232.

110 On the dynamics of iterated maps III: The individual molecules of the $\mathcal{M}$-set: self-similarity properties, the N^{-2} rule, and the N^{-2} conjecture; IV: The notion of "normalized radical" $\mathcal{R}$, and the fractal dimension of the boundary of $\mathcal{R}$; V: Conjecture that the boundary of the $\mathcal{M}$-set has a fractal dimension equal to 2; VI: Conjecture that certain Julia sets include smooth components, VII: Domain-filling ("Peano") sequences of fractal Julia sets, and an intuitive rationale for the Siegel discs, in: Chaos, Fractals and Dynamics, P. Fischer and W. Smith, eds. (Dekker, New York, 1985) pp. 213–253.

111 Self-affine fractals and fractal dimension, Physica Scripta 32 (1985) 257–260.

1986

112 Self-affine fractal sets, I: The basic fractal dimensions; II: Length and area measurements; III: Hausdorff dimension anomalies and their implications, in: Fractals in Physics (Trieste, 1985), L. Pietronero and E. Tosatti, eds. (North-Holland, Amsterdam, 1986) pp. 3–28.

113   Fractal measures (their infinite moment sequences and dimensions) and multiplicative chaos: early works and open problems, in: Dimensions and Entropies in Dynamical Systems (Pecos River NM, 1985), G. Mayer-Kress, ed. (Springer, New York, 1986) pp. 19–27.
– Letter to the Editor: Multifractals and fractals, Physics Today (September 1986) 11,12.

1987

114   Towards a second stage of indeterminism in science (preceded by historical reflections), Interdisciplin. Sci. Rev. 12 (1987) 117–127.

1988

115   Invariant multifractal measures in chaotic Hamiltonian systems, and related structures (with M.C. Gutzwiller), Phys. Rev. Lett. 60 (1988) 673–676.

116   Fractal landscapes without creases and with rivers, in: The Science of Fractal Images, H.-O. Peitgen and D. Saupe, eds. (Springer, New York, 1988) pp. 243–260.

117   An introduction to multifractal distribution functions, in: Fluctuations and Pattern Formation (Cargèse, 1988) H.E. Stanley and N. Ostrowsky, eds. (Kluwer, Dordrecht, 1988) pp. 345–360.

1989

118   The fractal range of the distribution of galaxies: crossover to homogeneity and multifractals, in: Large-scale Structure and Motions in the Universe (Trieste, 1988), F. Mardirossian, M. Mezzetti and D. Sciama, eds. (Kluwer, Dordrecht, 1989) pp. 259–279.

119   Temperature fluctuations: a well-defined and unavoidable notion, Physics Today (January 1989) 71–73.

120   Directed recursive models for fractal growth (with T. Vicsek), J. Phys. A 22 (1989) L377–L383.

121   Fractals in Geophysics (edited with H. Scholz), Pure Appl. Geophys. 131, Nos. 1, 2 (special issue) (1989) pp. 5ff.

122   A class of multifractal measures that exemplify negative (latent) values for the "dimension" $f(\alpha)$, in: Fractals (Erice, 1988), L. Pietronero, ed. (Plenum, New York, 1989).

123   Limit lognormal multifractal measures and their continuously variable embedding spaces, in: Frontiers of Physics: Landau Memorial Conference (Tel Aviv, 1988), E. Gotsman, ed. (Pergamon, New York, 1989).

LIST OF CONTRIBUTORS

Aharony, A. 1

Bak, P. 5
Ball, R.C. 13
Ben-Jacob, E. 16
Berry, M.V. 29
Binnig, G. 32
Blumen, A. 291
Blumenfeld, R. 93
Botet, R. 208
Bunde, A. 184

Chen, K. 5
Coniglio, A. 37
Courtens, E. 41

Desideri, J.P. 56
Domb, C. 64
Duplantier, B. 71

Edwards, S.F. 88
Entin-Wohlman, O. 93

Family, F. 98
Feder, J. 104
Fisher, M.E. 112

Garik, P. 16
Gefen, Y. 119

Giaever, I. 128
Goldburg, W.I. 134
Goldhirsch, I. 119
Goldstein, S. 141
Grey, F. 154
Guinea, F. 235
Gutzwiller, M.C. 160
Guyon, E. 172

Halperin, B.I. 179
Havlin, S. 184
Herrmann, H.J. 192
Hinrichsen, E.L. 104
Hulin, J.-P. 172
Hwa, T. 198
Hyosu, H. 299

Jensen, M.H. 203
Jøssang, T. 104
Jullien, R. 208

Kadanoff, L.P. 213
Kantor, Y. 215
Kardar, M. 198
Keese, C.R. 128
Kelly, K. 141
Kertész, J. 221
Kjems, J.K. 154

Landauer, R. 226
Lebowitz, J.L. 141
Legrand, O. 56
Lenormand, R. 230
Louis, E. 235
Lung, C.W. 242

Macon, L. 56
Måløy, K.J. 104
Masters, B.R. 98
Matsushita, M. 246
Meakin, P. 252
Meir, Y. 93
Mitescu, C.D. 172

Niklasson, G.A. 260

Oakeshott, R.B.S. 88
Orbach, R. 266
Ouchi, S. 246

Pak, H.K. 134
Peebles, P.J.E. 273
Pietronero, L. 279
Platt, D.E. 98
Prasad, R.R. 322

Redner, S. 287
Roux, S. 172

Roy, A.K. 291

Sapoval, B. 296
Sawada, Y. 299
Shlesinger, M.F. 304
Sinha, S.K. 310
Sivan, U. 93
Skjeltorp, A.T. 315
Sornette, D. 56
Sreenivasan, K.R. 322
Stanley, H.E. 330
Stauffer, D. 341
Stinchcombe, R.B. 345
Stoll, E. 41
Szasz, D. 141

Tong, P. 134

Vacher, R. 41
Vannimenus, J. 351
Vicsek, T. 356
Voss, R.F. 362
Vulpiani, A. 372

Wolf, D.E. 221

Zannetti, M. 37
Zhang, S.Z. 242
Ziff, R.M. 377

ANALYTIC SUBJECT INDEX

ac conductivity   260
Acoustics   41
Acoustic waves   56
Aerogels   41, 310
Aggregates   88, 154, 208, 310
Aggregation   13
Anderson localization   56
Angiogenesis   98
Anisotropic Kepler problems   160
Anisotropy   16, 192
Anomalous diffusion   184
Anomalous roughening   221
Arching   88
Asymmetric random walk   141

Ballistic deposition   252, 330
Ballistic deposition (at oblique incidence)   252
Ballistic deposition (spatially correlated)   252
Beam model   192
Besicovitch   64
Blood vessels   98
Boole   64
Boundary layer   203
Branched polymers (animals)   112
Branching   154
Branch orders   104
Brownian motion   71

Cantor set   141
Cantor spectrum   56
Cell motility   128
Clustering hierarchy   273
Coding off trajectories   160
Colloids   13, 310
Computer simulation   341
Conductivity   179
Conformations   291
Consolidation   13
Correlation functions   273, 345
Cosmology   273
Coupled maps   203
Criticality   5
Critical phenomena   56, 213
Cytotypes   112

Damping   296
Dendrites   16, 299
Deposition   315

Deterministic fractals   356
Deterministic growth   192
Dielectric breakdown model   279
Dielectric permittivity   260
Diffuse X-ray reflectivity   310
Diffusion length   299
Diffusion-limited aggregation   1, 98, 104, 154, 279, 330
Diffusion Stockgate processes   119
Directed clusters   356
Disorder   266
Dynamic critical phenomena   198
Dynamics   37

Eden model   330
Einstein relation   119
Elasticity   13, 192, 215
Elastic moduli   179
Electrochemical deposits   299
Energy transfer   291
Equilibrium   226
Evolution   32

Fluctuations   203
Fractal aggregates   208
Fractal dimensions   71, 104, 242, 291, 322
Fractal growth   356
Fractal shapes (variable)   112
Fractal structures   154, 260
Fractal time   260
Fractional Brownian motion   128, 362
Fracton dynamics   93
Fractons   41, 93, 266
Fracton scaling model   93
Fracture   192, 235
Fractured surface   242
Fracture of cracks   330
Fracture toughness   242
Free energy   345
Frustration   345

Galaxies   273
Game of life   5
Gel   330
Geometrical crossover   104
Grain size   242
Granular materials   88
Growth   37, 221
Growth patterns   235

Growth velocity   299
Guinier regime   310

Hausdorff   64
Hele-Shaw flow   16
Hopping conductivity   119
Hulls   377
Hurst exponent   128
Hydrodynamics   29

Inclusions   242
Interfaces   322
Intermittency   372
Intrinsic surface width   221
Invasion   172
Irreversible growth   279
Ising   345
Ising clusters   71

Kauffman model   341

Lacunary series   304
Lagrangian chaoticity   372
Large Schmidt numbers   322
Levy flights   304
Lifshitz singularity   287
Light scattering   134
Localization   41, 141, 266, 296
Localized modes   93

"Manhattan"   287
Membranes   112
Memory effects   192
Microspheres   315
Mixing of passive scalars   322
Multifractal measures   160
Multifractals   1, 37, 71, 184, 213, 356, 372

Neutron scattering   310
NeXt computer   330
Noise   226, 362
Nonequilibrium growth   16
Nonfractal shapes   112
Non-Newtonian fluids   172
Nuclear winter   29
Nyquist   226

$1/f$ noise   5, 198
Optical properties of fractals   208

Optics   208

Packing   88
Pair correlation function   310
Passive scalars   372
Pattern formation   98
Percolation   1, 5, 71, 88, 172, 179, 184, 215, 377
Perimeters   377
Permeability   172
Personal computers   330
Petroleum engineering   230
Phase transitions   351
Photon correlation spectroscopy   134
Physical laws   32
Plasticity   235
Polarization   208
Polymers   71, 215, 291, 351
Porod's law   310
Porous media   172, 230
Powders   88
Probability   330

Quasicrystal   56

R/S analysis   128
Random   266
Random Boolean networks   341
Random environment   141
Random fractals   184
Random networks   119
Random rough surfaces (growth of)   252
Random systems   215
Random velocities   287
Random walk   119, 230, 377
Renormalization   351
Resistor networks   1
Resonator   296
Retinal vessels   98
Riemann hypothesis   304

Sand   213
Sandpiles   198
Scaling   37, 134, 213, 310
Scaling analyses   112
Scattering   208
Self-affine fractal   310
Self-affine fractal surfaces   252
Self-affinity   221, 246, 362
Self-avoiding rings   112
Self-avoiding walks   291
Self-organization   5, 198
Self-similarity   310, 362
Sierpiński gasket   215, 351
Simulation   362
Singly connected bonds   179
Smart kinetic walk   377
Snowflake   330
Spinodal   37
Stability islands   160
Stress   88
Stress corrosion   192

Superdiffusive transport   287
Surface fractals   310

Thue–Morse sequence   141
Tip-splitting   16
Topography   246
Transmission   226
Tree-growth   154
Tree structure   104
Turbulence   5, 134, 203
Turbulent diffusion   304
Turbulent flows   322
Turing   64
Two-dimensional critical phenomena   71

Universality   213

Vesicles   112
Vibrational states of tenuous materials   93
Vibrations   13, 41
Viscoelasticity   192
Viscous fingering   104, 330

Water   330
Waves   296

X-ray scattering   310

ERRATA

Here are given some corrections and additions to the preceding text, which has been reprinted from Physica D, Vol. 38, Nos. 1-3 (1989).

Page 1: In the 2nd line of the abstract "nominated" must be changed into "dominated".

Page 2: In the 22th line of the right column "$(|i_{\min}|^{2q})_{av}$" must be changed into "$|i_{\min}|_{av}$".

Page 3: In the 9th line of the left column "negative a" must be changed into "negative q".

Page 4: The year of publication of ref. [20] is 1989 instead of 1980.

Page 114: In the line before eq. (4) "$R_0 \simeq 0.34a$" must be changed into "$R_0 \simeq 0.35a$".

In the 4th line before eq. (6) "$2\nu_a$" must be changed into "$2\nu_A$".

The value in eq. (6) should be "2.55 ± 0.05" instead of "2.5 ± 0.3" and in eq. (8) "0.40" instead of "0.41_2".

Page 115: The paragraph after eq. (12) regarding highly inflated vesicles reports an incorrect assertion of ref. [1]. The power in question should be N^3, as later simulations confirm: see A.C. Maggs, S. Leibler, M.E. Fisher and C.J. Camacho, to be published.

In the 12th line after eq. (13) the equation for ν^- should read "$\nu^- = \nu(1-\sigma)$", i.e. φ should be deleted.

Page 116: In the 1st line after eq. (16) "$a_0 \simeq 2.5a$" must be changed into "$a_0 \simeq 0.48a$".

Page 145: The 2nd line of eq. (12) should read:

$$(f_1 \square f_2)(x) = f_1(x_1) + f_2(x-x_1) \quad \text{for } x_1 \leq x \leq x_1 + x_2.$$

Page 177: P. Magnico is also acknowledged. His name should be added after "C. Leroy".

Page 178: Ref. [45] has been published now:

S. Roux and E. Guyon, J. Phys. A 22 (1989) 3693.

Page 229: In eq. (15) the sum should be over "nn'" instead of "m".

The book in ref. [1] has been published in 1989 and the contribution of R. Landauer can be found on p. 17 therein.

Page 236: In eq. (1) "δ" must be changed into "∂".

In the 7th line of the right column "[29]" must be changed into "[23]".

Page 238: In the 16th line of the left column "$i=1$" must be changed into "$i=l$".

In the 9th/10th line of section 3 the values for the fractal dimension should be "1.51" and "1.62" instead of "1.55" and "1.60".

Page 239: In the 2nd line of the left column "was reduced" must be changed into "is rescued".

In eq. (8) "δ" must be changed into "∂".

Page 243: In eq. (1) "$A_m(\epsilon)$" must be changed into "$A_n(\epsilon)$".

In fig. 3 the text near the vertical axis should read $\log P_n(\varepsilon)/P_0$ instead of $\log L_n(\varepsilon)/L_0$.

Page 248: In the caption to fig. 3 "ν_T" must be changed into "ν_t".

Page 249: In eq. (6) "$\exp(x^2/4Dt^{2H})$" must be changed into "$\exp(-x^2/4Dt^{2H})$".

In the 13th line from the bottom of the right column "$\nu_x \approx \nu_z \approx 0.73$" must be changed into "$\nu_x \approx \nu_y \approx 0.73$".

Errata

Page 253: In eq. (4) "$\langle h(x)-h(x+r)\rangle_{|r|=r}$" must be changed into "$\langle |h(x)-h(x+r)|\rangle_{|r|=r}$".
Eq. (5) should read "$C_h(r) \sim r^H$" instead of "$C_h(r) \sim r^{-H}$".

Page 256: In fig. 5 the horizontal width is 2000 diameters instead of 2000 lattice units".

Page 264: The name of the first author in ref. [11] should be Clerc instead of Clarc.

Page 288: In the 7th line of section 2 "$D_\perp t$" must be changed into "$(D_\perp t)^{1/2}$" and in the 1st line after eq. (1) "$D_\perp^{1/2}$" must be changed into "$(D_\perp t)^{1/2}$".

Page 289: In the 2nd line of eq. (6) "dt_w" must be changed into "dt_2".
In the 9th line after eq. (8) "$\langle P(x \sim t, t)\rangle$" must be changed into "$\langle P(x \sim t, t)\rangle_c$".

Page 290: In the 2nd line "$\langle P(x, t)\rangle$" must be changed into "$\langle P(x, t)\rangle_c$".

Page 319: In the 6th line from the bottom of the right column the porosity should be given by "$\varphi = 1 - \eta$" instead of "$\varphi = 1 - \beta$".

Page 322: In the right column a part of the text is missing between the 5th and 4th line from the bottom; the correct sentences now read:
"If σ is much smaller than unity (as in the stellar atmosphere) the smallest scalar scale is much larger than η. Of more common occurrence is the case of large η where the smallest scalar scale is the so-called Batchelor scale $\eta_b = \eta = \sigma^{-1/2}$ [7]."

Page 340: The following reference must be added to the list of references:
[14] A.D. Fowler, H.E. Stanley and G. Daccord, Diequilibrium silicate mineral textures: fractal and non-fractal features, Nature 341 (1989) 134–138.

Page 344: The following note should be added:
"For $L = 6976$ the threshold was between 0.304 and 0.3045 for one sample, consistent with fig. 1."

Page 349: In the 6th line of the right column "for the Ising model" must be changed into "for the two-dimensional Ising model".
In the 9th line from the bottom of the right column "with $\frac{1}{2}\nu_h$, ξ_h replaced by" must be changed into "with $\frac{1}{2}\nu_h$, ξ_h replaced respectively by".